7

SYNTHESIS AND OPTIMIZATION
OF DIGITAL CIRCUITS

McGraw-Hill Series in Electrical and Computer Engineering

Senior Consulting Editor

Stephen W. Director, Carnegie Mellon University

Circuits and Systems
Communications and Signal Processing
Computer Engineering
Control Theory
Electromagnetics
Electronics and VLSI Circuits
Introductory
Power and Energy
Radar and Antennas

Previous Consulting Editors

Ronald N. Bracewell, Colin Cherry, James F. Gibbons, Willis W. Harman, Hubert Heffner,
Edward W. Herold, John G. Linvill, Simon Ramo, Ronald A. Rohrer, Anthony E. Siegman,
Charles Susskind, Frederick E. Terman, John G. Truxal, Ernst Weber, and John R. Whinnery

Electronics and VLSI Circuits

Also Available from McGraw-Hill

Schaum's Outline Series in Electronics & Electrical Engineering

Most outlines include basic theory, definitions and hundreds of example problems solved in step-by-step detail, and supplementary problems with answers.

Related titles on the current list include:

Analog & Digital Communications
Basic Circuit Analysis
Basic Electrical Engineering
Basic Electricity
Basic Mathematics for Electricity & Electronics
Descriptive Geometry
Digital Principles
Electric Circuits
Electric Machines & Electromechanics
Electric Power Systems
Electromagnetics
Electronic Circuits
Electronic Communication
Electronic Devices & Circuits
Electronics Technology
Engineering Economics
Feedback & Control Systems
Introduction to Digital Systems
Microprocessor Fundamentals

Schaum's Solved Problems Books

Each title in this series is a complete and expert source of solved problems with solutions worked out in step-by-step detail.

Related titles on the current list include:

3000 Solved Problems in Calculus
2500 Solved Problems in Differential Equations
3000 Solved Problems in Electric Circuits
2000 Solved Problems in Electromagnetics
2000 Solved Problems in Electronics
3000 Solved Problems in Linear Algebra
2000 Solved Problems in Numerical Analysis
3000 Solved Problems in Physics

Available at most college bookstores, or for a complete list of titles and prices, write to:
Schaum Division
McGraw-Hill, Inc.,
Princeton Road, S-1
Hightstown, NJ 08520

SYNTHESIS AND OPTIMIZATION OF DIGITAL CIRCUITS

Giovanni De Micheli

Stanford University

McGraw-Hill, Inc.

New York   St. Louis   San Francisco   Auckland   Bogotá   Caracas
Lisbon   London   Madrid   Mexico City   Milan   Montreal
New Delhi   San Juan   Singapore   Sydney   Tokyo   Toronto

This book was set in Times Roman by Electronic Technical Publishing Services.
The editors were Anne Brown Akay, George T. Hoffman, and John M. Morriss;
the production supervisor was Elizabeth J. Strange.
The cover was designed by Initial Graphic Systems, Inc.
R. R. Donnelley & Sons Company was printer and binder.

SYNTHESIS AND OPTIMIZATION OF DIGITAL CIRCUITS

3 4 5 6 7 8 9 0 DOC DOC 9 0 9 8 7 6 5

ISBN 0-07-016333-2

Library of Congress Cataloging-in-Publication Data

De Micheli, Giovanni.
    Synthesis and optimization of digital circuits / Giovanni De
Micheli.
        p.   cm. — (McGraw-Hill series in electrical and computer
engineering. Electronics and VLSI circuits)
    Includes bibliographical references.
    ISBN 0-07-016333-2
    1. Digital integrated circuits—Design and construction—Data
processing.   2. Digital electronics—Data processing.   3. Computer-
aided design.   I. Title.   II. Series.
    TK7874.65.D4   1994
    621.39′5′028551—dc20   93-43595

ABOUT THE AUTHOR

Giovanni De Micheli is Associate Professor of Electrical Engineering and, by courtesy, of Computer Science at Stanford University. From 1984 to 1986 he worked at the IBM T. J. Watson Research Center, Yorktown Heights, where he was project leader of the Design Automation Workstation group. Previously he held positions at the Department of Electronics of the Politecnico di Milano, Italy and at Harris Semiconductor, Melbourne, Florida. He holds a Nuclear Engineer degree (Politecnico di Milano, 1979) and a Ph.D. degree in Electrical Engineering and Computer Science (University of California at Berkeley, 1983).

His research interests include several aspects of the computer-aided design of integrated circuits and systems, with particular emphasis on automated synthesis, optimization, and verification. Dr. De Micheli is a Fellow of IEEE. He was technical and general chairman of the IEEE International Conference on Computer Design (ICCD) in 1988 and 1989 respectively. He was also co-director of the Advanced Study Institute on Logic Synthesis and Silicon Compilation, held in L'Aquila, Italy, under the sponsorship of NATO, in 1986 and in 1987.

To all those who have taught me,
and to those who will learn from this book.

CONTENTS

5 Scheduling Algorithms

PREFACE

Computer-aided design (CAD) of digital circuits has been a topic of great importance in the last two decades, and the interest in this field is expected to grow in the coming years. Computer-aided techniques have provided the enabling methodology to design efficiently and successfully large-scale high-performance circuits for a wide spectrum of applications, ranging from information processing (e.g., computers) to telecommunication, manufacturing control, transportation, etc.

The book addresses one of the most interesting topics in CAD for digital circuits: *design synthesis* and *optimization*, i.e., the generation of detailed specifications of digital circuits from architectural or logic models and the optimization of some figures of merit, such as performance and area. This allows a designer to concentrate on the circuit specification issues and to let the CAD programs elaborate and optimize the corresponding representations until appropriate for implementation in some fabrication technology.

Some choices have been made in selecting the material of this book. First, we describe synthesis of digital synchronous circuits, as they represent the largest portion of circuit designs. In addition, the state of the art in synthesis for digital synchronous circuits is more advanced and stable than the corresponding one for asynchronous and analog circuits. Second, we present synthesis techniques at the architectural and logic level, and we do not report on physical design automation techniques. This choice is motivated by the fact that physical microelectronic design forms a topic of its own and has been addressed by other books. However, at present no textbook describes both logic and architectural synthesis in detail. Third, we present an algorithmic approach to synthesis based on a consistent mathematical framework.

In this textbook we combined CAD design issues that span the spectrum from circuit modeling with hardware description languages to cell-library binding. These problems are encountered in the design flow from an architectural or logic model of a circuit to the specification of the interconnection of the basic building blocks in any semicustom technology. More specifically, the book deals with the following issues:

- *Hardware modeling* in hardware description languages, such as VHDL, Verilog, UDL/I, and Silage.

- *Compilation techniques* for hardware models.
- *Architectural-level* synthesis and optimization, including scheduling, resource sharing and binding, and data-path and control generation.
- *Logic-level* synthesis and optimization techniques for combinational and synchronous sequential circuits.
- *Library binding* algorithms to achieve implementations with specific cell libraries.

These topics have been the object of extensive research work, as documented by the large number of journal articles and conference presentations available in the literature. Nevertheless, this material has never been presented in a unified cohesive way. This book is meant to bridge this gap. The aforementioned problems are described with their interrelations, and a unifying underlying mathematical structure is used to show the inherent difficulties of some problems as well as the merit of some solution methods.

To be more specific, the formulation of synthesis and optimization problems considered in this book is based upon graph theory and Boolean algebra. Most problems can be reduced to fundamental ones, namely coloring, covering, and satisfiability. Even though these problems are computationally intractable and often solved by heuristic algorithms, the exact problem formulation helps in understanding the relations among the problems encountered in architectural and logic synthesis and optimization. Moreover, recent improvements in algorithms, as well as in computer speed and memory size, have allowed us to solve some problems exactly for instances of practical size. Thus this book presents both exact and heuristic optimization methods.

This text is intended for a senior-level or a first-year graduate course in CAD of digital circuits for electrical engineers and/or computer scientists. It is also intended to be a reference book for CAD researchers and developers in the CAD industry. Moreover, it provides an important source of information to the microelectronic designers who use CAD tools and who wish to understand the underlying techniques to be able to achieve the best designs with the given tools.

A prerequisite for reading the book, or attending a course based on the book, is some basic knowledge of graph theory, fundamental algorithms, and Boolean algebra. Chapter 2 summarizes the major results of these fields that are relevant to the understanding of the material in the remaining parts of the book. This chapter can serve as a refresher for those whose knowledge is rusty, but we encourage the totally unfamiliar reader to go deeper into the background material by consulting specialized textbooks.

This book can be used in courses in various ways. It contains more material than could be covered in detail in a quarter-long (30-hour) or semester-long (45-hour) course, leaving instructors with the possibility of selecting their own topics. Sections denoted by an asterisk report on advanced topics and can be skipped in a first reading or in undergraduate courses.

Suggested topics for a graduate-level course in CAD are in Chapters 3–10. The instructor can choose to drop some detailed topics (e.g., sections with asterisks) in quarter-long courses. The text can be used for an advanced course in logic synthesis and optimization by focusing on Chapters 3 and 7–10. Similarly, the book can be used for an advanced course in architectural-level synthesis by concentrating on Chapters 3–6.

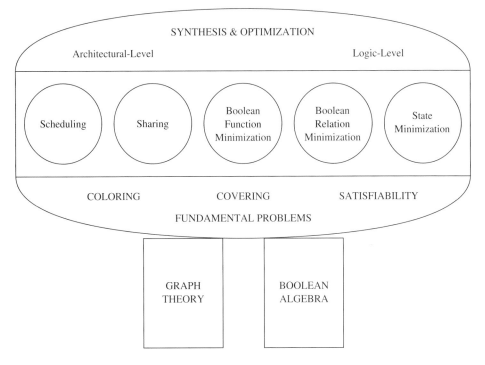

For senior-level teaching, we suggest refreshing the students with some background material (Chapter 2), covering Chapters 3–4 and selected topics in scheduling (e.g., Sections 5.2 and 5.3), binding (e.g., Section 6.2), two-level logic optimization (e.g., Section 7.2), multiple-level optimization (e.g., Sections 8.2 and 8.3), sequential synthesis (e.g., Sections 9.2 and 9.3.1), and library binding (e.g., Section 10.3.1). Most chapters start with a general introduction that provides valuable information without digging into the details. Chapter 4 summarizes the major issues in architectural-level synthesis.

The author would like to thank all those who helped him in writing and correcting the manuscript: Jonathan Allen, Massachusetts Institute of Technology; Gaetano Borriello, University of Washington; Donald Bouldin, University of Tennessee; Raul Camposano, Synopsys; Maciej Ciesielski, University of Massachusetts at Amherst; José Fortes, Purdue University; David Ku, Stanford University and Redwood Design Automation; Michael Lightner, University of Colorado at Boulder; Sharad Malik, Princeton University; Victor Nelson, Auburn University; Fabio Somenzi, University of Colorado at Boulder; and, Wayne Wolf, Princeton University. The author would also like to acknowledge enlightening discussions with Robert Brayton of the University of California at Berkeley, with Randal Bryant and Rob Rutenbar of Carnegie Mellon University, and with Alec Stanculescu of Fintronic. The graduate students of the CAD synthesis group at Stanford University provided invaluable help in improving the manuscript: Luca Benini, Claudionor Coelho, David Filo, Rajesh Gupta, Norris Ip, Polly Siegel, Thomas Truong, and Jerry Yang. Thanks also to those brave students of EE318 at Stanford, of ECE 560 at Carnegie Mellon and of CSE590F at the University of Washington who took courses in 1992 and in 1993 based on a prototype of this book and gave intelligent comments. Last but not least, I would like to thank my wife Marie-Madeleine for her encouragement and patience during the writing of this book. This work was supported in part by the National Science Foundation under a Presidential Young Investigator Award. Comments on this book can be sent to the author by electronic mail: book@galileo.stanford.edu. An instructor's manual, with solutions to the problems, is available from the publisher and foils for overhead projection are available from the author.

Giovanni De Micheli

SYNTHESIS AND OPTIMIZATION
OF DIGITAL CIRCUITS

PART
I

CIRCUITS
AND
MODELS

CHAPTER
1

INTRODUCTION

Wofür arbeitet ihr? Ich halte dafür, dass das einzige Ziel der Wissenschaft darin besteht, die Mühseligkeit der menschlichen Existenz zu erleichtern.

What are you working for? I maintain that the only purpose of science is to ease the hardship of human existence.

B. Brecht. Leben des Galilei.

1.1 MICROELECTRONICS

Microelectronics has been the enabling technology for the development of hardware and software systems in the recent decades. The continuously increasing level of integration of electronic devices on a single substrate has led to the fabrication of increasingly complex systems. The integrated circuit technology, based upon the use of semiconductor materials, has progressed tremendously. While a handful of devices were integrated on the first circuits in the 1960s, circuits with over one million devices have been successfully manufactured in the late 1980s. Such circuits, often called *Very Large Scale Integration* (VLSI) or *microelectronic* circuits, testify to the acquired ability of combining design skills with manufacturing precision (Figure 1.1).

Designing increasingly integrated and complex circuits requires a larger and larger capital investment, due to the cost of refining the precision of the manufacturing process so that finer and finer devices can be implemented. Similarly, the growth in

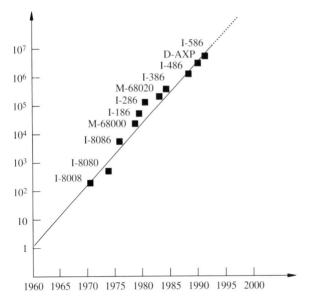

FIGURE 1.1
Moore's law showing the growth of microelectronic circuit size, measured in terms of devices per chip.

scale of the circuits requires larger efforts in achieving zero-defect designs. A particular feature of microelectronics is the near impossibility to repair integrated circuits, which raises the importance of designing circuits correctly and limiting the manufacturing defects.

The economics of VLSI circuits relies on the principle that replicating a circuit is straightforward and therefore the design and manufacturing costs can be recovered over a large volume of sales. The trend toward higher integration is economically positive because it leads to a reduction in the number of components per system, and therefore to a reduced cost in packaging and interconnection, as well as to a higher reliability of the overall system. More importantly, higher integration correlates to faster circuits, because of the reduction of parasitic effects in the electronic devices themselves and in the interconnections. As a result, the higher the integration level is, the better and cheaper the final product results.

This economic analysis is valid under the simplifying assumption that the volume of sales of a circuit (or of a system with embedded circuits) is large enough to recapture the manufacturing and design costs. In reality, only few circuit applications can enjoy a high volume of sales and a long life. Examples have been some general purpose microprocessors. Unfortunately, the improvement in circuit technology makes the circuits obsolete in a short time.

At present, many electronic systems require integrated dedicated components that are specialized to perform a task or a limited set of tasks. These are called *Application Specific Integrated Circuits*, or ASICs, and occupy a large portion of the market share. Some circuits in this class may not be produced in large volume because of the specificity of their application.

Thus other factors are important in microelectronic circuit economics. First, the use of particular design styles according to the projected volume of sales. Second, the reduction of design time, which has two implications. A reduction of design cost, associated with the designers' salaries and a reduction of the time-to-market. This last factor has been shown to be key to the profitability of many applications. Third, the quality of the circuit design and fabrication, measured in performance and manufacturing yield.

Computer-Aided Design (CAD) techniques play a major role in supporting the reduction of design time and the optimization of the circuit quality. Indeed, as the level of integration increases, carrying out a full design without errors becomes an increasingly difficult task to achieve for a team of human designers. The size of the circuit, usually measured in terms of the number of active devices (e.g., transistors), requires handling the design data with systematic techniques. In addition, optimizing features of large-scale circuits is even a more complex problem, because the feasible implementation choices grow rapidly in number with the circuit size.

Computer-aided design tools have been used since the inception of the integrated circuits. CAD systems support different facets of the design of microelectronic circuits. CAD techniques have reached a fairly good level of maturity in many areas, even though there are still niches with open problems. Moreover, the continuous growth in size of the circuits requires CAD tools with increased capability. Therefore continuous research in this field has targeted design and optimization problems for large-scale circuits, with particular emphasis on the design of scalable tools. The CAD industry has grown to be a substantial sector of the global electronic industry. Overall, CAD is as strategic for the electronic industry as the manufacturing technology is. Indeed, CAD is the codification of the design-flow technology.

In the following sections, we consider first the different microelectronic circuit technologies and design styles. We consider then the design flow and its abstraction as a series of tasks. We explain the major issues in synthesis and optimization of digital synchronous circuits, that are the subject of this book. Finally, we present the organization of the book and comment on some related topics.

1.2 SEMICONDUCTOR TECHNOLOGIES AND CIRCUIT TAXONOMY

Microelectronic circuits exploit the properties of semiconductor materials. The circuits are constructed by first patterning a substrate and locally modifying its properties and then by shaping layers of interconnecting wires. The fabrication process is often very complex. It can be classified in terms of the type of semiconductor material (e.g., silicon, gallium-arsenide) and in terms of the electronic device types being constructed. The most common circuit technology families for silicon substrates are *Complementary Metal Oxide Semiconductor* (CMOS), *Bipolar* and a combination of the two called *BiCMOS*. Even within a family, such as CMOS, different specific technologies have been developed. Examples for CMOS are: *single-well* (P or N), *twin-well* and *silicon on sapphire* (SOS). Specific technologies target particular markets: for example, SOS is less sensitive to radiation and proper for space applications.

Electronic circuits are generally classified as *analog* or *digital*. In the former class, the information is related to the value of a continuous electrical parameter, such as a voltage or a current. A power amplifier for an audio stereo system employs analog circuits. In digital circuits, the information is quantized. Most digital circuits are binary, and therefore the quantum of information is either a TRUE or a FALSE value. The information is related to ranges of voltages at circuit internal nodes. Digital circuits have shown to be superior to analog circuits in many classes of applications, primarily electronic computing. Recently, digital applications have pervaded also the signal processing and telecommunication fields. Examples are compact disc players and digital telephony. Today the number of digital circuits overwhelms that of analog circuits.

Digital circuits can be classified in terms of their mode of operations. *Synchronous* circuits are characterized by having information storage and processing synchronized to one (or more) global signal, called *clock*. *Asynchronous* circuits do not require a global clock signal. Different styles have been proposed for asynchronous design methods. When considering large-scale integration circuits, synchronous operation is advantageous because it is conceptually simpler; there is a broader common knowledge of design and verification methods and therefore the likelihood of final correct operation is higher. Furthermore, testing techniques have been developed for synchronous digital circuits. Conversely, the advantage of using asynchronous circuits stems from the fact that global clock distribution becomes more difficult as circuits grow in size, therefore invalidating the premise that designing synchronous circuits is simpler. Despite the growing interest in asynchronous circuits, most commercial digital designs use synchronous operation.

It is obvious that the choice of a semiconductor technology and the circuit class affect the type of CAD tools that are required. The choice of a semiconductor technology affects mainly the physical design of a circuit, i.e., the determination of the geometric patterns. It affects only marginally functional design. Conversely, the circuit type (i.e., analog, synchronous digital or asynchronous digital) affects the overall design of the circuit. In this book we consider CAD techniques for synchronous digital circuits, because they represent the vast majority of circuits and because these CAD design techniques have reached a high level of maturity.

1.3 MICROELECTRONIC DESIGN STYLES

The economic viability of microelectronic designs relies on some conflicting factors, such as projected volume, expected pricing and circuit performance required to be competitive. For example, instruction-set microprocessors require competitive performance and price; therefore high volumes are required to be profitable. A circuit for a space-based application may require high performance and reliability. The projected volume may be very low, but the cost may be of no concern because still negligible with respect to the cost of a space mission. A circuit for consumer applications would require primarily a low cost to be marketable.

Different design styles, often called methodologies, have been used for microelectronic circuits. They are usually classified as *custom* and *semicustom* design styles. In the former case, the functional and physical design are handcrafted, requiring an extensive effort of a design team to optimize each detailed feature of the circuit. In this case, the design effort and cost are high, often compensated by the achievement of high-quality circuits. It is obvious that the cost has either to be amortized over a large volume production (as in the case of processor design) or borne in full (as in the case of special applications). Custom design was popular in the early years of microelectronics. Today, the design complexity has confined custom design techniques to specific portions of a limited number of projects, such as execution and floating point units of some processors.

Semicustom design is based on the concept of restricting the circuit primitives to a limited number, and therefore reducing the possibility of fine-tuning all parts of a circuit design. The restriction allows the designer to leverage well-designed and well-characterized primitives and to focus on their interconnection. The reduction of the possible number of implementation choices makes it easier to develop CAD tools for design and optimization, as well as reducing the design time. The loss in quality is often very small, because fine-tuning a custom design may be extremely difficult when this is large, and automated optimization techniques for semicustom styles can explore a much wider class of implementation choices than a designer team can afford. Today the number of semicustom designs outnumbers custom designs. Recently, some high-performance microprocessors have been designed fully (and partially) using semicustom styles.

Semicustom designs can be partitioned in two major classes: *cell-based design* and *array-based design*. These classes further subdivide into subclasses, as shown in Figure 1.2. Cell-based design leverages the use of *library cells*, that can be designed once and stored, or the use of *cell generators* that synthesize macro-cell layouts from their functional specifications. In cell-based design the manufacturing process is not simplified at all with respect to custom design. Instead, the design process is simplified, because of the use of ready-made building blocks.

Cell-based design styles include *standard-cell design* (Figure 1.3). In this case, the fundamental cells are stored in a library. Cells are designed once, but updates are required as the progress in semiconductor technology allows for smaller geometries. Since every cell needs to be parametrized in terms of area and delay over ranges of temperatures and operating voltages, the library maintenance is far from a trivial task. The user of a standard-cell library must first conform his or her design to the available library primitives, a step called *library binding* or *technology mapping*. Then, cells are placed and wired. All these tasks have been automated. An extension is the *hierarchical standard-cell* style, where larger cells can be derived by combining smaller ones.

Macro-cell based design consists of combining building blocks that can be synthesized by computer programs, called *cell* or *module generators*. These programs vary widely in capabilities and have evolved tremendously over the last two decades. The first generators targeted the automatic synthesis of memory arrays and programmable-logic arrays (PLAs). Recently highly sophisticated generators have been able to synthesize the layout of various circuits with a device density and performance equal or superior to that achievable by human designers. The user of macro-cell generators has just to provide the functional description. Macro-cells are then placed and wired. Even though these steps have been automated, they are inherently more difficult and may be less efficient as compared to standard-cell placement and wiring, due to the irreg-

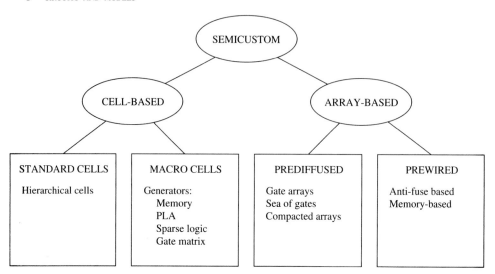

FIGURE 1.2
Semicustom design styles: a taxonomy.

ularity in size of the macro-cells. A major advantage of cell-based design (standard and macro-cells) is the compatibility with custom design. Indeed custom components can be added to a semicustom layout, and vice versa. The combination of custom design and cell-based design has been used often for microprocessor design, and also referred to as *structured custom* design (Figure 1.4).

Array-based design exploits the use of a matrix of uncommitted components, often called *sites*. Such components are then personalized and interconnected. Array-based circuits can be further classified as *prediffused* and *prewired*, also called *mask programmable* and *field programmable* gate arrays, respectively (MPGAs and FP-GAs). In the first case batches of wafers, with arrays of sites, are manufactured. The chip fabrication process entails programming the sites by contacting them to wires, i.e., by manufacturing the routing layers. As a result, only the metal and contact layers are used to program the chip, hence the name "mask programmable." Fewer manufacturing steps correlate to lower fabrication time and cost. In addition, the cost of the prediffused batches of wafers can be amortized over several chip designs.

There are several styles of prediffused wafers, that often go under the names of gate arrays, compacted arrays™ and sea-of-gates (Figure 1.5). The circuit designer that uses these styles performs tasks similar to those used for standard cells. He or she must first conform the design to the available sites, described by a set of library primitives, then apply placement and wiring, that are subject to constraints typical of the array image, e.g., slot constraints for the position of the sites. Also in this case, all these tasks have been automated.

Prewired gate arrays have been introduced recently and are often called "field programmable gate arrays" because these can be programmed in the "field," i.e., outside the semiconductor foundry. They consist of arrays of programmable modules, each having the capability of implementing a generic logic function (Figure 1.6).

FIGURE 1.3
One of AT&T's Application Specific Standard Product chips. The chip was designed and laid out using
AT&T CAD tools, with a standard cell design style. (Courtesy of AT&T.)

Programming is achieved in two ways. Wires, already present in the form of segments,
can be connected by programming antifuses.[1] Generic functions can be specialized by
connecting module inputs to voltage rails or together. Alternatively, memory elements
inside the array can be programmed to store information that relates to the module
configuration and interconnection.

---

[1]A fuse is a short circuit device that becomes an open circuit when traversed by an appropriate
current pulse. Conversely, an antifuse is an open circuit device that becomes a short circuit when traversed
by an appropriate current pulse.

FIGURE 1.4
Microphotograph of the Alpha AXP chip by Digital Equipment Corporation, using several macro-cells designed with proprietary CAD tools. (Courtesy of Digital Equipment Corporation.)

In a prewired circuit, manufacturing is completely carried out independently of the application, hence amortized over a large volume. The design and customization can be done on the field, with negligible time and cost. A drawback of this design style is the low capacity of these arrays and the inferior performance as compared to other design styles. However, due to the relative immaturity of the underlying technology, it is conceivable that capacity and speed of prewired circuits will improve substantially in the future. At present, prewired circuits provide an excellent means for prototyping.

We would now like to compare the different design styles, in terms of density, performance, flexibility, design and manufacturing times and cost (for both low and high volumes). Manufacturing time of array-based circuits is the time spent to customize them. The comparison is reported in the following table, which can serve as

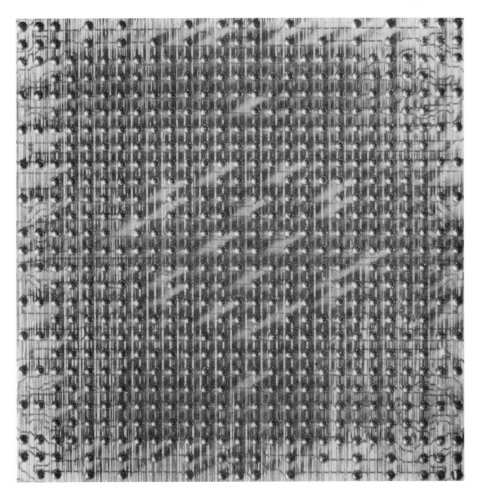

FIGURE 1.5
A mask programmable gate array from the IBM Enterprise System/9000 air-cooled processor technology.
(Courtesy and copyright of IBM, reprinted with permission.)

a guideline in choosing a style for a given application and market:

	Custom	*Cell-based*	*Prediffused*	*Prewired*
Density	Very High	High	High	Medium-Low
Performance	Very High	High	High	Medium-Low
Flexibility	Very High	High	Medium	Low
Design time	Very Long	Short	Short	Very Short
Manufacturing time	Medium	Medium	Short	Very Short
Cost - low volume	Very High	High	High	Low
Cost - high volume	Low	Low	Low	High

Even though custom is the most obvious style for general purpose circuits and semicustom is the one for application-specific circuits, processor implementations

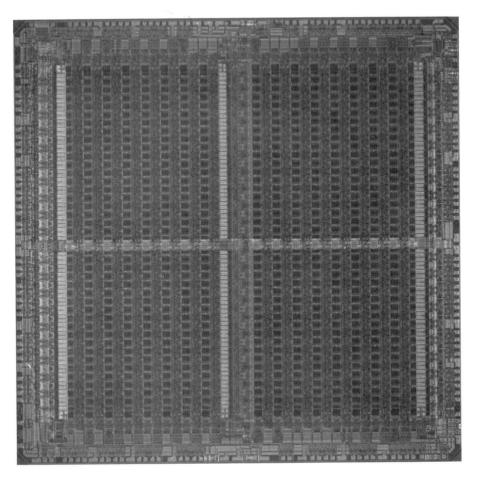

FIGURE 1.6
A field-programmable gate array: ACT^{TM} 2 by Actel Corporation. (Courtesy of ACTEL Corporation.)

have been done in semicustom and ASIC in custom, as in the case of some space applications. It is important to stress that ASIC is not a synonym for semicustom style, as often erroneously considered.

1.4 DESIGN OF MICROELECTRONIC CIRCUITS

There are four stages in the creation of an integrated circuit: *design, fabrication, testing* and *packaging* [9, 10] (Figure 1.7). We shall consider here the design stage in more detail, and we divide it into three major tasks: *conceptualization* and *modeling*, *synthesis* and *optimization*, and *validation*. The first task consists of casting an idea into a model, which captures the function that the circuit will perform. Synthesis consists of refining the model, from an abstract one to a detailed one, that has all the features

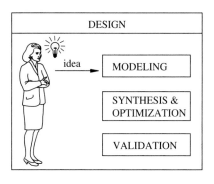
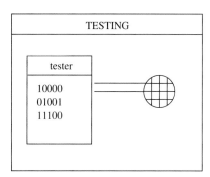
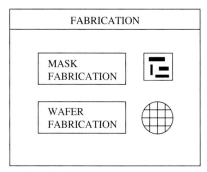
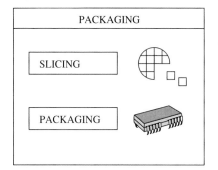

FIGURE 1.7
The four phases in creating a microelectronic chip.

required for fabrication. An objective of design is to maximize some figures of merit of the circuit that relate to its quality. Validation consists of verifying the consistency of the models used during design, as well as some properties of the original model.

Circuit models are used to represent ideas. Modeling plays a major role in microelectronic design, because it represents the vehicle used to convey information. Modeling must be rigorous as well as general, transparent to the designer and machine-readable. Today, most modeling is done using *Hardware Description Languages*, or HDLs. There are different flavors of HDLs, which are described in detail in Chapter 3. Graphic models are also used, such as flow diagrams, schematic diagrams and geometric layouts. The very large-scale nature of the problem forces the modeling style, both textual and graphical, to support hierarchy and abstractions. These allow a designer to concentrate on a portion of the model at any given time.

Circuit synthesis is the second creative process. The first is performed in the designer's mind when conceptualizing a circuit and sketching a first model. The overall goal of circuit synthesis is to generate a detailed model of a circuit, such as a geometric layout, that can be used for fabricating the chip. This objective is achieved by means of a stepwise refinement process, during which the original abstract model provided by the designer is iteratively detailed. As synthesis proceeds in refining the model, more information is needed regarding the technology and the desired design implementation style. Indeed a functional model of a circuit may be fairly independent from the implementation details, while a geometric layout model must incorporate all technology-dependent specifics, such as, for example, wire widths.

Circuit optimization is often combined with synthesis. It entails the selection of some particular choices in a given model, with the goal of raising one (or more) figures of merit of the design. The role of optimization is to enhance the overall quality of the circuit. We explain now in more detail what quality means. First of all, we mean circuit *performance*. Performance relates to the time required to process some information, as well as to the amount of information that can be processed in a given time period. Circuit performance is essential to competitive products in many application domains. Second, circuit quality relates to the overall *area*. An objective of circuit design is to minimize the area, for many reasons. Smaller circuits relate to more circuits per wafer, and therefore to lower manufacturing cost. The manufacturing yield[2] decreases with an increase in chip area. Large chips are more expensive to package too. Third, the circuit quality relates to the *testability*, i.e., to the ease of testing the chip after manufacturing. For some applications, the *fault coverage* is an important quality measure. It spells the percentage of faults of a given type that can be detected by a set of test vectors. It is obvious that testable chips are desirable, because earlier detection of malfunctions in electronic systems relates to lower overall cost.

Circuit validation consists of acquiring a reasonable certainty level that a circuit will function correctly, under the assumption that no manufacturing fault is present. Circuit validation is motivated by the desire to remove all possible design errors, before proceeding to the expensive chip manufacturing. It can be performed by *simulation* and by *verification* methods. Circuit simulation consists of analyzing the circuit variables over an interval of time for one specific set (or more sets) of input patterns. Simulation can be applied at different levels, corresponding to the model under consideration. Simulated output patterns must then conform to the expected ones. Even though simulation is the most commonly used way of validating circuits, it is often ineffective for large circuits except for detecting major design flaws. Indeed, the number of relevant input pattern sequences to analyze grows with the circuit size and designers must be content to monitor a small subset. Verification methods, often called *formal verification* methods, consist of comparing two circuit models, and to detect their consistency. Another facet of circuit verification is checking some properties of a circuit model, such as, for example, whether there are deadlock conditions.

1.5 COMPUTER-AIDED SYNTHESIS AND OPTIMIZATION

Computer-aided tools provide an effective means for designing microelectronic circuits that are economically viable products. Synthesis techniques speed up the design cycle

---

[2]The yield is the percentage of manufactured chips that operate correctly. In general, when fabrication faults relate to spots, the defect density per unit area is constant and typical of a manufacturing plant and process. Therefore the probability that one spot makes the circuit function incorrectly increases with the chip area.

and reduce the human effort. Optimization techniques enhance the design quality. At present, synthesis and optimization techniques are used for most digital circuit designs. Nevertheless their power is not yet exploited in full. It is one of the purposes of the book to foster the use of synthesis and optimization techniques, because they can be instrumental in the progress of electronic design.

We report in this book on circuit synthesis and optimization for digital synchronous circuits. We consider now briefly the circuit models and views and then comment on the major classification of the synthesis tasks.

1.5.1   Circuit Models

A model of a circuit is an abstraction, i.e., a representation that shows relevant features without associated details. Synthesis is the generation of a circuit model, starting from a less detailed one. Models can be classified in terms of *levels of abstraction* and *views*. We consider here three main abstractions, namely: *architectural*, *logic* and *geometrical*. The levels can be visualized as follows. At the *architectural* level a circuit performs a set of operations, such as data computation or transfer. At the *logic* level, a digital circuit evaluates a set of logic functions. At the *geometrical* level, a circuit is a set of geometrical entities. Examples of representations for architectural models are HDL models or flow diagrams; for logic models are state transition diagrams or schematics; for geometric models are floor plans or layouts.

> **Example 1.5.1.**   A simple example of the different modeling levels is shown in Figure 1.8. At the architectural level, a processor is described by an HDL model. A schematic captures the logic-level specification. A two-dimensional geometric picture represents the mask layout.
>
> Design consists of refining the abstract specification of the architectural model into the detailed geometrical-level model, that has enough information for the manufacturing of the circuit.

We consider now the views of a model. They are classified as: *behavioral*, *structural* and *physical*. *Behavioral views* describe the function of the circuit regardless of its implementation. *Structural views* describe a model as an interconnection of components. *Physical views* relate to the physical objects (e.g., transistors) of a design.

Models at different levels can be seen under different views. For example, at the architectural level, a behavioral view of a circuit is a set of operations and their dependencies. A structural view is an interconnection of the major building blocks. As another example, consider the logic-level model of a synchronous circuit. A behavioral view of the circuit may be given by a state transition diagram, while its structural view is an interconnection of logic gates. Levels of abstractions and views are synthetically represented by Figure 1.9, where views are shown as the segments of the letter Y. This axial arrangement of the views is often referred to as Gajski and Kuhn's Y-chart [3].

> **Example 1.5.2.**   Consider again the circuit of Example 1.5.1. Figure 1.10 highlights behavioral and structural views at both the architectural and the logic levels.

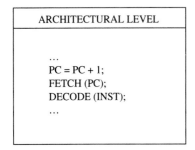

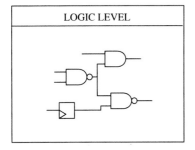

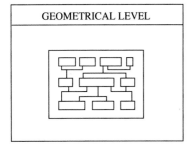

FIGURE 1.8
Three abstraction levels of a circuit representation.

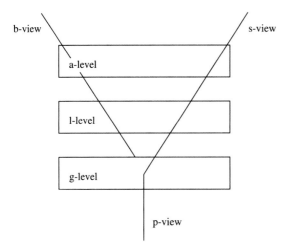

FIGURE 1.9
Circuit views and levels of abstractions.

BEHAVIORAL VIEW	STRUCTURAL VIEW	VIEWS / LEVELS

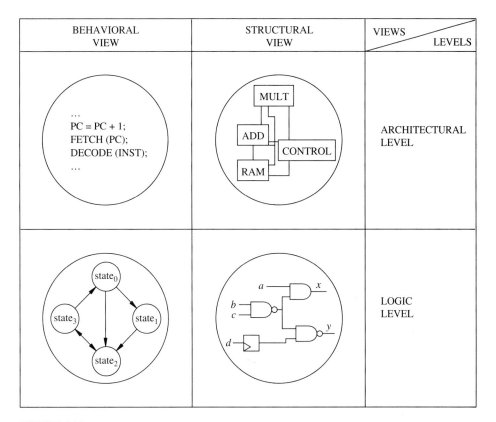

FIGURE 1.10
Levels of abstractions and corresponding views.

1.5.2 Synthesis

The model classification relates to a taxonomy of the synthesis tasks. Synthesis can be seen as a set of transformations between two axial views. In particular we can distinguish the synthesis subtasks at the different modeling levels as follows:

- *Architectural-level synthesis* consists of generating a structural view of an architectural-level model. This corresponds to determining an assignment of the circuit functions to operators, called *resources*, as well as their interconnection and the timing of their execution. It has also been called *high-level synthesis* or *structural synthesis*, because it determines the macroscopic (i.e., block-level) structure of the circuit. To avoid ambiguity, and for the sake of uniformity, we shall call it *architectural synthesis*.

- *Logic-level synthesis* is the task of generating a structural view of a logic-level model. *Logic synthesis* is the manipulation of logic specifications to create logic models as an interconnection of logic primitives. Thus logic synthesis determines the microscopic (i.e., gate-level) structure of a circuit. The task of transforming a

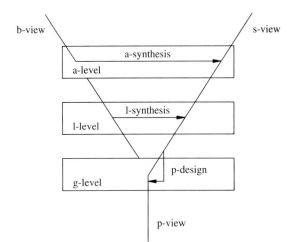

FIGURE 1.11
Levels of abstractions, views and
synthesis tasks.

logic model into an interconnection of instances of library cells, i.e., the back end
of logic synthesis, is often referred to as *library binding* or *technology mapping*.

- *Geometrical-level synthesis* consists of creating a physical view at the geometric
  level. It entails the specification of all geometric patterns defining the physical
  layout of the chip, as well as their position. It is often called *physical design*, and
  we shall call it so in the sequel.

The synthesis tasks are synthetically depicted in Figure 1.11. We now describe
these tasks in more detail. We consider them in the order that corresponds to their use
in a top-down synthesis system. This sequence is the converse of that corresponding
to their historical development and level of maturity.

ARCHITECTURAL SYNTHESIS. A behavioral architectural-level model can be abstracted as a set of operations and dependencies. Architectural synthesis entails identifying the hardware resources that can implement the operations, *scheduling* the execution time of the operations and *binding* them to the resources. In other words,
synthesis defines a structural model of a *data path*, as an interconnection of resources,
and a logic-level model of a *control unit*, that issues the control signals to the data
path according to the schedule.

The macroscopic figures of merit of the implementation, such as circuit area and
performance, depend heavily on this step. Indeed, architectural synthesis determines
the degree of parallelism of the operations. Optimization at this level is very important,
as mentioned in Section 1.5.3. Architectural synthesis is described in detail in Part II.

Example 1.5.3. We consider here first an example of a behavioral view of an architectural model. The example has been adapted from one proposed by Paulin and Knight
[6]. It models a circuit designed to solve numerically (by means of the *forward Euler
method*) the following differential equation: $y'' + 3xy' + 3y = 0$ in the interval $[0, a]$
with step-size dx and initial values $x(0) = x$; $y(0) = y$; $y'(0) = u$.

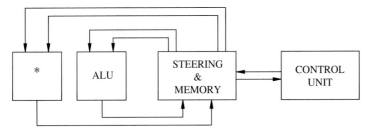

FIGURE 1.12
Example of structural view at the architectural level.

The circuit can be represented by the following HDL model:

```
diffeq {
        read ( x, y, u, dx, a );
        repeat  {
               x1  =    x + dx;
               u1  =    u - ( 3 * x * u * dx ) - ( 3 * y * dx );
               y1  =    y + u * dx;
               c   =    x1 <  a;
               x = x1 ; u = u1 ; y = y1;
               }
        until ( c );
        write ( y );
}
```

Let us now consider the architectural synthesis task. For the sake of simplicity, let us assume that the data path of the circuit contains two resources: a multiplier and an ALU, that can perform addition/subtraction and comparison. The circuit contains also registers, steering logic and a control unit.

A structural view of the circuit, at the architectural level, shows the macroscopic structure of the implementation. This view can be described by a block diagram, as in Figure 1.12, or equivalently by means of a structural HDL.

LOGIC SYNTHESIS. A logic-level model of a circuit can be provided by a state transition diagram of a *finite-state machine*, by a circuit schematic or equivalently by an HDL model. It may be specified by a designer or synthesized from an architectural-level model.

The logic synthesis tasks may be different according to the nature of the circuit (e.g., sequential or combinational) and to the starting representation (e.g., state diagram or schematic). The possible configurations of a circuit are many. Optimization plays a major role, in connection with synthesis, in determining the microscopic figures of merit of the implementation, as mentioned in Section 1.5.3. The final outcome of logic synthesis is a fully structural representation, such as a gate-level netlist. Logic synthesis is described in detail in Part III.

Example 1.5.4. Consider the control unit of the previous circuit. Its task is to sequence the operations in the data path, by providing appropriate signals to the resources. This is achieved by steering the data to resources and registers in the block called "Steering &

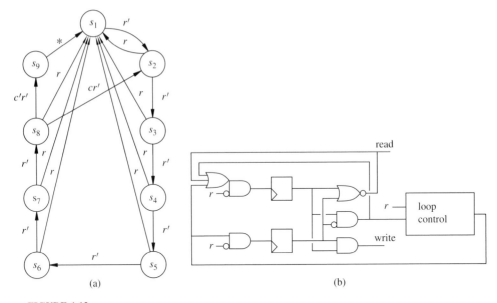

FIGURE 1.13
(a) Example of behavioral view at the logic level. (b) Example of structural view at the logic level.

Memory," in Figure 1.12. A behavioral view of the control unit at the logic level is given by a transition diagram, sketched in Figure 1.13 (a), where the signals to the "Steering & Memory" block are not shown for the sake of simplicity. The control unit uses one state for reading the data (reset state s_1), one state for writing the data (s_9) and seven states for executing the loop. Signal r is a reset signal.

A structural view of the control unit at the logic level is shown by the hierarchical schematic of Figure 1.13 (b), which shows the logic gates that implement the transitions among the states $\{s_1, s_2, s_9\}$ and that enable the reading and writing of the data. The subcircuit that controls the iteration is represented by the box labeled "loop control."

PHYSICAL DESIGN. Physical design consists of generating the layout of the chip. The layers of the layout are in correspondence with the masks used for chip fabrication. Therefore, the geometrical layout is the final target of microelectronic circuit design. Physical design depends much on the design style. On one end of the spectrum, for custom design, physical design is handcrafted by using layout editors. This means that the designer renounces the use of automated synthesis tools in the search for optimizing the circuit geometries by fine hand-tuning. On the opposite end of the spectrum, in the case of prewired circuits, physical design is performed in a virtual fashion, because chips are fully manufactured in advance. Instead, chip personalization is done by a *fuse map* or by a *memory map*.

The major tasks in physical design are *placement* and *wiring*, called also *routing*. *Cell generation* is essential in the particular case of macro-cell design, where cells are synthesized and not extracted from a library. These problems and their solutions are not described in this book. A brief survey is reported in Section 1.7.1.

1.5.3 Optimization

Circuit optimization is often performed in conjunction with synthesis. Optimization is motivated not only by the desire of maximizing the circuit quality, but also by the fact that synthesis without optimization would yield noncompetitive circuits at all, and therefore its value would be marginal. In this book, we consider the optimization of two quality measures, namely, *area* and *performance*. We shall comment also on the relation to *testability* in Sections 6.8, 7.2.4 and 8.5.

Circuit area is an *extensive* quantity. It is measured by the sum of the areas of the circuit components and therefore it can be computed from a structural view of a circuit, once the areas of the components are known. The area computation can be performed hierarchically. Usually, the fundamental components of digital circuits are logic gates and registers, whose area is known *a priori*. The wiring area often plays an important role and should not be neglected. It can be derived from a complete physical view, or estimated from a structural view using predictive or statistical models.

Circuit performance is an *intensive* quantity. It is not additive, and therefore its computation requires analyzing the structure and often the behavior of a circuit. To be more specific, we need to consider now the meaning of performance in more detail, according to the different classes of digital circuits.

The performance of a combinational logic circuit is measured by the input/output *propagation delay*. Often a simplifying assumption is made. All inputs are available at the same time and the performance relates to the delay through the *critical path* of the circuit.

The performance of a synchronous sequential circuit can be measured by its *cycle-time*, i.e., the period of the fastest clock that can be applied to the circuit. Note that the delay through the combinational component of a sequential circuit is a lower bound on the cycle-time.

When considering an architectural-level model of a circuit as a sequence of operations, with a synchronous sequential implementation, one measure of performance is circuit *latency*, i.e., the time required to execute the operations [4]. Latency can be measured in terms of clock-cycles. Thus the product of the *cycle-time* and *latency* determines the overall execution time. Often, *cycle-time* and *latency* are optimized independently, for the sake of simplifying the optimization problem as well as of satisfying other design constraints, such as interfacing to other circuits.

Synchronous circuits can implement a sequence of operations in a *pipeline* fashion, where the circuit performs concurrently operations on different data sets. An additional measure of performance is then the rate at which data are produced and consumed, called the *throughput* of the circuit. Note that in a non-pipelined circuit, the throughput is less than (or equal to) the inverse of the product of the cycle-time times the latency. Pipelining allows a circuit to increase its throughput beyond this limit. Maximum-rate pipelining occurs when the throughput is the inverse of the cycle-time, i.e., when data are produced and consumed at each clock cycle.

According to these definitions and models, performance optimization consists in minimizing the delay (for combinational circuits), the cycle-time and the latency (for synchronous circuits) and maximizing the throughput (for pipelined circuits).

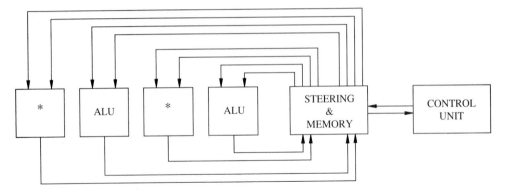

FIGURE 1.14
Alternative example of structural view at the architectural level.

Example 1.5.5. Consider the circuit of Example 1.5.3. For the sake of simplicity, we assume that all resources require one cycle to execute. The structure shown in Figure 1.12 has a minimal resource usage, but it requires seven cycles to execute one iteration of the loop. An alternative structure is shown in Figure 1.14, requiring a larger area (two multipliers and two ALUs) but just four cycles. (A detailed explanation of the reason is given in Chapters 4 and 5.) The two structures correspond to achieving two different optimization goals.

We consider here design optimization as the combined minimization of area and maximization of performance. Optimization may be subject to constraints, such as upper bounds on the area, as well as lower bounds on performance. We can abstract the optimization problem as follows. The different feasible structural implementations of a circuit define its *design space*. The design space is a finite set of design points. Associated with each design point, there are values of the area and performance evaluation functions. There are as many evaluation functions as the design *objectives* of interest, such as area, latency, cycle-time and throughput. We call *design evaluation space* the multidimensional space spanned by these objectives.

Example 1.5.6. Let us construct a simplified design space for the previous example, by just considering architectural-level modeling. For the sake of the example, we assume again that the resources execute in one cycle and the delay in the steering and control logic is negligible. Thus we can drop cycle-time issues from consideration.

The design space corresponds to the structures with a_1 multipliers and a_2 ALUs, where a_1, a_2 are positive integers. Examples for $a_1 = 1; a_2 = 1$ and for $a_1 = 2; a_2 = 2$ were shown in Figures 1.12 and 1.14, respectively.

For the sake of the example, let us assume that the multiplier takes five units of area, the ALU 1 unit, and the control unit, steering logic and memory and an additional unit. Then, in the design evaluation space, the first implementation has an area equal to seven and latency proportional to seven. The second implementation has an area equal to 13 and latency proportional to four.

When considering the portion of the design space corresponding to the following pairs for (a_1, a_2): $\{(1, 1); (1, 2); (2, 1); (2, 2)\}$, the design evaluation space is shown in

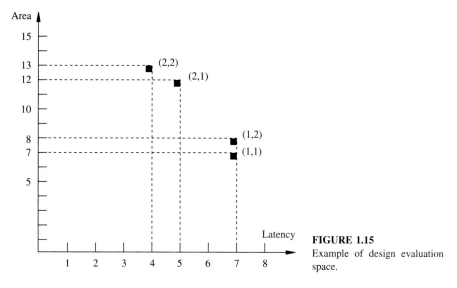

FIGURE 1.15
Example of design evaluation
space.

Figure 1.15. The numbers in parentheses relate to the corresponding points in the design space.

Circuit optimization corresponds to searching for the best design, i.e., a circuit configuration that optimizes all objectives. Since our optimization problem involves *multiple criteria*, special attention must be given to defining points of optimality. For the sake of simplicity and clarity, and without loss of generality, we assume now that the optimization problem corresponds to a minimization one. Note that maximizing the throughput corresponds to minimizing its complement.

A point of the design space is called a *Pareto point* if there is no other point (in the design space) with at least an inferior objective, all others being inferior or equal [1]. A Pareto point corresponds to a global optimum in a monodimensional design evaluation space. It is a generalization of this concept in the more general context. Note that there may be many Pareto points, corresponding to the design implementations not dominated by others and hence worth consideration.

The image of the Pareto points in the design evaluation space is the set of the optimal *trade-off* points. Their interpolation yields a *trade-off curve* or *surface*.

Example 1.5.7. Consider the design space and the design evaluation space of Example 1.5.6. The point (1, 2) of the design space is not a Pareto point, because it is dominated by point (1, 1) that has equal latency and smaller area. In other words, adding an ALU to an implementation with one multiplier would make its area larger without reducing the latency: hence this circuit structure can be dropped from consideration. The remaining points {(1, 1); (2, 1); (2, 2)} are Pareto points. It is possible to show that no other pair of positive integers is a Pareto point. Indeed the latency cannot be reduced further by additional resources.

We now consider examples of design (evaluation) spaces for different circuit classes.

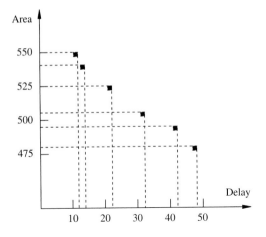

FIGURE 1.16
Design evaluation space: area/delay trade-off points for a 64-bit adder. (Area and delay are measured in terms of equivalent gates.)

COMBINATIONAL LOGIC CIRCUITS. In this case, the design evaluation space objectives are (*area, delay*). Figure 1.16 shows an example of the trade-off points in this space, for multiple-level logic circuit implementations. These points can be interpolated to form an area/delay trade-off curve.

> **Example 1.5.8.** A first example of area/delay trade-off is given by the implementation of integer adders. Ripple-carry adders have smaller area and larger delay than carry look-ahead adders. Figure 1.16 shows the variation in area and in delay of gate-level implementations of a 64-bit integer adder ranging from a two-level carry look-ahead to a ripple-carry style.
>
> As a second example, consider the implementation of a circuit whose logic behavior is the product of four Boolean variables (e.g., $x = pqrs$). Assume that the implementation is constrained to using two-input and/or three-input AND gates, whose area and delay are proportional to the number of inputs. Assume also that the inputs arrive simultaneously. Elements of the design space are shown in Figure 1.17, namely, four different logic structures.
>
> The corresponding design evaluation space is shown in Figure 1.18. Note that structures labeled c and d in Figure 1.17 are not Pareto points.

SEQUENTIAL LOGIC CIRCUITS. When we consider logic-level models for sequential synchronous circuits, the design evaluation space objectives are: (*area, cycle-time*), where the latter is bounded from below by the critical path delay in the combinational logic component. Therefore, the design evaluation space is similar in concept to that for combinational circuits, shown in Figures 1.16 and 1.18.

When considering nonpipelined architectural-level models, the design evaluation space objectives are: (*area, latency, cycle-time*), where the product of the last two yields the overall execution delay. An example of the design evaluation space is shown in Figure 1.19. Slices of the design evaluation space for some specific values of one objective are also of interest. An example is the (*area, latency*) trade-off, as shown in Figure 1.15.

Last, let us consider the architectural-level model of synchronous pipelined circuits. Now, performance is measured also by the circuit throughput. Hence, the design

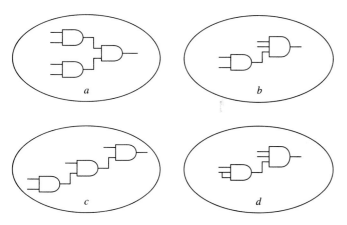

FIGURE 1.17
Example of design space: four implementations of a logic function.

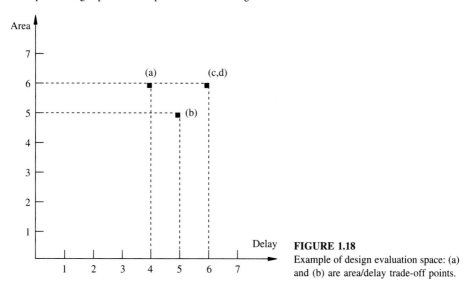

FIGURE 1.18
Example of design evaluation space: (a) and (b) are area/delay trade-off points.

evaluation space involves the quadruple (*area, latency, cycle-time, throughput*). Architectural trade-offs for pipelined circuit implementations will be described in Section 4.8, where examples will also be given.

GENERAL APPROACHES TO OPTIMIZATION. Circuit optimization involves multiple objective functions. In the previous examples, the optimization goals ranged from two to four objectives. The optimization problem is difficult to solve, due to the discontinuous nature of the objective functions and to the discrete nature of the design space, i.e., of the set of feasible circuit implementations.

In general, Pareto points are solutions to constrained optimization problems. Consider, for example, combinational logic optimization. Then the following two

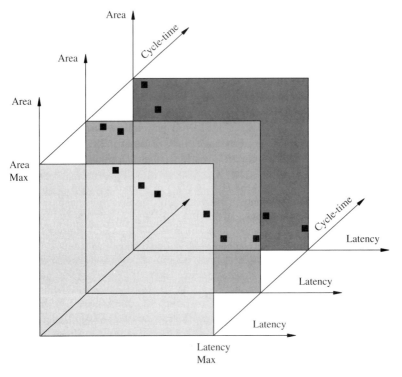

FIGURE 1.19
Design evaluation space: area/latency/cycle-time trade-off points.

problems are of interest:

- Minimize the circuit *area* under *delay constraints.*
- Minimize the circuit *delay* under *area constraints.*

Unfortunately, due to the difficulty of the optimization problems, only approximations to the Pareto points can be computed.

Consider next architectural-level models of synchronous circuits. Pareto points are solutions to the following problems, for different values of the *cycle-time*:

- Minimize the circuit *area* under *latency constraints.*
- Minimize the circuit *latency* under *area constraints.*

These two problems are often referred to as *scheduling* problems. Unfortunately again, the scheduling problems are hard to solve exactly in most cases and only approximations can be obtained. When considering pipelined circuits, Pareto points can be computed (or approximated) by considering different values of *cycle-time* and *throughput*, and by solving the corresponding scheduling problems.

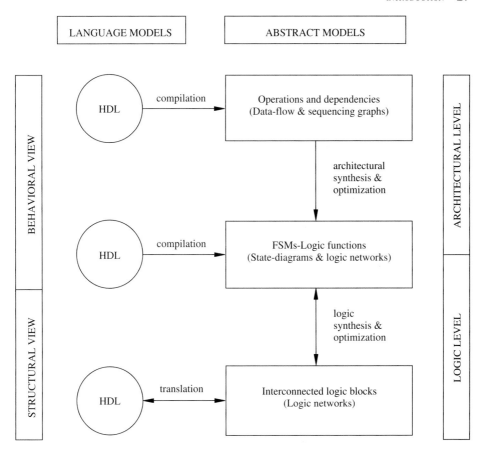

FIGURE 1.20
Circuit models, synthesis and optimization: a simplified view.

1.6 ORGANIZATION OF THE BOOK

This book presents techniques for synthesis and optimization of combinational and sequential synchronous digital circuits, starting from models at the architectural and logic levels. Figure 1.20 represents schematically the synthesis and optimization tasks and their relations to the models. We assume that circuit specification is done by means of HDL models. Synthesis and optimization algorithms are described as based on abstract models, that are powerful enough to capture the essential information of the HDL models, and that decouple synthesis from language-specific features. Example of abstract models that can be derived from HDL models by compilation, are: *sequencing* and *data-flow graphs* (representing operations and dependencies), *state transition diagrams* (describing finite-state machine behavior) and *logic networks* (representing interconnected logic blocks, corresponding to gates or logic functions).

This book is divided into four parts. Part I describes circuits and their models. Chapter 2 reviews background material on optimization problems and algorithms, as well as the foundation of Boolean algebra. Abstract circuit models used in this book are based on graph and switching theory. Chapter 3 describes the essential features of some HDLs used for synthesis, as well as issues related to circuit modeling with HDLs. Abstract models and their optimal derivation from language models are also presented in Chapter 3.

Part II is dedicated to architectural-level synthesis and optimization. Chapter 4 gives an overview of architectural-level synthesis, including data-path and control-unit generation. Operation scheduling and resource binding are specific tasks of architectural synthesis, that are described in Chapters 5 and 6, respectively. Architectural models that are constrained in their schedule and resource usage may still benefit from control-synthesis techniques (Chapter 4), that construct logic-level models of sequential control circuits.

Logic-level synthesis and optimization are described in Part III. Combinational logic functions can be implemented as two-level or multiple-level circuits. An example of a two-level circuit implementation is a PLA. The optimization of two-level logic forms (Chapter 7) has received much attention, also because it is applicable to the simplification of more general circuits. Multiple-level logic optimization (Chapter 8) is a rich field of heuristic techniques. The degrees of freedom in implementing a logic function as a multiple-level circuit make its exact optimization difficult, as well as its heuristic optimization multiform. Sequential logic models can also be implemented by two-level or multiple-level circuits. Therefore, some synthesis and optimization techniques for sequential circuits are extensions of those for combinational circuits. Specific techniques for synchronous design (Chapter 9) encompass methods for optimizing *finite-state machine* models (e.g., state minimization and encoding), as well as optimizing synchronous networks (e.g., retiming). Library binding, or technology mapping (Chapter 10), is applicable to both combinational and sequential circuits and is important to exploit at best the primitive cells of a given library.

In Part IV, Chapter 11 summarizes the present state of the art in this field, and reports on existing synthesis and optimization packages as well as on future developments.

We consider now again the synthesis and optimization tasks and their relation to the organization of this book. The flow diagram shown in Figure 1.21 puts the chapters in the context of different design flows. HDL models are shown at the top and the underlying abstract models at the bottom of the figure. Note that the order in which the synthesis algorithm is applied to a circuit does not match with the numbering of the chapters, for didactic reasons. We describe general issues of architectural synthesis before scheduling and binding also because some design flows may skip these steps. We present combinational logic optimization before sequential design techniques for the sake of a smooth introduction to logic design problems.

1.7 RELATED PROBLEMS

The complete design of a microelectronic circuit involves tasks other than architectural/logic synthesis and optimization. A description of all computer-aided methods and programs for microelectronic design goes beyond the scope of this book. We

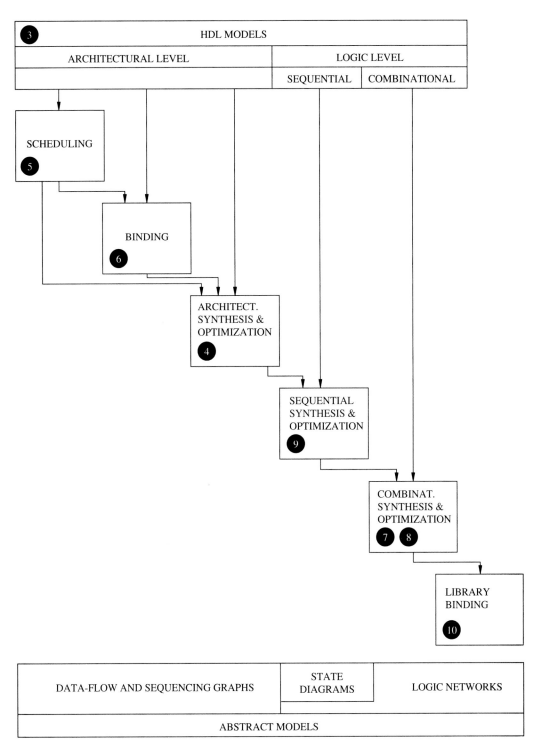

FIGURE 1.21
Chapters, models and design flows.

would like to mention briefly in Section 1.7.1 some physical design and optimization problems. Indeed, physical design is the back end of circuit synthesis, that provides the link between structural descriptions and the physical view, describing the geometric patterns required for manufacturing. We comment on those topics that are tightly related to architectural/logic synthesis, such as design validation by means of simulation and verification, and design for testability.

1.7.1 A Closer Look at Physical Design Problems

Physical design provides the link between design and fabrication. The design information is transferred to the fabrication masks from the geometrical layout, that is a representation of all geometries of a circuit. A layout is a set of layers, each being a two-dimensional representation. The relative position of the circuit geometries corresponds to the electronic devices, as well as to their interconnections.

A VLSI circuit layout involves a large amount of information. For this reason, appropriate data structures that exploit the notion of hierarchy are required to support the physical design algorithms. Designing an efficient data base for microelectronic circuits is a problem of its own, and it will not be described here. Instead, we concentrate on the constructive techniques for semicustom layout.

The design primitives, or cells, are generally available in a library, except in the case of macro-cell design style, where they are constructed automatically. Libraries of standard cells contain the full definition of the layout of the cells. In the case of prediffused arrays, libraries refer to those cells that can be obtained by wiring the sites. The richness of a library influences the flexibility of the design implementation. Library binding algorithms, as those described in Chapter 10, exploit the library cells in the search for an optimal implementation.

Macro-cell generation is performed by specialized programs, called *module generators*. These programs range from simple to complex ones. Simple generators may construct a personalized array of cells of a programmable-logic array, or the layout of a logic gate. Physical synthesis involves the transformation of the logic specification into a layout. Optimization can be done in different ways. For example, a PLA can be compacted by folding its rows and columns. As another example, the optimization of a logic CMOS gate corresponds to finding the linear arrangement of the transistors that maximizes connection by abutment, and thus reducing area and parasitic capacitance.

Complex macro-cell generators often use *procedural layout styles*, where the cells are the results of executing parametrized programs in specific layout languages. The parameters are related to the logic personality of the cell being constructed. A full circuit layout can be constructed by procedural layout programs. However, this approach may be inefficient for large circuits, due to the large amount of information required in the procedural layout program itself. Conversely, competitive physical implementations of data paths can be achieved with these methods. Other approaches to macro-cell synthesis involve a combination of techniques, ranging from the synthesis of basic cells to their placement and wiring in arrays with dedicated tools, that support different cell shapes and the use of privileged layers for interconnection. Optimization is related to sizing the cell geometrical features for maximal performance and arranging the cells to minimize the wiring and its associated delay penalty.

The two most common tasks in physical design are *placement* and *wiring*. Placement is the assignment of positions to the cells. In the case of standard cells, cells are organized in rows, while in the case of prediffused circuits, cell positions must correspond to some site locations in the array. For macro-cell design, placement has to deal with irregularly shaped cells. The term *floor plan* is often associated with a rough arrangement of the cells, that takes into account their relative, rather than absolute, positions. The concept of floor plan is useful to estimate the wiring area and delay, for any circuit design that is partitioned into macro-blocks, regardless of the implementation style.

The objective of placement is often to reduce the overall layout size; this correlates to the overall wiring length. Some distinguishing features of the different semicustom styles, with respect to placement, need to be highlighted. In the case of cell-based design, the position of the cells can be fine-tuned. Since wiring occupies often a large space, its minimization is of paramount importance. This is related to placing heavily interconnected cells close to each other. Timing-driven placement techniques have also been proposed; these aim at reducing the wire lengths along the critical paths. On the other hand, in the case of array-based design styles, the overall circuit area is fixed. Obviously the number of primitive cells must not outnumber the sites. Similarly, the wires must fit in the dedicated wiring space. This requirement is specifically stringent for gate arrays with fixed channels for wiring and for field-programmable gate arrays. In these cases, the placement has to be such that there exists a feasible wiring satisfying the channel capacity.

Wiring is the definition of the interconnections. *Global wiring*, called also *loose wiring*, determines the regions traversed by the wires. *Detailed wiring* determines the specific location of the wires, as well as their layers and contacts. Wiring strategies depend on the number of wiring layers available in a given technology. Early routers, called *river routers*, used only one layer and the interconnections were required to be shaped like the affluents of a river, i.e., without crossing. Multilevel routers use two, three or more levels. The objective in routing is related to reduce the wiring length, as well as the number of contacts between wire segments on different layers. Wiring length correlates to wiring area and delay; contacts correlate to delay.

Most physical design problems are computationally intractable. However, heuristic algorithms have shown to be effective in achieving good-quality solutions. Much work has been published on physical design automation. For this reason we do not report on the physical synthesis problem in this book. We refer the interested reader to the books by Lengauer [5], Preas and Lorenzetti [7], and Rubin [8].

1.7.2   Computer-Aided Simulation

Simulation has been the traditional way of validating a circuit. Simulation entails analyzing the circuit response to a set of input *stimuli* over a time interval. Simulation can be performed at different levels. At the geometrical level of abstraction, circuits can be modeled as an interconnection of electronic devices (e.g., transistors) and parasitics, such as resistance and capacitance.

Circuit-level simulation applies to such models and it corresponds to deriving the voltage levels at all (or some) circuit nodes as a function of time. This entails the formulation of a large set of differential equations and its solution. The sparsity of the equations makes the problem solvable for circuits with tenths of thousands of devices. Larger circuits are less amenable to circuit simulation, because of the memory and computing time requirements, as well as the difficulty in selecting a meaningful set of node voltages, from whose analysis the circuit correctness can be inferred. The major advantage of circuit-level simulation is the precise computation of delays, which incorporates the parasitics associated with the geometrical layout.

Logic-level simulation is the analysis of the functionality of a circuit in terms of logic variables. Very often, simulation of sequential circuits is referred to as *register-transfer level* simulation, because the variables being simulated are those stored in registers. The simulation iterates the evaluation of logic functions feeding registers at each clock-cycle. Today logic simulators can handle large-scale circuits.

Functional-level simulation corresponds to simulating HDL models of digital circuits. The different flavors of the modeling languages affect the way in which simulation is done. For example, standard programming languages can be used for modeling circuit behavior. In this case, simulation consists of compiling and executing the model.

Entire digital systems can be simulated, when modeled consistently. A designer can simulate a model of a circuit under development together with models of other previously designed (or off-the-shelf) circuits. In other words, simulation allows a designer to validate a circuit in its environment.

1.7.3 Computer-Aided Verification

Verification is the comparison of two models for consistency. There are two major classes of verification techniques. The first one relates to models that are intermediate steps in the generation of the circuit layout. These methods are often called *implementation verification* techniques. The second class involves the comparison of a circuit model at the architectural level against some abstract model, in order to prove some circuit property. These methods are called *design verification* methods. In other words, the former class of methods relates to verifying the correctness of the synthesis task; the latter class relates to the correctness of the conceptualization and modeling.

Implementation verification methods entail two representations at different levels. An example is the *extraction* of a circuit schematic from a layout and the comparison to another schematic, from which the layout was derived. The verification of congruency is not sufficient to assert the correctness of the overall design. However, the correctness of the layout implementation can be verified. Other examples relate to the logic level of abstraction. Comparisons of combinational and sequential logic functions are important to verify the correctness of the manipulations done by synthesis.

At a first glance, implementation verification should play a limited role when computer-aided synthesis methods are used. Indeed such methods should provide *correctness by construction*. This myth has fallen with the widespread use of synthesis systems. While synthesis algorithms have of course guaranteed properties of correctness, their software implementation (as well as the implementation of the operating systems and on which data bases they run) cannot be guaranteed to be error-free. Hence the need of implementation verification methods that support synthesis tools by performing validity checks.

Design verification methods are useful for verifying the correctness of a design, by proving assertions on the properties of the model. An example is the check as to whether an architectural model of a processor can deadlock. Automated theorem-proving techniques have been used in this perspective. Their present limitations relate to the requirement of extensive user intervention to drive the verification method toward its goal and to restrictions on the types and styles of models that are verifiable.

1.7.4   Testing and Design for Testability

Microelectronic circuits are tested after manufacturing to screen fabrication errors. Thus testing is the verification of correct manufacturing. There are several testing techniques and their relation to design is very important. Additional functions may be added to a circuit to make it more testable. For example, self-testing techniques can be implemented by adding pseudo-random pattern generators and signature compressors. These functions are often merged with the circuit primitives. Therefore circuit synthesis must be extended to cope directly with self-testing functions. Similarly, *level-sensitive scan design* (LSSD) techniques have been used to access the individual bits stored in the registers. The synthesis of circuits with full, or partial, scan features requires appropriate enhancement of the synthesis tools.

Circuit testability affects its quality. A circuit that is not fully testable, for some fault model, is less valuable than another one that is fully testable. Enhancing testability can be done by appropriate logic and architectural synthesis techniques. For some kind of fault models, such as the common *stuck-at* model, increasing the fault coverage is related to removing redundancies in the circuit. Logic synthesis techniques that support redundancy removal have been developed and used. Other faults, such as those modeled as *delay faults*, are more difficult to detect and remove, but promising recent research has shown viable ways. Design for enhanced testability of sequential synchronous circuits is also a very important topic, which is currently being researched. We refer the interested reader to reference [2] for further details.

1.8   SYNTHESIS AND OPTIMIZATION: AN HISTORIC PERSPECTIVE

Early efforts on logic synthesis and optimization algorithms date to the 1950s and 1960s. While this work has much historical importance, it has had less practical impact on CAD tools for microelectronic circuit design. Indeed most of the classical approaches to solving logic design problems were developed before the advent of very large-scale integration. Therefore those methods suffer from being not practical for medium/large-scale problems.

The earliest physical design tools were targeted at automatic layout systems for gate arrays, such as the ENGINEERING DESIGN SYSTEM of IBM, and for standard cells, like the LTX systems of AT&T Bell Laboratories. Algorithms for placement and routing flourished in the 1970s. Then, with the advent of low-cost, high-resolution color graphic workstations, physical design entry tools became popular. At the same time, methods for procedural design and symbolic layout were developed. We refer the reader to reference [7] for details on the development of physical design algorithms and tools.

As soon as *large-scale integration* (LSI) became possible from a technological standpoint, and LSI circuit design became practical because of the support provided by physical design tools, the research interest shifted to the logic level of abstraction. One major motivation was avoiding the design entry at the physical level, which is a tedious and error-prone task. The use of logic synthesis techniques required also a change in attitude toward problem solving. In the early 1970s the computational intractability theory was introduced. Researchers realized that it was not useful to extend the classical methods of switching theory, which tried to solve problems exactly. Instead, "good" engineering solutions, such as those constructed by heuristic algorithms, were the goal to pursue in order to cope with LSI.

An early example of "modern" logic optimization was IBM's program MINI, a heuristic minimizer for two-level logic forms implementable by PLAs. MINI could compute solutions whose quality was near optimal and could run in reasonable time on most circuits of interest. In the 1970s, logic synthesis tools targeted mainly two-level logic representations, and they were coupled to macro-cell generators for PLAs. Programs for PLA-based combinational and sequential logic design were then developed. The most prominent example was program ESPRESSO, developed at IBM jointly with the University of California at Berkeley.

In the early 1980s, as VLSI circuit technology matured, two requirements became important: the possibility of configuring logic circuits into multiple-level forms appropriate for cell-based and array-based styles and the necessity of using a higher, architectural level of abstraction in modeling and as a starting point for synthesis.

The former issue was related to the evolution of semicustom design. Technology mapping became the primary tool for redesigning *transistor-transistor-logic* (TTL)-based boards into CMOS integrated circuits. Technology mapping, also called library binding, was researched at AT&T Bell Laboratories, where the first algorithmic approach was implemented in program DAGON and at IBM, where the rule-based LOGIC SYNTHESIS SYSTEM (LSS) was introduced and used for designing mainframe computers. Multiple-level logic optimization algorithms and programs were then developed, such as the YORKTOWN LOGIC EDITOR (YLE) program at IBM and MULTIPLE-LEVEL INTERACTIVE SYSTEM (MIS) at the University of California at Berkeley, that could cope with the manipulation and reduction of large sets of logic equations. At the same time, the transduction method was also introduced at the University of Illinois and then used in the Fujitsu's logic design system.

As logic synthesis tools matured and companies started commercializing them, it became apparent that VLSI design could not be started by conceptualizing systems in terms of logic-level primitives and entering large circuits by means of schematic capture systems. Architectural (or high-level) synthesis techniques that traced back to the MIMOLA system at Honeywell and the ALERT systems at IBM became an important research subject, especially at Carnegie-Mellon University, the University of Southern California and Carleton University. The early architectural synthesis systems targeted instruction-set processor design and were not coupled to logic synthesis tools. The first fully integrated synthesis systems were IBM's YORKTOWN SILICON COMPILER (YSC), GTE's SILC system and the CATHEDRAL systems developed at the Catholic University of Leuven, Belgium.

At present, synthesis systems are common for circuit design at all levels. Commercial systems are available from vendors and research systems from universities and research centers. Some companies have developed their own internal synthesis systems, such as IBM, Digital Equipment Corporation, NEC, etc. Some synthesis systems are described in Chapter 11.

1.9 REFERENCES

1. R. Brayton and R. Spence, *Sensitivity and Optimization*, Elsevier, 1980.
2. S. Devadas, A. Ghosh, and K. Keutzer, *Logic Synthesis*, McGraw-Hill, 1994.
3. D. Gajski, Editor, *Silicon Compilers*, Addison-Wesley, Reading, MA, 1987.
4. J. Hennessy and D. Patterson, *Computer Architecture: A Quantitative Approach*, Morgan Kaufman, San Mateo, CA, 1990.
5. T. Lengauer, *Combinational Algorithms for Integrated Circuit Layout*, Wiley, New York, 1990.
6. P. Paulin and J. Knight, "Force-directed Scheduling for the Behavioral Synthesis of ASIC's," *IEEE Transactions on CAD/ICAS*, Vol. CAD-8, No. 6, pp. 661-679, July 1989.
7. B. Preas and M. Lorenzetti, *Physical Design Automation of VLSI Systems*, Benjamin Cummings, Menlo Park, CA, 1988.
8. S. Rubin, *Computer Aids for VLSI Design*, Addison-Wesley, Reading, MA, 1987.
9. N. Weste and K. Eshraghian, *Principles of CMOS VLSI Design: A Systems Perspective*, Addison-Wesley, Reading, MA, 1993.
10. W. Wolf, *Modern VLSI Design: A Systems Approach*, Prentice-Hall, Englewood Cliffs, NJ, 1994.

CHAPTER
2

BACKGROUND

On fait la science avec des faits, comme on fait une maison avec des pierres; mais une accumulation de faits n'est pas plus une science qu'un tas de pierres n'est une maison.
Science is built up with facts, as a house is with stones. But a collection of facts is no more a science than a heap of stones is a house.

H. Poincaré. La Science et l' Hypothèse.

The purpose of this chapter is to review some fundamental definitions, problems and algorithms that are necessary for the understanding of the material in the rest of this book. Sections 2.1, 2.2, 2.3 and 2.4 are meant to be a brief refresher for the reader; those who need more explanations should consult the related textbooks mentioned in the reference section.

2.1 NOTATION

A *set* is a collection of *elements*. We denote sets by uppercase letters and elements by lowercase ones. We use braces (i.e., { }) to denote unordered sets and parentheses [i.e., ( )] to denote ordered sets. For example, an unordered set of three elements is denoted as $S = \{a, b, c\}$. The *cardinality* of a set is the number of its elements, denoted by | |. Given a set S, a *cover* of S is a set of subsets of S whose union is S. A *partition* of S is a cover by disjoint subsets, called *blocks* of the partition. Set membership of an element is denoted by $\in$, set inclusion by $\subset$ or $\subseteq$, set union by $\cup$ and intersection by $\cap$. The symbol $\forall$ is the *universal quantifier*; the symbol $\exists$ the *existential quantifier*. Implication is denoted by $\Rightarrow$ and co-implication by $\Leftrightarrow$. The symbol ":" means "such that."

The set of real numbers is denoted by R, the set of integers is denoted by Z and the set of positive integers by Z^+. The set of binary values $\{0, 1\}$ is denoted by B. Vectors and matrices are ordered sets. They are denoted by lower- and uppercase bold characters, respectively. For example, **x** denotes a vector and **A** a matrix. We denote a vector with all 0 entries by **0** and one with all 1 entries by **1**.

The *Cartesian product* of two sets X and Y, denoted by $X \times Y$, is the set of all ordered pairs (x, y), such that $x \in X$ and $y \in Y$. A *relation* R between two sets X and Y is a subset of $X \times Y$. We write xRy when $x \in X$, $y \in Y$ and $(x, y) \in R$. An *equivalence* relation is *reflexive* [i.e., $(x, x) \in R$], *symmetric* [i.e., $(x, y) \in R \Rightarrow (y, x) \in R$] and *transitive* [i.e., $(x, y) \in R$ and $(y, z) \in R \Rightarrow (x, z) \in R$]. A *partial order* is a relation that is reflexive, anti-symmetric [i.e., $(x, y) \in R$ and $(y, x) \in R \Rightarrow x = y$] and transitive. A *partially ordered set* (or *poset*) is the combination of a set and a partial order relation on the set. A *totally* (or *linearly*) *ordered set* is a poset with the property that any pair of elements can be ordered such that they are members of the relation. A *consistent enumeration* of a partially ordered set S is an integer labeling $i(S)$ such that $s_a R s_b$ implies $i(s_a) < i(s_b)$, $\forall$ pair $(s_a, s_b) \in S$.

A *function* (or *map*) between two sets X and Y is a relation having the property that each element of X appears as the first element in one and only one pair of the relation. A function between two sets X and Y is denoted by $f : X \rightarrow Y$. The sets X and Y are called the *domain* and *co-domain* of the function, respectively. The function f assigns to every element $x \in X$ a unique element $f(x) \in Y$. The set $f(X) = \{f(x) : x \in X\}$ is called the *range* of the function. A function is *onto* or *surjective* if the range is equal to the co-domain. A function is *one-to-one* or *injective* if each element of its range has a unique element of the domain that maps to it, i.e., $f(x_1) = f(x_2)$ implies $x_1 = x_2$. In this case, the function has an *inverse*, $f^{-1} : f(X) \rightarrow X$. A function is *bijective* if it is both surjective and injective. Given a function $f : X \rightarrow Y$ and a subset of its domain $A \subseteq X$, the *image* of A under f is $f(A) = \{f(x) : x \in A\}$. Conversely, given a function $f : X \rightarrow Y$ and a subset of its co-domain $A \subseteq Y$, the *inverse image* of A under f is $f^{-1}(A) = \{x \in X : f(x) \in A\}$.

2.2 GRAPHS

A *graph* $G(V, E)$ is a pair (V, E), where V is a set and E is a binary relation on V [4, 13, 16]. We consider only finite graphs, i.e., those where set V is bounded. The elements of the set V are called *vertices* and those of the set E are called *edges* of the graph [Figure 2.1 (a)]. In a *directed graph* (or *digraph*) the edges are ordered pairs of vertices; in an *undirected graph* the edges are unordered pairs [Figure 2.1 (b)]. A directed edge from vertex $v_i \in V$ to $v_j \in V$ is denoted by (v_i, v_j) and an undirected edge with the same end-points by $\{v_i, v_j\}$. We also say that an edge (directed or indirected) is *incident* to a vertex when the vertex is one of its end-points. The *degree* of a vertex is the number of edges incident to it. A *hypergraph* is an extension of a graph where edges may be incident to any number of vertices [Figure 2.1 (c)].

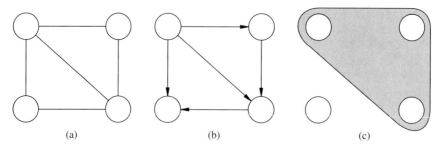

FIGURE 2.1
(a) Undirected graph. (b) Directed graph. (c) Hypergraph.

2.2.1 Undirected Graphs

We consider undirected graphs first. We say that a vertex is *adjacent* to another vertex when there is an edge incident to both of them. An edge with two identical end-points is called a *loop*. A graph is *simple* if it has no loops and no two edges link the same vertex pair. Otherwise it is called a *multi-graph*. Throughout the remainder of this book, we shall refer to simple graphs as graphs, unless explicitly stated otherwise.

A *walk* is an alternating sequence of vertices and edges. A *trail* is a walk with distinct edges, and a *path* is a trail with distinct vertices. A *cycle* is a closed walk (i.e., such that the two end-point vertices coincide) with distinct vertices. A graph is *connected* if all vertex pairs are joined by a path. A graph with no cycles is called an *acyclic* graph or a *forest*. A *tree* is a connected acyclic graph. A *rooted tree* is a tree with a distinguished vertex, called a *root*. Vertices of a tree are also called *nodes*. In addition, they are called *leaves* when they are adjacent to only one vertex each and they are distinguished from the root.

A *cutset* is a minimal set of edges whose removal from the graph makes the graph disconnected. Similarly, a *vertex separation set* is a minimal set of vertices whose removal from the graph makes the graph disconnected.

A *complete* graph is one such that each vertex pair is joined by an edge. The *complement* of a graph $G(V, E)$ is a graph with vertex set V, two vertices being adjacent if and only if they are not adjacent in $G(V, E)$. A *bipartite* graph is a graph where the vertex set can be partitioned into two subsets such that each edge has end-points in different subsets. Hypergraphs can be modeled by bipartite graphs, by associating their vertex and edge sets to the two vertex subsets of corresponding bipartite graphs. Edges in the bipartite graphs represent the incidence relation among vertices and edges of the hypergraphs.

A *subgraph* of a graph $G(V, E)$ is a graph whose vertex and edge sets are contained in the vertex and edge sets, respectively, of $G(V, E)$. Given a graph $G(V, E)$ and a vertex subset $U \subseteq V$, the subgraph *induced* by U is the maximal subgraph of $G(V, E)$ whose edges have end-points in U. A *clique* of a graph is a complete subgraph; it is maximal when it is not contained in any other clique [1, 4, 12, 13, 28]. (Some authors refer to maximal cliques as cliques [16, 19, 22].)

Every graph has one (or more) corresponding diagrams. A graph is said to be *planar* if it has a diagram on a plane surface such that no two edges cross. Two graphs are said to be *isomorphic* if there is a one-to-one correspondence between their vertex sets that preserves adjacency (Figures 2.2 and 2.3).

2.2.2 Directed Graphs

The definitions of the previous section can be extended to apply to directed graphs in a straightforward way. For any directed edge (v_i, v_j), vertex v_j is called the *head* of the edge and vertex v_i the *tail*. The *indegree* of a vertex is the number of edges where it is the head, the *outdegree* the number of edges where it is the tail. A walk is an alternating sequence of vertices and edges with the same direction. Trails, paths and cycles are defined similarly. The concept of head and tail are extended to trails and paths. Given a directed graph, the *underlying* undirected graph is one having the same vertex set and undirected edges replacing directed edges with the same end-points. Given an undirected graph, an *orientation* is a directed graph obtained by assigning a direction to the edges.

Directed acyclic graphs, also called *dags*, represent partially ordered sets. In a dag, a vertex v_j is called the *successor* (or *descendant*) of a vertex v_i if v_j is the head of a path whose tail is v_i. We say also that a vertex v_j is *reachable* from vertex

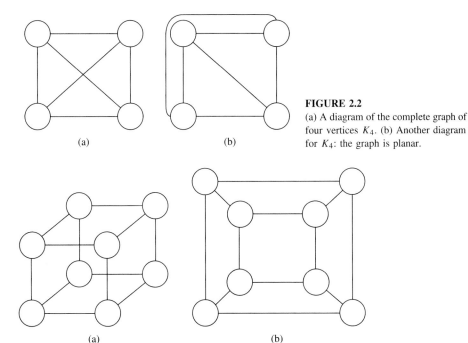

(a)   (b)

FIGURE 2.2
(a) A diagram of the complete graph of four vertices K_4. (b) Another diagram for K_4: the graph is planar.

(a)   (b)

FIGURE 2.3
(a) A diagram of the three-dimensional cube graph. (b) Another diagram of the cube graph: the graph is planar and bipartite.

v_i when v_j is a successor of v_i. Similarly, a vertex v_i is called the *predecessor* (or *ancestor*) of a vertex v_j if v_i is the tail of a path whose head is v_j. In addition, vertex v_j is a *direct successor* of, or *adjacent* to, vertex v_i if v_j is the head of an edge whose tail is v_i. Direct predecessors are similarly defined.

A *polar* dag is a graph having two distinguished vertices, a *source* and a *sink*, and where all vertices are reachable from the source and where the sink is reachable from all vertices [Figure 2.4 (a)].

Undirected and directed graphs can also be represented by matrices. The *incidence matrix* of a simple undirected graph is a matrix with as many rows as vertices and as many columns as edges. An entry in position (i, j) is 1 if the jth edge is incident to vertex v_i; else it is 0. For directed graphs, an entry in position (i, j) is 1 if vertex v_i is the head of the jth edge, -1 if it is its tail and otherwise 0 [Figure 2.4 (b)]. For both undirected and directed graphs, an *adjacency* matrix is a square matrix with dimensions equal to the vertex set cardinality. An entry in position (i, j) is 1 if vertex v_j is adjacent to vertex v_i; else it is 0 [Figure 2.4 (c)]. The adjacency matrix is symmetric only for undirected graphs, because the corresponding definition of adjacency relation is symmetric.

Directed and undirected graphs can be *weighted*. Weights can be associated with vertices and/or with edges, i.e., the graph can be *vertex weighted* and/or *edge weighted*.

2.2.3 Perfect Graphs

We consider here properties of undirected graphs. Each graph can be characterized by four numbers. The first number we consider is the *clique number* $\omega(G)$, which is the cardinality of its largest clique, called *maximum clique*.

A graph is said to be *partitioned into cliques* if its vertex set is partitioned into (disjoint) subsets, each one inducing a clique. Similarly, a graph is said to be *covered by cliques* when the vertex set can be subdivided into (possibly overlapping) subsets, each one inducing a clique. Note that a clique partition is a disjoint clique cover. The cardinality of a minimum clique partition is equal to the cardinality of a minimum clique cover, it is called *clique cover number* and it is denoted by $\kappa(G)$.

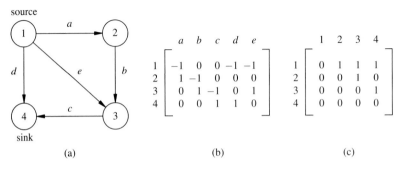

FIGURE 2.4
(a) A polar graph. (b) Its incidence matrix. (c) Its adjacency matrix.

A *stable set*, or *independent set*, is a subset of vertices with the property that no two vertices in the stable set are adjacent. The *stability number* $\alpha(G)$ of a graph is the cardinality of its largest stable set.

A *coloring* of a graph is a partition of the vertices into subsets, such that each is a stable set. The *chromatic number* $\chi(G)$ is the smallest number that can be the cardinality of such a partition. Visually, it is the minimum number of colors needed to color the vertices, such that no edge has end-points with the same color.

The size of the maximum clique is a lower bound for the chromatic number, because all vertices in that clique must be colored differently. Namely:

$$\omega(G) \leq \chi(G) \tag{2.1}$$

Similarly, the stability number is a lower bound for the clique cover number, since each vertex of the stable set must belong to a different clique of a clique cover. Thus:

$$\alpha(G) \leq \kappa(G) \tag{2.2}$$

A graph is said to be *perfect* when the inequalities can be replaced by equalities.

Example 2.2.1. Consider the graph of Figure 2.1 (a), reported again for convenience in Figure 2.5 (a). The size of the maximum clique $\{v_1, v_2, v_3\}$ is $\omega(G) = 3$. The graph can be partitioned into cliques $\{v_1, v_2, v_3\}$ and $\{v_4\}$. Alternatively, it can be covered by cliques $\{v_1, v_2, v_3\}$ and $\{v_1, v_3, v_4\}$. The clique cover number $\kappa(G) = 2$. The largest stable set is $\{v_2, v_4\}$. Thus, the stability number is $\alpha(G) = 2$. A minimum coloring would require three colors for $\{v_1, v_2, v_3\}$. Vertex v_4 can have the same color as v_2. Hence the chromatic number $\chi(G) = 3$. Therefore the graph is perfect.

Some special perfect graphs are worth considering. A graph is said to be *chordal*, or *triangulated*, if every cycle with more than three edges possesses a *chord*, i.e., an edge joining two non-consecutive vertices in the cycle.

A subclass of chordal graphs is the one of *interval graphs*. An interval graph is a graph whose vertices can be put in one-to-one correspondence with a set of *intervals*, so that two vertices are adjacent if and only if the corresponding intervals intersect.

A graph $G(V, E)$ is a *comparability graph* if it satisfies the *transitive orientation property*, i.e., if it has an orientation such that in the resulting directed graph $G(V, F)$, $\{(v_i, v_j) \in F$ and $(v_j, v_k) \in F\} \Rightarrow (v_i, v_k) \in F$.

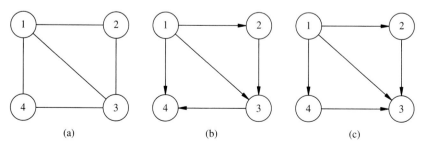

FIGURE 2.5
(a) Undirected graph. (b) An orientation. (c) A transitive orientation.

An important theorem, by Gilmore and Hoffman, relates comparability to interval graphs [15].

Theorem 2.2.1. An undirected graph is an interval graph if and only if it is chordal and its complement is a comparability graph.

Interval, chordal and comparability graphs play an important role in synthesis algorithms, because specialized efficient algorithms exist for coloring and clique partitioning.

Example 2.2.2. The graph of Figure 2.5 (a) is chordal. Figure 2.5 (b) shows an orientation that does not satisfy the transitive property (because there is no edge between vertices v_2 and v_4). Nevertheless, the orientation shown in Figure 2.5 (c) (obtained by reversing the direction of the edge between v_3 and v_4) has the transitive property. Hence the graph of Figure 2.5 (a) is a comparability graph. In addition, the complement of the graph of Figure 2.5 (a) is also a comparability graph, and by Gilmore's theorem the graph of Figure 2.1 (a) is an interval graph. Indeed, the graph can represent the intersection among a set of intervals, for example, $\{[1, 8], [1, 3], [2, 6], [5, 7]\}$ associated with v_1, v_2, v_3 and v_4, respectively.

2.3 COMBINATORIAL OPTIMIZATION

Computer-aided design of microelectronic circuits requires modeling some design problems in a precise mathematical framework, which supports problem analysis and development of solution methods. Most architectural and logic design problems for digital circuits are discrete in nature. It is therefore necessary to solve combinatorial decision and optimization problems.

2.3.1 Decision and Optimization Problems

A *decision problem* is a problem with a binary-valued solution: TRUE or FALSE. Such problems are related to testing some properties of a model, for example, questioning the existence of a vertex coloring with k colors in a given graph.

An *optimization problem* is a problem whose solution can be measured in terms of a *cost* (or *objective*) function and such that the cost function attains a maximum or minimum value. For example, the search for a vertex coloring with a minimum number of colors in a given graph is an optimization problem.

Optimization problems can be reduced to sequences of decision problems and are always at least as difficult to solve as the corresponding decision problems. For this reason, in problem complexity analysis, only decision problems are considered, because they provide a lower bound on the complexity of the corresponding optimization problem. Nevertheless, solution methods to optimization problems do not necessarily need to solve sequences of decision problems.

2.3.2   Algorithms

An *algorithm* is a computational procedure that has a set of *inputs* and *outputs*, has a *finite* number of unambiguously defined steps and terminates in a finite number of steps. Well-defined problems that can be solved by algorithms are called *decidable*. Some undecidable problems do exist, like the *halting problem* [28].

Algorithms can be described in natural languages or in software programming languages. In this book, we use a pseudo-code notation reminiscent of the C programming language. Algorithms can also be seen as the abstract models of the computational engines provided by computer programs, which are just descriptions of algorithms in appropriate executable languages. Algorithms can be evaluated according to two major criteria: (i) the quality of the solution and (ii) the computational effort required to achieve the solution. In order to quantify these two factors, some additional clarifications are needed.

Consider all instances of decision or optimization problems of a given type. An *exact* algorithm always provides the exact solution. It is obvious that exact algorithms are desirable and indeed they can be conceived for all decidable problems. Unfortunately, some exact algorithms may have such a high computational cost that their use can be prohibitive on problems of typical size. Hence there is a need for *approximation* algorithms, which are not guaranteed to find the exact solution to all instances of the problem but can provide good approximations to the exact solutions. These good but inexact solutions may be valid for practical applications. Approximation algorithms are often called *heuristic* algorithms, because they use problem-solving techniques based on experience.

The computational effort of an algorithm is measured in terms of time and memory-space consumption for both the *worst* and *average* cases. Often only the worst-case time complexity is reported, and therefore when we mention complexity, we shall be referring to worst-case time complexity unless otherwise specified. Complexity is measured in terms of elementary operations which are repeated in the algorithm. It is convenient to measure the growth in the number of elementary operations as a function of the problem size, i.e., as the size of some input to the algorithm. Let us denote by n the relevant input size. We say that the complexity is *of the order of* $f(n)$ if there exists a constant c such that $cf(n)$ is an upper bound on the number of elementary operations. We denote the order as $O(f(n))$, e.g., $O(n)$; $O(n \log n)$; $O(n^2)$; $O(2^n)$.

It is often believed that algorithms with polynomial complexity are viable, while exponential ones are not. This is generally true, but attention must also be paid to the constant c. For example, a cubic algorithm with a very large constant may not be viable for problems of large size and may perform worse than an exponential algorithm on problems of small size. It is also important to bear in mind that the average-case complexity may be much lower than the worst-case complexity.

The efficiency of an exact algorithm can be measured by comparing its complexity to that of the problem being solved. The complexity of the problem is a *lower bound* on the number of operations required to solve it. For example, searching for the largest element in an unordered set of integers of cardinality n requires $O(n)$

comparisons, because $n - 1$ elements must be discarded. An algorithm is said to be *optimal* when its complexity is of the same order as the problem complexity [3]. Thus, a searching algorithm with $O(n)$ complexity is optimal.

Note that algorithm optimality is a different concept than solution optimality. Indeed, we consider algorithm optimality only for exact algorithms which compute the *optimum* (called also *global optimum*) solution to a problem. In the case of approximation algorithms, the notion of a *local optimum* solution, or *optimal* solution, is often used. A solution is optimum with respect to some local property when there is no other solution with that property of lower/higher cost. A local optimum solution is often the result of algorithms that perform a *local search*, i.e., that search for a solution in a restricted neighborhood of the solution space.

> **Example 2.3.1.** A vertex cover of a graph is a subset of vertices such that each edge has an end-point in the set. A *minimum* cover is a cover of minimum cardinality. A vertex cover is *minimal* with respect to containment, when there is no vertex of the cover that can be removed, while preserving the covering property (see Figure 2.13).

> **Example 2.3.2.** Consider the problem of arranging n objects in a sequence while minimizing the associated cost. An *optimal* solution, with respect to a single pairwise swap, is an arrangement whose cost is lower than or equal to the cost of any other arrangement reachable from that arrangement by a single pairwise swap.
>
> As an example, consider the linear arrangement of $n = 3$ objects $\{a, b, c\}$. There are $3! = 6$ arrangements. Figure 2.6 shows a graph representing the solution space. Each vertex corresponds to an arrangement. A directed edge between two vertices denotes that the head vertex is an arrangement that can be derived from the tail arrangement by a pairwise swap.
>
> For the sake of the example, we assume an arbitrary cost of the arrangements, as shown in Figure 2.6. Edges are shown in Figure 2.6 with solid lines when a swap increases the cost and with dashed lines when it decreases the cost. It is straightforward to verify that configuration (a, b, c) is a global minimum, whereas configurations $\{(a, b, c), (c, a, b), (b, c, a)\}$ are local minima.
>
> Consider the class of algorithms which perform a local search and whose moves are pairwise swaps. Some of these algorithms can guarantee an optimal solution with respect to a single pairwise swap. For example, an algorithm may attempt to improve upon the cost of a current arrangement by considering all swaps decreasing its cost and choosing the local best, i.e., maximizing the cost decrease. If such an algorithm stops when no swap can decrease the cost, then the solution is a local optimum.
>
> On the other hand, consider an algorithm that chooses swaps at random and that stops when a move increases the cost. Then, the solution may not be optimal with respect to a single pairwise swap.

2.3.3 Tractable and Intractable Problems

Some decision problems can be solved by algorithms with polynomial complexity. This class of problems is often referred to as $\mathcal{P}$ or the class of *tractable* problems [14]. Unfortunately this class covers only a few of the relevant problems encountered in the synthesis and optimization of digital circuits.

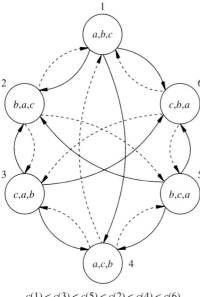

$$c(1) < c(3) < c(5) < c(2) < c(4) < c(6)$$

---- ▶ decreasing cost
——— ▶ increasing cost

FIGURE 2.6
Solution space for an optimum linear arrangement problem of $n = 3$ objects.

Other problems could be solved by polynomial algorithms on non-deterministic machines, i.e., on hypothetical computers that can perform guesses. This class of problems is called $\mathcal{NP}$. Obviously $\mathcal{P} \subseteq \mathcal{NP}$. The question of whether $\mathcal{P} = \mathcal{NP}$ has interested computer scientists for many years and is still unsolved.

It has been shown that there are problems with the property that if any one of such problems can be solved by a polynomial algorithm, then $\mathcal{P} = \mathcal{NP}$. The class of such problems is called $\mathcal{NP}$-hard, while the subclass that is contained in $\mathcal{NP}$ is called $\mathcal{NP}$-complete. The relation among these classes of problems is shown in Figure 2.7.

This implies that if a polynomial time algorithm is found for an $\mathcal{NP}$-complete or $\mathcal{NP}$-hard problem, then many other problems for which no polynomial algorithm has ever been known can be readily solved by polynomial algorithms. Therefore it is highly unlikely that such problems have polynomial algorithms and thus they are called *intractable*.

Algorithms with exponential (or higher) complexity exist for intractable problems, and often they perform well on some instances. Unfortunately, problem instances exist for which non-polynomial algorithms are impractical. Furthermore, it is always possible to set a threshold in problem size above which those algorithms cannot be used.

This motivates the search for heuristic algorithms that are conceived to have polynomial complexity with small exponents. Unfortunately such algorithms do not guarantee finding the global optimum solution, even though they often guarantee local optima with respect to some properties.

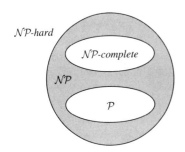

FIGURE 2.7
Relation between $\mathcal{P}$ and $\mathcal{NP}$.

2.3.4 Fundamental Algorithms

We now review some algorithms that are very general in nature and that can be applied to solve optimization and decision problems. We refer the reader to specialized textbooks, such as those described in Section 2.7, for a more complete and detailed description.

ALGORITHMS FOR LINEAR AND INTEGER PROGRAMS. Many problems can be cast as the optimization of a linear function under linear constraints. We consider first the general case, in which we search for a solution defined by an unknown vector $\mathbf{x}$. Such a formulation is referred to as a *linear program* (LP):

$$\min \mathbf{c}^T \mathbf{x} \text{ such that}$$

$$\mathbf{A}\,\mathbf{x} \geq \mathbf{b} \tag{2.3}$$

$$\mathbf{x} \geq \mathbf{0} \tag{2.4}$$

where $\mathbf{x} \in \mathsf{R}^n$; $\mathbf{c} \in \mathsf{R}^n$; $\mathbf{b} \in \mathsf{R}^m$ and $\mathbf{A} \in \mathsf{R}^{m \times n}$ has full rank.

Linear problems can be solved efficiently by the *simplex algorithm* [27, 28] in most cases, despite its complexity is above polynomial. The *ellipsoid* algorithm [28] was the first algorithm with polynomial complexity to solve linear programs. Unfortunately, its implementations are not efficient and they do not rival the simplex algorithm. The *projective* algorithm due to Karmarkar also has polynomial complexity. It is thought that this approach may compete with or be superior to the simplex algorithm in the future [27].

In the particular case of combinatorial optimization problems, the linear programming formulation can often be applied under the condition that the unknown is bound to integer values. Hence the name *integer linear program* (ILP):

$$\min \mathbf{c}^T \mathbf{x} \text{ such that}$$

$$\mathbf{A}\,\mathbf{x} \geq \mathbf{b} \tag{2.5}$$

$$\mathbf{x} \geq \mathbf{0} \tag{2.6}$$

$$\mathbf{x} \in \mathsf{Z}^n \tag{2.7}$$

The discrete nature of the problem makes it intractable. A lower bound on the solution of an ILP can be found by relaxing the integer constraint 2.7. Note that rounding a real solution vector does not guarantee the achievement of an optimum. This is especially true when ILPs model decision problems and the variables are restricted to be binary valued, i.e., 0 or 1. These ILPs are called 0-1-ILPs, or ZOLPs, or binary linear programs.

Integer linear programs are solved by a variety of algorithms [27, 28], including the branch-and-bound algorithm described in the next section. The solution of the linear program derived by relaxing the integer constraint is often used as a starting point for these algorithms. We refer the reader to references [27, 28] for details.

THE BRANCH-AND-BOUND ALGORITHM. The branch-and-bound algorithm is very general in nature. We shall introduce it here by considering its application to ZOLPs. For example, suppose we were to solve a ZOLP with n decision variables, i.e., $\mathbf{x} = [x_1, x_2, \ldots, x_n]^T$. A simple enumerative approach is to list the 2^n possible values for $\mathbf{x}$ and evaluate the objective function. A systematic approach would be to choose a variable, set it first to 1 and solve the corresponding reduced problem obtained by deleting that variable. Then, the variable would be set to 0 and the corresponding reduced problem solved. These steps can be visualized as a *decision tree*, whose leaves represent all possible solutions. Any algorithm that visits all leaves has exponential complexity in both the worst and average cases.

The branch-and-bound algorithm is based on the idea of visiting only a part of the decision tree. For each *branch*, i.e., local decision on the value of a variable, a *lower bound* is computed for all solutions in the corresponding subtree. If that bound is higher than the cost of any solution found so far, the subtree is *pruned* (or *killed*), because all its leaves would yield a solution of provably higher cost. For example, when solving an ILP, a lower bound can be computed by solving the corresponding LP after relaxing the integer constraints. When considering ZOLPs, other methods are usually more efficient to derive bounds and may depend on the application.

The worst-case complexity of the branch-and-bound algorithm is still exponential. However, its average complexity can be much less, and so the approach can be viable for practical problems. The features of a branch-and-bound algorithm applied to a specific problem are related to two issues: the branching selection and the bounding function. Heuristics that lead quickly to promising solutions can be used in the branching selection. If the cost function attains values close to the optimum for these promising solutions, then it is likely that a large part of the tree is pruned. Similarly, it is important that the bounding function computes a bound which is as sharp as possible. To be practical, the bounding function should also be quick to evaluate. Note that the heuristic selection in the branch-and-bound algorithm affects only the execution time and not the exactness of the solution.

We describe now the general branch-and-bound algorithm. A sequence of branching decisions is denoted by s. A sequence s may identify a solution, or a group of solutions when only a subset of decisions has been made. Let s_0 denote the sequence of zero length, i.e., corresponding to the initial state when no decisions have been made and equivalently corresponding to all possible solutions. The variable *Current_best* stores the best solution seen so far and variable *Current_cost* the corresponding cost. The initial value of *Current_best* can be any and its cost is

```
BRANCH_AND_BOUND {
    Current_best = anything;
    Current_cost = ∞;
    S = s₀;
    while (S ≠ ∅) do {
        Select an element in s ∈ S;
        Remove s from S;
        Make a branch based on s yielding sequences {sᵢ, i = 1, 2, ..., m};
        for ( i = 1 to m) {
            Compute the lower bound bᵢ of sᵢ;
            if (bᵢ ≥ Current_cost)
                Kill sᵢ;
            else {
                if (sᵢ corresponds to a complete solution ) {
                    Current_best = sᵢ;
                    Current_cost = cost of sᵢ;
                }
                else
                    Add sᵢ to set S;
            }
        }
    }
}
```

ALGORITHM 2.3.1

set to infinity. If a feasible solution to the problem is known, this can be used instead as the starting value for *Current_best* and its cost for *Current_cost* (Algorithm 2.3.1).

Often the decisions made by the branch-and-bound algorithms are binary, and thus $m = 2$. This corresponds to binary search trees. The algorithm can also be written recursively.

Example 2.3.3. Consider a problem with four possible solutions. The decision tree, shown in Figure 2.8 (a), visualizes the sequences of decisions. Thus the root corresponds to s_0 and the two branching decisions from the root lead to vertices a and b. Solutions correspond to the leaves of the tree. The number below each leaf in Figure 2.8 (a) is the cost of the corresponding solution.

Assume that the branch-and-bound algorithm visits first the subtree rooted in a. Since the *Current_cost* is infinity, the subtree is not killed. The algorithm then visits leaf x which has cost 5. Thus *Current_best* $= x$ and *Current_cost* $= 5$. Then it visits leaf y with cost 4 and *Current_best* is updated to y as well as *Current_cost* to 4.

The algorithm visits next vertex b. Assume that a lower bound for the subtree rooted in subtree b is 6. Then no solution in that subtree can be better than *Current_best* and the subtree is killed. The algorithm finishes, because all live solutions have been considered.

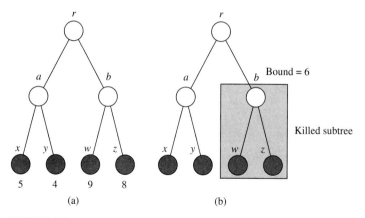

FIGURE 2.8
(a) Decision tree. (b) Pruned decision tree.

DYNAMIC PROGRAMMING. Dynamic programming is an algorithmic method that solves an optimization problem by decomposing it into a sequence of decisions. Such decisions lead to an optimum solution because the following *principle of optimality* is satisfied [19].

> **Definition 2.3.1.** An optimal sequence of decisions has the property that whatever the initial state and decisions are, the remaining decisions must constitute an optimal decision sequence with regard to the state resulting from the first decision.

Therefore, in order to solve a problem by dynamic programming, the problem itself must exhibit an *optimal substructure* [12], i.e., an optimum solution to the problem must contain optimum solutions to its subproblems. Each decision in dynamic programming thus involves the results of decisions on smaller subproblems. Adherence to the principle of optimality guarantees that suboptimal subsequences will not be generated. Thus, the decision sequence that stems from making the best decision to every subproblem leads to the optimum. The efficiency of the corresponding dynamic programming algorithm depends on the length of the decision sequence and of the complexity of solving the subproblems.

Some dynamic programming algorithms are particularly appealing, because the size of the sequence of decisions is of the order of the size of the problem and there exists a finite upper bound on the complexity of solving each subproblem. In this case, such algorithms can compute the optimum solution in linear time.

To be more concrete, we describe next a *tree covering* algorithm based on dynamic programming. Consider a rooted binary tree $T(V, E)$, called a *subject tree*, and a family of rooted *pattern trees*, each one associated with a non-negative cost. We assume that the family of pattern trees contains two *basic* trees: (i) a two-vertex tree and (ii) a three-vertex tree, where two of the vertices are children of the third. We call a *cover* of the subject tree a partition of the edge set E, where each block induces a tree isomorphic to a pattern tree. In other words, a set of instances of the pattern

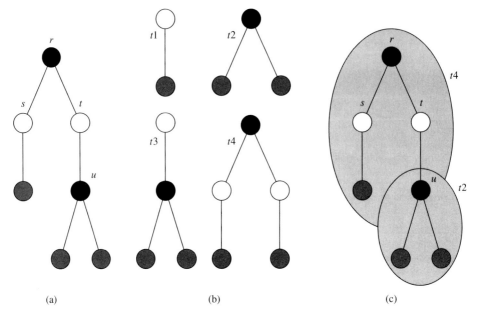

FIGURE 2.9
(a) Subject tree. (b) Pattern trees. (c) Cover of the subject tree.

trees cover the subject tree without covering any edge more than once. A *minimum cover* is one where the sum of the cost of the covering patterns is minimum.

> **Example 2.3.4.** Consider the subject graph in Figure 2.9 (a) and a set of pattern trees shown in Figure 2.9 (b) and labeled $\{t_1, t_2, t_3, t_4\}$. The first two patterns are the basic ones, and they are sufficient to cover any binary tree. Other patterns can yield lower cost solutions. An example of a cover is shown in Figure 2.9 (c).

A dynamic programming algorithm for covering binary trees can be defined as follows. There are as many decisions as the vertices of the subject tree, and they are considered according to a bottom-up traversal of the tree. A cost is assigned to each vertex; the cost of the leaves is zero. Each decision consists first of verifying whether a pattern tree is isomorphic to any subtree rooted at the corresponding vertex of the subject tree. Such pattern trees are called *matching* pattern trees. For the successful cases, the corresponding cost is equal to the sum of the cost of the matching pattern tree plus the cost of the vertices of the subject tree corresponding to the leaves of the matching pattern tree. The decision process involves selecting the pattern tree leading to minimum cost and assigning this cost to the vertex. Since every decision selects the optimum for the subproblem corresponding to the subtree rooted at that vertex, the sequence of decisions satisfies the optimality principle and leads to a minimum cover. The complexity of the algorithm is $O(|V|)$, because there are as many decisions as vertices and the complexity of each decision is bounded from above by the cardinality of the pattern tree set, which is constant. Note that including the basic trees in the pattern tree set guarantees that at least one matching pattern tree will be found at each vertex.

$TREE\_COVER(T(V, E))$ {

    Set the cost of the internal vertices to -1;

    Set the cost of the leaf vertices to 0;

    while (some vertex has negative weight) **do** {

        Select a vertex $v \in V$ whose children have all nonnegative cost;

        M = set of all matching pattern trees at vertex v;

        $$cost\,(v) = \min_{m \in M(v)} \left(cost\,(m) + \sum_{u \in L(m)} cost\,(u) \right);$$

    }

}

ALGORITHM 2.3.2

In Algorithm 2.3.2, we denote by $M(v)$ the set of matching pattern trees at any vertex $v \in V$ and by $L(m)$ the vertices of the subject tree corresponding to the leaves of a matching pattern tree $m \in M(v)$.

> **Example 2.3.5.** Consider again the subject and pattern trees of Figure 2.9 (a,b). Assume that the costs of the pattern trees $t_i, i = 1, 2, 3, 4$ are $(2, 3, 4, 5)$, respectively.
>
>   Vertex u can be matched by t_2 only. Its cost is 3. Vertex t can be matched by t_3, with cost 4, or by t_1, with cost equal to the cost of t_1 (i.e., 2) plus the cost of u (i.e., 3), yielding an overall cost of 5. The first choice is minimum and thus the cost of t is set to 4. Vertex s can be matched by t_1 only and its cost is 2. The root vertex v can be matched by t_2 with cost $3 + 2 + 4 = 9$. It can also be matched by t_4 with cost $5 + 3 = 8$. The second match is optimum. An optimum cover is shown in Figure 2.9 (c).

THE GREEDY ALGORITHM. The *greedy* algorithm solves a problem by making a sequence of local decisions, as with dynamic programming. Whereas in dynamic programming the decisions depend on the solutions to some subproblems, the greedy algorithm makes decisions based solely on local information, and then it solves the related subproblems. In other words, dynamic programming solves subproblems bottom up while the greedy strategy is usually top down [12].

The greedy algorithm yields an optimum solution to a problem when two conditions are satisfied. First, the problem must show an optimal substructure, as in the case of dynamic programming. Second, the greedy choice must be such that the problem substructure is exploited, by iteratively reducing the problem to subproblems similar in nature [12]. There is an interesting theory on the applicability of the greedy algorithm to solving exactly some classes of problems. It can be shown formally that the greedy algorithm returns optimum solutions on problems that have appropriate models in terms of computational structures called *matroids*. We refer the interested reader to specialized textbooks [12, 22] for a detailed description of the theory of the greedy algorithm.

In practice, the greedy algorithm is applied to many problems that do not exhibit any specific property. Thus the optimality of the solution is not guaranteed and the greedy algorithm is used as a heuristic method. Since the complexity of the greedy algorithm depends on the number of local decisions, it is often implemented so that its complexity is linear.

```
GREEDY_SCHEDULING(T) {
        i = 1;
    repeat {
            while ((Q = {unscheduled tasks with release time < i}) == ∅)  do
                    i = i + 1;
            if (∃ an unscheduled task  p : i + l(p) > d(p)) return(FALSE);
            Select q ∈ Q with smallest deadline;
            Schedule q at time i;
                i = i + l(q);
    } until (all tasks scheduled);
    return(TRUE);
}
```

ALGORITHM 2.3.3

We consider here as an example the *sequencing problem with release times and deadlines*. The problem can be stated as follows [14]. We are given a set of tasks T and for each task $t \in T$ a positive length $l(t) \in Z^+$, a release time $r(t) \in Z^+$ and a deadline $d(t) \in Z^+$. We would like to find an ordering of the tasks, i.e., a schedule on a single processor such that the release times and deadlines are satisfied. This decision problem can be transformed into an optimization problem by requiring the completion time of the last task to be minimum.

Example 2.3.6.   Consider a set of three tasks $T = \{a, b, c\}$ with release times $r(a) = 1; r(b) = 1; r(c) = 3$ and deadlines $d(a) = 4; d(b) = \infty; d(c) = 6$. There is only one sequence satisfying the deadlines: (a, c, b), as shown in Figure 2.10.

We consider now a greedy scheduling algorithm (Algorithm 2.3.3) for the sequencing problem that assigns a start time to the tasks. We assume (arbitrarily) that the tasks start at time 1.

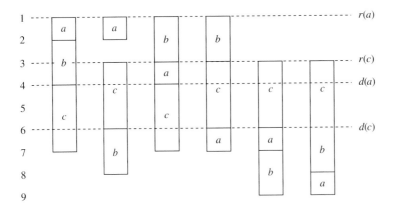

FIGURE 2.10
Six sequences of tasks. Only the second sequence meets the deadlines.

The algorithm increments the time counter i until there are tasks that can be scheduled. Then it schedules the most constrained task, i.e., the one with the smallest deadline. If at any time i we are left with unscheduled tasks whose deadline can no longer be met, the algorithm terminates, stating that no schedule exists. The algorithm has no backtracking capability, because scheduled tasks cannot be rescheduled. The complexity of the algorithm is $O(|T|)$.

Unfortunately the algorithm is not exact in the general case. It always constructs a schedule satisfying the release times and deadlines when it terminates by returning TRUE. Nevertheless, it may happen that it returns FALSE even though a solution to the problem exists. The algorithm just missed it, because of its greed. In addition, the computed sequence may not yield a minimum completion time for the last task. This result should not be surprising, because sequencing with release times and deadlines is intractable in the general case.

Example 2.3.7. Consider the tasks of Example 2.3.6. At time $i = 1$ the algorithm can schedule $Q = \{a, b\}$. It chooses a because it has the earliest deadline and it sets $i = 1 + l(a) = 2$. At time $i = 2$ the algorithm schedules $Q = b$. It sets $i = 2 + l(b) = 4$. At time $i = 4$, $Q = c$. Since $i + l(c) = 7 > d(c) = 6$, the algorithm terminates without completing a valid sequence.

In this case, the greedy algorithm is an example of a *heuristic* method that attempts to construct a solution with low computational cost. Most scheduling algorithms are based on a greedy strategy, like *list scheduling* algorithms which are described in detail in Section 5.4.3.

When considering the application of the greedy scheduling algorithm to sequencing problems without release times and/or with unit-length tasks, it is possible to show that the greedy algorithm is exact, i.e., it returns FALSE, if and only if no schedule exists that satisfies the given release times and deadlines. In addition, it constructs a sequence yielding a minimum completion time of the last task. In this case, the local best choice leads to a global optimum sequencing of the tasks. This can be proven formally on the basis of the fact that delaying the most urgent task (i.e., the one with the earliest deadline) leads to a subproblem whose solution is no more favorable than the one where that task is scheduled [23].

Example 2.3.8. Consider again the tasks of Example 2.3.6, but assume that all release times are 1. At time $i = 1$ the algorithm can schedule $Q = \{a, b, c\}$. It chooses again a because it has the earliest deadline and it sets $i = 1 + l(a) = 2$. At time $i = 2$ the algorithm can schedule $Q = \{b, c\}$. It chooses c because it has the earliest deadline. It sets $i = 2 + l(c) = 5$. At time $i = 5$ it schedules $Q = b$ and terminates successfully.

2.4 GRAPH OPTIMIZATION PROBLEMS AND ALGORITHMS

We briefly review here some important graph optimization problems and algorithms that are relevant to synthesis and optimization of digital circuits.

2.4.1 The Shortest and Longest Path Problems

The shortest and longest path problems are central to the modeling of many other problems. First we review the shortest path problem and some algorithms that solve it. We then extend this analysis to the longest path problem and to solving linear sets of inequalities.

We consider a connected and directed edge-weighted graph, with a given vertex with zero in-degree, called the *source*. The *single-source shortest path problem* is to determine a path of total minimum weight from the source to any other vertex. We shall refer to it as the shortest path problem for brevity. In the most general case, the problem can be difficult when negative cycles (closed paths with total negative weight) are present. Therefore we restrict the shortest path problem to graphs with no negative cycles, and we term *inconsistent* a problem that has negative cycles. Similarly, we define the longest path as a path of total maximum weight from the source, we restrict its search to graphs with no positive cycles and we term *inconsistent* a problem that has positive cycles.

We consider an edge-weighted graph $G(V, E, W)$, with vertex set $V = \{v_i, i = 0, 1, \ldots, n\}$, the source vertex being denoted as v_0. The weight set is $W = \{w_{ij}, i, j = 0, 1, \ldots, n\}$, where any vertex pair not joined by an edge has an infinite weight. We denote the weight of the shortest path from the source to each vertex by $\{s_i; i = 0, 1, \ldots, n\}$. Hence $s_0 = 0$.

The shortest path problem is characterized by Bellman's equations [22], which define the path weight to a vertex in terms of the path weight to its direct predecessors. Namely:

$$s_i = \min_{k \neq i} (s_k + w_{ki}); \qquad i = 1, 2, \ldots, n \tag{2.8}$$

$$= \min_{k:(v_k, v_i) \in E} (s_k + w_{ki}); \quad i = 1, 2, \ldots, n \tag{2.9}$$

The simplest case for solving Bellman's equations occurs when the graph is acyclic. Then the vertices can be ordered in a way consistent with the partial order represented by the dag, and the equations can be solved in that order. Finding a consistent vertex enumeration is called *topological sort* and can be achieved with complexity $O(|V| + |E|) \leq O(n^2)$ [1, 12].

When the graph has cycles, Bellman's equations are cross-coupled and harder to solve. In the special case when all weights are positive, the shortest path problem can be solved by the greedy algorithm proposed by Dijkstra [1]. The algorithm keeps a list of tentative shortest paths, which are then iteratively refined. Initially, all vertices are unmarked and $\{s_i = w_{0,i}, i = 1, 2, \ldots, n\}$. Thus, the path weights to each vertex are either the weights on the edges from the source or infinity. Then the algorithm iterates the following steps until all vertices are marked. It selects and marks the vertex that is head of a path from the source whose weight is minimal among those paths whose heads are unmarked. The corresponding tentative path is declared final. It updates the other (tentative) path weights by computing the minimum between the previous (tentative) path weights and the sum of the (final) path weight to the newly marked vertex plus the weights on the edges from that vertex (see Algorithm 2.4.1).

```
DIJKSTRA(G(V, E, W)) {
    s₀ = 0;
    for (i = 1 to n)                                    /* path weight initialization */
        sᵢ = w₀,ᵢ;
    repeat {
        Select an unmarked vertex vq such that sq is minimal;
        Mark vq;
        foreach (unmarked vertex vᵢ)
            sᵢ = min{sᵢ, (sq + wq,ᵢ)};
    }
    until (all vertices are marked);
}
```

ALGORITHM 2.4.1

```
BELLMAN_FORD(G(V, E, W)) {
    s₀¹ = 0;
    for (i = 1 to n)                                    /* path weight initialization */
        sᵢ¹ = w₀,ᵢ;
    for (j = 1 to n) {
        for (i = 1 to n) {
            sᵢ^(j+1) = min{sᵢ^j, (sₖ^j + wₖ,ᵢ)};          /* path weight update */
                      k≠i
        }
        if (sᵢ^(j+1) == sᵢ^j  ∀i) return (TRUE);        /* convergence */
    }
    return (FALSE);                                      /* inconsistent problem */
}
```

ALGORITHM 2.4.2

The greedy strategy pays off by providing an exact solution with computational complexity $O(|E| + |V| \log |V|) \leq O(n^2)$. Details and examples are provided in several textbooks [1, 22].

Let us consider now the general case where the graph can have cycles and the sign of the weights is not restricted. The Bellman-Ford algorithm, seen in Algorithm 2.4.2, can be used to detect if the problem is consistent and to compute the shortest paths for consistent problems. The Bellman-Ford algorithm solves Bellman's equations by relaxation. It initializes the shortest path weights to upper bounds provided by the weights on the edges from the source. It then refines all path weights iteratively. When the problem is consistent, the estimates of the path weights (denoted by superscripts) converge to stable values, corresponding to the weights of the shortest paths. If convergence is not achieved in $n = |V| - 1$ outer iterations, the problem is provably inconsistent.

The complexity of Bellman-Ford's algorithm is $O(|V||E|) \leq O(n^3)$.

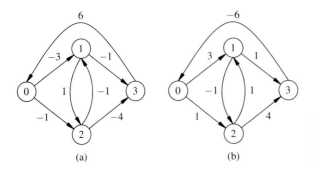

FIGURE 2.11
(a) Graph with cycles and mixed-sign weights. (b) Graph with opposite weights.

Example 2.4.1. Consider the graph of Figure 2.11 (a). The Bellman-Ford algorithm initializes the shortest paths to $s_0^1 = 0$; $s_1^1 = -3$; $s_2^1 = -1$; $s_3^1 = \infty$.
At the first iteration (i.e., $j = 1$), the path weights are updated as follows:

$$s_0^2 = \min\{s_0^1, s_3^1 + w_{3,0}\} = \min\{0, \infty + 6\} = 0$$

$$s_1^2 = \min\{s_1^1, s_2^1 + w_{2,1}\} = \min\{-3, -1 - 1\} = -3$$

$$s_2^2 = \min\{s_2^1, s_1^1 + w_{1,2}\} = \min\{-1, -3 + 1\} = -2$$

$$s_3^2 = \min\{s_3^1, s_1^1 + w_{1,3}, s_2^1 + w_{2,3}\} = \min\{\infty, -3 - 1, -1 - 4\} = -5$$

At the second iteration:

$$s_0^3 = \min\{s_0^2, s_3^2 + w_{3,0}\} = \min\{0, -5 + 6\} = 0$$

$$s_1^3 = \min\{s_1^2, s_2^2 + w_{2,1}\} = \min\{-3, -2 - 1\} = -3$$

$$s_2^3 = \min\{s_2^2, s_1^2 + w_{1,2}\} = \min\{-2, -3 + 1\} = -2$$

$$s_3^3 = \min\{s_3^2, s_1^2 + w_{1,3}, s_2^2 + w_{2,3}\} = \min\{-5, -3 - 1, -2 - 4\} = -6$$

At the third iteration:

$$s_0^4 = \min\{s_0^3, s_3^3 + w_{3,0}\} = \min\{0, -6 + 6\} = 0$$

$$s_1^4 = \min\{s_1^3, s_2^3 + w_{2,1}\} = \min\{-3, -2 - 1\} = -3$$

$$s_2^4 = \min\{s_2^3, s_1^3 + w_{1,2}\} = \min\{-2, -3 + 1\} = -2$$

$$s_3^4 = \min\{s_3^3, s_1^3 + w_{1,3}, s_2^3 + w_{2,3}\} = \min\{-6, -3 - 1, -2 - 4\} = -6$$

Since the path weights match those computed at the previous iteration, the algorithm terminates successfully.

We consider next the longest path problem. A longest path problem can be transformed into a shortest path problem by changing the sign of the weights. Thus it can be solved by the same family of algorithms described before. Alternatively, the shortest path algorithms can be transformed into longest path algorithms by replacing the *min* operator with the *max* operator. The following points are worth noticing. As with the shortest path problem, directed acyclic graphs are the simplest cases, because a consistent enumeration of the vertices can be computed by topological sort. Dijkstra's

```
LIAO_WONG(G(V, E ∪ F, W)) {
    for (j = 1 to |F| + 1) {
        foreach vertex vᵢ
            lᵢʲ⁺¹ = longest path in G(V, E, Wₑ);        /* compute longest path in G(V, E, Eᵥ) */
        flag = TRUE;
        foreach edge (vₚ, v_q) ∈ F {                    /* verify each feedback edge */
            if (l_qʲ⁺¹ < lₚʲ⁺¹ + wₚ,q){                  /* constraint violation */
                flag = FALSE;
                E = E ∪ (v₀, v_q);                       /* add edge */
                w₀,q = (lₚʲ⁺¹ + wₚ,q);
            }
        }
        if ( flag ) return (TRUE);                      /* convergence */
    }
    return (FALSE)                                      /* inconsistent problem */
}
```

ALGORITHM 2.4.3

algorithm can be applied to graphs where all weights are negative. CAD problems modeled by negative-weighted graphs are rare, and therefore Dijkstra's algorithm is not often used for computing the longest paths. The Bellman-Ford algorithm can be applied to both the shortest and longest path problems *mutatis mutandis*.

> **Example 2.4.2.** Consider the problem of computing the longest paths in the graph of Figure 2.11 (b). The shortest path algorithm can be applied to the graph of Figure 2.11 (a), where the signs are reversed. The weights of the longest paths in Figure 2.11 (b) are the opposite of the shortest paths of the graph of Figure 2.11 (a).

We describe now a special case of the longest path problem that has several practical applications in CAD. Let us consider a graph $G(V, E \cup F, W)$ where E is the subset of the edges with non-negative weights and F is the subset of the remaining ones, called *feedback* edges because each one closes a cycle. Moreover, let E induce dag $G(V, E, W_E)$ and let $|E| >> |F|$.

Liao and Wong [25] proposed a variation of the Bellman-Ford algorithm that is specific to this situation and that has lower computational complexity than the Bellman-Ford algorithm. The algorithm exploits the specific knowledge that the edges in F close the cycles. We denote in Algorithm 2.4.3 the set of longest path weights from the source to each vertex as $\{l_i; i = 1, 2, \ldots, n\}$.

The rationale of the algorithm is the following. It attempts to neglect the feedback edges and to derive a solution in terms of the forward edges only. If this solution happens to be correct when the feedback edges are reintroduced, then the algorithm terminates successfully. Otherwise, it corrects the problem by setting a lower bound on the weight of the incorrectly computed paths. The bound can be modeled by adding weighted edges from the source. The process is iterated $|F| + 1$ times.

Algorithm 2.4.3 computes the longest path weights by construction when it terminates favorably. Interestingly enough, Liao and Wong [25] were able to prove that the problem is inconsistent when the algorithm fails.

The algorithm performs at most $|F|+1$ longest path computations of the acyclic graph $G(V, E, E_W)$, where at most $|F|$ edges can be added. Hence the overall complexity is $O((|V|+|E|+|F|)|F|) \leq O(n^3)$. Note that when $|F| << |E|$, the algorithm is much more efficient than Bellman-Ford's.

> **Example 2.4.3.** Consider the longest path problem of Figure 2.11 (b) repeated again for convenience in Figure 2.12 (a). At the first iteration, the directed acyclic subgraph induced by the edges with positive weights is shown in Figure 2.12 (b). The longest paths from the sources in $G(V, E, W_E)$ are $l_1^2 = 3; l_2^2 = 1; l_3^2 = 5$. When the algorithm examines the feedback edges, it finds a constraint violation on edge (v_1, v_2). Namely $l_2^2 = 1 < l_1^2 + w_{1,2} = 3 - 1 = 2$. Thus, the algorithm adds edge (v_0, v_2) to E with weight $w_{0,2} = l_1^2 + w_{1,2} = 3 - 1 = 2$.
>
> At the second iteration the subgraph induced by the edges with positive weights is shown in Figure 2.12 (c). The longest path from the sources in $G(V, E, W_E)$ are $l_1^3 = 3; l_2^3 = 2; l_3^3 = 6$. When the algorithm examines the feedback edges, it does not find any constraint violation and terminates successfully.

It is interesting to relate the longest (or shortest) path problem to the solution of a linear program. Many design problems require the solution to a set of linear inequality constraints of the type:

$$x_j \geq x_i + w_{ij}; \quad i, j = 0, 1, \ldots, n - 1 \tag{2.10}$$

These constraints can be written in matrix form as:

$$\mathbf{A}^T \mathbf{x} \geq \mathbf{b} \tag{2.11}$$

where $\mathbf{x} \in \mathbf{R}^n; \mathbf{b} \in \mathbf{R}^m$, m equals the number of constraints and the entries of $\mathbf{b}$ are $\{w_{ij}\}$ for the values of i, j where a constraint is specified; $\mathbf{A} \in \mathbf{R}^{n \times m}$. Thus, a set of

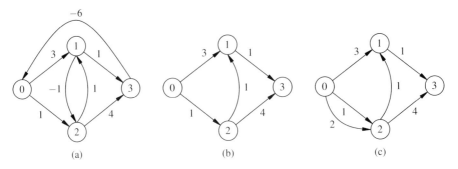

(a)   (b)   (c)

FIGURE 2.12
(a) Graph with cycles and mixed-sign weights. (b) Directed acyclic graph obtained by deleting the feedback edges. (c) Directed acyclic graph obtained by deleting the feedback edges and adding an edge resolving the constraint violation.

linear inequalities can be expressed as the feasibility constraint of a linear program. Matrix A can be viewed as the incidence matrix of a directed graph with n vertices and m edges, each one weighted by the corresponding entry in b. Such a graph is called the *constraint graph* of the problem.

Example 2.4.4. Consider the set of linear inequalities:

$$x_1 \geq x_0 + 3$$
$$x_1 \geq x_2 + 1$$
$$x_2 \geq x_0 + 1$$
$$x_2 \geq x_1 - 1$$
$$x_3 \geq x_1 + 1$$
$$x_3 \geq x_2 + 4$$
$$x_0 \geq x_3 - 6$$

Let:

$$A = \begin{bmatrix} -1 & 0 & -1 & 0 & 0 & 0 & +1 \\ +1 & +1 & 0 & -1 & -1 & 0 & 0 \\ 0 & -1 & +1 & +1 & 0 & -1 & 0 \\ 0 & 0 & 0 & 0 & +1 & +1 & -1 \end{bmatrix}$$

and $b = [3, 1, 1, -1, 1, 4, -6]^T$. Then the set of inequalities can be written as $A^T x \geq b$. Note that A is the incidence matrix of the graphs shown in Figure 2.11.

The longest path problem can then be modeled as a solution to the minimization of $c^T x$, where $c = 1$ (i.e., a vector of 1s), under the constraint 2.11. Note that one vertex of the graph must be chosen as the source and the corresponding entry in x be set to zero to have a unique solution. Conversely, a shortest path problem can be modeled by a set of inequalities with the reversed directions and solved by maximizing $c^T x$.

It is often important to know whether a set of inequalities 2.11 is satisfiable. The following key theorem relates the satisfiability of a set of linear inequalities to the properties of the corresponding *constraint graph*, which has A as the incidence matrix and b as weights.

Theorem 2.4.1. A set of linear inequalities $A^T x \geq b$ is satisfiable if and only if the corresponding constraint graph has no positive cycles.

The proof to this theorem is reported in several textbooks [12]. Intuitively, a set of linear inequalities is satisfiable if and only if the corresponding longest path problem is consistent.

2.4.2 Vertex Cover

A *vertex cover* of an undirected graph $G(V, E)$ is a subset of the vertices such that each edge in E has at least one end-point in that subset. The vertex covering decision problem is to determine if a given graph $G(V, E)$ has a vertex cover of cardinality smaller than (or equal to) a given integer. The corresponding optimization problem is the search for a vertex cover set of minimum cardinality. The vertex cover decision problem is intractable [14].

Heuristic and exact algorithms have been proposed to solve the minimum covering problem. Some heuristic algorithms can guarantee only that the cover is *minimal with respect to containment*, i.e., that no vertex in the cover is redundant and therefore no vertex can be removed while preserving the covering property. Such a cover is often termed *irredundant*.

> **Example 2.4.5.** Consider the graphs of Figure 2.13 (a). Minimum covers are shown in Figure 2.13 (b) and non-minimum irredundant covers in Figure 2.13 (c). Redundant covers are shown in Figure 2.13 (d).

We consider two heuristic algorithms based on a greedy strategy. Algorithm 2.4.4 constructs a cover by iteratively selecting vertices

Two remarks are important. First, selecting the vertex with largest degree corresponds to covering the largest edge subset with a single vertex. Second, the deletion of a vertex corresponds to the removal of the vertex itself and of all edges incident to it.

This simple algorithm may perform well in some cases, but there are examples where the cardinality of the computed cover exceeds twice that of a minimum

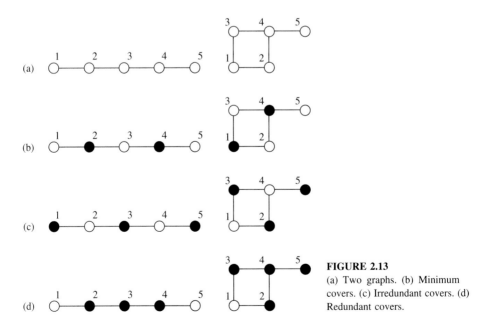

FIGURE 2.13
(a) Two graphs. (b) Minimum covers. (c) Irredundant covers. (d) Redundant covers.

```
VERTEX_COVER_V(G(V, E)) {
    C = ∅;
    while (E ≠ ∅) do {
        Select a vertex v ∈ V;
        Delete v from G(V, E);          /* remove vertex u and edges incident to u */
        C = C ∪ {v};
    }
}
```

ALGORITHM 2.4.4

```
VERTEX_COVER_E(G(V, E)) {
    C = ∅;
    while (E ≠ ∅) do {
        Select an edge {u, v} ∈ E;
        C = C ∪ {u} ∪ {v};
        Delete u and v from G(V, E);
    }
}
```

ALGORITHM 2.4.5

cover [28]. Algorithm 2.4.5, which has a better worst-case performance, can be obtained by iteratively selecting edges instead of vertices.

In this case, the algorithm has the property that the cardinality of the computed cover is at most twice that of the minimum cover [28]. Unfortunately the covers computed by algorithms $VERTEX\_COVER\_V$ and $VERTEX\_COVER\_E$ may be redundant. For example, a possible sequence of vertex selections by algorithm $VERTEX\_COVER\_V$ for the first graph of Figure 2.13 (a) is (v_2, v_3, v_4). The computed cover is redundant, as shown in Figure 2.13 (d). A simple way to make a vertex cover irredundant is to add a post-processing step that considers each vertex of the cover in turn and removes it from the cover if the remaining vertices still cover all edges.

The vertex covering problem can be solved exactly by the branch-and-bound algorithm for general covering problems that is described in Section 2.5.3. It is interesting to note that the *edge covering* problem in a graph can be solved exactly in polynomial time by matching algorithms [22].

2.4.3 Graph Coloring

A *vertex coloring* of an undirected graph $G(V, E)$ is a labeling of the vertices such that no edge in E has two end-points with the same label. The vertex coloring decision problem is to determine if a given graph $G(V, E)$ has a chromatic number smaller than (or equal to) a given integer. The corresponding optimization problem is the search for a vertex coloring with a minimum number of colors. The vertex coloring decision problem is intractable [14].

```
VERTEX_COLOR(G(V, E)) {
    for (i = 1  to  |V|) {
        c = 1;
        while ( ∃ a vertex adjacent to vᵢ with color c ) do {
            c = c + 1;
        }
        Label vᵢ with color c;
    }
}
```

ALGORITHM 2.4.6

Most algorithms for graph coloring are based on a sequential scan of the vertex set, where vertices are colored one at a time. In Algorithm 2.4.6 we consider a simple heuristic algorithm for minimum coloring first [3]. Colors are represented by positive integers.

The total number of colors is the maximum value that c attains. It is larger than (or equal to) the chromatic number $\chi(G(V, E))$. Unfortunately it may be much larger than $\chi(G(V, E))$, and it is sensitive to the order in which the vertices are colored. The algorithm can be modified to cope with the coloring decision problem by providing an exit (with negative response) when the color counter c exceeds a given threshold. Unfortunately, due to the heuristic nature of the algorithm, a negative response can be given even if a proper coloring exists with c colors.

Some improvements to this basic scheme have been proposed. An example is to allow color interchange between two (already colored) vertices when this helps in not increasing the number of colors. We refer the interested reader to reference [3] for details and examples.

Exact algorithms for coloring have been proposed with computation complexity above polynomial. Most techniques are enumerative in nature. For example, the algorithm to solve the coloring decision problem reported in reference [19] is also based on a sequential scan and coloring of the vertices. When the number of possible colors is exhausted, the algorithm backtracks and recolors previously colored vertices. In the worst case, all color assignments to the vertices are considered.

Example 2.4.6. Consider the graph of Figure 2.14 (a). A minimum coloring is shown in Figure 2.14 (b), where three colors are used.

Consider now the application of algorithm $VERTEX\_COLOR$ to this graph, with the vertices sorted according to their identifiers. The first four vertices are labeled with alternating colors. A third color is required for v_5 and a fourth for v_6. Thus the solution, shown in Figure 2.14 (c), is not optimum.

An optimum solution would require some recoloring of previously colored vertices. For example, if we want to color the graph with three colors and we fail at coloring v_6 properly, then we can backtrack and try to recolor v_5 (which is not possible) or v_4 (which is possible). Once v_4 is recolored, then a solution with three colors can be achieved, as shown in Figure 2.14 (b).

A heuristic alternative to backtracking is to swap colors in colored vertex pairs. In this case, the colors of v_4 and v_5 should have been swapped.

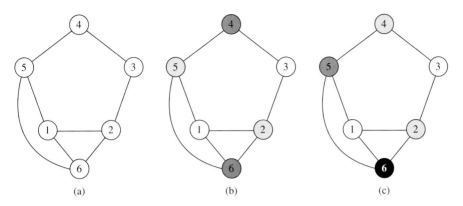

FIGURE 2.14
(a) Graph. (b) Minimum coloring. (c) Non-minimum coloring.

The coloring problem is solvable in polynomial time for chordal graphs, which are perfect graphs. Vertex coloring is tractable in this case, because chordal graphs have a *perfect vertex elimination scheme* [15]. Briefly, a perfect vertex elimination scheme is a linear order of the vertices, denoted by σ, such that, for each vertex $\{v_i; i = 1, 2, \ldots, |V|\}$, the vertex subset $\{v_j \in V : \{v_i, v_j\} \in E$ and $\sigma(v_j) > \sigma(v_i)\}$ induces a clique. For example, when considering the graph of Figure 2.5 (a), the sequence (v_2, v_1, v_3, v_4) is perfect while (v_1, v_2, v_3, v_4) is not.

Coloring chordal graphs can be done by considering the vertices according to a perfect elimination scheme. This specific order guarantees that no backtracking is necessary to find an optimum coloring, and thus optimum coloring can be achieved by algorithms with polynomial complexity. In particular, checking whether a graph is chordal, computing a perfect elimination scheme and coloring can be done in $O(|V| + |E|)$ time. Moreover, scanning the vertices according to a perfect elimination scheme allows us to determine all maximal cliques with no additional effort, and hence we can identify a clique cover. We refer the interested reader to Golumbic's book [15] for details.

Let us now consider interval graphs, which model many relevant circuit optimization problems. Since interval graphs are chordal, coloring and searching for all maximal cliques can be done by exploiting a perfect elimination scheme, as mentioned before. Very often, problems are directly specified by the sets of intervals $I = \{[l_j, r_j]; \ j = 1, 2, \ldots, |I|\}$ that induce the interval graph. Since intervals correspond to vertices, coloring the intervals is equivalent to coloring the vertices.

A well-known algorithm for coloring interval graphs is the *LEFT_EDGE* algorithm (Algorithm 2.4.7), which was proposed by Hashimoto and Stevens for channel routing [17], where intervals correspond to wire segments to be arranged in a channel. The name stems from the selection of the intervals based on their left-edge coordinate, i.e., the minimum value of the interval. The algorithm first sorts the intervals by their left edge. It then considers one color at a time and assigns as many intervals as possible to that color by scanning a list of intervals in ascending order of the left edge before incrementing the color counter.

```
LEFT_EDGE(I) {
        Sort elements of I in a list L in ascending order of l_i;
        c = 0;
        while (some interval has not been colored ) do {
                S = ∅;
                r = 0;                                 /* initialize coordinate of rightmost edge in S */
                while ( ∃ an element in L whose left edge coordinate is larger than r) do{
                        s = First element in the list L with l_s > r;
                        S = S ∪ {s};
                        r = r_s;                       /* update coordinate of rightmost edge in S */
                        Delete s from L;
                }
                c = c + 1;
                Label elements of S with color c;
        }
}
```

ALGORITHM 2.4.7

The algorithm yields a provably minimum coloring of $G(V, E)$, i.e., the maximum value of c is the chromatic number $\chi(G(V, E))$. The complexity of the algorithm is $O(|V| \log |V|)$ [17].

> **Example 2.4.7.** Consider the set of intervals specified in the following table. Each interval is associated with a vertex, and the corresponding interval graph is shown in Figure 2.15 (a). Recall that edges correspond to intersections of intervals, according to the definition of the interval graph:

Vertex	Left edge	Right edge
v_1	0	3
v_2	3	5
v_3	6	8
v_4	0	7
v_5	7	8
v_6	0	2
v_7	2	6

The intervals, sorted by ascending order of the left edge, are shown in Figure 2.15 (b). The algorithm will assign the first color to v_1, v_2 and v_3; the second to v_6, v_7 and v_5 and the last one to v_4. The colored graph is shown in Figure 2.15 (c). The intervals can then be packed into as many windows (tracks) as the chromatic number, as shown in Figure 2.15 (d).

2.4.4 Clique Covering and Clique Partitioning

Clique partitioning and clique covering are closely related. A clique partition is a disjoint clique cover. A clique cover that is not disjoint can be made disjoint by selecting a set of disjoint cliques each contained in a corresponding clique of the

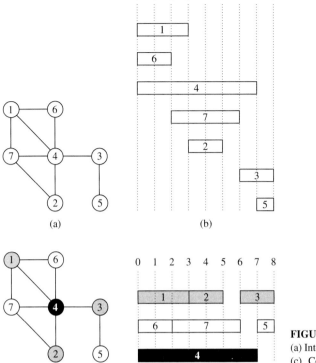

FIGURE 2.15
(a) Interval graph. (b) Sorted intervals. (c) Colored graph. (d) Packed intervals.

cover. Finding a clique cover, or partition, with bounded cardinality is an intractable problem [14].

A partition into cliques can be obtained by coloring the complement of the graph. Indeed each clique identifies a set of pairwise adjacent vertices, which are not adjacent in the graph complement, and thus can be colored with the same color. Therefore the clique cover number of a graph corresponds to the chromatic number of its complement. As a result, heuristic and exact algorithms for coloring can be applied to the search of minimum-cardinality clique partitions.

Example 2.4.8. The complement of the graph of Figure 2.14 (a), repeated again for convenience in Figure 2.16 (a), is shown in Figure 2.16 (b). Emboldened edges correspond to cliques. Three cliques cover the graph.

When directly solving the clique partitioning problem, a solution can be determined by iteratively searching for the maximum clique of the graph and then deleting it from the graph until there are no more vertices, as seen in Algorithm 2.4.8.

Therefore the problem is reduced to that of finding the maximum clique, which is also intractable [14]. A maximum clique corresponds to a maximum independent set in the complement of the graph. A maximum independent set is the subset of

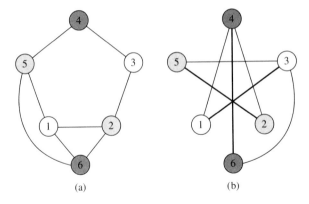

(a)   (b)

FIGURE 2.16
(a) Colored graph. (b) Clique partitioning of the complement of the graph.

```
CLIQUE_PARTITION(G(V, E)) {
    Π = ∅;                                    /* Initialize partition */
    while (G(V, E) not empty ) do {
        C = MAX_CLIQUE(G(V, E));              /* Compute maximum clique C ⊆ V in G(V, E) */
        Π = Π ∪ C;
        Delete C from G(V, E);
    }
}
```

ALGORITHM 2.4.8

vertices not included in a vertex cover. Thus, exact and heuristic algorithms for vertex cover can be applied to the maximum clique problem.

Example 2.4.9. Consider the graph of Figure 2.17 (a). Its complement is shown in Figure 2.17 (b), and it can be covered by vertices $\{v_2, v_4\}$, which is a minimum vertex cover. Thus the maximum independent set $\{v_1, v_3, v_5\}$ identifies a maximum clique in the original graph.

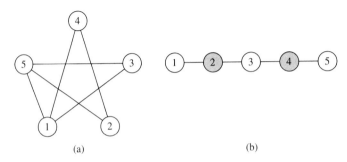

(a)   (b)

FIGURE 2.17
(a) Graph. (b) Vertex cover of the complement of the graph.

```
MAX_CLIQUE(G(V, E)) {
    C = vertex with largest degree;
    repeat {
        repeat {
            U = {v ∈ V  : v ∉ C and adjacent to all vertices of C};
            if (U = ∅)
                return(C);
            else {
                Select vertex v ∈ U;
                C = C ∪ {v};
            }
        }
    }
}
```

ALGORITHM 2.4.9

Algorithm 2.4.9 is a simple heuristic algorithm that can be used to compute a maximal clique as an approximation to the maximum clique.

The clique partitioning problem can be solved exactly in $O(|V| + |E|)$ time for chordal graphs [15], and therefore for interval graphs as well. It can also be solved in polynomial time for comparability graphs by transforming the problem into a minimum-flow computation of a related network [15].

2.5 BOOLEAN ALGEBRA AND APPLICATIONS

An *algebraic system* is the combination of a set and one or more operations. A *Boolean algebra* is defined by the set $B \supseteq \mathsf{B} \equiv \{0, 1\}$ and by two operations, denoted by $+$ and $\cdot$, which satisfy the *commutative* and *distributive* laws and whose identity elements are 0 and 1, respectively. In addition, any element $a \in B$ has a complement, denoted by a', such that $a + a' = 1$ and $a \cdot a' = 0$. These axioms, which define a Boolean algebra, are often referred to as Huntington's postulates [18]. The major properties of a Boolean algebra can be derived from Huntington's postulates and are shown in Table 2.1. Note that a Boolean algebra differs from an ordinary algebra in that distributivity applies to both operations and because of the presence of a complement.

There are many examples of Boolean algebraic systems, for example set theory, propositional calculus and arithmetic Boolean algebra [9]. We consider in this book only the binary Boolean algebra, where $B = \mathsf{B} \equiv \{0, 1\}$ and the operations $+, \cdot$ are the *disjunction* and *conjunction*, respectively, often called *sum* and *product* or OR and AND. The multi-dimensional space spanned by n binary-valued Boolean variables is denoted by B^n. It is often referred to as the n-dimensional *cube*, because it can be graphically represented as a hypercube. A point in B^n is represented by a binary-valued vector of dimension n.

TABLE 2.1
Some properties of Boolean algebraic systems.

$a + (b + c) = (a + b) + c$	Associativity
$a(bc) = (ab)c$	Associativity
$a + a = a$	Idempotence
$aa = a$	Idempotence
$a + (ab) = a$	Absorption
$a(a + b) = a$	Absorption
$(a + b)' = a'b'$	De Morgan
$(ab)' = a' + b'$	De Morgan
$(a')' = a$	Involution

When binary variables are associated with the dimensions of the Boolean space, a point can be identified by the values of the corresponding variables. A *literal* is an instance of a variable or of its complement. A product of n literals denotes a point in the Boolean space: it is a zero-dimensional cube. Often, products of literals are called cubes.

Example 2.5.1. Let us consider a three-dimensional Boolean space, as shown in Figure 2.18 with the associated variables a, b, c. A vertex in the Boolean space can be expressed by the product of $n = 3$ literals, for example, $ab'c'$, or equivalently by the row vector [100]. An arbitrary subspace can be expressed by a product of literals, for example, ab, or by the row vector [11∗], where ∗ means that the third variable can take any value.

2.5.1 Boolean Functions

A *completely specified* Boolean function is a mapping between Boolean spaces. An n-input, m-output function is a mapping $\mathbf{f} : \mathbf{B}^n \to \mathbf{B}^m$. It can be seen as an array of m scalar functions over the same domain, and therefore we use the vector

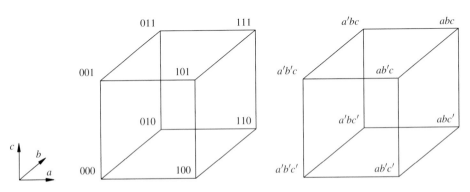

FIGURE 2.18
The three-dimensional Boolean space.

notation. An *incompletely specified* scalar Boolean function is defined over a sub-set of B^n. The points where the function is not defined are called *don't care* con-ditions. They are related to the input patterns that can never occur and to those for which the output is not observed. In the case of multiple-output functions (i.e., $m > 1$), *don't care* points may differ for each component, because different out-puts may be sampled under different conditions. Therefore, incompletely specified functions are represented as $\mathbf{f} : B^n \rightarrow \{0, 1, *\}^m$, where the symbol $*$ denotes a *don't care* condition. For each output, the subsets of the domain for which the func-tion takes the values 0, 1 and $*$ are called the *off set*, *on set*, and *dc set*, respec-tively.

A *Boolean relation* is a relation between Boolean spaces. It can be seen as a generalization of a Boolean function, where a point in the domain can be associated with more than one point in the co-domain. Boolean relations and their optimization play an important role in multiple-level logic synthesis and optimization, as shown in Chapter 8.

We review some definitions and properties of Boolean functions. For the sake of simplicity, we consider single-output functions. Let $f(x_1, x_2, \ldots, x_n)$ be a Boolean function of n variables. The set $\{x_1, x_2, \ldots, x_n\}$ is called the *support* of the function.

> **Definition 2.5.1.** The **cofactor** of $f(x_1, x_2, \ldots, x_i, \ldots, x_n)$ with respect to variable x_i is $f_{x_i} = f(x_1, x_2, \ldots, 1, \ldots, x_n)$. The **cofactor** of $f(x_1, x_2, \ldots, x_i, \ldots, x_n)$ with respect to variable x_i' is $f_{x_i'} = f(x_1, x_2, \ldots, 0, \ldots, x_n)$.

In this book, we shall use mainly the short notation for cofactors as in the definition, e.g., f_{x_i}. In Chapter 8 we shall use an extended notation with a bar and an assignment, e.g., $f|_{x_i=1}$, to avoid confusion with the labeling of a function.

Boole's expansion of a function f, often called Shannon's expansion, is shown by the following theorem, whose proof is reported in several textbooks, e.g., refer-ence [9].

> **Theorem 2.5.1.** Let $f : B^n \rightarrow B$. Then $f(x_1, x_2, \ldots, x_i, \ldots, x_n) = x_i \cdot f_{x_i} + x_i' \cdot f_{x_i'} = (x_i + f_{x_i'}) \cdot (x_i' + f_{x_i}) \; \forall i = 1, 2, \ldots, n$.

Any function can be expressed as a *sum of products* of n literals, called the *minterms* of the function, by recursive expansion. Alternatively, it can be represented as a *product of sums* of n literals, called *maxterms*. We use mainly *sum of products* forms and minterms in the sequel, all considerations being applicable to *product of sums* forms and maxterms *mutatis mutandis*.

As a result of a complete expansion, a Boolean function can be interpreted as the set of its minterms. Operations and relations on Boolean functions over the same domain can be viewed as operations and relations on their minterm sets. In particular, the sum and the product of two functions are the union ($\cup$) and intersection ($\cap$) of their minterm sets, respectively. Implication between two functions corresponds to the containment ($\subseteq$) of their minterm sets. Sometimes set operators are used instead of Boolean operators.

Example 2.5.2. Consider the majority function $f = ab + bc + ac$. Its cofactor with respect to (w.r.t.) a is computed by setting $a = 1$. Therefore $f_a = b + c$. Similarly, its cofactor w.r.t. a' is computed by setting $a = 0$, yielding $f_{a'} = bc$. The function can be re-written as $f = af_a + a'f_{a'} = a(b + c) + a'bc = ab + ac + a'bc = ab + bc + ac$. By applying the expansion further (for example, by expanding f_a and $f_{a'}$ w.r.t. b), we obtain $f_{ab} = 1$, $f_{ab'} = c$, $f_{a'b} = c$ and $f_{a'b'} = 0$. Thus $f = ab + ab'c + a'bc$. By repeating the expansion w.r.t. c, we obtain $f = abc + abc' + ab'c + a'bc$, which is a representation of the function in terms of four minterms.

Definition 2.5.2. A function $f(x_1, x_2, \ldots, x_i, \ldots, x_n)$ is (positive/negative) **unate in variable** x_i if $f_{x_i} \supseteq f_{x_i'}$ ($f_{x_i} \subseteq f_{x_i'}$). Otherwise it is binate (or mixed) in that variable.
   A function is (positive/negative) **unate** if it is (positive/negative) unate in all support variables. Otherwise it is binate (or mixed).

Note that stating $f_{x_i} \supseteq f_{x_i'}$ is equivalent to say that the set of minterms of f_{x_i} includes the set of minterms of $f_{x_i'}$, or $f_{x_i} \geq f_{x_i'}$, for all possible assignments to the other variables $x_j : j \neq i$, $j = 1, 2, \ldots, n$.

Example 2.5.3. The function $f = a + b + c'$ is positive unate with respect to variable a, because $f_a = 1 \geq f_{a'} = b + c'$ for any assignment to variables b and c. Alternatively, the set of minterms associated with f_a, i.e., $\{bc, b'c, bc', b'c'\}$, contains the set of minterms associated with $f_{a'}$, i.e., $\{bc, bc', b'c'\}$. By a similar argument, f is positive unate with respect to variable b and negative unate with respect to variable c.

Three operators are important in Boolean algebra: *Boolean difference*, *consensus* (universal quantifier) and *smoothing* (existential quantifier).

Definition 2.5.3. The **Boolean difference**, or **Boolean derivative**, of $f(x_1, x_2, \ldots, x_i, \ldots, x_n)$ with respect to variable x_i is $\partial f / \partial x_i = f_{x_i} \oplus f_{x_i'}$.

The Boolean difference w.r.t. x_i indicates whether f is sensitive to changes in input x_i. When it is zero, then the function does not depend on x_i. In that case, x_i is said to be unobservable. The concept of Boolean derivative is important in testing. Indeed, any test applied to x_i must produce effects observable at the circuit outputs to be useful.

Definition 2.5.4. The **consensus** of $f(x_1, x_2, \ldots, x_i, \ldots, x_n)$ with respect to variable x_i is $\mathcal{C}_{x_i}(f) = f_{x_i} \cdot f_{x_i'}$.

The consensus of a function with respect to a variable represents the component that is independent of that variable. The consensus can be extended to sets of variables, as an iterative application of the consensus on the variables of the set.

Definition 2.5.5. The **smoothing** of $f(x_1, x_2, \ldots, x_i, \ldots, x_n)$ with respect to variable x_i is $\mathcal{S}_{x_i}(f) = f_{x_i} + f_{x_i'}$.

The smoothing of a function with respect to a variable corresponds to dropping that variable from further consideration. Informally, it corresponds to deleting all appearances of that variable.

Example 2.5.4. Consider again the function $f = ab + bc + ac$ with $f_a = b + c$ and $f_{a'} = bc$ [Figure 2.19 (a)]. The Boolean difference w.r.t. a is $\partial f/\partial a = f_a \oplus f_{a'} = (b + c) \oplus bc = b'c + bc'$. When $b'c + bc'$ is true, the value of f changes as a changes [Figure 2.19 (b)].

The consensus of f w.r.t. a is $\mathcal{C}_a(f) = f_a \cdot f_{a'} = (b+c)bc = bc$, and it represents the component of the function independent of a [Figure 2.19 (c)].

The smoothing of f w.r.t. a is $\mathcal{S}_a(f) = f_a + f_{a'} = (b + c) + bc = b + c$, and it represents the function when we drop the dependence of f on a [Figure 2.19 (d)].

It is important to notice that Boole's expansion is a special case of an expansion using an orthonormal basis [9]. Let $\phi_i, i = 1, 2, \ldots, k$, be a set of Boolean functions such that $\sum_{i=1}^{k} \phi_i = 1$ and $\phi_i \cdot \phi_j = 0$, $\forall i \neq j \in \{1, 2, \ldots, k\}$. An orthonormal expansion of a function f is $f = \sum_{i=1}^{k} f_{\phi_i} \phi_i$. The term f_{ϕ_i} is called the *generalized cofactor* of f w.r.t. $\phi_i, \forall i$. Note that in the special case of Boole's expansion on a variable x_i, the orthonormal functions are just the literals x_i and x'_i and the generalized

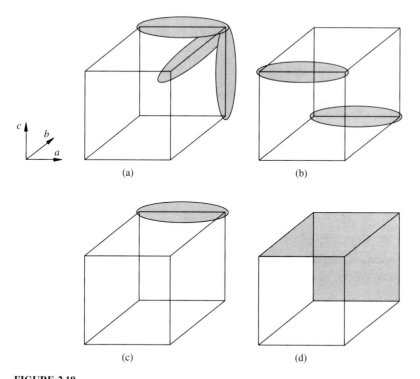

(a) (b)

(c) (d)

FIGURE 2.19
(a) The function $f = ab + bc + ac$. (b) The Boolean difference $\partial f/\partial a$. (c) The consensus $\mathcal{C}_a(f)$. (d) The smoothing $\mathcal{S}_a(f)$.

cofactor is the usual cofactor. The generalized cofactor may not be unique and satisfies the following bounds [9]: $f \cdot \phi_i \subseteq f_{\phi_i} \subseteq f + \phi_i'$.

Example 2.5.5. Consider function $f = ab + bc + ac$. Let $\phi_1 = ab$ and $\phi_2 = a' + b'$. Then $ab \subseteq f_{\phi_1} \subseteq 1$ and $a'bc + ab'c \subseteq f_{\phi_2} \subseteq ab + bc + ac$. By choosing $f_{\phi_1} = 1$ and $f_{\phi_2} = a'bc + ab'c$, we can write:

$$f = \phi_1 f_{\phi_1} + \phi_2 f_{\phi_2}$$
$$= ab1 + (a' + b')(a'bc + ab'c)$$
$$= ab + bc + ac$$

The following theorem has many applications to the manipulation of Boolean functions.

Theorem 2.5.2. Let f, g be two Boolean functions expressed by an expansion with respect to the same orthonormal basis $\phi_i, i = 1, 2, \ldots, k$. Let $\odot$ be an arbitrary binary operator representing a Boolean function of two arguments. Then:

$$f \odot g = \sum_{i=1}^{k} \phi_i \cdot (f_{\phi_i} \odot g_{\phi_i}) \tag{2.12}$$

A proof of this theorem is reported by Brown [9].

Example 2.5.6. Let $f = ab + bc + ac$ and $g = c$. Let $\phi_1 = ab$ and $\phi_2 = a' + b'$. Choose $f_{\phi_1} = 1$, $f_{\phi_2} = a'bc + ab'c$, $g_{\phi_1} = abc$, $g_{\phi_2} = (a' + b')c$.
Then, $fg = bc + ac$ can be computed as:

$$fg = \phi_1(f_{\phi_1} g_{\phi_1}) + \phi_2(f_{\phi_2} g_{\phi_2})$$
$$= ab(1abc) + (a' + b')((a'bc + ab'c)(a' + b')c)$$
$$= ab(abc) + (a' + b')(a'bc + ab'c)$$
$$= abc + a'bc + ab'c$$
$$= bc + ac$$

The following corollary is widely used, and it relates to restricting the orthonormal expansion to Boole's expansion.

Corollary 2.5.1. Let f, g be two Boolean functions with support variables $\{x_i, i = 1, 2, \ldots, n\}$. Let $\odot$ be an arbitrary binary operator, representing a Boolean function of two arguments. Then:

$$f \odot g = x_i \cdot (f_{x_i} \odot g_{x_i}) + x_i' \cdot (f_{x_i'} \odot g_{x_i'}), \quad \forall i = 1, 2, \ldots, n \tag{2.13}$$

The extension of these definitions and properties to multiple-output functions can be done in a straightforward way by representing the function and its cofactors as m-dimensional vectors, where each entry corresponds to an output of the function itself.

2.5.2 Representations of Boolean Functions

There are different ways of representing Boolean functions, which can be classified into *tabular forms*, *logic expressions* and *binary decision diagrams*.

TABULAR FORMS. Tabular forms are two-dimensional tables. The tables can be partitioned into two parts, corresponding to the inputs and outputs of the function. Tabular forms are also called *personality matrices* of a function.

The simplest tabular form is the *truth table*. A truth table is a complete listing of all points in the Boolean input space and of the corresponding values of the outputs. Therefore the input part is the set of all row vectors in B^n and it can be omitted if a specific ordering of the points of the domain is assumed. The output part is the set of corresponding row vectors in $\{0, 1, *\}^m$. Since the size of a truth table is exponential in the number of inputs, truth tables are used only for functions of small size.

The most common tabular form is a list of *multiple-output implicants*. A multiple-output implicant has input and output parts. The input part represents a cube in the domain and is a row vector in $\{0, 1, *\}^n$. The output part has a symbol in correspondence to each output, denoting whether the corresponding input implies a value in the set $\{0, 1, *\}$ or not. There are different formats possible for multiple-output implicants; those relevant to logic optimization will be described in detail in Chapter 7. Since each implicant can represent a subset of the domain, the overall number of implicants needed to represent a function can be made smaller than 2^n.

Example 2.5.7. Consider a three-input two-output function. The first output is $x = ab + a'c$ and the second is $y = ab + bc + ac$. The function is represented by the following truth table:

abc	xy
000	00
001	10
010	00
011	11
100	00
101	01
110	11
111	11

We show next a list of implicants for the same function that represents the *on set* of each output. Thus a 1 entry in the output part denotes that the corresponding input implies a TRUE value of that output.

abc	xy
001	10
*11	11
101	01
11*	11

Operations on logic functions represented by lists of implicants are described in detail in Chapter 7 in conjunction with two-level logic optimization.

EXPRESSION FORMS. Scalar Boolean functions can be represented by expressions of literals linked by the $+$ and $\cdot$ operators. (Note that the $\cdot$ operator is often omitted.) Expressions can be nested by using parentheses.

The number of *levels* of an expression form often refers to the number of primitive operators applied to two or more arguments. The definition of primitive operator may depend on the context and may be ambiguous. To remove the ambiguity, we consider here only $+$ and $\cdot$ as primitive operators.

Single-level forms use only one operator. Therefore they are restricted to expressing only the sum or product of literals and cannot be used to represent arbitrary Boolean functions.

Standard *two-level* forms are *sum of products* of literals and *product of sums* of literals. All Boolean functions can be represented by such forms. In particular, *sum of minterms* and *product of maxterms* are canonical forms. A truth table or an implicant table can be represented by a *sum of products* expression, where the products (of literals) correspond to the minterms or implicants, respectively. The same applies to *product of sums* forms, *mutatis mutandis*.

Most early work on switching theory was applied to functions in two-level forms, because of its generality and because of the belief that the number of levels relates to the propagation delay of a corresponding implementation. The advent of microelectronic circuits brought more complex relations between delay and logic forms. As a result, multiple-level forms became of interest.

Multiple-level forms involve arbitrary nesting of Boolean operators by means of parentheses. A particular multiple-level form is the *factored form* [7].

Definition 2.5.6. A factored form is one and only one of the following:

- A literal.
- A sum of factored forms.
- A product of factored forms.

In other words, factored forms are arbitrary parenthesized expressions with the only constraint being that complementation can be applied only to single variables. Note that factored forms include *sum of products* and *product of sums* and therefore they can represent arbitrary functions.

Example 2.5.8. Let us consider again the majority function $f = ab + ac + bc$. The function cannot be expressed in single-level form, but it can of course be represented in a two-level form as *sum of products* (as above) or as *product of sums* [e.g., $f = (a+b)(a+c)(b+c)$]. An example of a factored form representation is $f = a(b+c)+bc$. Note that many factored form representations of the same function may exist.

The sum and product of logic functions represented by factored forms are straightforward. Complementation of a Boolean function can be achieved by applying De Morgan's law to its expression form. It is often important to perform operations on two-level forms that yield results in two-level forms. In this case, it is best to transform the two-level representations into implicant tables first, operate on the tables and then translate the result back into a two-level expression form.

BINARY DECISION DIAGRAMS. A *binary decision diagram* represents a set of binary-valued decisions, culminating in an overall decision that can be either TRUE or FALSE. Decision diagrams can be represented by trees or rooted dags, where decisions are associated with vertices. Examples of binary decision diagrams are shown in Figure 2.20.

Binary decision diagrams (or BDDs) were proposed by Lee [24] and later by Akers [2] to represent scalar Boolean functions. Many variations on this theme have been proposed. In Aker's model, the decisions are the evaluation of a variable. Bryant [10] showed that efficient algorithms can manipulate decision diagrams under a mild assumption on the ordering of the decision variables. For this reason, Bryant's diagrams are called *ordered binary decision diagrams*, or OBDDs. Karplus [21] investigated diagrams where each decision is the evaluation of a function. These diagrams are called *if-then-else dags*, or ITE dags, and are more general than OBDDs in nature. As a result, the representation of some functions is more compact with ITE dags than with OBDDs.

OBDDs have more practical applications than other types of decision diagrams for two reasons [10, 11]. First, OBDDs can be transformed into canonical forms that uniquely characterize the function. (This applies also to ITE dags, but the conditions for canonicity are more complex [21].) Second, operations on OBDDs can be done in polynomial time of their size, i.e., the vertex set cardinality. This may lead to the false conclusion that OBDDs are a panacea and can be used to efficiently solve intractable problems. Unfortunately, the size of the OBDDs depends on the ordering of the variables. For example, OBDDs representing adder functions are very sensitive to variable ordering and can have exponential size in the worst case and linear size in the best case. Other functions, such as arithmetic multiplication, have OBDDs with exponential size regardless of the variable order. In practice, for most common and

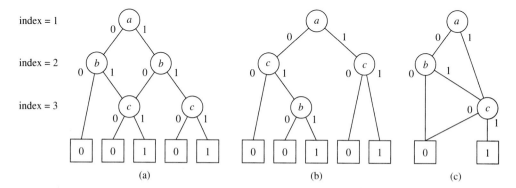

FIGURE 2.20
Binary decision diagrams for $f = (a + b)c$: (a) OBDD for the variable order (a, b, c). (b) OBDD for the variable order (a, c, b). (c) ROBDD for the variable order (a, b, c).

reasonable examples of logic functions, a variable order can be found such that the size of the OBDD is tractable. Therefore, for most practical cases, OBDDs provide a computationally efficient means of manipulating Boolean functions.

There have been many additions to the OBDD model. We consider first the original model presented by Bryant [10], then its extension to the tabular representation supporting multiple-output functions and last other additional features used by software programs supporting OBDD models and computations.

To review the properties of OBDDs and to introduce operations on OBDDs, a formal definition is required [10].

Definition 2.5.7. An **OBDD** is a rooted directed graph with vertex set V. Each non-leaf vertex has as attributes a pointer $index(v) \in \{1, 2, \ldots, n\}$ to an input variable in the set $\{x_1, x_2, \ldots, x_n\}$, and two children $low(v), high(v) \in V$. A leaf vertex v has as an attribute a value $value(v) \in \mathsf{B}$.

For any vertex pair $\{v, low(v)\}$ (and $\{v, high(v)\}$) such that no vertex is a leaf, $index(v) < index(low(v))$ (and $index(v) < index(high(v))$).

Note that the restriction on the variable ordering guarantees that the graph is acyclic. A function can be associated with an OBDD as follows.

Definition 2.5.8. An OBDD with root v denotes a function f^v such that:

- If v is a leaf with $value(v) = 1$, then $f^v = 1$.
- If v is a leaf with $value(v) = 0$, then $f^v = 0$.
- If v is not a leaf and $index(v) = i$, then $f^v = x_i' \cdot f^{low(v)} + x_i \cdot f^{high(v)}$.

Example 2.5.9. An ordered binary decision diagram is shown in Figure 2.20 (a). The diagram corresponds to the function $f = (a + b)c$. The vertices are ordered in a way that the top level corresponds to variable a, the second to b and the third to c. For the sake of clarity, the vertices are labeled in the figure with the corresponding variables.

A rigorous interpretation of the definition would require us to label the vertices as $\{v_1, v_2, v_3, v_4, v_5\}$ and the variables as $x_1 = a, x_2 = b, x_3 = c$. Assume that v_1 is the root. Then $index(v_1) = 1$, which just means that vertex v_1 is related to the first variable in the order, i.e., to $x_1 = a$. And so on.

Figure 2.20 (b) shows the OBDD for a different variable order, namely (a, c, b).

Two OBDDs are *isomorphic* if there is a one-to-one mapping between the vertex sets that preserves adjacency, indices and leaf values. Thus two isomorphic OBDDs represent the same function. Whereas an OBDD uniquely defines a Boolean function, the converse is not true. To make an OBDD canonical, any redundancy in the OBDD must be removed.

Definition 2.5.9. An OBDD is said to be a **reduced** OBDD (or **ROBDD**) if it contains no vertex v with $low(v) = high(v)$, nor any pair $\{v, u\}$ such that the subgraphs rooted in v and in u are isomorphic.

Example 2.5.10. Consider again function $f = (a + b)c$. Figure 2.20 (c) shows the ROBDD corresponding to the OBDD of Figure 2.20 (a). Note that redundancies have been eliminated from the diagram. This representation is unique for that function and the variable order (a, b, c).

Bryant proved that ROBDDs are canonical forms [10] and proposed an algorithm to reduce an OBDD, which is described next. The algorithm visits the OBDD bottom up, and labels each vertex $v \in V$ with an identifier $id(v)$. An ROBDD is identified by a subset of vertices with different identifiers.

The OBDD traversal is done as follows. Let us assume that the portion of the OBDD corresponding to vertices with index larger than i has been processed and the identifiers assigned. Let us consider the subset of vertices with index i, called $V(i)$. There are two conditions for redundancy. If $id(low(v)) = id(high(v))$, then vertex v is redundant and we set $id(v) = id(low(v))$. Similarly, if there are two vertices u and v such that $id(low(v)) = id(low(u))$ and $id(high(v)) = id(high(u))$, then the two vertices are roots of isomorphic graphs and we set $id(v) = id(u)$. In the remaining cases, a different identifier is given to each vertex at level i. The algorithm terminates when the root is reached.

In Algorithm 2.5.1 we present a simplified form of Bryant's reduction algorithm [10] that extends the previous ideas for the sake of efficiency. Namely, instead of considering all pairs of vertices at each level and checking whether their children match, a *key* is formed using the children's identifiers. The elements of $V(i)$ are then sorted by these keys. If a key matches the key of a previously considered vertex, the vertex is discarded. Otherwise, the identifier (stored in variable *nextid*) is incremented and assigned to the current vertex. At this point the vertex can be saved as part of the ROBDD. The complexity of the algorithm is dominated by sorting the vertex subsets, i.e., $O(|V| \log |V|)$.

Example 2.5.11. Let us apply the reduction algorithm to the OBDD of Figure 2.20 (a), repeated for convenience in Figure 2.21 (a). Figure 2.21 (b) shows the same OBDD with numerically labeled vertices, so that we can refer to them in this example. The identifiers $id(v)$, $\forall v \in V$, are also shown.

First, the algorithm labels the leaves as 1 and 2 and initializes the ROBDD with two distinguished leaves. Then, the vertices with $index = 3$ are visited, i.e., those corresponding to variable c and labeled v_4 and v_5. A key is constructed for these vertices from the labels of the children. The key for both vertices is $(1, 2)$, because both have the same right (i.e., high) and left (i.e., low) child. Since the key is the same, the first vertex being sorted, say v_4, is assigned the identifier $id = 3$ and is added to the ROBDD with edges to the leaves. The other vertex (e.g., v_5) gets the same identifier but is not part of the ROBDD, because it is redundant. (Its key matches the key of the previously sorted vertex.)

Next, the algorithm visits the vertices with $index = 2$, i.e., those corresponding to variable b and labeled v_2 and v_3. The left and the right children of v_3 have the same identifier. Therefore v_3 inherits their identifier, i.e., $id = 3$, and it is discarded. Vertex v_2 receives identifier $id = 4$ and is added to the ROBDD, with edges to the 0-leaf and to v_4, because the identifiers of the children of v_2 are 1 and 3.

Finally, the vertex with $index = 1$ (i.e., the root) is visited. Since its children have different identifiers, the root is added to the ROBDD, with edges to v_2 and to v_4, because

```
REDUCE(OBDD){
        Set id(v) = 1 to all leaves v ∈ V with value(v) = 0;
        Set id(v) = 2 to all leaves v ∈ V with value(v) = 1;
        Initialize ROBDD with two leaves with id = 1 and id = 2 respectively;
        nextid = 2;                              /* nextid is the next available identifier value */
        for (i = n to 1 with i = i − 1){
                V(i) = {v ∈ V : index(v) = i};
                foreach (v ∈ V(i)){                              /* consider vertices at level i */
                        if (id(low(v)) = id(high(v))){
                                id(v) = id(low(v));                      /* redundant vertex */
                                Drop v from V(i);
                        }
                        else
                                key(v) = id(low(v)), id(high(v));
                                                /* define key(v) as the identifier pair of v's children */
                }
                oldkey = 0, 0;                    /* initial key that cannot be matched by any vertex */
                foreach v ∈ V(i) sorted by key(v) {
                        if (key(v) = oldkey)                      /* graph rooted at v is redundant */
                                id(v) = nextid
                        else {                    /* nonredundant vertex to receive new identifier value */
                                nextid = nextid + 1;
                                id(v) = nextid;
                                oldkey = key(v);
                                Add v to ROBDD with edges to vertices in ROBDD
                                        whose id equal those of low(v) and high(v);
                        }
                }
        }
}
```

ALGORITHM 2.5.1

the identifiers of the children of v_1 are 4 and 3. Figure 2.21 (c) shows the completed ROBDD.

It is important to construct ROBDDs for Boolean functions directly, thus avoiding the reduction step and memory overflow problems when the unreduced OBDDs are large. This is possible by using a hash table called the *unique table*, which contains a key for each vertex of an ROBDD. The key is a triple, composed of the corresponding variable and the identifiers of the left and right children. The function associated with each vertex is uniquely identified by the key within the unique table. The unique table is constructed bottom up. When considering the addition of a new vertex to the OBDD being constructed, a look-up in the table can determine if another vertex in the table implements the same function, by comparing the keys. In this case, no

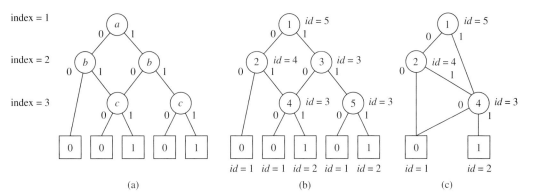

FIGURE 2.21
Binary decision diagrams for $f = (a + b)c$: (a) OBDD for the variable order (a, b, c). (b) OBDD with identifiers. (c) ROBDD for the variable order (a, b, c).

redundant vertex is added to the table. As a result, the unique table represents an ROBDD.

The unique table has been shown to be a *strong canonical form*[1] [5, 6]. Each key can be labeled by a unique identifier, and two Boolean expressions (cast in this form and with the same variable order) can be checked for equivalence by comparing the corresponding identifiers. Strong canonicity is important for developing efficient algorithms that operate on Boolean functions. These algorithms are designed to keep the ROBDDs in reduced form, thus preserving canonicity. The unique table can also be used to represent multiple-output functions, because it can model multiple ROBDDs with shared subgraphs. These are also called multi-rooted decision diagrams or shared ROBDDs.

> **Example 2.5.12.** Consider first the ROBDD of Figure 2.22 (a), representing function $f = (a + b)c$ with the variable order (a, b, c), associated with the vertex with $id = 5$. Assuming that constants 0 and 1 have identifiers 1 and 2, respectively, the unique table is:

		Key	
Identifier	**Variable**	**Left child**	**Right child**
5	a	4	3
4	b	1	3
3	c	1	2

> Consider now the function pair $f = (a + b)c$; $g = bcd$ with the variable order (d, a, b, c), where f is associated with the vertex with $id = 5$, and g is associated

---

[1] A strong canonical form is a form of pre-conditioning which reduces the complexity of an equivalence test between elements in a set. A unique identifier is assigned to each element in the set, so that an equivalence check is just a test on the equality of the identifiers.

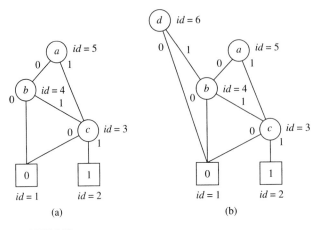

FIGURE 2.22

(a) ROBDD for $f = (a+b)c$ with variable order (a, b, c). (b) Multi-rooted ROBDD for $f = (a+b)c$; $g = bcd$ with variable order (d, a, b, c).

with the vertex with $id = 6$. A multi-rooted decision diagram for f and g is shown in Figure 2.22 (b). The unique table is:

		Key	
Identifier	**Variable**	**Left child**	**Right child**
6	d	1	4
5	a	4	3
4	b	1	3
3	c	1	2

The ROBDD construction and manipulation can be done with the ite operator, which takes its name from its operative definition. Given three scalar Boolean functions, f, g and h, $ite(f, g, h) = f \cdot g + f' \cdot h$ which is equivalent to $if\ f\ then\ g\ else\ h$. In particular, functions f, g and h can be Boolean variables or constants.

We assume that all variables are ordered and we call the *top variable* of a function (or set of functions) the first variable in the support set (or in the union of the support sets). Let $z = ite(f, g, h)$ and let x be the top variable of functions f, g and h. Then function z is associated with the vertex whose variable is x and whose children implement $ite(f_x, g_x, h_x)$ and $ite(f_{x'}, g_{x'}, h_{x'})$. The reason can be explained by considering the identity:

$$z = xz_x + x'z_{x'} \tag{2.14}$$

$$= x(fg + f'h)_x + x'(fg + f'h)_{x'} \tag{2.15}$$

$$= x(f_x g_x + f'_x h_x) + x'(f_{x'} g_{x'} + f'_{x'} h_{x'}) \tag{2.16}$$

$$= ite(x, ite(f_x, g_x, h_x), ite(f_{x'}, g_{x'}, h_{x'})) \tag{2.17}$$

TABLE 2.2

Boolean functions of two arguments and equivalent representation in terms of the ite operator.

Operator	Equivalent ite form
0	0
$f \cdot g$	$ite(f, g, 0)$
$f \cdot g'$	$ite(f, g', 0)$
f	f
$f'g$	$ite(f, 0, g)$
g	g
$f \oplus g$	$ite(f, g', g)$
$f + g$	$ite(f, 1, g)$
$(f + g)'$	$ite(f, 0, g')$
$f \overline{\oplus} g$	$ite(f, g, g')$
g'	$ite(g, 0, 1)$
$f + g'$	$ite(f, 1, g')$
f'	$ite(f, 0, 1)$
$f' + g$	$ite(f, g, 1)$
$(f \cdot g)'$	$ite(f, g', 1)$
1	1

and by the fact that x is the top variable of f, g and h. The relevant terminal cases of this recursion are $ite(f, 1, 0) = f$, $ite(1, g, h) = g$, $ite(0, g, h) = h$, $ite(f, g, g) = g$ and $ite(f, 0, 1) = f'$. Other terminal cases can be reduced to these, because they are equivalent [5, 6].

The usefulness of the ite operator stems from the fact that all Boolean functions of two arguments can be represented in terms of ite, as shown in Table 2.2. Thus, the ite operator can be used to construct the ROBDD of a function, which is represented as the application of binary operators to variables.

Example 2.5.13. Note first that $f \cdot g = ite(f, g, 0)$ and that $f + g = ite(f, 1, g)$ for any function pair f, g, as shown in Table 2.2.

Consider the construction of the ROBDD of the function $f = ac + bc$ with variable order (a, b, c). Figures 2.23 (a,b,c) show elementary ROBDDs related to variables a, b and c, respectively. The ROBDD of ac can be computed as $ite(a, c, 0)$ and is shown in Figure 2.23 (d). Similarly, the ROBDD of bc can be computed as $ite(b, c, 0)$ and is shown in Figure 2.23 (e). Finally, the ROBDD of $f = ac + bc$ can be computed as $ite(ac, 1, bc)$. Since $ite(ac, 1, bc) = ite(a, ite(c, 1, bc), ite(0, 1, bc)) = ite(a, c, bc)$, the ROBDD has a as top variable, the ROBDD of bc [Figure 2.23 (e)] as the left child of the root and the ROBDD of c [Figure 2.23 (c)] as the right child of the root. Moreover, since a vertex corresponding to variable c is already present within the ROBDD representing bc, which is now the left child of the root, this vertex is used also as a right child of the root. Therefore the OBDD is reduced.

It is interesting to note that the same ROBDD will be constructed from other equivalent representations of the same function, e.g., $f = (a + b)c$, provided that the same order of the variables is used.

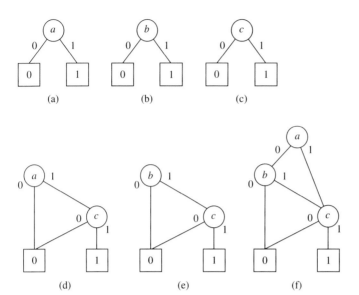

FIGURE 2.23
(a) ROBDD of a. (b) ROBDD of b. (c) ROBDD of c. (d) ROBDD of ac. (e) ROBDD of bc. (f) ROBDD of $ac + bc$.

The *ITE* algorithm seen in Algorithm 2.5.2 implements the *ite* operator and allows us to construct an ROBDD for arbitrary sets of Boolean functions over an ordered set of variables. The *ITE* algorithm uses two tables. The first one is the unique table, which stores the ROBDD information in a strong canonical form. A second table, called the *computed table*, is used to improve the performance of the algorithm. The computed table stores the mapping of any triple (f, g, h) to the vertex implementing $ite(f, g, h)$ once this has been computed. This allows the algorithm to use results from previous invocations of *ITE* in subsequent calls.

The worst-case complexity of the *ITE* algorithm can be derived under the assumption that look-up and insertion in the unique and computed tables can be done in constant time. *ITE* will be called at most once for each distinguished triple of vertices of the ROBDDs of f, g and h and therefore the complexity is of the order of the product of the sizes (i.e., vertex cardinalities) of the corresponding graphs. The average complexity is much lower.

The *ITE* algorithm can be used to apply Boolean operators (e.g., AND, OR, *et cetera*) to ROBDDs. For example, it can be used to check the implication of two functions, e.g., $f \rightarrow g$ or, equivalently, if $f' + g$ is a tautology. This test can be done by verifying whether $ite(f, g, 1)$ has an identifier equal to that of the leaf with value 1. Alternatively, a specialized algorithm can be used which checks directly for tautology of $ite(f, g, h)$. Such an algorithm is based on the following observation. The function associated with an ROBDD vertex is a tautology if both its children represent a tautology. Hence the algorithm can return a negative answer as soon as this condition is not satisfied at some vertex [5, 6].

```
ITE(f, g, h){
    if (terminal case)
        return (r = trivial result);
    else {                                              /* exploit previous information */
        if (computed table has entry {(f, g, h), r})
            return (r from computed table);
        else {
            x = top variable of f, g, h;
            t = ITE(fₓ, gₓ, hₓ);
            e = ITE(fₓ', gₓ', hₓ');
            if ( t == e )                               /* children with isomorphic OBDDs */
                return (t);
            r = find_or_add_unique_table(x, t, e);      /* add r to unique table if not present */
            Update computed table with {(f, g, h), r};
            return (r);
        }
    }
}
```

ALGORITHM 2.5.2

The *ITE* algorithm can also be used for functional composition, i.e., for replacing a variable of a Boolean expression by another expression, because $f_{|x=g} = f_x g + f_{x'} g' = ite(g, f_x, f_{x'})$. A more efficient implementation of this operation can be achieved by another specialized algorithm, called *COMPOSE* [6, 10].

Whereas the consensus and the smoothing operations can be computed by *ITE*, there is also a specific algorithm for these tasks [5, 6]. This algorithm is called here *QUANTIFY* and has a structure similar to *ITE*. The algorithm can perform both universal and existential quantification (i.e., consensus and smoothing) of a function f with respect to variables in a list *varlist*. The algorithm calls procedure *OP(t,e)*, which performs either the conjunction of its arguments in the case of consensus or their disjunction in the case of smoothing. Procedure *OP(t,e)* is implemented as *ITE(t, e, 0)* in the former case and as *ITE(t, 1, e)* in the latter.

The algorithm is recursive and makes use of a computed table that stores the quantification results. Thus, the table is checked before the actual computation is performed. If the table does not contain the desired entry, the top variable x in f is selected and the cofactors with respect to this variable are computed. (This step is trivial, because the cofactors are just represented by the children of the vertex representing f.)

Next the algorithm computes recursively the quantification of the cofactors, which are named t and e, respectively (see Algorithm 2.5.3). If x is included in *varlist*, then either $r = t \cdot e$ or $r = t + e$ is computed according to the goal of using *QUANTIFY* to derive either the consensus or the smoothing of f. If x is not included in *varlist*, the result is expressed by *ITE(x, t, e)*.

```
QUANTIFY(f, varlist){
    if (f is constant)
        return (f);
    else {
        if (computed table has entry {(f, varlist), r})        /* exploit previous information */
            return (r from computed table);
        else {
            x = top variable of f;
            g = f_x;
            h = f_x';
            t = QUANTIFY(g, varlist);        /* compute quantification recursively on g */
            e = QUANTIFY(h, varlist);        /* compute quantification recursively on h */
            if (x is in varlist)
                r = OP(t, e);
            else
                r = ITE(x, t, e);
            Update computed table with {(f, varlist), r};
            return (r);
        }
    }
}
```

ALGORITHM 2.5.3

Example 2.5.14. Consider function $f = ab + bc + ac$ and let us compute $C_a(f)$. Then $varlist = a$. Assume the variables are ordered as (a, b, c). When $QUANTIFY(f, a)$ is invoked, the top variable is a and the cofactors are computed as $g = f_a = b + c$ and $h = f_{a'} = bc$. The recursive calls return $t = g = b + c$ and $e = h = bc$, because g and h do not depend on a. Since a is in $varlist$, $r = OP(t, e) = ITE(t, e, 0) = ITE(b + c, bc, 0) = bc$. Thus the result is $C_a(f) = bc$, as shown in Figure 2.19 (c).

Several extensions to ROBDD have been proposed. *Complemented edges* are flags added to edges to denote local complementation. As an example, this feature can be useful for representing a function and its complement with a multi-rooted ROBDD, with no extra cost in terms of vertices for the complement. In general, complemented edges can contribute to reducing the size of ROBDDs and provide a means of complementing functions in constant time. To maintain a canonical form, restrictions are placed on where the complemented edges can be used. Namely, for each vertex v, the edge $\{v, high(v)\}$ must not be complemented. Note that only one terminal vertex is needed when using complemented edges.

Another extension to ROBDDs is the addition of a *don't care* leaf to represent incompletely specified functions. Furthermore, multi-terminal ROBDDs have been proposed having a finite number of leaves, each associated with a different value (e.g., positive integers). Some specific additions to BDDs are motivated by particular applications.

There have been several implementations of software programs supporting ROBDDs for a wide set of purposes. Brace, Rudell and Bryant [5] developed a portable package for ROBDD manipulation. The package exploits the tabular representation of ROBDDs described before, in conjunction with other techniques to render the ROBDD manipulation efficient. In particular, vertex identifiers in the unique table are pointers to memory locations and thus comparison is very fast. The computed table is implemented as a hash-based *cache*, which takes advantage of a high locality of reference. Entries are created for every non-trivial call to *ITE*, but newer entries are allowed to overwrite older ones. The package also takes advantage of grouping triples into equivalence classes, when they yield the same result under the *ite* operator, e.g., $ite(f, f, g) = ite(f, 1, g) \; \forall f, g$. The package defines a *standard triple* in each equivalence class, which is used for look-up or for writing the computed table. Finally, the package supports an efficient *garbage collection* procedure to recycle memory freed by unused entries. Overall, the package uses about 22 bytes per vertex, and so ROBDDs with millions of vertices can be represented with present computers.

There are still many interesting open research problems related to BDDs. A major issue is searching for the variable order which minimizes the ROBDD size. Some simple heuristic rules have been shown effective for ROBDDs representing combinational circuits, and they are based on the following consideration. Variables related to signals that feed deep cones of logic should come first in the order, because they are important decision makers. In general, ROBDDs have been used for modeling disparate problems and such rules may not be applicable. Furthermore, as manipulations are performed on ROBDDs, the variable order which minimizes their size may change, and the order which was originally chosen may no longer be appropriate. Methods for automatic *dynamic reordering* of variables are very appealing for handling large problems and are currently the subject of investigation.

2.5.3   Satisfiability and Cover

Many synthesis and optimization problems can be reduced to a fundamental one: *satisfiability*. A Boolean function is satisfiable if there exists an assignment of Boolean values to the variables that makes the function TRUE. The most common formulation of the satisfiability problem requires the function to be expressed in a *product of sums* form. The sum terms in the *product of sums* form are called *clauses*. Thus the satisfiability decision problem is the quest for a variable assignment that makes all clauses TRUE.

> **Example 2.5.15.** Consider the following *product of sums* form: $(a + b)(a' + b' + c)$. The expression can be satisfied by choosing $a = 1; b = 1; c = 1$. Note that the assignment is not the only one that satisfies the expression.

The satisfiability problem was shown to be intractable by Cook's theorem [14] and it is considered the root of all intractable problems. Indeed, the intractability of a problem can be shown by a transformation (or a chain of transformations) of satisfiability into that problem. For example, the proof of intractability of the ZOLP can be done by formulating a satisfiability problem as a ZOLP.

Example 2.5.16. Consider again the *product of sums* form: $(a + b)(a' + b' + c)$. The corresponding satisfiability problem can be modeled as a ZOLP:

$$a + b \geq 1$$

$$(1 - a) + (1 - b) + c \geq 1$$

$$a, b, c \in \mathsf{B}$$

Whereas formulating satisfiability as a ZOLP allows us to use the corresponding algorithms, this approach is inefficient and seldom used.

It is worth mentioning that there are several variations of the satisfiability problem. The *3-satisfiability* problem is a restriction of the satisfiability problem to clauses with three literals. It is also intractable. On the other hand, the *2-satisfiability* problem, which is defined similarly, can be solved in polynomial time. The *tautology* decision problem is the complementary problem to satisfiability. Given an arbitrary Boolean expression, we question if it is always satisfied. Note that while satisfiability questions the *existence* of a truth assignment, tautology questions *universality* of a truth assignment. The non-tautology problem, i.e., the question of the existence of an assignment that makes the expression FALSE, is intractable. Nevertheless there are efficient algorithms for tautology, such as those based on ROBDDs, and those described in Section 7.3.4.

A *minimum-cost satisfiability* problem is an optimization problem whose feasibility is expressed by a set of decision variables $\mathbf{x} \in \mathsf{B}^n$ and a *product of sums* form to be satisfied. We restrict our attention to linear minimum-cost satisfiability problems, where the cost function is a weighted sum of the decision variables. Thus we express the cost function as $\mathbf{c}^T \mathbf{x}$ where $\mathbf{c}$ is a weight vector.

Given a collection C of subsets of a finite set S, a minimum *covering* problem is the search for the minimum number of subsets that cover S. We call the elements of C *groups*, to distinguish them from the elements of S. When groups are weighted, the corresponding covering problem is weighted. An example of an unweighted covering problem is the vertex cover problem described in Section 2.4.2. In general, covering problems can be modeled by matrices. Let $\mathbf{A} \in \mathsf{B}^{n \times m}$, where the set of rows is in one-to-one correspondence with S ($n = |S|$) and the set of columns is in one-to-one correspondence with C ($m = |C|$). Then, a cover corresponds to a subset of columns, having at least a 1 entry in all rows of $\mathbf{A}$. Equivalently, it corresponds to selecting $\mathbf{x} \in \mathsf{B}^m$, such that $\mathbf{Ax} \geq \mathbf{1}$. A minimum-weighted cover corresponds to selecting $\mathbf{x} \in \mathsf{B}^m$, such that $\mathbf{Ax} \geq \mathbf{1}$ and $\mathbf{c}^T \mathbf{x}$ is minimum.

Example 2.5.17. Consider the second graph of Figure 2.13 (a), reported again for convenience in Figure 2.24 (a), where edges are labeled $\{a, b, c, d, e\}$. The corresponding

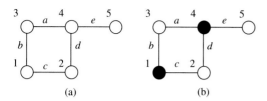

FIGURE 2.24

(a) Graph. (b) Minimum vertex cover.

(a)   (b)

(vertex/edge) incidence matrix is:

$$\mathbf{A}_I = \begin{bmatrix} 0 & 1 & 1 & 0 & 0 \\ 0 & 0 & 1 & 1 & 0 \\ 1 & 1 & 0 & 0 & 0 \\ 1 & 0 & 0 & 1 & 1 \\ 0 & 0 & 0 & 0 & 1 \end{bmatrix}$$

In the vertex cover problem, we search for the minimum number of vertices covering all the edges. Thus the edge set corresponds to S and the vertex set to C. Note that each vertex set can be identified by the group of edges incident to it. Therefore $\mathbf{A} = \mathbf{A}_I^T$ and $\mathbf{c} = \mathbf{1}$.

Examples of covers are represented by the following vectors: $\mathbf{x}^1 = [10010]^T$, $\mathbf{x}^2 = [01101]^T$ and $\mathbf{x}^3 = [01111]^T$. Note that $\mathbf{A}\,\mathbf{x} \geq \mathbf{1}$ for $\mathbf{x} = \mathbf{x}^1, \mathbf{x}^2, \mathbf{x}^3$. Vector $\mathbf{x}^1$ is a minimum cover that is shown in Figure 2.24 (b).

Example 2.5.18. Consider the hypergraph shown in Figure 2.25 (a), with five vertices $\{v_i, i = 1, 2, \ldots, 5\}$ and five edges $\{a, b, c, d, e\}$. The corresponding (vertex/edge) incidence matrix is:

$$\mathbf{A}_I = \begin{bmatrix} 1 & 0 & 1 & 0 & 0 \\ 1 & 1 & 0 & 0 & 1 \\ 0 & 1 & 1 & 0 & 1 \\ 0 & 0 & 0 & 1 & 0 \\ 0 & 1 & 1 & 1 & 0 \end{bmatrix}$$

We search now for the minimum number of edges that cover all the vertices. Thus the vertex set corresponds to S and the edge set to C of the covering problem formulation. Therefore $\mathbf{A} = \mathbf{A}_I$ and $\mathbf{c} = \mathbf{1}$. Edge covers satisfy $\mathbf{A}\mathbf{x} \geq \mathbf{1}$.

An example of a minimum cover is given by edges $\{a, b, d\}$, or equivalently by vector $\mathbf{x} = [11010]^T$. The cover is shown in Figure 2.25 (b).

Let us extend the problem by associating weights with each edge. Let $\mathbf{c} = [1, 2, 1, 1, 1]^T$. Then, it is easy to verify that a minimum-cost cover is $\{a, c, d\}$, corresponding to vector $\mathbf{x} = [10110]^T$ with cost 3.

Note that a minimum vertex cover problem can be defined in an analogous way.

Covering problems can be cast as minimum-cost satisfiability problems, by associating a selection variable with each group (element of C) and a clause with each element of S. Each clause represents those groups that can cover the element, i.e., it is the disjunction of the variables related to the corresponding groups. It is important to note that the product of the clauses is an unate expression, and thus this problem is often referred to as *unate cover*.

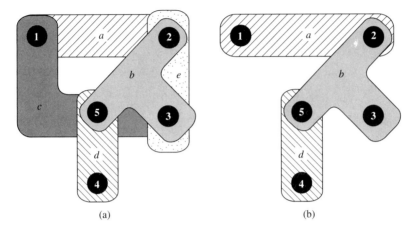

FIGURE 2.25
(a) Hypergraph. (b) Minimum-edge cover.

Example 2.5.19. Consider the covering problem of Example 2.5.18. The first covering clause $(x_1 + x_3)$ denotes that vertex v_1 must be covered by edge a or c. The overall *product of sums* expression to be satisfied is:

$$(x_1 + x_3)(x_1 + x_2 + x_5)(x_2 + x_3 + x_5)(x_4)(x_2 + x_3 + x_4)$$

It is easy to verify that $\mathbf{x} = [11010]^T$ satisfies the *product of sums* expression above.

The covering problem can be generalized by assuming that the choice of a group implies the choice of another one. This additional constraint can be represented by an implication clause. For example, if the selection of group a implies the choice of b, then the clause $(a' + b)$ is added. Note that the implication clause makes the *product of sums* form binate in variable a, because a (uncomplemented) is also part of some covering clauses. Therefore this class of problems is referred to as *binate covering*, or *covering with closure*. Binate covering problems are minimum-cost satisfiability problems. Nevertheless, some minimum-cost satisfiability problems exist that do not represent covering problems even with implications (e.g., when all clauses have variables with mixed polarity).

Example 2.5.20. Consider again the covering problem of Example 2.5.18. Assume that the selection of edge a implies the selection of edge b. Then the clause $x_1' + x_2$ must also be considered. The overall *product of sums* expression to be satisfied is:

$$(x_1 + x_3)(x_1 + x_2 + x_5)(x_2 + x_3 + x_5)(x_4)(x_2 + x_3 + x_4)(x_1' + x_2)$$

It is easy to verify that $\mathbf{x} = [11010]^T$ is a minimum cover and that $\mathbf{x} = [10110]^T$ is not a cover.

Binate covering problems can still be characterized by matrices. Let $\mathbf{A} \in \{-1, 0, +1\}^{n \times m}$ represent a problem with n clauses and m variables. Each row corresponds to a clause depending on variables with non-zero entries in the corresponding column. A -1 entry corresponds to a complemented variable and a 1 entry to an uncomplemented one. (Note that an unate problem has only non-negative entries in **A**.)

The binate covering problem corresponds to selecting a subset of the columns, such that all rows have at least a 1 entry in that subset or a -1 entry in the complement subset.

Example 2.5.21. Consider again the expression:

$$(x_1 + x_3)(x_1 + x_2 + x_5)(x_2 + x_3 + x_5)(x_4)(x_2 + x_3 + x_4)(x_1' + x_2)$$

The corresponding matrix formulation is:

$$\mathbf{A} = \begin{bmatrix} 1 & 0 & 1 & 0 & 0 \\ 1 & 1 & 0 & 0 & 1 \\ 0 & 1 & 1 & 0 & 1 \\ 0 & 0 & 0 & 1 & 0 \\ 0 & 1 & 1 & 1 & 0 \\ -1 & 1 & 0 & 0 & 0 \end{bmatrix}$$

Vector $[11010]^T$ denotes a solution, because the submatrix identified by the first, second and fourth columns has at least a 1 in all rows. Vector $[10110]^T$ is not a solution, because the last clause is not satisfied. Indeed, there is no 1 in the last row in correspondence to the first, third and fourth columns, and no -1 in correspondence with the second and fifth columns.

Whereas both unate and binate covering problems are intractable, binate covering is intrinsically more difficult than unate covering. Indeed, the addition of an element to a feasible solution may make it infeasible if its implications are not satisfied. A classic example of unate covering is that related to exact two-level optimization of logic functions (Section 7.2.2). Notable examples of binate covering are those related to the exact minimization of Boolean relations (Section 7.6), to state minimization for finite-state machines (Section 9.2.1) and to cell-library binding (Chapter 10).

AN EXACT ALGORITHM FOR THE COVERING PROBLEM. We consider first a branch-and-bound algorithm for the unweighted unate covering problem. We shall extend it later to cope with weighted and binate covering problems as well as with minimum-cost satisfiability problems. The algorithm described here is based upon an algorithm used in exact logic minimization [29]. We present here the essential features of the algorithm and we refer the reader to reference [29] for details.

We assume in the sequel that a covering problem is specified by matrix **A**, with rows corresponding to the elements of the set to be covered and columns corresponding to groups of elements. For the sake of visualizing the definitions and the algorithm, we consider the covering problem as an edge covering problem of a hypergraph, having **A** as the incidence matrix. The following definitions are useful in characterizing and simplifying the covering problem.

An *essential* column is a column having the only 1 entry of some row. Graphically, it corresponds to an essential edge, i.e., to an edge which is the only one incident to a vertex. Essential columns (edges) must be part of any cover.

A column *dominates* another column if the entries of the former are larger than or equal to the corresponding entries of the latter. Equivalently, an edge dominates another edge when it is incident to at least all vertices incident to the other. Dominated columns (edges) can be discarded from consideration.

A row *dominates* another row if the entries of the former are larger than or equal to the corresponding entries of the latter. Equivalently, a vertex dominates another vertex when it is incident to at least all edges incident to the other. Dominant rows (vertices) can be neglected, because any cover of the dominated ones is also a cover of the complete set.

> **Example 2.5.22.** Consider matrix $\mathbf{A}$ of Example 2.5.18. The fourth column is essential, because it is the only one with a 1 entry in the fourth row. Equivalently, edge d is essential, because it is the only one incident to vertex v_4. Hence it must be part of a cover.
>
> The second column ($[01101]^T$) dominates the fifth column ($[01100]^T$). Equivalently, selecting edge b corresponds to covering all vertices covered by edge e. Thus edge e can be dropped from consideration.
>
> The fifth row ($[01110]$) dominates the fourth row ($[00010]$). Equivalently, vertex v_5 dominates vertex v_4, because it is incident to a superset of the edges incident to v_4. Thus, v_5 can be dropped from consideration, because any solution covering v_4 covers v_5 as well.
>
> Thus, any optimum solution to the covering problem specified by A has $x_4 = 1; x_5 = 0$.

The covering algorithm, as seen in Algorithm 2.5.4, is recursive. Binary vector $\mathbf{x}$ contains the current solution and binary vector $\mathbf{b}$ the best solution seen so far. The algorithm is invoked with arguments matrix $\mathbf{A}$, describing the problem, $\mathbf{x} = \mathbf{0}$ and $\mathbf{b} = \mathbf{1}$. Matrix $\mathbf{A}$ is recursively reduced in size. We assume that pointers (not described here for simplicity) relate the columns of the reduced $\mathbf{A}$ to the corresponding elements in $\mathbf{x}$ and $\mathbf{b}$, whose sizes are invariant. The algorithm returns a binary vector indicating the cover.

At the first step, the reduction of matrix $\mathbf{A}$ is done as follows. First, if matrix $\mathbf{A}$ is reducible (i.e., it can be reordered into a block-diagonal form, because it represents two or more disjoint subproblems), $EXACT\_COVER$ is invoked on its partition blocks and the current solution $\mathbf{x}$ is set to the disjunction of the return values. Second, dominated columns, dominant rows and essential columns are iteratively removed until there is no change. For every essential column detected and deleted, the corresponding entry in $\mathbf{x}$ is set to 1 and the rows to which it is incident are deleted as well.

> **Example 2.5.23.** Consider the covering problem specified in Example 2.5.18. As shown in Example 2.5.22, the fourth column is essential and the fifth is dominated. The fifth row is dominant. Thus the fourth element of $\mathbf{x}$ is set to 1, the fifth element of $\mathbf{x}$ is set to

$EXACT\_COVER(\mathbf{A}, \mathbf{x}, \mathbf{b})$ {

    Reduce matrix $\mathbf{A}$ and update corresponding $\mathbf{x}$;

    if $(Current\_estimate \geq |\mathbf{b}|)$ **return**($\mathbf{b}$);

    if ($\mathbf{A}$ has no rows) **return** ($\mathbf{x}$);

    Select a branching column c;

    $x_c = 1$;

    $\widetilde{\mathbf{A}} = \mathbf{A}$ after deleting column c and the rows incident to it;

    $\widetilde{\mathbf{x}} = EXACT\_COVER(\widetilde{\mathbf{A}}, \mathbf{x}, \mathbf{b})$;

    if ($|\widetilde{\mathbf{x}}| < |\mathbf{b}|$)

        $\mathbf{b} = \widetilde{\mathbf{x}}$;

    $x_c = 0$;

    $\widetilde{\mathbf{A}} = \mathbf{A}$ after deleting column c;

    $\widetilde{\mathbf{x}} = EXACT\_COVER(\widetilde{\mathbf{A}}, \mathbf{x}, \mathbf{b})$;

    if ($|\widetilde{\mathbf{x}}| < |\mathbf{b}|$)

        $\mathbf{b} = \widetilde{\mathbf{x}}$;

    return ($\mathbf{b}$);

}

ALGORITHM 2.5.4

0 and the corresponding reduced covering matrix is:

$$\mathbf{A} = \begin{bmatrix} 1 & 0 & 1 \\ 1 & 1 & 0 \\ 0 & 1 & 1 \end{bmatrix}$$

The bounding is done by the evaluation of $Current\_estimate$, which is a lower bound of the cardinality of all covers that can be derived from the current $\mathbf{x}$. The estimate is the sum of two components: the number of 1s in $\mathbf{x}$ (i.e., the number of columns already committed) plus a bound on the size of the cover of the current $\mathbf{A}$. Such a bound can be given by the size of a maximally independent set of rows, i.e., rows that have no 1s in the same column. Pairwise independence can be easily identified and a graph can be constructed whose vertices are in one-to-one correspondence with the rows (vertices in the original hypergraph) and whose edges denote independence. The bound is the clique number in this graph. Computing an exact bound is difficult, because identifying the maximum clique is as complex as solving the covering problem itself. However, any lesser approximation (that can be computed by a fast heuristic algorithm) is valuable. The worse the approximation is, the less pruning the algorithm does. In any case, the algorithm is still exact.

Example 2.5.24. Consider the covering matrix $\mathbf{A}$ of Example 2.5.18. The fourth row is independent from the first, second and third, but the first three rows are dependent. Thus the maximal set of mutually independent rows has cardinality 2. Hence the bound is 2.

    Consider the reduced covering matrix $\mathbf{A}$ of Example 2.5.23. There are no independent rows. Thus a lower bound on covering that matrix is 1. A lower bound on covering the original matrix is 1 plus $|\mathbf{x}| = 1$, i.e., 2. When applying the algorithm to the matrix of Example 2.5.18, the bound is less than $|\mathbf{b}| = |\mathbf{1}| = 5$ and the search is not killed.

The branch-and-bound algorithm explores the solution space recursively, by assuming that each column (edge) is included or excluded from the cover. In the first case, the recursive call has as an argument a reduced matrix $\widetilde{\mathbf{A}}$, where the column being included is deleted as well as the rows incident to that column. In the second case, the recursive call has as an argument a reduced matrix $\widetilde{\mathbf{A}}$, where the column being excluded is deleted. Obviously the complexity of the algorithm is exponential, but pruning and matrix reduction make it viable for problems of interesting size.

Example 2.5.25. Consider the reduced covering matrix $\mathbf{A}$ shown in Example 2.5.23. Let us assume that the branching column c is the first one. Then $x_1 = 1$, $\mathbf{x} = [10010]^T$ and:

$$\widetilde{\mathbf{A}} = [1\ 1]$$

Consider now the recursive call to $EXACT\_COVER$. Once a dominated column is deleted, the other column is essential. Assume it corresponds to x_2. Then $\mathbf{x} = [11010]^T$. After removing the row, the matrix has no rows and the call returns $\widetilde{\mathbf{x}} = \mathbf{x}$. Now $|\widetilde{\mathbf{x}}| = 3 < |\mathbf{b}| = 5$ and thus the best solution seen so far is updated to $\mathbf{b} = [11010]^T$.

Then the algorithm tries to exclude the first column. Thus $x_1 = 0$, $\mathbf{x} = [00010]^T$, and:

$$\widetilde{\mathbf{A}} = \begin{bmatrix} 0 & 1 \\ 1 & 0 \\ 1 & 1 \end{bmatrix}$$

Now both columns are essential. Then $\mathbf{x} = [01110]^T$. After removing the rows incident to the essential columns, the matrix has no rows and the call returns $\widetilde{\mathbf{x}} = \mathbf{x}$. Now $|\widetilde{\mathbf{x}}| = 3 = |\mathbf{b}| = 3$ and $\mathbf{b}$ is not updated. The algorithm returns then $[11010]^T$.

The algorithm can be modified to cope with weighted covering problems as follows. First, the cost of a cover specified by $\mathbf{x}$ is $\mathbf{c}^T\mathbf{x}$, which replaces $|\mathbf{x}|$ in the algorithm. Similar considerations apply to the cost of the best solution seen so far, $\mathbf{b}$. Second, the current estimate computation must reflect the cost of covering a maximal independent sets of rows. Last, column dominance must also reflect the weights, i.e., an additional condition for column dominance is that the dominant column must not have larger weight than the dominated one.

Example 2.5.26. Consider the covering matrix $\mathbf{A}$ of Example 2.5.18, with weight vector $\mathbf{c} = [1, 2, 1, 1, 1]^T$. Then, the initial best solution has cost 6. A lower bound on the cost of the cover can be derived by considering the two independent sets of rows that would require one unit-weight column each to be covered in the best case. Hence the bound is 2, less than 6, and the search is not killed.

In this case, the fifth column is not dominated by the second, because its cost is lower. Hence it should not be discarded. Indeed, once the essential column is detected and the rows incident to it are deleted, the second column can be dropped, because it is dominated by the fifth column.

At this point, the problem can be represented by a reduced matrix similar to that used in Example 2.5.25, with the only difference that the columns are related to edges $\{a, e, c\}$. All columns have unit weight. A minimum-cost cover is $\{a, c, d\}$, represented by vector $\mathbf{x} = [10110]^T$.

Let us consider now a minimum-cost satisfiability problem that is modeled by a matrix $\mathbf{A} \in \{-1, 0, 1\}^{n \times m}$. This problem can also be solved by the $EXACT\_COVER$ algorithm with some modifications. We highlight here the major differences from the unate case, and we refer the reader to references [8, 20] for details.

First, note that an instance of a minimum-cost satisfiability problem (in particular a binate covering problem) may not have a solution, whereas unate covering problems always have a solution. Second, the matrix reduction step has to account for three-valued entries. Whereas an essential column is a column having the only nonzero entry of some row, an essential must be a part of a solution when the entry is 1, and it must not be a part of a solution when the entry is -1. Column and row dominance can be defined in a similar way. In particular, a row dominates another row if the corresponding clause is satisfied when the latter is satisfied. Third, the computation of the bounding function can still be based on the search for a maximal independent set of rows. In the general case, two rows are independent when they cannot be satisfied simultaneously by setting a single variable to 1 [8].

Example 2.5.27. Consider the minimum-cost satisfiability problem specified by the following *product of sums* expression:

$$(x_1' + x_2)(x_1' + x_2 + x_3 + x_4)(x_1 + x_3 + x_4')(x_5)(x_6')$$

It can be described by the following matrix:

$$\mathbf{A} = \begin{bmatrix} -1 & 1 & 0 & 0 & 0 & 0 \\ -1 & 1 & 1 & 1 & 0 & 0 \\ 1 & 0 & 1 & -1 & 0 & 0 \\ 0 & 0 & 0 & 0 & 1 & 0 \\ 0 & 0 & 0 & 0 & 0 & -1 \end{bmatrix}$$

Variables x_5 and x_6 are essential: the former must be set to 1 and the latter to 0. Therefore the last two rows and columns can be dropped from consideration. Variable x_3 dominates x_4: thus x_4 can be set to 0, satisfying the clause expressed by the third row. The second row dominates the first one and can be discarded, because $x_1' + x_2$ implies $x_1' + x_2 + x_3 + x_4$. The problem is thus reduced to satisfying $x_1' + x_2$, which can be done by setting both x_1 and x_2 to 0. A solution is $\mathbf{x} = [000010]^T$.

In general, a minimum-cost satisfiability problem can be represented and solved using an OBDD that models the product of all clauses. Thus each vertex of an OBDD represents the selection of a group to be placed in the cover. By weighting each edge corresponding to the TRUE decision by the corresponding weight (by 1 in the unweighted covering case) and each edge corresponding to the FALSE decision by 0, it is possible to associate a cost with each path in the OBDD. A minimum-cost solution corresponds to a shortest path from the root to the TRUE leaf [26]. Indeed such a path identifies a minimum-cost assignment to the decision variables that make all clauses TRUE. A shortest path in a dag can be computed by topological sort very efficiently.

Example 2.5.28. Consider the problem of Example 2.5.27. Figure 2.26 shows a fragment of the corresponding OBDD, with all paths to the TRUE leaf. The emboldened path in the figure has minimum weight. It identifies solution $\mathbf{x} = [000010]^T$.

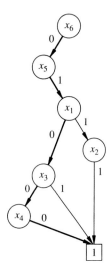

FIGURE 2.26
OBDD fragment modeling a minimum-cost satisfiability problem.

Whereas this simple formulation is appealing because it leverages the OBDD manipulation algorithms, it is often the case that constructing and storing an OBDD for large problems is not efficient. A recently proposed method has combined the branch-and-bound approach with the OBDD-based approach, where the overall problem is modeled as a *product of OBDDs* [20]. Moreover various heuristic algorithms have been developed for the covering problem. At present, efficient algorithms for solving the minimum-cost satisfiability problem are the subject of ongoing research. Thus the problem is still considered open to improvements.

2.6   PERSPECTIVES

Graph theory and Boolean algebra form the basis for solving complex decision and optimization problems related to circuit synthesis. Despite the large number of results available in these fields, new algorithms for solving (or approximating) fundamental problems are continuously sought for. Examples are methods for solving minimum-cost satisfiability and for manipulating Boolean functions with BDDs.

Since Bryant's seminal paper [10] on OBDDs, a large amount of technical papers have tackled the solution of different problems that can be modeled by BDDs. Examples range from digital circuit design and verification to sensitivity and probabilistic analysis of digital circuits and to finite-state system analysis [11]. While OBDDs provide today an efficient data structure to solve problem instances previously out of reach, the search for improving BDD models and algorithms is still ongoing.

Overall, CAD researchers and developers have used several classical results of mathematics and computer science. The size of the problems to be solved has led them to explore heuristic algorithms that are often dedicated to solving specific applications. Nevertheless, some results achieved in the CAD domain, like BDD manipulation and the minimization of two-level logic forms, have had tremendous impact in other areas of science.

2.7 REFERENCES

Graph theory is extensively described in the textbooks by Deo [13], Bondy and Murty [4], and Harary [16]. Perfect graphs and algorithms that exploit the properties of perfect graphs are well described in Golumbic's book [15].

Fundamental algorithms are presented in detail in the recent book by Cormen *et al.* [12]. A classical reference is Lawler's book [22]. Linear programs and network algorithms are also described in detail in Nemhauser and Wolsey's text [27] and surveyed in the more recent book by Papadimitriou and Steiglitz [28]. Another good reference for fundamental algorithms is the textbook by Horowitz and Sahni [19]. Garey and Johnson [14] have written an excellent book on tractable and intractable problems with an extended list of problems and their taxonomy.

Boolean algebra and its applications have been the subject of several textbooks, e.g., Hill and Peterson's [18]. A good reference for formal properties of Boolean algebra is Brown's book [9]. Binary decision diagrams have been described in several papers [2, 5, 10, 24]. Most applications of BDDs are based on the model popularized by Bryant [10, 11]. Recent applications of BDDs are often supported by the BDD package developed by Brace *et al.* [5], who introduced the use of the *ITE* algorithm and the tabular notation, also credited to Karplus and Bryant.

1. A. Aho, J. Hopcroft and J. Ullman, *Data Structures and Algorithms*, Addison-Wesley, Reading, MA, 1983.
2. S. Akers, "Binary Decision Diagrams," *IEEE Transactions on Computers*, Vol. C-27, pp. 509–516, June 1978.
3. S. Baase, *Computer Algorithms*, Addison-Wesley, Reading, MA, 1978.
4. J. Bondy and U. Murty, *Graph Theory with Applications*, North-Holland, New York, 1976.
5. K. Brace, R. Rudell and R. Bryant, "Efficient Implementation of a BDD Package," *DAC, Proceedings of the Design Automation Conference*, pp. 40–45, 1990.
6. K. Brace, "Ordered Binary Decision Diagrams for Optimization in Symbolic Switch-level Analysis of MOS Circuits," Ph.D. Dissertation, Carnegie-Mellon University, Pittsburgh, PA, 1993.
7. R. Brayton, R. Rudell, A. Sangiovanni and A. Wang, "MIS: A Multiple-Level Logic Optimization System," *IEEE Transactions on CAD/ICAS*, Vol. CAD-6, No. 6, pp. 1062–1081, November 1987.
8. R. Brayton and F. Somenzi, "An Exact Minimizer for Boolean Relations," *ICCAD, Proceedings of the International Conference on Computer Aided Design*, pp. 316–320, 1989.
9. F. Brown, *Boolean Reasoning*, Kluwer Academic Publishers, Boston, MA, 1990.
10. R. Bryant, "Graph-based Algorithms for Boolean Function Manipulation," *IEEE Transactions on Computers*, Vol. C-35, No. 8, pp. 677–691, August 1986.
11. R. Bryant, "Symbolic Boolean Manipulation with Ordered Binary-Decision Diagrams," *ACM Computing Surveys*, Vol. 24, No. 3, September 1992.
12. T. Cormen, C. Leiserson and R. Rivest, *Introduction to Algorithms*, McGraw-Hill, New York, 1990.
13. N. Deo, *Graph Theory with Applications to Engineering and Computer Science*, Prentice-Hall, Englewood Cliffs, NJ, 1974.
14. M. Garey and D. Johnson, *Computers and Intractability*, Freeman, New York, 1979.
15. M. Golumbic, *Algorithmic Graph Theory and Perfect Graphs*, Academic Press, San Diego, CA, 1980.
16. F. Harary, *Graph Theory*, Addison-Wesley, Reading, MA, 1972.
17. A. Hashimoto and J. Stevens, "Wire Routing by Optimizing Channel Assignment within Large Apertures," *Proceedings of 8th Design Automation Workshop*, pp. 155–163, 1971.
18. F. Hill and G. Peterson, *Introduction to Switching Theory and Logical Design*, Wiley, New York, 1981.
19. E. Horowitz and S. Sahni, *Fundamentals of Computer Algorithms*, Computer Science Press, New York, 1978.
20. S. Jeong and F. Somenzi, "A New Algorithm for the Binate Covering Problem and Its Application to the Minimization of Boolean Relations," *ICCAD, Proceedings of the International Conference on Computer Aided Design*, pp. 417–420, 1992.
21. K. Karplus, "Representing Boolean Functions with If-Then-Else Dags," University of California at Santa Cruz, Report CRL-88-28, November 1988.

22. E. Lawler, *Combinatorial Optimization: Networks and Matroids*, Holt, Rinehart and Winston, New York, 1976.
23. E. Lawler, "Optimal Sequencing of a Single Machine Subject to Precedence Constraints," *Management Science*, Vol. 19, No. 5, pp. 544–546, January 1973.
24. C. Lee, "Representation of Switching Circuits by Binary-Decision Programs," *Bell System Technical Journal*, Vol. 38, pp. 985–999, July 1959.
25. Y. Liao and C. Wong, "An Algorithm to Compact a VLSI Symbolic Layout with Mixed Constraints," *IEEE Transactions on CAD/ICAS*, Vol. CAD-2, No. 2, pp. 62–69, April 1983.
26. B. Lin and F. Somenzi, "Minimization of Symbolic Relations," *ICCAD, Proceedings of the International Conference on Computer Aided Design*, pp. 88–91, 1990.
27. G. Nemhauser and L. Wolsey, *Integer and Combinatorial Optimization*, Wiley, New York, 1988.
28. C. Papadimitriou and K. Steiglitz, *Combinatorial Optimization*, Prentice-Hall, Englewood Cliffs, NJ, 1982.
29. R. Rudell and A. Sangiovanni-Vincentelli, "Multiple-valued Optimization for PLA Optimization," *IEEE Transactions on CAD/ICAS*, Vol. CAD-6, No. 5, pp. 727–750, September 1987.

2.8 PROBLEMS

1. Show that the adjacency matrix of a dag can be cast into lower triangular form by a row and column permutation.

2. Compute the longest path weights of the graph of Figure 2.11 (b) by a modified Bellman-Ford algorithm where the *min* operator is replaced by a *max* operator. Show all steps.

3. Prove that the $LEFT\_EDGE$ algorithm is exact.

4. Consider the function $f = ab + bc + ac$. Compute $\partial f/\partial b$, $\mathcal{C}_b(f)$ and $\mathcal{S}_b(f)$. Represent the function, the Boolean difference, the consensus and the smoothing on the three-dimensional Boolean cube.

5. Consider the function $f = ab + bc + ac$. Compute an expansion on the orthonormal basis $\{\phi_1 = a; \ \phi_2 = a'b; \ \phi_3 = a'b'\}$.

6. Consider the function $f = ab + cd + ef$. Determine the variable orders that minimize and maximize the size of the corresponding OBDDs.

7. Consider functions $f = ab + bc$ and $g = ac$. Draw the corresponding OBDDs and determine the ROBDD corresponding to $f \oplus g$. (Use order (a, b, c).)

8. Design an algorithm for detecting if $ite(f, g, h)$ is a constant. Explain why such an algorithm is preferable to the regular *ITE* algorithm.

CHAPTER

3

HARDWARE MODELING

Die Grenzen meiner Sprache bedeuten die Grenzen meiner Welt.
The limits of my language mean the limits of my world.

L.Wittgenstein. Tractatus Logico-Philosophicus.

3.1 INTRODUCTION

A model of a circuit is an abstraction, i.e., a representation that shows relevant features without associated details. Models are used to specify circuits, to reason about their properties and as means of transferring the information about a design among humans as well as among humans and computer-aided design tools. Circuit specifications are models describing circuits to be implemented, which are often accompanied by constraints on the desired realization, e.g., performance requirements.

Formal models have a well-defined syntax and semantics. Hence they provide a way of conveying the information about a circuit in a consistent way that can be unambiguously interpreted. Thus automated tools can be designed to read, process and write such models. Conversely, informal circuit specifications, such as textual descriptions of the *principles of operations* of a computer in a natural language, have limited applications when CAD methods are used. In addition, informal descriptions of large-scale circuits or systems may be sources of misunderstanding among humans, because it is often impossible to check their completeness and consistency. In the sequel we shall refer to formal models only, and we shall drop the adjective "formal" for brevity.

A circuit can be modeled differently according to the desired abstraction level (e.g., architectural, logic, geometric), view (e.g., behavioral, structural, physical) and to the modeling means being used (e.g., language, diagram, mathematical model). Abstraction levels and views were described in Section 1.5. We comment now on the modeling means.

In recent years, there has been a trend toward using *hardware description languages* (HDLs) for circuit specification. Conceiving an HDL model has similarities to writing a software program. The conciseness of HDL models has made them preferable to the corresponding flow, state and logic diagrams, even though some diagram models [7] are more powerful in visualizing the circuits' functions. We shall not describe circuit specifications in terms of diagrams, because the information that most of them convey can be expressed in equivalent form by HDL models.

This book does not advocate the use of any specific hardware language. The synthesis techniques that we present have general value and are not related to the specifics of any particular language. For this reason, we shall consider also *abstract models* for circuits at both the architectural and logic levels. Abstract models are mathematical models based on graphs and Boolean algebra. At the architectural level, the circuit behavior can be abstracted by a set of *operations* (called also *tasks*) and their *dependencies*. The operations can be general in nature, ranging from arithmetic to logic functions. The behavior of a sequential logic circuit is abstracted by a finite-state machine that degenerates to a Boolean function in the combinational case. Structural views are abstracted as interconnections of logic blocks or gates (at the logic level) or resources (at the architectural level).

Abstract models are powerful enough to capture the essential features described by HDL and diagram models. At the same time, they are simple enough that properties of circuit transformations can be proven. We show in Figure 3.1 the relations among HDL models, abstract models and the synthesis tasks.

The outline of this chapter is as follows. First, we shall address the motivation and features of HDLs and review some of the most important ones. Second, we shall consider abstract models that capture hardware representations at different levels and with different views and show the relation between the languages and these models. Eventually we shall present synthesis and optimization of abstract models from HDL models.

3.2 HARDWARE MODELING LANGUAGES

Hardware description languages are primarily motivated by the need of specifying circuits. Several HDLs exist, with different features and goals. Even though some evolved from programming languages, like AHPL, which was based upon APL and VHDL, which was derived from ADA, the specific nature of hardware circuits

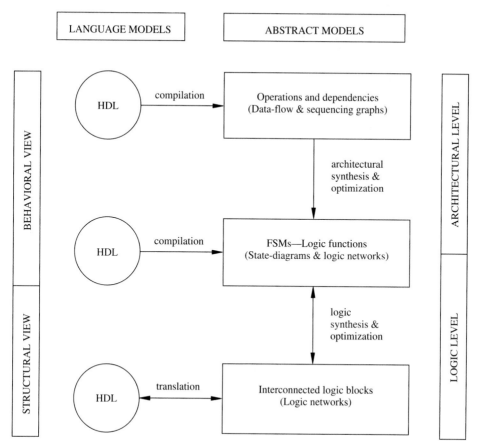

FIGURE 3.1
Circuit models, synthesis and optimization: a simplified view.

makes them fairly different from the commonly used software programming languages.

To appreciate the difference, let us compare the most distinguishing features between the objects that hardware and software languages describe. First, hardware circuits can execute operations with a wide degree of concurrency. Conversely, software programs are most commonly executed on uni-processors and hence operations are serialized. In this respect, hardware languages are closer to programming languages for parallel computers. Second, the specifications of hardware circuits entail some structural information. For example, the interface of a circuit with the environment may require the definition of the input/output ports and the data formats across these ports. In addition, a designer may want to express hints (or constraints) on some partitions of the circuit. Hence, HDLs must support both behavioral and structural views, to be used efficiently for circuit specification. Third, detailed timing of the operations is very important in hardware, because of interface requirements of the circuit being described. On the other hand, the specific execution time frames of the operations in software programs are of less concern, with some exceptions in the case of real-time applications.

Circuits can be modeled under different views, and consequently HDLs with the corresponding flavors have been developed. In particular, when considering architectural and logic level modeling, *behavioral* or *structural* views are used. Some languages support combined views, thus allowing a designer to specify implementation details for desired parts of the circuit as well as performing stepwise refinement of behavioral models into structural models. Synthesis tools support the computer-aided transformation of behavioral models into structural ones, at both the architectural and logic levels. HDLs serve also the purpose of exchange formats among tools and designers.

Hardware languages are not only motivated by hardware specification reasons. Circuit models require validation by simulation or verification methods. Synthesis methods use HDL models as a starting point. As a result, several goals must be fulfilled by HDLs. Modeling language design has been driven by different priorities in achieving these goals. As an example, VHDL [15, 18, 20] was developed with the objective of documenting large systems. Verilog HDL®[22], called here Verilog for brevity, was designed to support simulation efficiently. The objectives of UDL/I [10] were a standard representation format that could be used for synthesis, simulation and verification.

The multiple goals of HDLs cannot be achieved by programming languages applied to hardware specification. Standard programming languages have been used for functional modeling of processors that can be validated by compiling and executing the models. Nevertheless, such models cannot be used for synthesis, because of the lack of the features described before. The enhancement of the C programming language for simulation and synthesis has led to new HDL languages, such as ESIM [6] and HardwareC [11].

We use the term HDL *model* as the counterpart of *program* in software. Similarly, we refer to *model call* as the invocation of a submodel, corresponding to procedure call. Hierarchical hardware models achieved by means of model calls can be used to increase modularity of the representation by encapsulating parts of a model as well as by expressing multiple shared submodels.

3.2.1   Distinctive Features of Hardware Languages

A language can be characterized by its *syntax*, *semantics* and *pragmatics*. The syntax relates to the language structure and it can be specified by a grammar. The semantics relates to the meaning of a language. The semantic rules associate actions to the language fragments that satisfy the syntax rules. The pragmatics relate to the other aspects of the language, including implementation issues. There are formal methods for describing syntax and semantics, reported by any major textbook on languages [1, 24].

Languages can be broadly classified as *procedural* and *declarative* languages [24]. Procedural programs specify the desired action by describing a sequence of steps whose order of execution matters. Conversely, declarative models express the problem to be solved by a set of declarations without detailing a solution method. Therefore the order of the statements is not important in such languages.

Alternative classifications of languages exist. Languages with an *imperative* semantics are those where there is an underlying dependence between the assignments and the values that variables can take. Languages with an *applicative* semantics are those based on function invocation. A precise classification of programming and hardware languages on this basis is often difficult, because languages sometimes exploit features of different semantic paradigms. Most commonly used languages (for both hardware and software) are procedural with an imperative semantics.

Languages for hardware specification are classified on the basis of the description view that they support (e.g., physical, structural or behavioral). For example, languages that support physical design are characterized by having geometric primitives and by supporting operations on those primitives. In this book, we concentrate on the structural and behavioral views and we consider the corresponding languages. Most HDLs support both structural and behavioral views, because circuit specifications often require both.

Hardware description languages have often been developed in conjunction with simulators, and simulators have influenced some choices in HDL design. Execution speed is a major requirement for simulators, so that large circuits can be analyzed in relatively short time. Event-driven simulation algorithms have been widely used, because they allow a simulator to minimize the number of required computations and hence its run time.

It is important to consider the hardware model as seen by most simulators to understand the *discrete-event* timing semantics of many modern HDLs. In particular, we consider the paradigm and terminology for VHDL simulation [15, 18, 20]. A digital circuit is a discrete system that can be thought of as a set of *processes* communicating by means of *signals*. Each process can be executing or suspended. In this last case, it can be reactivated. Usually, reactivation is triggered by *events*, i.e., a change of values of some signals placed onto a *sensitivity* list. In VHDL, sensitivity lists are described by *wait* statements.

Simulation consists of evaluating signal values over a specified time interval, partitioned into time frames. At each time frame, a simulation cycle consists of the following steps. First, signal values are propagated and updated. Second, all processes that are sensitized to some event are executed until they are suspended again (by executing a wait statement). Last, when all processes are suspended, the simulator time frame is advanced and another simulation cycle is performed.

This modeling style is very general and supports modeling of synchronous and asynchronous circuits. In the synchronous case, all processes may be forced to execute at each clock cycle by including the clock in the corresponding sensitivity list. When the circuit internal delays are known, they can be specified as attributes to the signals, and thus the simulation can model exactly the timing behavior of the circuit.

Problems arise when delays are unknown, or their specification allows for a wide interval of uncertainty, because the interpretation of the model cannot yield the actual timing of the operations. This is the case when a language is used to specify a circuit yet to be synthesized. Then, the designer either renounces to extract precise timing information from the behavioral models and relies for this on those models produced by synthesis or imposes constraints to the behavioral model to force the execution of the operations in pre-determined time frames. *Synthesis policies* have been proposed to provide a precise timing interpretation of circuit models, and they are described later in this chapter.

3.2.2   Structural Hardware Languages

Models in structural languages describe an interconnection of components. Hence their expressive power is similar to that of circuit schematics, even though specific language constructs can provide more powerful abstractions. Hierarchy is often used to make the description modular and compact. The basic features of structural languages place them close to the declarative class, even though some structural languages have also procedural features. Variables in the language correspond to ports of components.

> **Example 3.2.1.**   We consider here a half-adder circuit, as shown in Figure 3.2. The example is in the VHDL language, using its structural modeling capability. Note that `architecture` is a keyword in VHDL, encapsulating the internals of a model (i.e., excluding the interface). It does not relate to the architectural level of abstraction. Conversely, the word STRUCTURE is an arbitrary identifier.

```
architecture STRUCTURE of HALF_ADDER is
        component AND2
                port (x, y: in bit; o: out bit);
        component EXOR2
                port (x, y: in bit; o: out bit);
        begin
                G1: AND2
                        port map (a, b, carry);
                G2: EXOR2
                        port map (a, b, sum  );
end STRUCTURE;
```

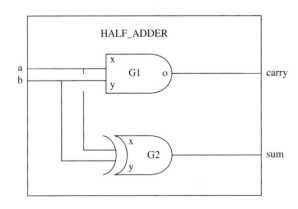

FIGURE 3.2
Structural representation of a half-adder circuit.

The model contains two declarations of other models, AND2 and EXOR2, as well as two instantiations of the models, called G1 and G2. Additional information on the components AND2 and EXOR2 is provided elsewhere, for example, in a standard logic library package. This description of the half-adder is not complete, because it lacks the definition of the interface, to be shown in Example 3.2.9.

Other types of variables, often called *metavariables*, are used to make circuit models compact. An example of a metavariable is an index of an array. Metavariables do not represent directly hardware entities and are eliminated from the model by expansion in the first steps of compilation.

Example 3.2.2. We model in VHDL an array of 32 inverters between two busses, as shown in Figure 3.3. The keyword generate relates to the multiple instantiation with metavariable i.

```
architecture STRUCTURE of BUS_INV is
        component INVERTER
                port (i1: in bit; o1: out bit);
        end component;
        begin
                G: for i in 1 to 32 generate
                        INV: INVERTER port map (inputs(i), outputs(i));
                end generate;
end STRUCTURE;
```

3.2.3 Behavioral Hardware Languages

We consider behavioral modeling for circuits in increasing levels of complexity. Combinational logic circuits can be described by a set of ports (inputs/outputs) and a set of

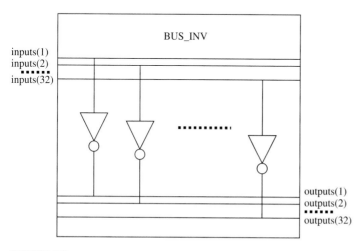

FIGURE 3.3
Structural representation of an array of inverters.

equations that relate variables to logic expressions. The declarative paradigm applies best to combinational circuits, which are by definition memoryless. Indeed they can be seen as an interconnection (i.e., a structure) of operators, each operator evaluating a logic function. These models differ from structural models in that there is not a one-to-one correspondence between expressions and logic gates, because for some expressions there may not exist a single gate implementing it.

Procedural languages can be used to describe combinational logic circuits. Most procedural hardware languages, except for the early ones, allow for multiple assignments to variables. Thus an unambiguous interpretation of multiple assignments to variables is required. There are various *resolution mechanisms*. For example, the last statement can override the others. Other languages, like YLL (an APL-based HDL used by the YORKTOWN SILICON COMPILER [12]) resolved multiple assignments by assigning to a variable the disjunction of the corresponding expressions.

Example 3.2.3. We consider again the half-adder circuit in the VHDL language, using its behavioral modeling capability.

```
architecture BEHAVIOR of HALF_ADDER is
process
        begin
                carry <= ( a and b ) ;
                sum   <= ( a xor b ) ;
        end process;
end BEHAVIOR;
```

In VHDL add and or are predefined operators. The keywords process and end process denote a set of statements to be executed sequentially. They can be avoided when modeling combinational logic.

Note that in this particular simple example, the two assignments correspond to the component instantiations of the related structural representations shown in Example 3.2.1. This is not true in general.

Let us now consider modeling of synchronous logic circuits. The modeling style is heavily affected by the timing semantics of the chosen language. Different paradigms have been proposed and used. We show first how modeling combinational circuits by declarative methods can be extended to sequential circuits. Sequential circuits can be described by a set of assignments of expressions. The arguments in the expressions are variables with synchronous delays. This modeling style applies well to synchronous data paths and DSPs.

Example 3.2.4. In the applicative language Silage, delayed variables are denoted by the @ sign, followed by a timing annotation that can be general in nature. In particular, Silage can be used to represent synchronous circuits by using synchronous delays as

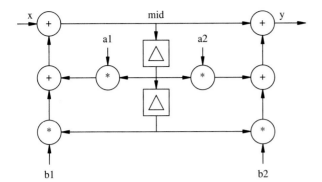

FIGURE 3.4
Recursive filter.

the timing annotation. A simple recursive filter, reported by Hilfinger [8] and shown in Figure 3.4, can be described by the following expressions:

```
function IIR ( a1, a2 , b1, b2, x: num)
        /* returns */ y: num =
begin
        y = mid + a2 * mid@1 + b2 * mid@2;
        mid = x + a1 * mid@1 + b1 * mid@2;
end
```

We consider now sequential circuit modeling with procedural languages. Finite-state machines can be described by procedural models where a variable keeps the state information. Then, the operations of the finite-state machine can be described as an iteration (synchronized to the clock), with branching to the fragments corresponding to the present state.

Example 3.2.5. We describe here a finite-state machine that recognizes two or more consecutive 1s in an input data stream. The model is again in the VHDL language:

```
architecture BEHAVIOR of REC is
        type STATE_TYPE is (STATE_ZERO, STATE_ONE);
        signal STATE : STATE_TYPE := STATE_ZERO;
        process
                begin
                wait until clock'event and clock='1';
                if ( in = '1' ) then
                        case STATE is
                                when => STATE_ZERO
                                        STATE <= STATE_ONE;
                                        out <= '0';
                                when => STATE_ONE
                                        STATE <= STATE_ONE;
                                        out <= '1';
                        end case;
                else
                                        STATE <= STATE_ZERO;
                                        out <= '0';
                end if;
        end process;
end BEHAVIOR;
```

In this model signal STATE is an enumerated type and stores the state of the finite-state machine. Note that the second assignment STATE <= STATE_ONE; is superfluous. VHDL has a discrete-event timing semantics and the timing interpretation of the model is the following. The process is executed every time the clock becomes true. The wait statement synchronizes the model to the clock.

Example 3.2.6. We show now a model of the previous finite-state machine in the UDL/I language using a particular construct called automaton. Note that the automaton statement defines the clock signal explicitly, which is used to synchronize the operations for simulation and synthesis purposes. Thus there is no need for a *wait* statement.

```
NAME:           REC;
PURPOSE:        synthesis;
LEVEL:          chip;
INPUTS:         clock, rst, in;
OUTPUTS:        o1;
BEHAVIOR_SECTION;
    BEGIN
        ->> a1**t1; "this automaton works unconditionally"
    END;
    automaton:a1(t1) : .rst : rise(.clock) ;
      STATE_ZERO : begin
              .o1 := 1B0;
              if .in then
                 -> STATE_ONE;
              end_if;
           end;
      STATE_ONE : begin
              if .in then
                 .o1 := 1B1;
              else
                 .o1 := 1B0;
                 -> STATE_ZERO;
              end_if;
           end;
    end_auto;
END_SECTION;
END; "end of module"
CEND;"end of design file (chip)"
```

We consider next behavioral modeling at the architectural level, with specific reference to procedural languages. A common modeling paradigm is to sequence assignments to variables by means of control-flow constructs, such as *branching*, *iteration* and *model call*. This explains why we can abstract the behavior of a circuit as a set of operations and dependencies. Operations correspond to assignments, groups of assignments or model calls. The data-flow and control-flow dependencies define constraints on the potential operation concurrency of an implementation. Additional constraints in the circuit specifications may be used, if desired, to prescribe the exact time frames of the operations, hence specifying the concurrency in the implementation.

Example 3.2.7. We consider here a fragment of a model describing a set of operations in a computer.

```
...
ir <= fetch(pc);
case ir is
        when => AND
                acc <= rega and regb;
        when => OR
                acc <= rega or regb;
        when => XOR
                acc <= rega xor regb;
        when => ADD
                acc <= rega + regb;
end case;
pc <= pc + 1;
...
```

The fragment describes three tasks:

- A call to function `fetch`, which returns the value of the instruction register `ir`.
- A multi-way branch, which sets the accumulator `acc` to the result of four mutually exclusive operations applied to the values of registers `rega` and `regb`.
- An increment of the program counter `pc`.

An analysis of the data dependencies can show us that the last task can be executed in parallel with one of the other two. Alternatively, the three tasks can be performed in series.

Behavioral models provide some degrees of freedom in interpreting the time frame of execution of the operations. Architectural synthesis and optimization tools exploit this freedom in the search for optimal implementations. On the other hand, the simulation plots of the initial behavioral models may not match those of the synthesized structural models, because the original models ignore the optimization choices. Thus it is important to provide means for representing the timing behaviors of those implementations that are compatible with the original behavioral model.

Synthesis policies have been developed for different languages and tools to provide an interpretation of the timing behavior. In addition, policies relate to the language subsets that can be synthesized, e.g., disallowing specific constructs for simulation, such as `assert` and `print`.

Example 3.2.8. Consider the finite-state machine of Example 3.2.5. A synthesis policy for VHDL [4] is to synchronize the operations to the *wait* statement in the circuit synthesized from that model. Since the model has only one `wait`, all operations will execute in one clock cycle.

Let us consider now Example 3.2.7 and let us add a `wait` statement before the first operation. A synthesis policy could be to assume that the three tasks are synchronized to the clock and that the second task follows the first, because of a data dependency. The difference from the finite-state machine example lays in the fact that all tasks may not be completed in a single cycle. The delays of the tasks may even be unknown before their synthesis is performed. Assuming that in the synthesized implementation each task takes one cycle to execute, the overall implementation of the fragment would execute in two cycles. On the other hand, a simulator would execute the three tasks in a single cycle.

In order to avoid the mismatch between the timing of the behavioral simulation and of its implementation, additional `wait` statements can be introduced. For example, consider the following model:

```
...
wait until clock'event and clock='1';
ir <= fetch(pc);
wait until clock'event and clock='1';
case ir is
        when => AND
                acc <= rega and regb;
        when => OR
                acc <= rega or regb;
        when => XOR
                acc <= rega xor regb;
        when => ADD
                acc <= rega + regb;
end case;

pc <= pc + 1
...
```

The model forces the simulator to delay the second and third tasks. A consistent synthesis policy of interpretation of `wait` statements is considering them as boundaries for the operation time frames. Thus all start times of the operations can be predetermined. This policy has been used in the CALLAS system [16].

Several synthesis policies have been proposed and used. As far as the timing interpretation of behavioral models, the spectrum of solutions that they propose range from considering just the partial order of the operations in each block to defining a complete order on them. In the sequel, we shall not refer to any specific policy but consider mainly the two limiting cases. Thus we interpret behavioral models as either partial or linear orders of tasks.

3.2.4   HDLs Used for Synthesis

We describe in this section some of the major languages in use today. In particular, we focus on the procedural languages VHDL, Verilog and UDL/I and on the applicative language Silage. We do not intend to describe the syntax of these languages. We refer the interested reader to the user's manuals and to the related literature [2, 8, 10, 15, 18, 20, 22]. Instead we survey here the major features of the languages and the major semantic differences.

VHDL. VHDL was developed as a part of the VHSIC project sponsored by the U.S. Department of Defense. It became an IEEE standard in 1987 and it is now very widespread. VHDL borrowed some features from the ADA language. It was born as a language for specifying large systems. Readability was preferred to writability, and consequently the language is fairly verbose.

A VHDL model of a circuit is a *design entity*. The entity consists of two components, the *interface description* and the *architectural body*. The interface specifies the circuit ports, the architecture its contents.

Example 3.2.9. A half-adder circuit would be described by the following interface:

```
entity HALF_ADDER is
        port(a, b: in bit; sum,carry: out bit);
end HALF_ADDER
```

Architectural bodies for the half-adder were shown in Examples 3.2.1 and 3.2.3.

The architectural body can have three different flavors: *structural, behavioral* and *data flow*. A structural architectural body describes an interconnection. A behavioral architectural body includes *processes* that are encapsulations of sequential statements with an imperative semantics. A data-flow architectural body is an additional style to model concurrent signal assignments. An example of a data-flow architectural body can be derived from Example 3.2.3, where the process statement is removed. All assignments are then concurrent, while respecting data dependencies. A VHDL model can include bodies with different flavors.

VHDL supports several data types, and it is a *strongly typed* language, in that all variables must be explicitly declared [1]. The user can also define its own types. The language supports constants, variables and signals. Variables are not directly related to hardware; they can be used internally in behavioral models. Signals are a generalization of variables. They relate to electrical signals that can traverse ports and have a time dimension, i.e., signal delays can be specified. The language supports the usual control-flow constructs, such as function and procedure call, branching and iteration. The wait statement is used to synchronize the execution to events, and therefore it is useful in conjunction with concurrent process modeling. VHDL is highly modular. Blocks (i.e., statements encapsulated by the begin and end keywords) can have a *guard* that enables certain types of statements inside the block. Logical, relational and arithmetic operators are provided, but the arithmetic operators apply only to the integer and real data types. Therefore the user has to provide his or her own procedures to perform arithmetic operations on bit vectors. VHDL allows the *overloading* of the operators, i.e., the redefinition of an operator according to the type of the operands.

In VHDL, data types, constants and subprograms can be defined inside entity declarations and inside architecture bodies. They are not visible elsewhere. To obviate the problem, VHDL provides also a `package` mechanism to encapsulate declarations and subprograms that can be referenced elsewhere. This enables the design team to build libraries of commonly used declarations and functions into packages. This feature enhances the modularity and reusability of the models, and it is especially useful for large programs. For example, there exists a common standard logic package that defines the logic types, primitives and functions.

Example 3.2.10. We present here a complete example of the VHDL behavioral model of the circuit integrating a differential equation proposed in Example 1.5.3. This model shows explicitly that the ports are read after signal `start` raises and that each loop iteration is synchronized to the `clock` signal. Note that the model could be described in different but equivalent ways.

```
package mypack is
  subtype bit8 is integer range 0 to 255;
end mypack;

use work.mypack.all;
entity DIFFEQ is
port(
        dx_port, a_port, x_port,  u_port : in  bit8;
        y_port : inout  bit8;
        clock, start  : in bit );
end diffeq;

architecture BEHAVIOR of DIFFEQ is
begin
process
        variable x, a, y, u, dx, x1, u1, y1:  bit8;
begin
        wait until start'event and start = '1';
        x := x_port; y := y_port; a := a_port;
        u := u_port; dx := dx_port;
        DIFFEQ_LOOP:
        while ( x < a ) loop
                wait until clock'event and clock = '1';
                x1 := x + dx;
                u1 := u - (3 * x * u * dx) - (3 * y * dx);
                y1 := y + (u * dx);
                x  := x1; u := u1 ; y := y1;
        end loop DIFFEQ_LOOP;
        y_port <=  y;
end process;
end BEHAVIOR;
```

VERILOG. Verilog was developed as a proprietary language first and then was opened to the public in the late 1980s. The language was conceived for modeling and simulating logic circuits at the *functional* (i.e., architectural), *logic* and *switch* levels. In contrast with VHDL, the language is concise in its syntax. It is also more limited in capabilities than VHDL, but it supports all major features for circuit modeling and stepwise refinement. We consider here the behavioral and structural views, which can be intermixed. We begin by reproposing Examples 3.2.1 and 3.2.3 in Verilog. Verilog does not require a distinction between the interface and the body specification.

Example 3.2.11. This is a structural representation of a half-adder in Verilog:

```
module   HALF_ADDER (a , b , carry , sum);
         input    a , b;
         output   carry, sum;
         and
                 G1 (carry, a , b);
         xor
                 G2 (sum,    a , b);
endmodule
```

In Verilog `and` and `xor` are language primitives.

Example 3.2.12. The following is a behavioral representation of a half-adder in Verilog:

```
module   HALF_ADDER (a , b , carry , sum);
         input    a , b;
         output   carry, sum;

                 assign carry = a & b ;
                 assign sum   = a ^ b ;
endmodule
```

Verilog supports two major kinds of assignments. The *continuous assignment*, with the keyword `assign`, denotes an immediate assignment. The *procedural assignment*, without the keyword `assign`, is an assignment that happens when the statement is executed.

Verilog supports two major data types: `register` and `net`. The language supports the usual control-flow statements, such as branching and iteration, as well as model calls of two types: function and subroutine, the latter named `task`. The `wait` statement serves the purpose of synchronizing operations or concurrent processes. Behavioral models are divided into an initialization portion, preceded by the keyword `initial` and a sequence of statements whose execution repeats forever, preceded by the keyword `always`. Also Verilog has a discrete-event timing semantics. Hence most considerations for synthesis of VHDL models apply to Verilog models too.

Example 3.2.13. We present here a Verilog behavioral model of the circuit proposed in Example 1.5.3. This model shows that the computation is delayed until signal *start* raises and that each loop iteration is synchronized to the *clock* signal.

```
module DIFFEQ (xp,yp,up,dx,a,clock,start);
input [7:0] a,dx,xp,up;
inout [7:0] yp;
input      clock, start;
reg   [7:0] xl,ul,yl,x,y,u;
always  @(posedge start)
begin
    x = xp;  y = yp;  u = up;
    while (x < a)
    begin
     xl = x + dx;
     ul = u - (3 * x * u * dx) - (3 * y * dx);
     yl = y + (u * dx);
     @(posedge clock);
     x = xl; u = ul; y = yl;
    end
end
assign yp = y;
endmodule
```

It is instructive to contrast the model to the VHDL model of Example 3.2.10.

UDL/I. The Unified Design Language for Integrated circuits (UDL/I) was developed in Japan with the goal of providing a common interchange format among manufacturers and to motivate common CAD tool development [10]. The language objectives included defining circuit specifications that could be used for simulation, synthesis and verification. Hence, a separation has been defined between the simulation constructs and the others.

The language supports *behavioral, structural* and *primitive* views, the last confined to the definition of function primitives. Structural descriptions involve a listing of the components and a listing of the nets. Delays can be added as attributes to nets.

Example 3.2.14. Let us consider the structural description of the half-adder.

```
...
inputs: a,b;
outputs: sum, carry;
types: and  (I1#inputs, I2#inputs, T1#outputs),
       xnor (I1#Inputs, I2#Inputs, T1#outputs);
and:  G1;
xnor: G2;

net_section;

        N1 = from (.a)    to (G1.I1);
        N2 = from (.b)    to (G1.I2);
        N3 = from (.a)    to (G2.I1);
        N4 = from (.b)    to (G2.I2);
        N5 = from (G1.T1) to (.carry);
        N6 = from (G2.T1) to (.sum);

end_section;
...
```

Note that all nets are listed. Primary inputs and outputs are referred to as starting with a dot. Pins on instances are labeled by the name of the instance and the local terminal.

The behavioral view supports several features, including the common ones like the assignments of expression to variables and the branching constructs. There is no iteration construct in UDL/I. The *at* statement can bind a specific action to an event, like a clock transition. In addition, in UDL/I it is possible to refer directly to hardware facilities [e.g., registers, latch, random-access memory (RAM), read-only memory (ROM), PLA, terminal]. A specific construct, called `automaton`, can be used to model finite-state machines, as shown by Example 3.2.6.

The formal semantics of UDL/I is defined by a small subset of the language, called *core subset*. Formal rules are provided to translate each construct of the language into the subset. The subset is related to three types of circuit primitives: (i) a facility, i.e., a memory array or a port; (ii) a combinational function; and (iii) a storage element, like a register, latch or memory. The transformation rules make the interpretation of UDL/I models unambiguous.

> **Example 3.2.15.** Let us consider now the differential equation integrator. Iteration is modeled by sensitizing the model to *clock*. Bidirectional signals are described by the BUS construct:

```
NAME: DIFFEQ;
PURPOSE: Synthesis;
LEVEL: end;
INPUTS: start, a[7:0], dx[7:0];
BUS: x<7:0>, y<7:0>, u<7:0>;
CLOCK: clock;

BEHAVIOR_SECTION
REGISTER: xl<7:0>, yl<7:0>, ul<7:0>;
    BEGIN

        AT clock & start & (x < a) DO
            xl := x + dx;
            ul := u - (3 * x * u * dx) - (3 * y * dx);
            yl := y + (u * dx);
            x := xl; u := ul ; y := yl;
        END_DO;
        END_SECTION;
    END;
CEND;
```

SILAGE. Silage is an applicative language developed by Hilfinger [8]. The basic object in Silage is a *signal* that is a vector with as many elements as the time points of interest. The fundamental operation is the function call, similar to that represented in Example 3.2.4. Each call is an expression of delayed values of the variables, and it expresses what happens during a single sample interval. The order of the calls is irrelevant. A Silage description can be made equivalent to a signal flow graph. Therefore the language is well suited for DSP applications. Multi-rate descriptions are possible using the primitive functions `decimate` and `interpolate`.

The language supports also branching and iteration on the metavariables describing signal arrays.

Example 3.2.16. We present again the differential equation integrator, proposed in Example 1.5.3, in Silage. The model consists of a function invocation. In the model, x, y, u are vectors, the second representing the value of the function being integrated at each time step.

The iterative construct, whose syntax is (j:lower_bound..upper_bound) :: assignment, has a fixed number of iterations NMAX. A branching construct breaks the loop when the integration interval has been covered, i.e., when x[i] >= a:

```
#define W fix<18,0>
#define NMAX 1000 /* Maximum number of iterations */

function diffeq ( Xinport, DXport, Aport, Yinport, Uinport : W)
    Xoutport, Youtport, Uoutport : W =
begin
   x[0] = Xinport; a = Aport; dx = DXport;
   y[0] = Yport; u[0] = Uinport;

   ( i : 1 .. NMAX ) ::
   begin
      x[i] = x[i-1] + dx;
      u[i] = u[i-1] - (3*x[i-1]*u[i-1]*dx) - (3*y[i-1]*dx);
      y[i] = y[i-1] + (u[i-1] * dx);

      (Xoutport, Youtport, Uoutport) =
      exit (x[i] >= a)
         -> (x[i], y[i], u[i])   || (0,0,0)   /* export variables */
      tixe;
   end;
end;
```

The language supports an annotation construct, called `pragma`, that can be used to describe constraints. For example, `pragma` statements can associate operations with the hardware resources to be used in an implementation. As a result, constraints on the operation concurrency in a synthesized implementation can be modeled by the use of pragmas.

SOME COMPARISONS OF HDLS USED FOR SYNTHESIS. Hardware description languages have to satisfy multiple criteria, and hence it is meaningless to search for the "best" language. Valid measures of the value of a language are hardware expressing power, support for computer-aided simulation, synthesis and verification and distribution.

As far as hardware expressing power is concerned, different languages are suited to the design of different classes of circuits. Whereas VHDL and Verilog support general circuit design, Silage is better suited toward describing DSP data paths and filters. UDL/I supports modeling in terms of interconnected finite-state machines. Circuits with complex control functions can be best described by procedural languages.

Availability of CAD tools for processing hardware models is a key factor for the widespread use of HDLs. At present, few CAD tools have been developed for UDL/I. Silage has been used mainly in the specific domain of DSPs, where architectural-level synthesis tools have been fairly successful. Synthesis and simulation tools from VHDL and Verilog have been developed and used. Synthesis policies have been proposed and used with restrictions to subsets of the languages, especially in the case of VHDL, because of the complexity of the full language.

VHDL and Verilog are widespread worldwide. Many designers are using VHDL because of economic factors related to marketing their products. UDL/I is used by some academic and industrial sites in Japan.

3.3 ABSTRACT MODELS

We present in this section some abstract models that are used to represent different circuit views at the logic and architectural levels. They are based on graphs.

3.3.1 Structures

Structural representations can be modeled in terms of *incidence* structures. An incidence structure consists of a set of *modules*, a set of *nets* and an incidence relation among modules and nets. A simple model for the structure is a *hypergraph*, where the vertices correspond to the modules and the edges to the nets. The incidence relation is then represented by the corresponding incidence matrix. Note that a hypergraph is equivalent to a bipartite graph, where the two sets of vertices correspond to the modules and nets. An alternative way of specifying a structure is to denote each module by its terminals, called *pins* (or *ports*), and to describe the incidence among nets and pins.

Often the incidence matrix is sparse, and *netlists* are a more efficient means of description. A netlist enumerates all nets of a given module (module-oriented netlist) or all modules of a given net (net-oriented netlist).

Example 3.3.1. Consider the example of Figure 3.5 (a). There are three modules, three nets and seven pins. The module-net incidence matrix is:

$$\begin{bmatrix} 1 & 1 & 1 \\ 1 & 1 & 0 \\ 0 & 1 & 1 \end{bmatrix}$$

The corresponding hypergraph and bipartite graphs are shown in Figures 3.5 (b) and (c), respectively. A module-oriented netlist is the following:

$$m1: n1, n2, n3$$
$$m2: n1, n2$$
$$m3: n2, n3$$

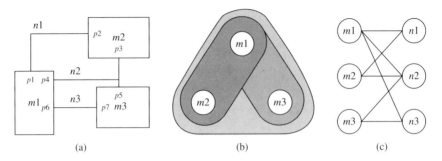

FIGURE 3.5
Example of a structure: (a) modules, nets and pins; (b) hypergraph; (c) bipartite graph.

Incidence structures can be made hierarchical in the following way. A *leaf* module is a primitive with a set of pins. A non-leaf module is a set of modules, called its submodules, a set of nets and an incidence structure relating the nets to the pins of the module itself and to those of the submodules.

> **Example 3.3.2.** Consider the previous example. We assume now that the structure is hierarchical and that module $m2$ has submodules. Figure 3.6 shows the details of module $m2$, which consists of submodules $m21$ and $m22$, subnets $n21$ and $n22$ and internal pins $p21$, $p22$, $p23$, $p24$ and $p25$.

3.3.2 Logic Networks

A *generalized logic network* is a structure, where each leaf module is associated with a combinational or sequential logic function. While this concept is general and powerful, we consider here two restrictions to this model: the combinational logic network and the synchronous logic network.

The combinational logic network, called also *logic network* or *Boolean network*, is a hierarchical structure where:

- Each leaf module is associated with a multiple-input, single-output combinational logic function, called a *local function*.
- Pins are partitioned into two classes, called *inputs* and *outputs*. Pins that do not belong to submodules are also partitioned into two classes, called *primary inputs* and *primary outputs*.

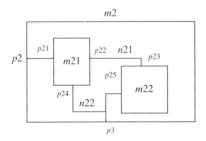

FIGURE 3.6
Example of a hierarchical structure: details of $m2$.

- Each net has a distinguished terminal, called a source, and an orientation from the source to the other terminals. The source of a net can be either a primary input or a primary output of a module at the inferior level. (In particular, it may correspond to the output of a local function.)
- The relation induced by the nets on the modules is a partial order.

Example 3.3.3. Figure 3.7 shows a logic network with three primary inputs, two modules and two primary outputs.

Logic networks are usually represented by graphs. We consider here the non-hierarchical case, for the sake of simplicity and because its extension is straightforward. A *logic network graph*, $G_n(V, E)$, is a directed graph, whose vertex set V is in one-to-one correspondence with the primary inputs, local functions (i.e., modules) and primary outputs. The set of directed edges E represents the decomposition of the multi-terminal nets into two terminal nets. Note that the graph is acyclic by definition, because the nets induce a partial order on the modules.

Example 3.3.4. A logic network graph is shown in Figure 3.8, corresponding to the structure of Figure 3.7. The graph has three input vertices, labeled v_a, v_b and v_c; two output vertices (v_x and v_y) and two internal vertices (v_p and v_q) corresponding to the logic functions. Note that the three-terminal net, whose source is the output of the module corresponding to vertex v_p, is represented by two edges in the graph.

In general, a logic network is a hybrid structural/behavioral representation, because the incidence structure provides a structure while the logic functions denote the terminal behavior of the leaf modules. Indeed some of these functions could not be implemented as such, because of technology-dependent fan-in limitations. For example, a 64-input AND function could be represented by a local function of a logic network. The actual implementation, where the AND function is split into a tree of AND gates, would correspond then to a different structure. In the particular case where a logic network denotes an interconnection of logic gates (possibly represented by

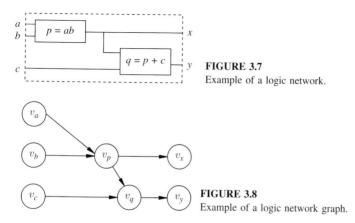

FIGURE 3.7
Example of a logic network.

FIGURE 3.8
Example of a logic network graph.

their functions), it is a purely structural representation. It is called a *bound* or *mapped network* in jargon.

In most cases, logic networks are used to represent multiple-input/output logic functions in a structured way. Indeed, logic networks have a corresponding unique input/output combinational logic function that can be obtained by combining the local functions together to express the primary outputs in terms of the primary inputs. Often the input/output functions cannot be easily represented in standard forms, such as *sum of products* or binary decision diagrams, because of the size. This is one reason for using the logic network model. Note though that this model is not a unique representation of a combinational function. Combinational logic networks are described in detail in Section 8.2.

The *synchronous* logic network model is a generalization of the combinational logic network model, to support the description of sequential synchronous circuits. In this model, leaf modules can implement multiple-input single-output combinational logic functions or *synchronous delay* elements. Nets are not required to induce a partial order on the modules. Nevertheless, the subset of the nets whose source is not a delay element must induce a partial order on the modules to model the requirement that in a synchronous circuit there are no combinational feedbacks. The synchronous network model and the corresponding synchronous network graph are described in detail in Section 9.3.1.

3.3.3 State Diagrams

The behavioral view of sequential circuits at the logic level can be expressed by finite-state machine transition diagrams. A finite-state machine can be described [9] by:

- A set of primary input patterns, X.
- A set of primary output patterns, Y.
- A set of states, S.
- A *state transition* function, $\delta : X \times S \to S$.
- An *output function*, $\lambda : X \times S \to Y$ for Mealy models or $\lambda : S \to Y$ for Moore models.
- An initial state.

The *state transition table* is a tabulation of the state transition and output functions. Its corresponding graph-based representation is the *state transition diagram*.

The state transition diagram is a labeled directed multi-graph $G_t(V, E)$, where the vertex set V is in one-to-one correspondence with the state set S and the directed edge set E is in one-to-one correspondence with the transitions specified by δ. In particular, there is an edge (v_i, v_j) if there is an input pattern $x \in X$ such that $\delta(x, s_i) = s_j \ \forall i, j = 1, 2, \ldots, |S|$. In the Mealy model, such an edge is labeled by $x/\lambda(x, s_i)$. In the Moore model, that edge is labeled by x only; each vertex $v_i \in S$ is labeled by the corresponding output function $\lambda(s_i)$.

Example 3.3.5. We describe here a Mealy-type finite-state machine that acts as a synchronizer between two signals. The primary inputs are a and b, plus the reset signal r. There is one primary output o that is asserted when both a and b are simultaneously true or when one is true and the other was true at some previous time. The finite-state machine has four states. A reset state s_0. A state memorizing that a was true while b was false, called s_1, and a similar one for b, called s_2. Finally, a state corresponding to both a and b being, or having been, true, called s_3. The state transitions and the output function are annotated on the diagram in Figure 3.9.

It is sometimes convenient to represent finite-state machine diagrams in a *hierarchical* way, by splitting them into subdiagrams. Each subdiagram, except the root, has an *entry* and an *exit* state and is associated with one (or more) *calling* vertex of a diagram at the higher level in the hierarchy. Each transition to a calling vertex is equivalent to a transition into the entry state of the corresponding finite-state machine diagram. A transition into an exit state corresponds to return to the calling vertex.

Hierarchical diagrams are used in synthesis for assembling the finite-state machine transition diagram in a modular way. They are also used for specification, to manage the size of large representations.

Example 3.3.6. A hierarchical state transition diagram is shown in Figure 3.10. There are two levels in the diagram: the top level has three states, the other four. A transition into calling state s_{01} is equivalent to a transition to the entry state of the lower level of the hierarchy, i.e., into s_{10}. Transitions into s_{13} correspond to a transition back to s_{01}. In simple words, the dotted edges of the diagrams are traversed immediately.

3.3.4 Data-flow and Sequencing Graphs

We consider here models that abstract the information represented by procedural HDLs with imperative semantics. Abstract models of behavior at the architectural

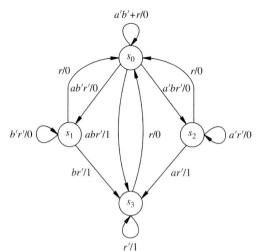

FIGURE 3.9
Example of a state transition diagram of a synchronizer finite-state machine.

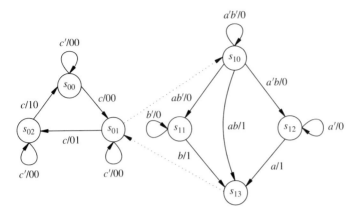

FIGURE 3.10
Example of a hierarchical state transition diagram.

level are in terms of *tasks* (or *operations*) and their *dependencies*. Tasks may be *No-Operations* (NOPs), i.e., fake operations that execute instantaneously with no side effect. Dependencies arise from several reasons. A first reason is availability of data. When an input to an operation is the result of another operation, the former operation depends on the latter. A second reason is serialization constraints in the specification. A task may have to follow a second one regardless of data dependency. A simple example is provided by the two following operations: loading data on a bus and raising a flag. The circuit model may require that the flag is raised after the data are loaded. Last, dependencies may arise because two tasks share the same resource that can service one task at a time. Thus one task has to perform before the other. Note, though, that, in general, dependencies due to resource sharing are not part of the original circuit specification, because the way in which resources are exploited is related to the circuit implementation.

Data-flow graphs represent operations and data dependencies. Let the number of operations be n_{ops}. A data-flow graph $G_d(V, E)$ is a directed graph whose vertex set $V = \{v_i; \; i = 1, 2, \ldots, n_{ops}\}$ is in one-to-one correspondence with the set of tasks. We assume here that operations require one or more operands and yield one or more results (e.g., an addition with two addends yielding a result and an overflow flag). The directed edge set $E = \{(v_i, v_j); \; i, j = 1, 2, \ldots, n_{ops}\}$ is in correspondence with the transfer of data from an operation to another one. Data-flow graphs can be extended by adding other types of dependencies, such as task serialization, that are represented by additional directed edges.

The data-flow graph model implicitly assumes the existence of *variables* (or *carriers*), whose values store the information required and generated by the operations. Each variable has a *lifetime* that is the interval from its *birth* to its *death*, where the former is the time at which the value is generated as an output of an operation and the latter is the latest time at which the variable is referenced as an input to an operation. The model assumes that the values are preserved during their lifetime, i.e., that some sort of storage (registers in case of synchronous implementations) is provided.

Example 3.3.7. Consider the following model fragment, from Example 1.5.3, describing a set of computations.

$$xl = x + dx;$$
$$ul = u - (3 * x * u * dx) - (3 * y * dx);$$
$$yl = y + u * dx;$$
$$c = xl < a;$$

The four assignments in the model can be broken down into a set of 11 simple operations such as addition, subtraction, multiplication and comparison. The data-flow graph representation of these tasks is shown in Figure 3.11. There are 11 vertices, labeled by numeric identifiers. The input and output data, represented by variables $\{x, y, u, dx, a, 3\}$ and by $\{xl, yl, ul, c\}$, respectively, are explicitly marked on the edges.

Note that the program fragment could have been represented by different data-flow graphs by exploiting the commutativity and associativity of addition and multiplication.

Control-flow information, related to *branching* (or *conditional*) and *iteration* (or *loop*) constructs, can also be represented graphically. Many different models have been proposed to represent *control/data-flow graphs* (CDFGs) [5, 17, 23]. The simplest approach is to extend further data-flow graphs by introducing branching vertices that represent operations that evaluate conditional clauses. A branching vertex is the tail of a set of alternative paths, corresponding to the possible branches. Iteration can be modeled as a branch based on the iteration exit condition. The corresponding vertex is the tail of two edges, one modeling the exit from the loop and the other the return to the first operation in the loop.

We consider in this book one particular abstract model for tasks subject to data- and control-flow dependencies. This model will be used consistently to describe synthesis algorithms and the proof of their properties. The model is called *sequencing*

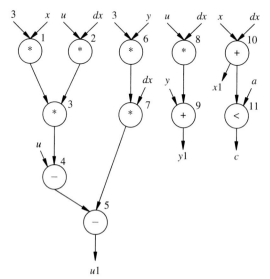

FIGURE 3.11
Example of a data-flow graph.

graph and it is a hierarchical control/data-flow graph, where control-flow primitives such as branching and iteration are modeled through the hierarchy, whereas data-flow and serialization dependencies are modeled by graphs. In addition, the hierarchical model supports a *model call*, i.e., the encapsulation of subsets of operations and their dependencies into blocks that can be multiply invoked.

To be more specific, we call *sequencing graph* $G_s(V, E)$ a hierarchy of directed graphs. A generic element in the hierarchy is called a *sequencing graph entity*, when confusion is possible between the overall hierarchical model and a single graph. A sequencing graph entity is an extended data-flow graph that has two kinds of vertices: *operations* and *links*, the latter linking other sequencing graph entities in the hierarchy. Sequencing graph entities that are leaves of the hierarchy have no link vertices.

Let us consider first a sequencing graph entity that has only operation vertices, e.g., a non-hierarchical model or an entity that is a leaf of the hierarchy. The vertex set V is in one-to-one correspondence with the operations. The edge set E models the dependencies due to data flow or serialization. The graph has two major properties. First, it is *acyclic*. Hence it models a partial order on the tasks. Acyclic dependencies suffice because iteration is modeled outside the graph entity, as shown later. Second, the graph is *polar*; i.e., there are two vertices, called *source* and *sink*, that represent the first and last tasks. Since there may not be a unique task to be executed first (or last) in the circuit specification, the source and sink vertices always model *No-Operations*. The source is the tail of directed edges to all those vertices representing the initial operations and the sink is the head of directed edges from all those representing the final tasks.

The source and sink vertices are labeled by v_0 and v_n, respectively. Therefore the graph has $n_{ops} + 2$ vertices and subscript n is interchangeable with $n_{ops} + 1$. The vertex set is then $V = \{v_i;\ i = 0, 1, \ldots, n\}$ and the edge set $E = \{(v_i, v_j);\ i, j = 0, 1, \ldots, n\}$ represents those operation pairs such that operation v_j follows directly v_i for any $i, j = 0, 1, \ldots, n$. We say that vertex v_i is a predecessor (or direct predecessor) of v_j when there is a path (or an edge) with tail v_i and head v_j. Similarly, we say that vertex v_i is a successor (or direct successor) of v_j when there is a path (or an edge) with tail v_j and head v_i. Note that paths in the graph represent *concurrent* (and not alternative) streams of operations.

Example 3.3.8. Let us consider again the program fragment of the previous example (with no serialization requirements). We show in Figure 3.12 the corresponding sequencing graph. Note that we represent explicitly the source and sink vertices v_0 and v_n. Vertices $\{v_3, v_4, v_5\}$ are successors of v_1. Vertex v_3 is the direct successor of v_1.

Let us consider now a generic sequencing graph entity. The same considerations presented above still apply, with the exception that some vertices are links to other sequencing graph entities, representing *model call*, *branching* and *iteration* constructs.

A model call vertex is a pointer to another sequencing graph entity at a lower level in the hierarchy. It models a set of dependencies from its direct predecessors to the source vertex of the called entity and another set of dependencies from the corresponding sink to its direct successors.

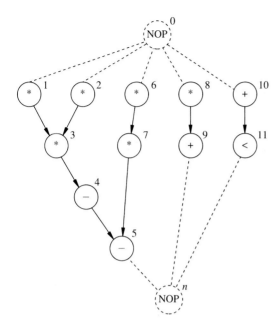

FIGURE 3.12
Example of a sequencing graph.

Example 3.3.9. An example of a hierarchical sequencing graph is shown in Figure 3.13, corresponding to the following code sequence: $x = a * b; y = x * c; z = a + b;$ $submodel(a, z)$, where $submodel(m, n)\{p = m + n; q = m * n\}$. Vertex a,4, modeling the call to $submodel$, links the two graph entities.

Branching constructs can be modeled by a *branching clause* and *branching bodies*. A branching body is a set of tasks that are selected according to the value of the branching clause. There are as many branching bodies as the possible values of the branching clause, and their execution is mutually exclusive. Branching is modeled by associating a sequencing graph entity with each branching body and a link vertex with the branching clause. The link vertex also models the operation of evaluating the clause and taking the branching decision. The selection of a branch body is then modeled as a selective model call to the corresponding sequencing graph.

Example 3.3.10. Figure 3.14 shows an example of a hierarchical sequencing graph corresponding to the following code sequence: $x = a * b; y = x * c; z = a + b;$ if $(z \geq 0)\{p = m + n; q = m * n\}$. The branching clause $z \geq 0$ is evaluated at vertex a,4. There are two bodies of the branch, corresponding to the cases in which the branching clause is TRUE or FALSE. Since no action is required when the clause is FALSE, the second body is represented by *No-Operations*.

Iterative constructs are modeled by an *iteration clause* and an *iteration body*. An iteration (or loop) body is a set of tasks that are repeated as long as the iteration clause

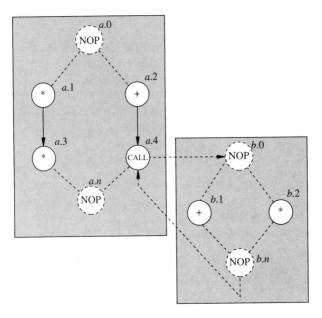

FIGURE 3.13
Example of a hierarchical sequencing graph.

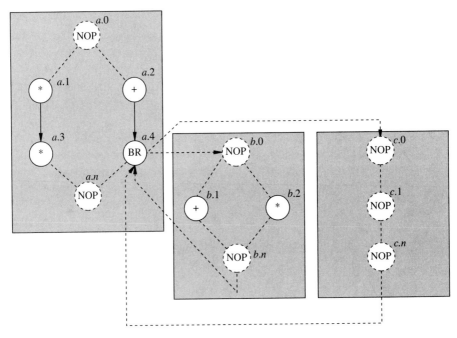

FIGURE 3.14
Example of a hierarchical sequencing graph with a branching construct.

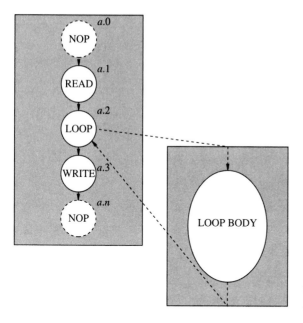

FIGURE 3.15
Example of a hierarchical sequencing graph with an iteration construct.

is true. Iteration is modeled in sequencing graphs through the use of the hierarchy, thus preserving the acyclic nature of the graph. Iteration is represented as a repeated model call to the sequencing graph entity modeling the iteration body. The link vertex models the operation of evaluating the iteration clause.

Example 3.3.11. We consider again the differential equation integrator circuit of Example 1.5.3 [19]. Even though syntactically different representations exist, as shown in Section 3.2, the circuit performs essentially three tasks: (i) reading input data; (ii) iterating the evaluation of a set of statements (shown in Example 3.3.7); and (iii) writing the result to a port.

The sequencing graph is shown in Figure 3.15. The loop body indicated in the figure is the sequencing graph entity shown in Figure 3.12. The vertex labeled LOOP evaluates the iteration clause c.

Also in this case, other equivalent graph models could be derived. Our particular choice is motivated by the desire of keeping this example similar to what has been presented in the literature [19].

The semantic interpretation of the sequencing graph model requires the notion of *marking* the vertices. A marking denotes the state of the corresponding operation, which can be (i) waiting for execution; (ii) executing; and (iii) having completed execution. *Firing* an operation means starting its execution. Then, the semantics of the model is as follows: *an operation can be fired as soon as all its direct predecessors have completed execution.*

We assume that a model can be reset by marking all operations as waiting for execution. Then, the model can be fired (i.e., executed) by firing the source vertex. The model has completed execution when the sink has completed execution. Pipelined circuits can be modeled by sequencing graphs where the source is fired before the sink has completed execution.

Some attributes can be assigned to the vertices and edges of a sequencing graph model, such as measures or estimates of the corresponding area or delay cost. In general, the delay of a vertex can be *data independent* or *data dependent*. Only data-independent delays can be estimated before synthesis. Examples of operations with data-dependent delay are *data-dependent delay branching* and *iteration*. For example, a branch may involve bodies with different delays, where in the limit one branch body can be a *No-Operation* (e.g., a floating point data normalization requiring conditional data alignment). An example data-dependent iteration is given by an arithmetic divisor, based on an iterative algorithm.

Data-dependent delays can be *bounded* or *unbounded*. The former case applies to data-dependent delay branching, where the maximum and minimum possible delays can be computed. It applies also to some iteration constructs, where the maximum and minimum number of iterations is known. The latter case is typical of some other iteration constructs, as well as modeling external synchronization.

A sequencing graph model with data-independent delays can be characterized by its overall delay, called *latency*. Graphs with bounded delays (including data-independent ones) are called *bounded-latency* graphs. Else they are called *unbounded-latency* graphs, because the latency cannot be computed.

We close this section by mentioning that the circuit behavior at the architectural level can be modeled by *Petri nets*. A wide literature on Petri nets is available [21]. Petri nets can be cast as directed bipartite graphs, where vertices correspond to *places* and *transitions*. The former can be thought of as states and the latter as operations. The Petri net model is more general and powerful, and it sets the framework to state properties of general concurrent circuits. On the other hand, modeling and synthesizing synchronous circuits can be achieved with simpler models, e.g., sequencing graphs or equivalent control/data-flow graphs. Indeed, just one type of vertices suffices, and there is no need to distinguish *places* from *transitions*. Petri nets fit better the modeling requirements of asynchronous circuits, which are beyond the scope of this book.

3.4   COMPILATION AND BEHAVIORAL OPTIMIZATION

We explain in this section how circuit models, described by HDL programs, can be transformed in the abstract models that will be used as a starting point for synthesis in the following chapters. Most hardware compilation techniques have analogues in software compilation. Since hardware synthesis followed the development of software compilers, many techniques were borrowed and adapted from the rich field of compiler design [1]. Nevertheless, some behavioral optimization techniques are applicable only to hardware synthesis. We shall briefly survey general issues on compilation, where the interested reader can find a wealth of literature, and we shall concentrate on the specific hardware issues.

A software compiler consists of a *front end* that transforms a program into an *intermediate form* and a *back end* that translates the intermediate form into the machine code for a given architecture. The front end is language dependent, and the back end varies according to the target machine. Most modern optimizing compilers improve the intermediate form, so that the optimization is neither language nor machine dependent.

Similarly, a hardware compiler can be seen as consisting of a front end, an optimizer and a back end (Figure 3.16). The back end is much more complex than a software compiler, because of the requirements on timing and interface of the internal operations. The back end exploits several techniques that go under the generic names of architectural synthesis, logic synthesis and library binding. We describe the front end in Section 3.4.1 and the optimization techniques in Section 3.4.2. The back end is described in the remaining chapters of this book.

3.4.1 Compilation Techniques

The front end of a compiler is responsible for *lexical* and *syntax analysis*, *parsing* and creation of the intermediate form. A lexical analyzer is a component of a compiler that reads the source model and produces as output a set of *tokens* that the parser then uses for syntax analysis. A lexical analyzer may also perform ancillary tasks, such as stripping comments and expanding macros. Metavariables can be resolved at this point.

A parser receives a set of tokens. Its first task is to verify that they satisfy the syntax rules of the language. The parser has knowledge of the grammar of the language and it generates a set of *parse trees*. A parse tree is a tree-like representation of the syntactic structure of a language. An example is shown in Figure 3.17. Syntactic errors, as well as some semantic errors (such as an operator applied to an incompatible operand), are detected at this stage. The error recovery policy depends on the

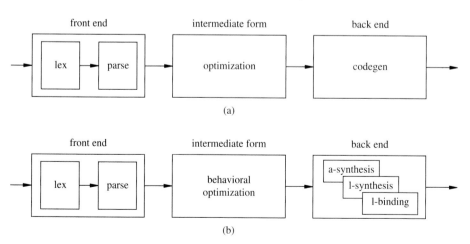

FIGURE 3.16
Anatomies of software and hardware compilers.

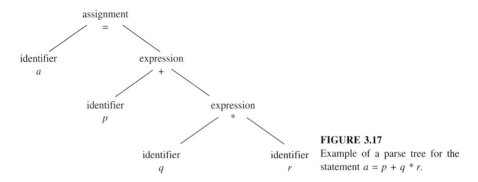

FIGURE 3.17
Example of a parse tree for the statement $a = p + q * r$.

compiler and on the gravity of the error. Software tools can be used to create lexical analyzers and parsers. Commonly used programs are *lex* and *yacc* provided with the *UNIX*™operating system.

Whereas the front ends of a compiler for software and hardware are very similar, the subsequent steps may be fairly different. In particular, for hardware languages, diverse strategies are used according to their semantics and intent.

Let us consider structural languages first. An assignment in such languages expresses a relation between pins (or modules) and nets. It is semantically equivalent to an element of an incidence structure, as described in Section 3.3.1. Similarly, there is a semantic equivalence between the notion of module call and hierarchical structure. Therefore the parse trees can be readily transformed into (possibly hierarchical) netlists. Netlist specifications are usually preferred to matrices, because they are more compact. In general, no optimization is done while compiling structural languages.

Let us turn now to languages that model combinational logic circuits. The simplest case is that of applicative languages that model the circuit by a set of Boolean equations. Then, the semantics of the model corresponds exactly to that of a logic network that is an interconnection of modules, each one characterized by a logic equation. The compilation of a logic network is then straightforward, as in the case of structural models. A more complex case is the one where the HDL has procedural semantics, possibly with branching statements. Multiple assignments to a variable must be resolved by means of some mechanism. This can be an interpretation policy, such as those mentioned in Section 3.2, or an explicitly modeled resolution function. For example, in the VHDL and Verilog language, signals can have different strengths, therefore allowing us to model three-state operation, among others.

Branching constructs can be used to model logic networks. A common way to exploit branching is by means of conditional assignments to a variable. A branching construct can be replaced by one logic expression, representing the disjunction of the possible assignments in conjunction with the test on the clause. When the branching construct does not specify an assignment for all values of the clause, the missing assignments represent *don't care* conditions on that variable, unless there is another assignment to that variable. The problem becomes more involved in the presence of nested branching constructs. We refer the interested reader to [3] for details.

Example 3.4.1. Consider the following model fragment of a combinational circuit:

```
if ( q ) {
        x = a + b;
        y = a + c;
}
else
        x = a b;
```

Assume there are no other assignments to x and y. Then the model can be expanded to $x = q(a + b) + q'ab$ and $y = q(a + c)$ with q' as a *don't care* condition for y, i.e., y can take any value when q is false. *Don't care* conditions can be used to simplify the circuit model in different ways, e.g., by simplifying the expression for y to $y = a + c$.

Often, branching constructs test the value of a variable with an enumerated type. A binary encoding of that variable is needed to generate a logic-level description, such as that captured by a logic network. In some cases, it is convenient to represent the values of such a variable as the state of the circuit and postpone the encoding to the logic synthesis stage.

A sequential model of a finite-state machine is characterized by a set of actions in coincidence with states and input patterns. In general, the state is declared by means of a variable with an enumerated type. Then, the possible values of that variable can be put in one-to-one correspondence with the finite-state machine states. The set of actions are the bodies of a branching construct whose clause relates to the present state and to the input values. They are in general combinational logic assignments. Therefore, the compilation of finite-state machine models entails just the state recognition and the processing of the combinational assignment statements.

Example 3.4.2. Consider the finite-state machine model of Example 3.2.5. The variable STATE can take two values, called STATE_ZERO and STATE_ONE, that are in one-to-one correspondence with the states. The assignment within the bodies specifies the following state transition table:

Present state	In	Next state	Out
STATE_ZERO	1	STATE_ONE	0
STATE_ONE	1	STATE_ONE	1
STATE_ZERO or STATE_ONE	0	STATE_ZERO	0

The compilation of hardware models at the architectural level involves a full *semantic analyses* that comprises *data-flow* and *control-flow* analyses and *type checking*. Semantic analysis is performed on the parse trees in different ways, for example, by transforming the parse trees into an intermediate form. Type checking has some peculiarities when compiling HDLs. Operations on vectors of Boolean variables are checked for operand compatibility. Vectors may be padded with 1s or 0s to achieve compatibility in some cases.

The overloading of the arithmetic and relational operators has to be resolved at this stage. Operations on Boolean vectors have to be mapped to hardware operators that do the corresponding function. For example, the sum of two Boolean vectors has to be resolved as a link to an adder circuit. Similarly, the comparison of two integers, to be implemented in the hardware by Boolean vectors, has to be mapped to a link to a hardware comparator. Since the mapping to hardware resources is not always uni-valent, because different hardware implementations have different area/performance parameters, abstract hardware operators are used at this stage and the binding to hardware resources is deferred to a later optimization stage (described in Chapter 6).

The semantic analysis of the parse trees leads to the creation of the intermediate form, which represents the implementation of the original HDL program on an abstract machine. Such a machine is identified by a set of operations and dependencies, and it can be represented graphically by a sequencing graph. The hardware model in terms of an abstract machine is virtual in the sense that it does not distinguish the area and delay costs of the operations. Therefore, behavioral optimization can be performed on such a model while abstracting the underlying circuit technological parameters.

We consider here procedural HDLs and we assume, for the sake of explanation and uniformity, that the sequencing graph model of Section 3.3.4 is used as interme-diate form. Note that other intermediate models could be used, with similar meaning but different aspects. Similarly we assume here, for the sake of simplicity, that struc-tured programming constructs are used (e.g., no unrestricted `goto` statements are employed) and that each model has a single entry and a single exit point. This allows us to interpret the hardware model as a sequencing graph that abides the definition given in Section 3.3.4.

Control-flow and data-flow analyses determine the hierarchical structure of the sequencing graph and the topology of its entities. The parse trees for each assignment statement identify the operations corresponding to the vertices of each graph entity. The edges are inferred by considering data-flow and serialization dependencies. Each leaf entity of a hierarchical sequencing graph corresponds to a *basic block* in compiler jargon.

Data-flow analysis comprises several tasks, and it is used as a basis for be-havioral optimization. It includes the derivation of the variable lifetimes. Note that sequencing graphs do not model explicitly the fact that variables need storage dur-ing their lifetime, with a corresponding cost in terms of circuit implementation. When considering hardware models with imperative semantics, multiple assignments to vari-ables may occur. Variables preserve their values until the next assignment. Hence variables may correspond to registers in the hardware implementation. Alternatively, they may correspond to wires, when the information they carry is readily consumed. The hardware implementation of variables is decided upon at later stages of architec-tural synthesis. It is often convenient to rename instances of variables, so that each instance has a single assignment and, of course, to resolve the references appropriately. A scheme for variable renaming is presented in reference [11].

An important issue for hardware synthesis is to propagate the hardware con-straints specified implicitly or explicitly in the HDL models. Let us consider first the timing interpretation of the model. Some synthesis policies interpret `wait` state-ments as boundaries on the time frame of the operations. In this case, the sequencing graph derived from the HDL model has a timing annotation for the start time of the operations. It is called a *scheduled sequencing graph*. Alternatively, explicit timing constraints may be specified by labeling operations and providing the means for expressing minimum and maximum timing separations between the start times of pairs of operations. Such constraints are annotated to the model and provide constraints for architectural optimization (Section 5.3.3). In this case, as well as when no timing constraints are specified, the schedule of the sequencing graph can be optimized by specialized algorithms, such as those described in Chapter 5.

Constraints may also link operations to hardware operators. For example, the model may require the use of a particular adder implementation for a specific addition. Such constraints can be seen as hints to be followed by hardware synthesis tools. Constraints will be described in detail in Chapter 4.

3.4.2 Optimization Techniques

Behavioral optimization is a set of semantic-preserving transformations that minimize the amount of information needed to specify the partial order of tasks. No knowledge about the circuit implementation style is required at this stage. The latitude of applying such optimization depends on the freedom to rearrange the intermediate code. Therefore, models that are highly constrained to adhere to a time schedule or to an operator binding may benefit very little from the following techniques.

Behavioral optimization can be implemented in different ways. It can be applied directly to the parse trees, or during the generation of the intermediate form, or even on the intermediate form itself, according to the different cases. For the sake of explanation, we consider here these transformations as applied to sequences of statements, i.e., as program-level transformations.

We review here the major techniques that are applicable to a broad class of circuits. We do not consider those methods that are specific of particular applications, such as concurrency-enhancing transformations for DSP design. Algorithms for behavioral optimization of HDL models can be classified as data-flow and control-flow oriented. The former group resembles most the transformations applied in software compilers. They rely on global data-flow analysis of the intermediate form.

Before delving into the description of the transformations, it is important to stress the different impact of assignments that use arithmetic versus logic operators. Hardware arithmetic operators (e.g., adders, multipliers) have non-negligible cost, and very often they accept at most two operands. Therefore, arithmetic assignments with more than two inputs have to be split into two-input ones, as shown below. Conversely, complex logic gates can implement expressions with several logic operators and inputs. As a result, logic assignments can be treated as single entities. Optimization of logic expression is best performed by logic-level optimization algorithms, such as those presented in Chapter 8.

DATA-FLOW-BASED TRANSFORMATIONS. These transformations are dealt with in detail in most books of software compiler design [1, 14, 24].

Tree-height reduction. This transformation applies to the arithmetic expression trees and strives to achieve the expression split into two-operand expressions, so that the parallelism available in hardware can be exploited at best. It can be seen as a local transformation, applied to each compound arithmetic statement, or as a global transformation, applied to all compound arithmetic statements in a basic block. Enough hardware resources are postulated to exploit all parallelism. If this is not the case, the gain of applying the transformation is obviously reduced.

> **Example 3.4.3.** Consider the following arithmetic assignment: $x = a + b + c + d$; this can be trivially split as $x = a + b$; $x = x + c$; $x = x + d$;. It requires three additions in series. Alternatively, the following split can be done: $p = a + b$; $q = c + d$; $x = p + q$;, where the first two additions can be done in parallel if enough resources are available (in this case, two adders). The second choice is better than the first one, because the corresponding implementation cannot be inferior for any possible resource availability.

Tree-height reduction was studied in depth as an optimization scheme for software compilers [14]. It is used in hardware compilers mainly as a local transformation, because of the limited parallelism in basic blocks. Tree-height reduction exploits some properties of the arithmetic operations, with the goal of balancing the expression tree as much as possible. In the best case, the tree height is $O(\log n_{ops})$ for n_{ops} operations, and the height is proportional to a lower bound on the overall computation time.

The simplest reduction algorithm uses the commutativity and associativity of the addition and multiplication. It permutes the operands to achieve subexpressions involving the same operator, which can be reduced by using the associative property.

> **Example 3.4.4.** Consider the following arithmetic assignment: $x = a + b * c + d$;. By using commutativity of addition, we get $x = a + d + b * c$;, and by associativity $x = (a + d) + b * c$;. The transformation is shown in Figure 3.18.

A further refinement can be achieved by exploiting the distributive property, possibly at the expense of adding an operation.

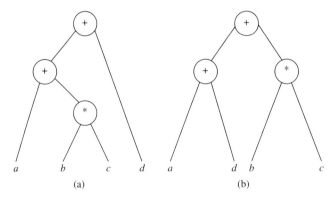

FIGURE 3.18
Example of tree-height reduction using commutativity and distributivity.

Example 3.4.5. Consider the following arithmetic assignment: $x = a*(b*c*d+e)$;. By using commutativity of addition, no reduction in height is possible. By using distributivity instead, we can write $x = a*b*c*d + a*e$;, which has a tree of height 3 and one additional operation. The transformation is shown in Figure 3.19. Note that two multipliers are necessary for reducing the computation time by this transformation.

Other tree-height reduction algorithms can exploit expression factorization techniques. We refer the interested reader to Kuck's textbook [14] for the computation of bounds on the tree height and for further details.

Constant and variable propagation.   Constant propagation, also called constant folding, consists of detecting constant operands and pre-computing the value of the operation with that operand. Since the result may be again a constant, the new constant can be propagated to those operations that use it as input.

Example 3.4.6. Consider the following fragment: $a = 0$; $b = a + 1$; $c = 2*b$;. It can be replaced by $a = 0$; $b = 1$; $c = 2$;.

Variable propagation, also called copy propagation, consists of detecting the *copies* of variables, i.e., the assignments like $x = y$, and using the right-hand side in the following references in place of the left-hand side. Data-flow analysis permits the identification of the statements where the transformation can be done. In particular, the propagation of y cannot be done after a different reassignment to x. Variable propagation gives the opportunity to remove the copy assignment. Note that copy assignments may have been introduced by other transformations.

Example 3.4.7. Consider the following fragment: $a = x$; $b = a + 1$; $c = 2*a$;. It can be replaced by $a = x$; $b = x + 1$; $c = 2*x$;. Statement $a = x$; may then be removed by dead code elimination, if there are no further references to a.

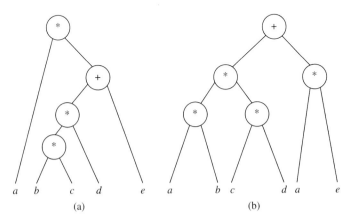

(a)   (b)

FIGURE 3.19
Example of tree-height reduction using the distributive property.

Common subexpression elimination. The search for common logic subexpressions is best done by logic-level optimization algorithms, described in Section 8.3. The search for common arithmetic subexpressions relies in general on finding isomorphic patterns in the parse trees. This step is greatly simplified if the arithmetic expressions are reduced to two-input ones. Then, this transformation consists of selecting a target arithmetic operation and searching for a preceding one of the same type and with the same operands. Operator commutativity can be exploited. Again, data-flow analysis is used in the search to ensure that in any matching expression the operands always take the same values. When a preceding matching expression is found, the target expression is replaced by a copy of the variable which is the result of that preceding matching expression.

> **Example 3.4.8.** Consider the following fragment: $a = x + y$; $b = a + 1$; $c = x + y$;. It can be replaced by $a = x + y$; $b = a + 1$; $c = a$;. Note that a variable copy has been introduced for variable a that can be propagated in the subsequent code.

Dead code elimination. Dead code consists of all those operations that cannot be reached, or whose result is never referenced elsewhere. Such operations are detected by data-flow analysis and removed. Obvious cases are those statements following a procedure `return` statement. Less obvious cases involve operations that just precede a `return` statement and whose results are not parameters of the procedure or do not affect any of its parameters.

> **Example 3.4.9.** Consider the following fragment: $a = x$; $b = x + 1$; $c = 2 * x$;. If variable a is not referenced in the subsequent code, the first assignment can be removed.

Operator strength reduction. Operator strength reduction means reducing the cost of implementing an operator by using a simpler one. Even though in principle some notion of the hardware implementation is required, very often general considerations apply. For example, a multiplication by 2 (or by a power of 2) can be replaced by a shift. Shifters are always faster and smaller than multipliers in many implementations.

> **Example 3.4.10.** Consider the following fragment: $a = x^2$; $b = 3 * x$;. It can be replaced by $a = x * x$; $t = x << 1$; $b = x + t$;.

Code motion. Code motion often applies to loop invariants, i.e., quantities that are computed inside an iterative construct but whose values do not change from iteration to iteration. The goal is to avoid the repetitive evaluation of the same expression.

> **Example 3.4.11.** Consider the following iteration construct: `for` $(i = 1; i \leq a * b)\{ \ldots \}$ where variables a and b are not updated in the loop. It can be transformed to $t = a * b$; `for` $(i = 1; i \leq t)\{ \ldots \}$.

CONTROL-FLOW-BASED TRANSFORMATIONS. The following transformations are typical of hardware compilers. In some cases these transformations are automated, in others they are user driven.

Model expansion. Writing structured models by exploiting subroutines and functions is useful for two main reasons: modularity and re-usability. Modularity helps in highlighting a particular task (or set of tasks). Often, models are called only once.

Model expansion consists in flattening locally the model call hierarchy. Therefore the called model disappears, being swallowed by the calling one. A possible benefit is that the scope of application of some optimization techniques (at different levels) is enlarged, yielding possibly a better final circuit. If the expanded model was called only once, there is no negative counterpart. Nevertheless, in the case of multiple calls, a full expansion leads to an increase in the size of the intermediate code and to the probable loss of the possibility of hardware sharing.

> **Example 3.4.12.** Consider the following fragment: $x = a + b;$ $y = a * b;$ $z = foo(x, y);$, where $foo(p, q)\{t = q - p;$ `return`$(t); \}$. Then by expanding foo, we have $x = a + b;$ $y = a * b;$ $z = y - x;$.

Conditional expansion. A conditional construct can be always transformed into a parallel construct with a test in the end. Under some circumstances this transformation can increase the performance of the circuit. For example, this happens when the conditional clause depends on some late-arriving signals. Unfortunately this transformation precludes some possibilities for hardware sharing, because the operations in all bodies of the branching construct have to be performed.

A special case applies to conditionals whose clauses and bodies are evaluations of logic functions. Then, the conditional expansion is favorable because it allows us to expand the scope of application of logic optimization.

> **Example 3.4.13.** Consider the following fragment: $y = ab;$ `if` (a) $\{x = b + d; \}$ `else` $\{x = bd; \}$. The conditional statement can be flattened to $x = a(b+d)+a'bd;$ and by some logic manipulation, the fragment can be rewritten as $y = ab;$ $x = y+d(a+b);$.

Loop expansion. Loop expansion, or unrolling, applies to an iterative construct with data-independent exit conditions. The loop is replaced by as many instances of its body as the number of operations. The benefit is again in expanding the scope of other transformations. Needless to say, when the number of iterations is large, unrolling may yield a large amount of code.

> **Example 3.4.14.** Consider the following fragment: $x = 0;$ `for` $(i = 1; i \leq 3; i++)$ $\{x = x + a[i]; \}$. The loop can be flattened to $x = 0;$ $x = x+a[1];$ $x = x+a[2];$ $x = x + a[3];$ and then transformed to $x = a[1] + a[2] + a[3]$ by propagation.

Other transformations on loops are possible, such as moving the evaluation of the iterative clause from the top to the bottom of the loop [23].

Block-level transformations. Branching and iterative constructs segment the intermediate code into basic blocks. Such blocks correspond to the sequencing graph entities. Trickey [23] studied the possibility of manipulating the size of the blocks by means of block-level transformations that include block merging and expansions of conditionals and loops. Even though he did not consider model expansion, the extension is straightforward. He assumed that operations in different blocks cannot overlap execution and that concurrency is limited only by the amount of hardware resources available for parallel execution in each block.

Therefore, collapsing blocks may provide more parallelism and enhance the average performance. To find the optimum number of expansions to be performed, he proposed five transformations, with rules to measure the expected performance improvement. The rules can then be applied bottom up in the hierarchy induced by the control-flow hierarchy of the model. He proposed a linear-time dynamic programming algorithm that returns an optimum block-level structure. Unfortunately, the optimality is weakened by the assumptions on the model and on the transformation rules. We refer the interested reader to reference [23] for further details.

3.5 PERSPECTIVES

Hardware modeling by means of HDLs has changed the way designers think of circuits. Schematic entry tools have been replaced by CAD systems that support HDL specifications and synthesis. In other words, hardware design looks today closer to software programming. The move from gate-level to architectural-level modeling can be paralleled to the change from programming in assembly code to high-level software languages. Object-oriented methods for hardware specification have been proposed, and the unifying trend between hardware and software design is unavoidable, especially when considering that system design requires specifying both hardware and software components.

The HDLs reviewed in this chapter serve the purpose of providing a focal point for the transfer of information between designers and CAD systems. Unfortunately this focal point is not sharp enough, due to the lack of a fully specified hardware semantics of most languages. Whereas users of CAD systems would like to see a standardization of languages and formats, as well as of synthesis policies, CAD researchers and providers are interested in the competitive edge given by supporting increasingly advanced features of modeling languages. This issue applies especially well to the VHDL language, which is very broad in scope.

It is conceivable that subsets of presently used HDLs will become standard formats for information exchange at the register-transfer level, while new languages will emerge to capture hardware behavior and be used as the starting point for synthesis. More importantly, languages and environments will be developed to support design conceptualization and verification. In other words, such languages and environments will have to make designs transparent to their conceivers, so that they are sure that what they have modeled is what they want. Transparent and efficient modeling is a co-requisite, together with effective synthesis tools, for rapid design of high-quality circuits.

3.6   REFERENCES

There are several sources of information for hardware description languages, including the original manuals. We suggest the books by Lipsett *et al.* [15], Navabi [18], Perry [20] and Armstrong [2] on VHDL, by Thomas and Moorby [22] on Verilog and the original paper by Hilfinger [8] for Silage. Silage is also described in a chapter of Gajski's edited book on silicon compilation [12]. Karatsu [10] describes UDL/I.

General issues on compilers are described in Aho *et al.*'s textbook [1] and in Watson's [24]. Internal models for hardware synthesis are described by several sources [e.g., 13]. The Value Trace was one of the first data structures used for architectural synthesis and is described by McFarland [17]. Other intermediate models are described in reference [5]. Techniques for behavioral-level optimization are fairly standard and are borrowed from the world of software compilers. A good reference is again reference [1]. Tree-height reduction is described extensively in Kuck's book [14].

1. A. Aho, R. Sethi and J. Ullman, *Compilers: Principles, Techniques and Tools*, Addison-Wesley, Reading, MA, 1988.
2. J. R. Armstrong, *Chip-Level Modeling with VHDL*, Prentice-Hall, Englewood Cliffs, NJ, 1989.
3. G. Colon-Bonet, E. Schwartz, D. Bostick, G. Hachtel and M. Ligthner, "On Optimal Extraction of Combinational Logic and Don't Care Sets from Hardware Description Languages," *ICCAD, Proceedings of the International Conference on Computer Aided Design*, pp. 308–311, 1989.
4. S. Carlson, *Introduction to HDL-based Design Using VHDL*, Synopsys, Mountain View, CA, 1990.
5. S. Hayati, A. Parker and J. Granacki, "Representation of Control and Timing Behavior with Applications to Interface Synthesis," *ICCD, Proceedings of the International Conference on Computer Design*, pp. 382–387, 1988.
6. E. Frey, "ESIM: A Functional-Level Simulation Tool," *ICCAD, Proceedings of the International Conference on Computer Aided Design*, pp. 48–50, 1984.
7. D. Harel, "Statecharts: A Visual Formalism for Complex Systems," *Science of Computer Programming*, No. 8, pp. 231–274, 1987.
8. P. Hilfinger, "A High-Level Language and Silicon Compiler for Digital Signal Processing," *CICC, Proceedings of the Custom Integrated Circuit Conference*, pp. 213–216, 1985.
9. F. J. Hill and G. R. Peterson, *Switching Theory and Logical Design*, Wiley, New York, 1981.
10. O. Karatsu, "VLSI Design Language Standardization in Japan," *DAC, Proceedings of the Design Automation Conference*, pp. 50–55, 1989.
11. D. Ku and G. DeMicheli, *High-Level Synthesis of ASICs under Timing and Synchronization Constraints*, Kluwer Academic Publishers, Boston, MA, 1992.
12. D. Gajski, Editor, *Silicon Compilation*, Addison-Wesley, Reading, MA, 1988.
13. D. Gajski, N. Dutt, A. Wu and S. Lin, *High-Level Synthesis*, Kluwer Academic Publishers, Boston, MA, 1992.
14. D. Kuck, *The Structure of Computers and Computation*, Wiley, New York, 1978.
15. R. Lipsett, C. Schaefer and C. Ussery, *VHDL: Hardware Description and Design*, Kluwer Academic Publishers, Boston, MA, 1991.
16. P. Michel, U. Lauther and P. Duzy, *The Synthesis Approach to Digital System Design*, Kluwer Academic Publishers, Boston, MA, 1992.
17. M. J. McFarland, "Value Trace," Carnegie-Mellon University Internal Report, Pittsburgh, PA, 1978.
18. Z. Navabi, *VHDL: Analysis and Modeling of Digital Systems*, McGraw-Hill, New York, 1993.
19. P. Paulin and J. Knight, "Force-directed Scheduling for the Behavioral Synthesis of ASIC's," *IEEE Transactions on CAD/ICAS*, Vol. CAD-8, No. 6, pp. 661–679, July 1989.
20. D. Perry, *VHDL*, McGraw-Hill, New York, 1991.
21. J. Peterson, *Petri Net Theory and Modeling of Systems*, Prentice-Hall, Englewood Cliffs, NJ, 1981.
22. D. Thomas and P. Moorby, *The Verilog Hardware Description Language*, Kluwer Academic Publishers, Boston, MA, 1991.
23. H. Trickey, "Flamel: A High-Level Hardware Compiler," *IEEE Transactions on CAD/ICAS*, Vol. CAD-6, No. 2, pp. 259–269, March 1987.
24. D. Watson, *High-level Languages and their Compilers*, Addison-Wesley, Reading, MA, 1989.

3.7   PROBLEMS

1. Design a model of a 1-bit full-adder in structural VHDL, incorporating the half-adder model of Example 3.2.1. Include the interface description.

2. Design a model of a 4-bit adder in behavioral and in data-flow VHDL. Include the interface description.

3. Design a model of the recursive filter of Example 3.2.4 in behavioral VHDL.

4. Consider a circuit that returns the integer addition of three 8-bit numbers, i.e., whose behavior is $x = a + b + c$. Do not consider the carry. You have available only 1-bit full-adder cells. Design a structural model of the circuit using 15 full-adders, and model it in your favorite HDL using metavariables. Show a block diagram and draw a bipartite graph representing the module/net structure.

5. Consider the following model fragment:

$$x = ac + de$$

$$y = a + b$$

$$w = p + a$$

$$z = q + b$$

$$p = ac + ad + e$$

$$q = ap + b$$

Draw the corresponding logic network graph, while considering a, b, c, d, e as primary input variables and x, y, w, z as primary output variables.

6. Consider a circuit that solves numerically (by means of the *backward Euler method*) the following differential equation: $y'' + 3xy' + 3 = 0$ in the interval $[0, a]$ with step size dx and initial values $x(0) = x$; $y(0) = y$; $y'(0) = u$. Design a behavioral model in the VHDL language. Then draw the data-flow graph of the inner loop. Eventually, sketch the sequencing graph for the entire model.

7. Design a model of a circuit (in your favorite HDL) that reads two numbers, computes their greatest common divisor by Euclid's algorithm and returns this value. The circuit should operate every time a signal `start` is raised. Then draw the corresponding sequencing graph.

8. Consider the data-flow graph of Figure 3.11. Apply the following optimization techniques: (i) operator strength reduction; (ii) common subexpression elimination. Draw the optimized data-flow graph.

PART
II

ARCHITECTURAL-
LEVEL
SYNTHESIS
AND
OPTIMIZATION

This part deals with synthesis and optimization of circuits at the architectural level. In particular, it describes techniques for transforming an abstract model of circuit behavior into a data path and a control unit. The data path is an interconnection of resources whose execution times and input/output data are determined by the control unit according to a schedule.

We survey the major tasks of architectural-level optimization in Chapter 4 as well as the data path and control synthesis. We devote Chapters 5 and 6 to the detailed description of algorithms for scheduling and resource sharing and binding. Thus Chapter 4 serves the purposes of both introducing architectural-level synthesis and optimization and summarizing it to the reader who intends to skip Chapters 5 and 6.

CHAPTER

4

ARCHITECTURAL SYNTHESIS

Et verbum caro factum est.
And the Word was made flesh.

Gospel according to St. John.

4.1 INTRODUCTION

Architectural synthesis means constructing the macroscopic structure of a digital circuit, starting from behavioral models that can be captured by data-flow or sequencing graphs. The outcome of architectural synthesis is both a structural view of the circuit, in particular of its data path, and a logic-level specification of its control unit. The data path is an interconnection of *resources* (implementing arithmetic or logic functions), *steering logic circuits* (e.g., multiplexers and busses), that send data to the appropriate destination at the appropriate time and *registers* or *memory arrays* to store data. An example of one possible macroscopic structure of the differential equation integrator of Example 1.5.3 is shown in Figure 4.1.

Architectural synthesis may be performed in many ways, according to the desired circuit implementation style. Therefore a large variety of problems, algorithms and tools have been proposed that fall under the umbrella of architectural synthesis. To

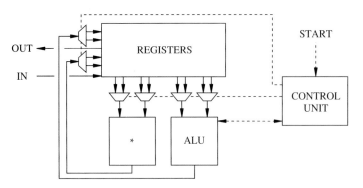

FIGURE 4.1
Structural view of the differential equation integrator with one multiplier and one ALU.

be more specific, we address here synthesis problems for synchronous, mono-phase digital circuits.

Circuit implementations are evaluated on the basis of the following objectives: *area, cycle-time* (i.e., the clock period) and *latency* (i.e., the number of cycles to perform all operations) as well as *throughput* (i.e., the computation rate) in the case of pipelined circuits. Worst-case bounds on area and on performance may be imposed to exclude undesirable implementations, for example, low-performance or large circuits. In addition, the circuit structure may be constrained to using some pre-specified units for some operations or specified ports for input/output (I/O) data. Note that area and performance measures can only be estimated at the architectural level, because just the macroscopic structure of the circuit is dealt with. In general, area and performance depend on the resources as well as on the steering logic, storage circuits, wiring and control. A common simplification is to consider area and performance as depending only on the resources. Circuits for which this assumption holds are called *resource-dominated* circuits and are typical of some application fields, such as digital signal processing (DSP).

The *design space*, introduced in Chapter 1, is the collection of all feasible structures corresponding to a circuit specification. The Pareto points are those points of the design space that are not dominated by others in all objectives of interest. Hence they represent the spectrum of implementations of interest to designers. Their image determines the trade-off curves in the design evaluation space. An example of a design evaluation space spanned by the objectives (*area, latency, cycle-time*) was reported in Chapter 1 as Figure 1.19. Other examples in this chapter are Figures 4.9, 4.10 and 4.11. Figure 4.23 shows the design evaluation space for pipelined circuits.

Realistic design examples have trade-off curves that are not smooth [13], because of two reasons. First, the design space is a finite set of points, since the macroscopic structure of a circuit has a coarse grain. For example, a hypothetical circuit may have one or two multipliers, and its area would jump in correspondence of this choice. Second, there are several non-linear effects that are compounded in determining the objectives as a function of the structure of the circuit. Due to the lack of compactness of the design space and of smoothness of the design evaluation space, architectural optimization problems are hard and their solution relies in general on solving some related subproblems.

Architectural exploration consists of traversing the design space and providing a spectrum of feasible non-inferior solutions, among which a designer can pick the desired implementation. Exploration requires the solution of constrained optimization problems. Architectural synthesis tools can select an appropriate design point, according to some user-specified criterion, and construct the corresponding data path and control unit.

It is the goal of this chapter to give an overview of the problems in architectural synthesis and optimization. We consider first circuit modeling in more detail and then architectural optimization problems for non-pipelined circuits, including scheduling and resource sharing. We comment on the interdependency between these problems, but we defer the detailed descriptions of the algorithms for scheduling and resource binding to Chapters 5 and 6, respectively. We describe then the synthesis of data paths and control units in detail. Eventually we consider architectural synthesis for pipelined circuits.

4.2 CIRCUIT SPECIFICATIONS FOR ARCHITECTURAL SYNTHESIS

Specifications for architectural synthesis include behavioral-level circuit models, details about the resources being used and constraints. Behavioral models are captured by sequencing graphs that were described in Section 3.3.4. Thus we comment here on resources and constraints in detail.

4.2.1 Resources

Resources implement different types of functions in hardware. They can be broadly classified as follows:

- *Functional resources* process data. They implement arithmetic or logic functions and can be grouped into two subclasses:
  - *Primitive resources* are subcircuits that are designed carefully once and often used. Examples are arithmetic units and some standard logic functions, such as encoders and decoders. Primitive resources can be stored in libraries. Each resource is fully characterized by its area and performance parameters.
  - *Application-specific resources* are subcircuits that solve a particular subtask. An example is a subcircuit servicing a particular interrupt of a processor. In general such resources are the implementation of other HDL models. When synthesizing hierarchical sequencing graph models bottom up, the implementation of the entities at lower levels of the hierarchy can be viewed as resources.
- *Memory resources* store data. Examples are registers and read-only and read-write memory arrays. Requirement for storage resources are implicit in the sequencing graph model. In some cases, access to memory arrays is modeled as transfer of data across circuit ports.

• *Interface resources* support data transfer. Interface resources include busses that may be used as a major means of communication inside a data path. External interface resources are I/O pads and interfacing circuits.

The major decisions in architectural synthesis are often related to the usage of functional resources. As far as formulating architectural synthesis and optimization problems, there is no difference between primitive and application-specific functional resources. Both types can be characterized in terms of area and performance and used as building blocks. Nevertheless, from a practical point of view, the second class requires either full top-down synthesis to derive the area/performance parameters or the use of estimation techniques. For example, the area and performance can be estimated from the complexity of the logic description of the corresponding circuit, while neglecting the wiring space.

Some circuit models require performing standard arithmetic operations with application-specific word lengths. Examples are bit-parallel digital filters, requiring arithmetic units with non-standard size. It would not be practical to store in libraries primitive resources for several choices of word lengths. However, due to the regularity of some implementations of arithmetic functions, *module generators* can be used to derive the layout of the resources and to evaluate their performance. This approach allows us to characterize precisely a class of application-specific resources.

When architectural synthesis targets synchronous circuit implementations, as is often the case and as considered in this book, it is convenient to measure the performance of the resources in terms of cycles required to execute the corresponding operation, which we call the *execution delay* of the resource. While this datum is part of the specification for any sequential resource, it can be easily computed for any combinational resource by rounding up the quotient of the *propagation delay* to the cycle-time. It is obvious that the execution delay of all resources depends on the cycle-time. We shall also refer to the execution delay of an operation as the execution delay of a resource performing that operation.

It is interesting to note that DSP circuits frequently employ primitive resources, such as arithmetic units. Often DSP circuits are resource dominated, because the area and performance of the resources dominate those of the steering and control logic as well as registers and wiring. Therefore good estimates of the entire structure can be inferred by the resource usage. In this case, architectural synthesis is mainly related to finding the best interconnection among few, well-characterized primitives.

Conversely, some ASICs exploit several application-specific resources. In the limit, each operation may require a different type of resource. The area and delays of the steering and control logic may be comparable to those of the resources. Then, the estimation of the area and performance is of paramount importance in the synthesis of good macroscopic structures.

Some circuit models can be implemented by using one general purpose ALU to perform all operations. In this case, synthesis is bound to one single resource and consists of determining the control sequence to the ALU. In this case architectural synthesis reduces to control synthesis and it is similar to *micro-compilation*.

4.2.2 Constraints

Constraints in architectural synthesis can be classified into two major groups: *interface constraints* and *implementation constraints.*

Interface constraints are additional specifications to ensure that the circuit can be embedded in a given environment. They relate to the format and timing of the I/O data transfers. The data format is often specified by the interface of the model. The timing separation of I/O operations can be specified by timing constraints that can ensure that a synchronous I/O operation follows/precedes another one by a prescribed number of cycles in a given interval. Such timing constraints are described in Section 5.3.3. Timing constraints can also specify *data rates* for pipelined systems.

Implementation constraints reflect the desire of the designer to achieve a structure with some properties. Examples are area constraints and performance constraints, e.g., *cycle-time* and/or *latency* bounds.

A different kind of implementation constraint is a resource binding constraint. In this case, a particular operation is required to be implemented by a given resource. These constraints are motivated by the designer's previous knowledge, or intuition, that one particular choice is the best and that other choices do not need investigation. Architectural synthesis with resource binding constraints is often referred to as *synthesis from partial structure* [9]. Design systems that support such a feature allow a designer to specify a circuit in a wide spectrum of ways, ranging from a full behavioral model to a structural one. This modeling capability may be useful to leverage previously designed components.

> **Example 4.2.1.** A processor design may require several addition operations, e.g., in the instruction fetch unit, in the ALU, in the memory address computation unit, etc. Assume that the instruction fetch unit always adds 4 to the operand and that the performance of this addition is critical. Then, the designer may want to identify a specific fast adder for this task, with 2 bits less than the word size. Such a resource may be pre-bound to the instruction fetch unit, and not used anywhere else in the circuit, so that no penalty for data steering is compounded to the adder delay.

In the case of resource-dominated circuits, the area is determined only by the resource usage. Hence bounding the number of resources is equivalent to imposing an upper bound on the area. For this reason architectural optimization algorithms, such as scheduling under resource constraints, are often used to explore the design space.

Sometimes resource usage constraints are used in synthesis and optimizations of circuits that are not resource dominated. Then, even though such constraints may not ensure bounds on area, they can still be valuable as heuristics in steering the solution toward a desired goal or bound.

> **Example 4.2.2.** A circuit design must execute an algorithm involving a stream of bit-parallel double-precision floating point additions and multiplications. For the sake of this example, we consider the circuit as not resource dominated. The fastest implementation is sought for, e.g., the minimal-latency structure for a given cycle-time that fits the effective chip area (excluding I/Os). To achieve the required performance level in a

given technology, a multiplier requires slightly more than half of the effective chip area. Then an obvious resource constraint is to limit the multipliers to one.

In a scaled-down technology, such a multiplier requires slightly more than one-third of the effective chip area. While reasonable multiplier bounds are 1 and 2, a designer may choose 1 as a bound, thinking that it is unlikely that a chip with two multipliers and the associated wiring may fit the area bound and perform better than a circuit with one multiplier. The designer's experience suggests that such systems have no more than one multiplier. It is very likely that he or she is correct, but there is no guarantee.

4.3 THE FUNDAMENTAL ARCHITECTURAL SYNTHESIS PROBLEMS

We consider now the fundamental problems in architectural synthesis and optimization. We assume that a circuit is specified by:

- A sequencing graph.
- A set of functional resources, fully characterized in terms of area and execution delays.
- A set of constraints.

We assume for now that storage is implemented by registers and interconnections by wires. Usage of internal memory arrays and busses will be described in Sections 6.2.4 and 6.2.5, respectively.

For the sake of explanation, we shall consider next non-hierarchical graphs with operations having bounded and known execution delays and present then extensions to hierarchical models and unbounded delays in Sections 4.3.3 and 4.3.4, respectively. We assume that there are n_{ops} operations. Sequencing graphs are polar and acyclic, the source and sink vertices being labeled as v_0 and v_n, respectively, where $n = n_{ops} + 1$. Hence the graph $G_s(V, E)$ has vertex set $V = \{v_i; \ i = 0, 1, \ldots, n\}$ in one-to-one correspondence with the set of operations and edge set $E = \{(v_i, v_j); \ i, j = 0, 1, \ldots, n\}$ representing dependencies. An example is shown in Figure 4.2.

Architectural synthesis and optimization consists of two stages. First, placing the operations in time and in space, i.e., determining the time interval for their execution and their binding to resources. Second, determining the detailed interconnections of the data path and the logic-level specifications of the control unit.

We show now that the first stage is equivalent to annotating the sequencing graph with additional information.

4.3.1 The Temporal Domain: Scheduling

We denote the execution delays of the operations by the set $D = \{d_i; \ i = 0, 1, \ldots, n\}$. We assume that the delay of the source and sink vertices is zero. We define the *start time* of an operation as the time at which the operation starts its execution. The start times of the operations, represented by the set $T = \{t_i; \ i = 0, 1, \ldots, n\}$, are attributes of the vertices of the sequencing graph. *Scheduling* is the task of determining the start times, subject to the precedence constraints specified by the sequencing graph. The

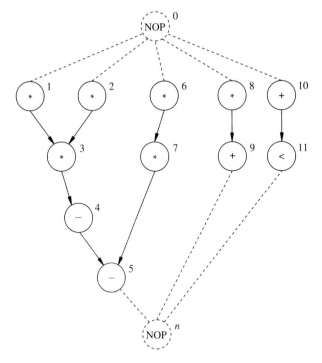

FIGURE 4.2
Sequencing graph.

latency of a scheduled sequencing graph is denoted by λ, and it is the difference between the start time of the sink and the start time of the source, i.e., $\lambda = t_n - t_0$.

Definition 4.3.1. A **schedule** of a sequencing graph is a function $\varphi : V \to Z^+$, where $\varphi(v_i) = t_i$ denotes the operation start time such that $t_i \geq t_j + d_j, \forall \, i, j : (v_j, v_i) \in E$.

A *scheduled sequencing graph* is a vertex-weighted sequencing graph, where each vertex is labeled by its start time. A schedule may have to satisfy timing and/or resource usage constraints. Different scheduling algorithms have been proposed, addressing unconstrained and constrained problems. They are described in detail in Chapter 5.

Example 4.3.1. An example of a sequencing graph, modeling the inner loop of the differential equation integrator, was shown in Figure 3.12. It is reported again in Figure 4.2 for convenience. All operations are assumed to have unit execution delay. A scheduled sequencing graph is shown in Figure 4.3. The start time of the operations is summarized by the following table. The latency of the schedule is $\lambda = t_n - t_0 = 5 - 1 = 4$.

Operation	Start time
$v_1, v_2, v_6, v_8, v_{10}$	1
v_3, v_7, v_9, v_{11}	2
v_4	3
v_5	4

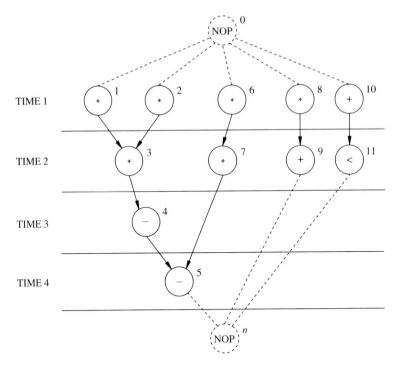

FIGURE 4.3
Scheduled sequencing graph.

Example 4.3.2. Consider again the sequencing graph of Figure 4.2, where all operations have unit execution delay. A schedule with a bound on the resource usage of one resource per type is the following:

Operation		Start time
Multiply	**ALU**	
v_1	v_{10}	1
v_2	v_{11}	2
v_3	-	3
v_6	v_4	4
v_7	-	5
v_8	v_5	6
-	v_9	7

The scheduled sequencing graph is shown in Figure 4.4. The latency of the schedule is $\lambda = t_n - t_0 = 8 - 1 = 7$.

The scheduling formulation can be extended by considering the propagation delays of the combinational resources instead of the integer execution delays. Thus two (or more) combinational operations in a sequence can be *chained* in the same execution cycle if their overall propagation delay does not exceed the cycle-time. This approach can be further extended to chains of resources whose overall delay spans

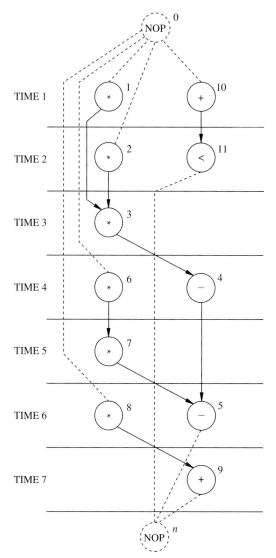

FIGURE 4.4
Scheduled sequencing graph under resource constraints.

more than one cycle. *Scheduling with chaining* can provide tighter schedules in some cases, but constrained schedules are harder to compute.

Example 4.3.3. Consider again the sequencing graph of Figure 4.2, where the propagation delay of a multiplication is 35 nsec and that of the other operations is 25 nsec. Assume a 50-nsec cycle-time.

The usual scheduling formulation is based on computing the execution delays as the rounded-up quotient of the propagation delays to the cycle-time. Thus, all operations have unit execution delay, and an unconstrained schedule is the same as that shown in Figure 4.3, with a latency of 4 cycles.

However, by allowing the schedule to chain operations $\{v_4, v_5\}$ and $\{v_{10}, v_{11}\}$ in a single cycle, a latency of 3 cycles can be achieved.

4.3.2 The Spatial Domain: Binding

Let us consider now the relations among operations and resources. We define the *type* of an operation as the type of computation it performs. It may be an arithmetic operation, such as addition or multiplication, or an application-specific operation, such as evaluating a Boolean function. We can extend the notion of type to functional resources. A covering relation can be defined among types to represent the fact that a resource type can implement more than one operation type. For example, the resource-type ALU may cover operation types {*addition, subtraction, comparison*}. Obviously, a feasible implementation requires that the resource types available cover the operation types.

We call a *resource-type set* the set of resource types. For the sake of simplicity, we identify the resource-type set with its enumeration. Thus, assuming that there are n_{res} resource types, we denote the resource-type set by $\{1, 2, \ldots, n_{res}\}$. The function $\mathcal{T} : V \rightarrow \{1, 2, \ldots, n_{res}\}$ denotes the resource type that can implement an operation. It is obvious that *No-Operations* do not require any binding to any resource. Therefore, when referring to a binding, we drop them from consideration, including the source and sink vertices.

It is interesting to note that there may be more than one operation with the same type. In this case, *resource sharing* may be applied, as described later in this section. Conversely, the binding problem can be extended to a *resource selection* (or *module selection*) problem by assuming that there may be more than one resource applicable to an operation (e.g., a ripple-carry and a carry-look-ahead adder for an addition). In this case $\mathcal{T}$ is a one-to-many mapping. We shall describe the module selection problem in Section 6.6. We assume for now that $\mathcal{T}$ is a single-valued function.

A fundamental concept that relates operations to resources is *binding*. It specifies which resource implements an operation.

Definition 4.3.2. A **resource binding** is a mapping $\beta : V \rightarrow R \times Z^+$, where $\beta(v_i) = (t, r)$ denotes that the operation corresponding to $v_i \in V$, with type $\mathcal{T}(v_i) = t$, is implemented by the rth instance of resource type $t \in R$ for each $i = 1, 2, \ldots, n_{ops}$.

A simple case of binding is a *dedicated resource*. Each operation is bound to one resource, and the resource binding β is a one-to-one function.

Example 4.3.4. Consider the scheduled sequencing graph of Figure 4.3. There are 11 operations. Assume that 11 resources are available. In addition, assume that the resource types are {*multiplier, ALU*}, where the ALU can perform addition, subtraction and comparisons. We label the multiplier as type 1, the ALUs as type 2. Thus $\mathcal{T}(v_1) = 1$, $\mathcal{T}(v_4) = 2$, etc. We need six instances of the multiplier type and five instances of the

ALU type. The following table shows the binding function:

$\beta(v_1)$	$(1,1)$
$\beta(v_2)$	$(1,2)$
$\beta(v_3)$	$(1,3)$
$\beta(v_4)$	$(2,1)$
$\beta(v_5)$	$(2,2)$
$\beta(v_6)$	$(1,4)$
$\beta(v_7)$	$(1,5)$
$\beta(v_8)$	$(1,6)$
$\beta(v_9)$	$(2,3)$
$\beta(v_{10})$	$(2,4)$
$\beta(v_{11})$	$(2,5)$

A resource binding may associate one instance of a resource type to more than one operation. In this case, that particular resource is *shared* and binding is a many-to-one function. A necessary condition for a resource binding to produce a valid circuit implementation is that the operations corresponding to a shared resource do not execute concurrently.

Example 4.3.5. It is obvious that the resource usage of the previous example is not efficient. Indeed only four multipliers and two ALUs are required by the scheduled sequencing graph of Figure 4.3. This is shown in Figure 4.5. The tabulation of the binding is the following:

$\beta(v_1)$	$(1,1)$
$\beta(v_2)$	$(1,2)$
$\beta(v_3)$	$(1,2)$
$\beta(v_4)$	$(2,1)$
$\beta(v_5)$	$(2,1)$
$\beta(v_6)$	$(1,3)$
$\beta(v_7)$	$(1,3)$
$\beta(v_8)$	$(1,4)$
$\beta(v_9)$	$(2,1)$
$\beta(v_{10})$	$(2,2)$
$\beta(v_{11})$	$(2,2)$

A resource binding can be represented by a *labeled hypergraph*, where the vertex set V represents operations and the edge set E_β represents the binding of the operations to the resources. In the hypergraph model, we assume that E_β defines a partition of the vertex set, i.e., the edges are disjoint (because resource sharing is transitive) and they cover all vertices (excluding *No-Operations*), because we assume that all operations are bound. A concise representation of a bound sequencing graph can be achieved by adding the edge set E_β to the sequencing graph $G_s(V, E)$.

Example 4.3.6. Consider the example of Figure 4.5. The shaded areas represent the edges $E_\beta = \{\{v_1\}, \{v_2, v_3\}, \{v_6, v_7\}, \{v_8\}, \{v_{10}, v_{11}\}, \{v_4, v_5, v_9\}\}$.

Consider now the example of dedicated resources. Then, the edge set E_β corresponds to the vertex set V.

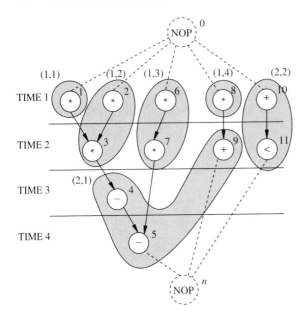

FIGURE 4.5
Scheduled sequencing graph with resource binding.

When binding constraints are specified, a resource binding must be compatible with them. In particular, a partial binding may be part of the original specification, as described in Section 4.2.2. This corresponds to specifying a binding for a subset of the operations $U \subseteq V$. A resource binding is *compatible* with a partial binding when its restriction to the operations U is identical to the partial binding itself.

Example 4.3.7. Consider again the sequencing graph of Figure 4.2. A hypothetical partial binding requires that operations v_6 and v_8 be performed by the same multiplier. Then, the operations v_6 and v_8 cannot execute concurrently and the corresponding schedule differs from that shown in Figure 4.5. The partial binding and the corresponding schedule are shown in Figure 4.6.

Common constraints on binding are upper bounds on the resource usage of each type, denoted by $\{a_k; \; k = 1, 2, \ldots, n_{res}\}$. These bounds represent the *allocation*[1] of instances for each resource type. A resource binding satisfies resource bounds $\{a_k; \; k = 1, 2, \ldots, n_{res}\}$ when $\beta(v_i) = (t, r)$ with $r \leq a_t$ for each operation v_i; $i = 1, 2, \ldots, n_{ops}$.

Scheduling and binding provide us with an annotation of the sequencing graph that can be used to estimate the area and performance of the circuit, as described in detail in Section 4.4. In general, scheduling and binding are interrelated problems and constraints may complicate the search for a solution. This section has shown a simple

---

[1]The term allocation has often been misused in the literature. Some authors refer to binding as allocation. We prefer to use the terms "resource bounds" and "binding" in this book, and we shall not use the term "allocation" at all.

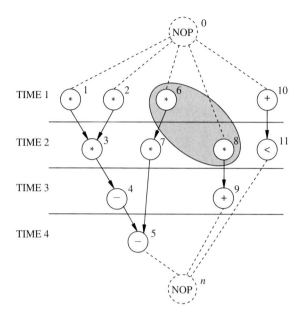

FIGURE 4.6

Example of partial binding of operations v_6 and v_8 to a single resource.

approach of unconstrained scheduling followed by binding, for the sake of introducing the fundamental concepts. Different approaches to solving the architectural synthesis and optimization problems will be reviewed in Section 4.5.

4.3.3 Hierarchical Models

When hierarchical graphs are considered, the concepts of scheduling and binding must be extended accordingly.

A hierarchical schedule can be defined by associating a start time to each vertex in each graph entity. The start times are now relative to that of the source vertex in the corresponding graph entity. The start times of the link vertices denote the start times of the sources of the linked graphs.

The latency computation of a hierarchical sequencing graph, with bounded delay operations, can be performed by traversing the hierarchy bottom up. The execution delay of a *model call* vertex is the latency of the corresponding graph entity. The execution delay of a *branching* vertex is the maximum of the latencies of the corresponding bodies. The execution delay of an *iteration* vertex is the latency of its body times the maximum number of iterations.

A hierarchical binding can be defined as the ensemble of bindings of each graph entity, restricted to the operation vertices. Operations in different entities may share resources. Whereas this can be beneficial in improving the area and performance of the circuit, it complicates the solution of the corresponding binding and resource-constrained scheduling problems. In addition, multiple model calls can be interpreted as viewing the called model as a shared application-specific resource. We refer the interested reader to reference [10] for details.

Example 4.3.8. Consider again the hierarchical sequencing graph shown in Figure 3.13. A binding with resource sharing across the hierarchy is shown by the following table:

$\beta(v_{a.1})$	(1, 1)
$\beta(v_{a.2})$	(2, 1)
$\beta(v_{a.3})$	(1, 2)
$\beta(v_{b.1})$	(2, 1)
$\beta(v_{b.2})$	(1, 1)

The bound sequencing graph is now shown in Figure 4.7.

4.3.4 The Synchronization Problem

There are operations whose delay is *unbounded* and not known at synthesis time. Examples are external synchronization and data-dependent iteration. (See Section 3.3.4.)

Scheduling unbounded-latency sequencing graphs cannot be done with traditional techniques. Different methods can be used. The simplest one is to modify the sequencing graph by isolating the unbounded-delay operations and by splitting the graph into bounded-latency subgraphs. Then, these subgraphs can be scheduled. Techniques for isolating the synchronization points and implementing them in the control unit are shown in Section 4.7.4. Alternatively, relative scheduling can be used, as described in Section 5.3.4.

Example 4.3.9. A sequencing graph with an unbounded-delay operation (v_a) is shown in Figure 4.8 (a). The graph can be modified by isolating the unbounded-delay operation, as shown in Figure 4.8 (b). The shaded subgraphs can be scheduled. Note that one

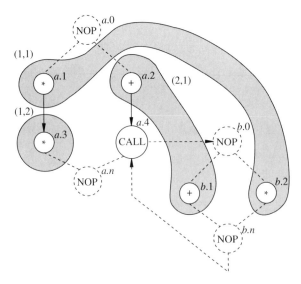

FIGURE 4.7
Example of a hierarchical sequencing graph.

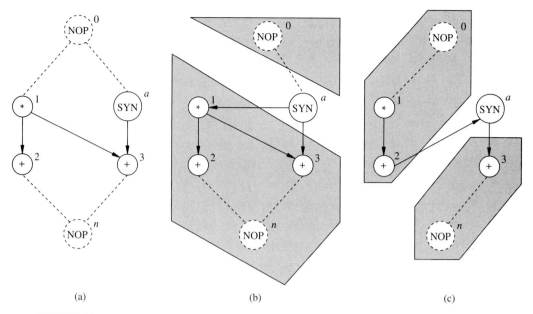

(a)   (b)   (c)

FIGURE 4.8
(a) A sequencing graph with a synchronization operation. (b) Isolation of bounded-latency components. (c) Alternative isolation of bounded-latency components. The shaded areas show subgraphs that can be independently scheduled.

subgraph is void in this case. Alternatively, the graph can be modified as in Figure 4.8 (c). Note that in the first graph the edge (v_a, v_1) has been added, while in the second the edge (v_2, v_a) has been added and (v_1, v_3) deleted, to make polar the subgraphs (together with the synchronization vertex). The edge addition and deletion are compatible with the serialization of the unbounded-delay vertex with the other operations.

The isolation of unbounded-delay operations can be done in different ways. Note that the graph of Figure 4.8 (b) has only one non-trivial component to be scheduled, as opposed to two in the case of the graph of Figure 4.8 (c). However, in the second graph, the operations v_1 and v_2 can start executing before the synchronization signal arrives, yielding possibly a better average performance of the resulting circuit.

4.4 AREA AND PERFORMANCE ESTIMATION

Accurate area and delay estimation is not a simple task. On the other hand, the solutions to architectural synthesis and optimization problems require estimating area and performance variations as a result of architectural-level decisions.

A schedule provides the latency λ of a circuit for a given cycle-time. A binding provides us with information about the area of a circuit. Therefore these two objectives can be evaluated for scheduled and bound sequencing graphs. Unfortunately, measures of area and performance are also needed while computing a schedule and a binding. Therefore it is important to be able to forecast, with some degree of accuracy, the impact of scheduling and binding decisions on area and performance. We defer to Section 4.5 a discussion about strategies for architectural optimization and we focus here on area and delay estimation.

Estimation is much simpler in the case of resource-dominated circuits, which we consider next.

4.4.1   Resource-Dominated Circuits

The area and the delay of the resources are known and dominate. The parameters of the other components can be neglected or considered a fixed overhead.

Without loss of generality, assume that the area overhead is zero. The area estimate of a structure is the sum of the areas of the bound resource instances. Equivalently, the total area is a weighted sum of the resource usage. A binding specifies fully the total area, but it is not necessary to know the binding to determine the area. Indeed, it is just sufficient to know how many instances of each resource type are used. If the area overhead is not zero, it can just be added to the overall resource area.

For a given cycle-time, the execution delays of the operations can be derived. If the delay overhead is not zero, this can be subtracted from the cycle-time. The latency of a circuit can thus be determined by its schedule.

> **Example 4.4.1.** Consider again the model of the loop of the differential equation integrator, shown in Figure 4.2. Assume that the circuit can be considered resource dominated, that the areas of the multiplier and ALU are 5 units and 1 unit, respectively, with an overhead of 1 unit, and that their propagation delays are 35 and 25 nsec, respectively. For a cycle-time of 40 nsec, all operations require one cycle to execute.
>
> An implementation with dedicated resources would require $6*5 + 5*1 + 1 = 36$ units of area and a latency of 4 cycles. Conversely, an implementation with one resource of each type requires then an area of $5 + 1 + 1 = 7$ units, independent of binding. A schedule would yield a latency of at least 7 cycles.

4.4.2   General Circuits

The area estimate is the sum of the areas of the bound resource instances plus the area of the steering logic, registers (or memories), control unit and wiring area. All these components depend on the binding, which in turn depends on the schedule.

The latency of the circuit still depends on the schedule, which in turn must take into account the propagation delays on paths between register boundaries. Thus performance estimation requires more detailed information. We consider the components of the data path first and then we comment on the control unit.

REGISTERS. All data transferred from a resource to another across a cycle boundary must be stored into some register. An upper bound on the register usage can then be derived by examining a scheduled sequencing graph. This bound is in general loose, because the number of registers can be minimized, as shown in Section 6.2.3. The binding information is needed for evaluating and/or performing the register optimization. Therefore, the accurate estimation of the number of registers requires both scheduling and binding.

The effect of registers on the evaluation of the cycle-time is easy to compute. In fact, their *set-up* and *propagation delay* times must be added to the propagation delays of the combinational logic. It is convenient to consider a reduced cycle-time in all computations that already discounts set-up times and propagation delays.

STEERING LOGIC. Steering logic affects the area and the propagation delay. While the area of multiplexers can be easily evaluated, their number requires the knowledge of the binding. Similarly, multiplexers add propagation delays to the resources.

Busses can also be used to steer data. In general, circuits may have a fixed number of busses that can support several multiplexed data transfers. Hence there are two components to take into account in the area and delay estimation. One corresponds to the busses themselves and can be considered as a fixed overhead. The second component relates to the bus drivers and receivers, which can be thought of as distributed multiplexers and modeled accordingly.

WIRING. Wiring contributes to the overall area and delay. The wiring area overhead can be estimated from the structure, once a binding is known, by using models that are appropriate for the physical design style of the implementation. The propagation delay on the wires may not be negligible, and it is proportional to the wiring length. Unfortunately, estimating the wiring area and length requires the knowledge of the structure (i.e., the binding) as well as the placement of the physical implementation of the resources. Fast floor planners or statistical placement models have been used. Alternatively, statistical wiring models have been used. In this case, it has been shown that the average interconnection length is proportional to the total number of blocks to the α power, where $0 \leq \alpha \leq 1$. The wiring delay and area track with the average interconnection length. We refer the interested reader to reference [14] for a detailed analysis of the wiring area and delay estimation.

CONTROL UNIT. The control circuit contributes to the overall area and delay, because some control signals can be part of the critical path. Recently, the interest in synthesizing control-dominated circuits, such as some communication ASICs, has exposed the importance and difficulty of the problem. Simple models for estimating the size of the control circuit can be based on the latency. Consider bounded-latency non-hierarchical sequencing graphs. Read-only-memory-based implementations of the control units require an address space wide enough to accommodate all control steps and a word length commensurate to the number of resources being controlled. Hard-wired implementations can be modeled by finite-state machines with as many states as the latency. Unfortunately, these models may provide loose bounds, because many optimization techniques can be applied to the controller, such as word-length reduction by encoding for microcode-based units or state minimization and encoding for hard-wired units. In addition, general models for sequencing graphs, including for example data-dependent delay operations, require more complex control units, as shown in Section 4.7. This complicates further the area and delay estimation.

Example 4.4.2. Consider again the model of the loop of the differential equation integrator, shown in Figure 4.2. The assumptions of Example 4.4.1 still hold, except that we do not consider the circuit as resource dominated, because we are now interested in a finer-grain analysis.

Let us consider registers first. There are 7 intermediate variables, 3 loop variables (x, y, u) and 3 loop invariants $(a, 3, dx)$, yielding a total of 13 variables to store. (Note that x is the output of operation v_{10}.) Nevertheless, according to the schedule, some registers may hold more than 1 variable when their lifetimes are disjoint intervals (see Section 6.2.3). For example, in the case of the schedule of Figure 4.3, 4 registers are enough to hold the intermediate variables (instead of 7), while 2 registers suffice in the case of the schedule of Figure 4.4.

An implementation with dedicated resources would not require multiplexers to steer data, whereas one with shared resources would. In particular, there are 5 possible operand pairs for both the multiplier and the ALU. An implementation with shared registers for the temporary variables would require each register to have a multiplexer to select the results to be stored. Thus, an implementation with two resources and two registers (for intermediates) would require 2 two-input multiplexers. Hence the overall requirement would be 4 five-input and 2 two-input multiplexers that must steer as many bits as the desired word size. A finer analysis shows that the multiplexer cost could be reduced further. Indeed, when using two registers (for intermediates), the result of the ALU could be stored always in the same register. Thus only 1 two-input multiplexer is required. Similarly, operands and results could be assigned to the registers (by exploiting also commutativity of addition and multiplication), to reduce the inputs of the multiplexers feeding the resources. A sketch of such a data path is shown later in Figure 4.12.

As far as the control unit for the loop is concerned, 4 states are required for the implementation with dedicated resources and 7 states for that with one resource of each type.

The overhead in terms of area and delay can be derived from the above analysis by empirical formulae.

4.5 STRATEGIES FOR ARCHITECTURAL OPTIMIZATION

Architectural optimization comprises scheduling and binding. Complete architectural optimization is applicable to circuits that can be modeled by sequencing (or equivalent) graphs without a start time or binding annotation. Thus the goal of architectural optimization is to determine a scheduled sequencing graph with a complete resource binding that satisfies the given constraints and optimizes some figure of merit.

Partial architectural optimization problems arise in connection with circuit models that either fully specify the timing behavior or fully characterize the resource usage. Equivalently, the initial circuit specification is either a scheduled sequencing graph or a sequencing graph with complete binding. It is obvious that any circuit model in terms of a scheduled and bound sequencing graph does not require any optimization at all, because the desired point of the design space is already prescribed. In all cases, a data path and control synthesis are needed to construct a circuit representation in terms of logic blocks. We consider the full architectural problem first, and we show how partial synthesis can be seen as solving some tasks of the full problem.

Architectural optimization consists of determining a schedule φ and a binding β that optimize the objectives (*area, latency, cycle-time*). The optimization of these multiple objectives reduces to the computation of the Pareto points in the design space and in the evaluation (or estimation) of the corresponding objective functions.

Architectural exploration is often done by exploring the (*area/latency*) trade-off for different values of the *cycle-time*. This approach is motivated by the fact that the cycle-time may be constrained to attain one specific value, or some values in an interval, because of system design considerations. The (*area/latency*) trade-off can then be explored by solving appropriate scheduling problems, as described later.

Other important approaches are the search for the (*cycle-time/latency*) trade-off for some binding or the (*area/cycle-time*) trade-off for some schedules. They are important for solving partial exploration when the input specification is bound to the resources or is scheduled, as well as when considering scheduling after binding or *vice versa*.

Unfortunately, the (*cycle-time/latency*) trade-off for some values of area as well as the (*area/cycle-time*) trade-off for some values of latency are complex problems to solve, because several bindings correspond to a given area and several schedules to a latency value.

4.5.1 Area/Latency Optimization

Let us consider resource-dominated circuits first. Given the cycle-time, the execution delays can be determined. Since the area depends on the resource usage (and not on any particular binding), scheduling problems provide the framework for determining the (*area/latency*) trade-off points. Indeed, solutions to the *minimum-latency* scheduling problem and to the *minimum-resource* scheduling problem provide the extreme points of the design space. Intermediate solutions can be found by solving *resource-constrained minimum-latency* scheduling problems or *latency-constrained minimum-resource* scheduling problems for different values of the constraints.

The (ideal) trade-off curve in the area/latency plane is monotonically decreasing in one parameter as a function of the other. Nevertheless, the intractability of the constrained scheduling problems often requires the use of heuristic algorithms that yield approximate solutions. The computed trade-off curve may lack monotonicity and solutions to the dual constrained scheduling problems may not coincide in practice.

The problem becomes more complicated when applied to general circuits, with more complex dependency of area and delay on the circuit structure. Assume first that control logic and wiring have a negligible impact on area and delay. Then only registers and steering logic must be taken into account, jointly with the resources, for the computation of the objectives. Area and latency can be determined by binding and scheduling, but the two problems are now deeply interrelated. Binding is affected by scheduling, because the amount of resource sharing depends on the concurrency of the operations. Registers and steering logic depend on the binding. Their delays must be compounded with the resource propagation delays when determining the execution delays, because the cycle-time is fixed. Thus the schedule may be affected.

This shows a circular dependency among scheduling, binding and estimation of the execution delays. An obvious solution is to solve these problems jointly, including the evaluation of registers and steering logic (Section 6.4). Alternatively, iterative methods can be used. The problem difficulty is exacerbated when the cost of the control unit and wiring is considered, because also the physical position of the resources and the controller must be taken into account. Again, by combining scheduling, binding, module generation, floor planning and wiring estimation in a single step, an optimal structure could be derived with an accurate area and performance model. This is a formidable task. Even though global algorithms for architectural optimization have been proposed, they have been shown to be practical for small-scale circuits only.

In practice, CAD systems for architectural optimization perform either scheduling followed by binding or *vice versa*. Area and delay are estimated before these tasks and are verified *a posteriori*. Most approaches to architectural synthesis perform scheduling before binding. Such an approach fits well with processor and DSP designs, because circuits often are resource dominated or close to being resource dominated. Examples of synthesis systems using this strategy are EMERALD/FACET [17], the SYSTEM'S ARCHITECT WORKBENCH [16] and the CATHEDRAL-II [8] systems.

Performing binding before scheduling permits the characterization of the steering logic and the more precise evaluation of the delays. No resource constraints are required in scheduling, because the resource usage is determined by binding. In this case, resource sharing requires that no operation pair with shared resources executes concurrently. This requirement can be embedded in the sequencing graph model by serializing the operations with a shared resource, i.e., by adding appropriate edges to the sequencing graph.

This approach best fits the synthesis of those ASIC circuits that are control dominated and where the steering logic parameters can be comparable to those of some application-specific resource. Since the unconstrained scheduling problem can be efficiently solved, the computational complexity is dominated by the binding task. This strategy has been used by programs BUD [12], CADDY/CALLAS [5] and HEBE [6].

Example 4.5.1. Consider again the model of the loop of the differential equation integrator under the assumptions of Example 4.4.1. For the sake of this example, we consider the circuit as resource dominated.

We consider the following upper bounds as constraints on the implementation: the area has to be smaller than 20 units and the latency less than 8 cycles. Recall that the ALU has a propagation delay of 25 nsec and the multiplier of 35 nsec. For a cycle-time of 40 nsec, all resources have unit execution delay. Hence, the area/latency trade-off curve is as shown in Figure 1.15. For a cycle-time of 30 nsec, the multiplier has an execution delay of 2 units, while the adder has an execution delay of 1 unit. A structure with one multiplier has latency equal to 13, violating the given bound. A structure with two multipliers and one adder has latency equal to 8 and area equal to 12. (Recall that we assign 1 unit of area to the control unit, as done in previous examples.) A structure with three multipliers and one adder has latency equal to 7 and area equal to 17. A structure with three multipliers and two adders has latency equal to 6 and area equal to 18. A structure with more than three multipliers violates the area bound. The trade-off curve is shown in Figure 4.9 for the specific values of the cycle-time considered here.

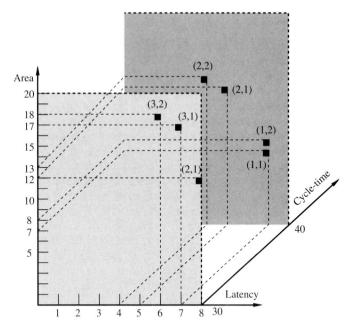

FIGURE 4.9
Design evaluation space: area/latency trade-off points for two values of the cycle-time. Note that (1,2) is not a Pareto point.

4.5.2   Cycle-Time/Latency Optimization

We consider bound sequencing graphs that are either representative of initial circuit specifications or derived by binding.

Let us consider resource-dominated circuits first. For each value of the cycle-time of interest, the corresponding execution delays of the operations can be derived and a minimum-latency schedule computed. Alternatively, *scheduling with chaining* can be performed by considering the propagation delays of the resources.

Let us consider now the case in which we are interested in the minimum cycle-time compatible with a given latency. When the resources are combinational in nature, the problem reduces to determining the register boundaries that optimize the cycle-time. This problem has been referred to as *retiming*, and it is dealt with in Section 9.3.1. The formulation and its solution can be extended to cope with sequential resources by modeling them as interconnections of a combinational component and register.

These considerations can be applied to general circuits when the binding of the multiplexers and registers is specified and fixed and when the wiring and control-unit area and delay can be approximated as constant or neglected. Under these assumptions the overall area is constant. Then, the multiplexers can be considered as additional combinational resources with their appropriate delays. The cycle-time/latency trade-off points can be determined again by scheduling with chaining or retiming.

Example 4.5.2. Consider again the model of the loop of the differential equation integrator under the assumptions of Example 4.4.1. For the sake of this example, we consider the circuit as resource dominated. We assume that the cycle-time is constrained to be in the interval 20-50 nsec and latency below 8 cycles.

Let us consider first an implementation with dedicated resources (area estimate 36). For a cycle-time larger than or equal to 35 nsec, all resources have unit execution delay. Hence latency is 4 cycles. For a cycle-time larger than or equal to 25 nsec and smaller than 35 nsec, the ALU has unit execution delay, but the multiplier has 2-cycle execution delay. Now the latency is 6 cycles. For a cycle-time larger than or equal to 20 nsec and smaller than 25 nsec, all resources have 2-cycle execution delay. Hence latency is 8 cycles.

A more accurate analysis can be performed by scheduling with chaining. In this case, chaining is possible only for a cycle-time of 50 nsec: operations $\{v_4, v_5\}$ as well as $\{v_{10}, v_{11}\}$ can execute together in a single cycle. Hence latency is 3 cycles.

Next we consider an implementation with one resource of each type (area estimate 7). For a cycle-time larger than or equal to 35 nsec, all resources have unit execution delay, leading to a latency of 7 cycles. For smaller values of the cycle-time, the latency constraint is violated.

The trade-off curve is shown in Figure 4.10, in correspondence with the two extreme values of the area. Points corresponding to other bindings are not shown.

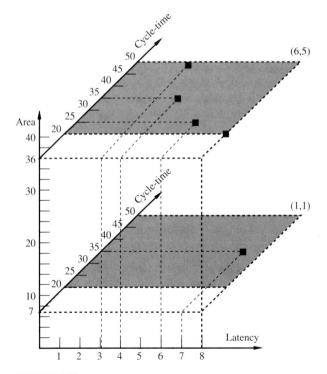

FIGURE 4.10
Design evaluation space: some cycle-time/latency trade-off points for two bindings.

4.5.3 Cycle-Time/Area Optimization

We consider now scheduled sequencing graphs where latency is fixed. We can envision the case in which we are solving either a partial synthesis problem or the binding problem after scheduling.

This problem is not relevant for resource-dominated circuits, because changing the binding does not affect the cycle-time. Conversely, it is important for general circuits, where the cycle-time is bounded from below by the delays in the resources, steering logic, etc. We assume here that only delays in steering logic matter. Note that the delay overhead due to registers is always constant.

With this approach, the schedule provides the latitude for applying resource sharing, because it defines the operation concurrency. The choice of a particular binding, and as a result the resource sharing configuration, affects the number and size of the multiplexers and hence the cycle-time. Minimal cycle-time binding and minimal area binding subject to cycle-time constraints can be computed as shown in Section 6.3.2.

> **Example 4.5.3.** Consider again the model of the loop of the differential equation integrator under the assumptions of Example 4.4.1. For the sake of this example, we consider the circuit cycle-time and area as dependent only on the resources and steering logic and we do not consider register sharing.
>
> Consider first the schedule shown in Figure 4.3. We consider two bindings: one with dedicated resources and one as shown in Figure 4.5. In the former case no multiplexer is needed. Thus, area is 36 units and cycle-time is bounded from below by the multiplier delay, i.e., 35. In the latter case, multiplexers are needed only for shared resources: namely one three-input and three two-input multiplexers. Hence, when computing the area, their contribution must be considered. Assuming a cost of 0.2 and 0.3 units for two-input and three-input multiplexers, respectively, the total area is now $4*5 + 2*1 + 1 + 1*0.3 + 3*0.2 = 23.9$ units. (Recall that we assign 1 unit of area to the control unit, as done in previous examples.) Similar considerations apply to delay. Assuming that the delay of two-input and three-input multiplexers is 2 and 3 nsec, respectively, the cycle-time is bound from below by 37 nsec. The two implementations considered so far have (*area, cycle-time*) = (36, 35) and (23.9, 37), respectively, and a latency of 4 cycles. Other implementations can be found with this latency, corresponding to different bindings.
>
> Consider now the schedule shown in Figure 4.4. We consider only one binding, corresponding to one resource per type. Two five-input multiplexers are required. Assuming that their area is 0.5 unit and their delay is 5 nsec, such an implementation would require an area of 8 units and cycle-time of 40 nsec. Figure 4.11 shows the points in the design evaluation space considered above.

4.6 DATA-PATH SYNTHESIS

Data-path synthesis is a generic term that has often been abused. We distinguish here data-path synthesis techniques at the physical level from those at the architectural level. The former exploit the regularity of data-path structures and are specific to data-path layout. The latter involve the complete definition of the structural view of the data path, i.e., refining the binding information into the specification of all interconnections. To avoid confusion, we call this task *connectivity synthesis*.

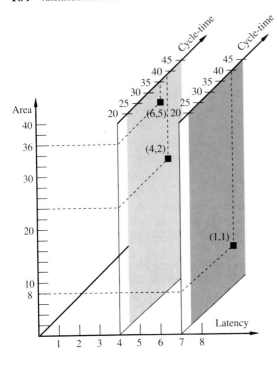

FIGURE 4.11

Design evaluation space: some cycle-time/latency trade-off points for two schedules.

Physical synthesis algorithms go beyond the scope of this book. We would like to mention that different approaches have been used for data-path design, according to different design styles, namely *bus-oriented*, *macro-cell-based* or *array-based* data paths. In the first case, a data-path generator constructs the data path as a stack of bit slices according to a predefined pattern. An example is the bus-oriented data path synthesized by the SYCO compiler [7], which has an architecture similar to the M68000 processor. Macro-cell-based data paths are typical of DSP circuits. Module generators are used to synthesize the resources that then need to be placed and wired. This method has been used by the CATHEDRAL-II compiler [8]. This approach is more flexible than using bus-oriented data-path synthesis with respect to the choice of a resource set, especially when application-specific resources are needed (e.g., arithmetic operators with non-standard word lengths). Unfortunately, this style leads often to a less efficient wiring distribution. Eventually, in the case of array-based data paths, logic and physical synthesis techniques are applied to the data path. Thus, the data path is treated no differently than other portions of the design. In general, bit-sliced data paths consume less area and perform better than data paths designed in a macro-cell-based or array-based style. The difference in performance may be thought to be small, as compared to manual design, when data-path optimization is used.

Data-path connectivity synthesis consists of defining the interconnection among resources, steering logic circuits (multiplexers or busses), memory resources (registers and memory arrays), input/output ports and the control unit. Therefore a complete binding is required. Problems related to the optimal binding of resources, registers, internal memory arrays and busses are described in Chapter 6. Connectivity synthesis refines the binding information by providing the detailed interconnection among the blocks. For example, the inputs, outputs and controls of the multiplexers have to be specified and connected.

> **Example 4.6.1.** Figure 4.12 shows a refined view of the data path of the differential equation integrator with one multiplier and one ALU. It shows explicitly the interconnection of the multiplexers.
>
> Note that the data format needs to be specified at this step to associate a word size with the registers, multiplexers and their interconnections. For simplicity, we have used one register to store the constant 3. Obviously, once the data format is chosen, this register can be changed into a hard-wired implementation.
>
> The connections to the input/output ports are not shown. Additional two-input multiplexers are required to load registers x, y, u, dx, a with input data. The output port is connected to register y.
>
> The connections to the control unit are the enable signals for all registers, the selectors of the multiplexers and a control signal for the ALU selecting the operation to be performed among $(+, -, <)$. The data path returns signal c to the control unit, detecting the completion of the iteration.

Data-path connectivity synthesis specifies also the interconnection of the data path to its environment through the input/output ports. Interfacing problems arise when the environment is not synchronous with the data path being synthesized. Notable examples are interfaces to asynchronous circuits and to synchronous circuits with different cycle-times. Borriello proposed models and algorithms for the synthesis of

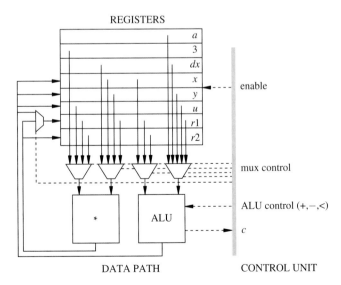

FIGURE 4.12
Structural view of the differential equation integrator with one multiplier and one ALU.

interface transducers. He specified (timing-constrained) interface problems by means of *event graphs*, showing dependencies among operations and events. We refer the reader to references [3, 4] for details.

The data path may include multi-port memory arrays, called also register files. Such arrays are dealt with as macro-cells, characterized by the word size and by the number of ports for reading and writing. An optimal usage of a memory array requires assigning data transfers to and from the array to the ports. Algorithms for this task are described in Section 6.2.4. Connectivity synthesis consists of specifying the interconnection between the ports of the memory array and the other data-path elements, e.g., resources and multiplexers. External memories are accessed through the circuit ports. The specific nature of the memory chip being used dictates the nature of the interface circuit required.

The interface to the control circuit is provided by the signals that enable the registers and that control the steering circuits (i.e., multiplexers and busses). Sequential resources require a *start* (and sometimes a *reset*) signal. Hence the execution of each operation requires a set of *activation* signals.

In addition, the control unit receives some *condition* signals from the data path that evaluate the clauses of some branching and iterative constructs. Condition signals provided by data-dependent operations are called *completion* signals. The ensemble of these control points must be identified in data-path synthesis.

Example 4.6.2. Figure 4.12 shows the interconnection between the data path and the control unit. Specifically, activation signals are the register enables, the ALU control, and multiplexer controls. The completion signal c is provided by the ALU.

Consider, for example, the execution of operation v_{10} in the sequencing graph of Figure 4.2, computing $x + dx$ and storing the result in register r_1. It requires controlling the multiplexers feeding the ALU and register r_1, setting the ALU control signal to perform an addition and enabling register r_1. The ensemble of this signal constitutes the activation signals for operation v_{10}.

4.7 CONTROL-UNIT SYNTHESIS

We consider in this section synthesis of control units. We assume that there are n_{act} *activation* signals to be issued by the control unit and we do not distinguish among their specific function (e.g., enable, multiplexer control, etc). From a circuit implementation point of view, we can classify the control-unit model as microcode based or hard wired. The former implementation style stores the control information into a read-only memory (ROM) array addressed by a counter, while the latter uses a hard-wired sequential circuit consisting of an interconnection of a combinational circuit and registers. From a logic standpoint, synchronous implementation of control can be modeled as a finite-state machine. Both implementation styles can be modeled as such, because a read-only memory and a synchronous counter behave as a finite-state machine as well as an interconnection of combinational logic gates and synchronous registers. It is important to remark that a synthesized microcoded control unit is not reprogrammable, i.e., the microcode-based implementation is a way of storing the control information in an organized fashion (e.g., in a memory array) but without support for modifying it.

The techniques used for control synthesis vary. We comment here on the major schemes that have been used. We consider first non-hierarchical sequencing graphs with data-independent delay operations. We extend then the techniques to cope with hierarchical graphs and eventually to unbounded-latency graphs. The reason for the choice of this sequence in presenting the control synthesis issues is as follows. Control synthesis for non-hierarchical graphs with data-independent delays requires the specification of the *activation* signals only. Hierarchical graphs, modeling branching and iteration, must also take into account the branching *condition* signals. Control units for unbounded-delay operations require handling the *completion* signals as well as the others. Therefore we shall analyze control synthesis methods for increasingly complex models.

It is important to note that control units for sequencing graphs with data-dependent bounded delays can be synthesized by considering the delays as data-independent and equal to their upper bounds. Then, the corresponding sequencing graph entities can be padded with finite-delay *No-Operations* to fit the upper bounds. Alternatively, if the operations provide a *completion* signal, they can be handled as unbounded delay operations. The second choice corresponds to a more complex controller but also to a circuit with better average performance.

> **Example 4.7.1.** Consider, for example, a branch into two equally probable tasks, with delays equal to 1 and 2 units, respectively. Assume the branch is on the critical path. If the branch is considered as a data-independent delay operation and the upper bound on the delays is used, then the branch will always take 2 units of delay. Conversely, if the immediate successors to the branch can start upon the assertion of its *completion* signal, then the branch will take on average 1.5 units.

4.7.1 Microcoded Control Synthesis for Non-Hierarchical Sequencing Graphs with Data-Independent Delay Operations

A microcoded implementation can be achieved by using a memory that has as many words as the latency λ. Each word is in one-to-one correspondence with a schedule step. Therefore the ROM must have as many address bits as $n_{bit} = \lceil \log_2 \lambda \rceil$. A synchronous counter with n_{bit} bits is used to address the ROM. The counter has a reset signal that clears the counter so that it can address the first word in memory corresponding to the first operations to be executed. When the sequencing graph models a set of operations that must be iterated, the last word of the schedule clears the counter. The counter runs on the system clock. The only external control signal provided by the environment is the counter *reset*. By raising that signal, the overall circuit halts and resets. By lowering it, it starts execution from the first operation.

There are different ways of implementing a microcoded memory array. The simplest way is to associate the *activation* signals of each resource to 1 bit of the word. This scheme is called *horizontal microcode*, because the word length n_{act} is usually much larger than λ and the ROM has a width larger than its height.

> **Example 4.7.2.** Consider the scheduled sequencing graph of Figure 4.3 and assume a binding with dedicated resources and registers. Hence the *activation* signals control the register enables. (The ALU's control signals are fixed, because the resources are dedicated.) Therefore we can assume that there are as many *activation* signals as operations, i.e., $n_{act} = n_{ops} = 11$. Then, a 2-bit counter driving a ROM with four words of 11 bits suffices, as shown in Figure 4.13.

A fully *vertical microcode* corresponds to the choice of encoding the n_{act} *activation* signals with $\lceil \log_2 n_{act} \rceil$ bits. This choice reduces drastically the width of the ROM. Two problems arise in conjunction with choosing a vertical microcode scheme. First, decoders are needed at the ROM output to decode the *activation* signals. Such decoders can be implemented using other ROMs and the scheme is then called two-stage control store (micro-ROM and nano-ROM). The second problem relates to the operation concurrency. A word corresponding to a schedule step may need to activate two or more resources. But this may be impossible to decode, unless code words are reserved for all possible n-tuples of concurrent operations. Vertical control schemes can be implemented by lengthening the schedule (i.e., renouncing to some concurrency) or by assuming that multiple ROM words can be read in each schedule step.

> **Example 4.7.3.** Consider again the scheduled sequencing graph of Figure 4.3 and assume a dedicated binding. For the sake of simplicity, assume again that there are as many *activation* signals as operations and that they are encoded with the 4-bit binary encoding of the operation identifier. Figure 4.14 can be interpreted in two ways. First, the operations are serialized and the latency is now 11. Second, multiple words can be read in a cycle. For example, the first five words could be read during the first control step.

4.7.2 Microcoded Control Optimization Techniques*

It is possible to encode the *activation* signals in ways that span the spectrum of choices between the horizontal and vertical schemes. A common microcode optimization approach is to search for the shortest encoding of the words, such that full concurrency is preserved. This problem is referred to as a *microcode compaction*

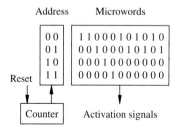

FIGURE 4.13
Example of horizontal microcode.

Microwords

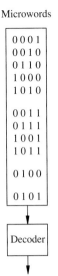

Activation signals

FIGURE 4.14
Example of vertical microcode.

problem and is intractable. This problem can be approached by assuming that the words are partitioned into fields and that operations are partitioned into corresponding groups (i.e., each group is assigned to a field). Operations in each group are vertically encoded. The partition into groups is done so that no pair of operations in each group is concurrent. Therefore full concurrency is preserved while retaining local vertical encoding.

A heuristic approach to find the optimal partition is to minimize the number of groups. Then, the partition can be done by considering a *conflict* graph, where the vertices correspond to the operations and the edges represent concurrency. A minimum *vertex coloring* of that graph yields the minimum number of fields that are needed. Minimum coloring, though intractable, can be computed by the algorithms shown in Section 2.4.3. This method, proposed by Schwartz, minimizes the number of fields and equivalently the number of decoders, but not necessarily the number of word bits.

An alternative approach, proposed by Grasselli and Montanari, is to model the problem as a *compatibility* graph, which is the complement of a conflict graph. Then a solution method can be cast in terms of a weighted *clique partitioning* problem, where the cliques represent classes of compatible operations and the weight relates to the number of bits required. By minimizing the total weight, instead of the clique cover number, a minimum-width ROM can be derived. We refer the reader to reference [1] for the details.

Example 4.7.4. Consider again the scheduled sequencing graph of Figure 4.3 and assume a dedicated binding and as many *activation* signals as operations. Let us partition the microword into five fields and the operations into four groups: $\{v_1, v_3, v_4\}$; $\{v_2\}$; $\{v_6, v_7, v_5\}$; $\{v_8, v_9\}$; $\{v_{10}, v_{11}\}$. Note that the operations inside each group are not concurrent and

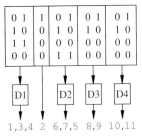

Microword format

Microwords

1,3,4 2 6,7,5 8,9 10,11

Activation signals

FIGURE 4.15

Example of encoded microwords using five fields.

therefore every group can be encoded vertically, as shown in Figure 4.15. The encoding is as follows:

Field	Operation	Code
A	v_1	01
A	v_3	10
A	v_4	11
B	v_2	1
C	v_6	01
C	v_7	10
C	v_5	11
D	v_8	01
D	v_9	10
E	v_{10}	01
E	v_{11}	10

Note also that null fields (i.e., all 0 entries in a field) denote that no operation in the group is activated in that particular control step. This encoding requires 9 bits (instead of 11 of the horizontal encoding and 4 of the vertical encoding) but preserves full concurrency.

4.7.3 Hard-Wired Control Synthesis for Non-Hierarchical Sequencing Graphs with Data-Independent Delay Operations

The synthesis of a Moore-type finite-state machine from a scheduled sequencing graph is straightforward. Indeed, such a machine has as many states as the latency λ (i.e., schedule length), and the state set $S = \{s_l; \ l = 1, 2, \ldots, \lambda\}$ is in one-to-one correspondence with the schedule steps. State transitions are unconditional and only among state pairs $(s_l, s_{l+1}); \ l = 1, 2, \ldots, (\lambda - 1)$. If the schedule requires a repetitive execution, then an unconditional transition (s_λ, s_1) is added. Conditional transitions into s_1 from all other states, controlled by a *reset* signal, provide the start and reset capability. (Alternatively, an idle state s_0 and transitions $(s_\lambda, s_0),(s_0, s_1)$ can be added, the latter controlled by a *start* signal. Transitions into s_0 are controlled by a different *reset* signal.)

The output function of the finite-state machine in each state s_l; $l = 1, 2, \ldots, \lambda$, activates those operations whose start time is t_l. More specifically the *activation* signal for the control point k, $k = 1, 2, \ldots, n_{act}$, in state s_l; $l = 1, 2, \ldots, \lambda$, is δ_{l,t_k}, where $\delta_{i,j}$ denotes a Kronecker delta function. A hard-wired control unit can be obtained by synthesizing the finite-state machine model using standard techniques (see Section 9.2) and in particular by encoding the states and by implementing the combinational logic in the appropriate style (e.g., sparse logic, PLA).

Example 4.7.5. Consider again the scheduled sequencing graph of Figure 4.3. The state transition diagram of the finite-state machine implementing a hard-wired control unit is shown in Figure 4.16. The numbers by the vertices of the diagram refer to the *activation* signals.

4.7.4 Control Synthesis for Hierarchical Sequencing Graphs with Data-Independent Delay Operations

Hierarchical sequencing graphs represent model calls, branching and iteration through the hierarchy. In this section, we assume that the graphs have bounded latency, and therefore each vertex has a known, fixed execution delay.

Let us consider first model calls and their control implementations. We assume that every sequencing graph entity in the hierarchy has a corresponding local control unit, called a *control block*. Each control block has its own *activation* signal, which controls the stepping of the counter (in microcode-based implementations) or the finite-state machine transitions as well as enables the *activation* signals at the corresponding level in the hierarchy. Therefore, raising the *activation* signal for a control block corresponds to executing the related operations. Lowering the *activation* signal corresponds to halting all related operations. We assume that each control block resets itself after having executed the last operation. All control blocks can be reset.

The hierarchical control implementation can be achieved as follows. The execution of a link vertex corresponding to a model call is translated to sending an *activation* signal to the corresponding control block. That signal is asserted for the

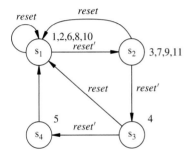

FIGURE 4.16
Example of state diagram for hard-wired control.

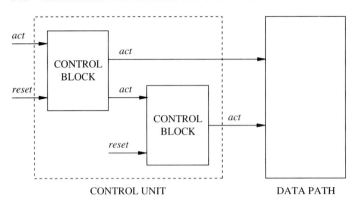

FIGURE 4.17
Example of interconnecting a hierarchical control structure.

duration of execution of the called model, i.e., as long as its local latency. Note that the control block of the calling model continues its execution, because the model call is in general concurrent with other operations. An example is shown in Figure 4.17.

The interconnection of the control blocks corresponding to the different sequencing graph entities in the hierarchy can be done regardless of the implementation style (e.g., microcode or hard wired) as long as the *activation* signal is provided. The *activation* signal of the root block can be used to start the overall circuit. Alternatively, this signal can always be asserted and the circuit can be started by pulsing the *reset* line. Note that the call to (and return from) a model does not require an additional control step with this scheme.

Let us consider now branching operations. A branch is represented in the hierarchical sequencing graph model by a selective model call, controlled by the branch condition clause. Therefore, a straightforward implementation can be achieved by activating the control blocks corresponding to the branch bodies by the conjunction of the *activation* signal with the branch *condition* clause value, as shown in Figure 4.18. For this control scheme to be correct, we must assume that the branching clause value does not change during the execution of the branch itself. Therefore, the value of the clause may have to be temporarily stored.

The control for an iteration link vertex can be done in a similar way. The loop body can be seen as a model call that is repeated a finite and known number of times. (Data-dependent iteration will be considered in Section 4.7.5.) Since we already assume that each control block resets itself when all operations have finished execution, it suffices to assert the *activation* signal for the loop body control block as long as the iteration has to last. Recall that the latency of an iteration link vertex is the product of the loop body latency times the number of executions. This number, which in this case is known at synthesis time, is the duration of the *activation* signal.

The overall control unit may be lumped into a single ROM or finite-state machine, according to the desired implementation style.

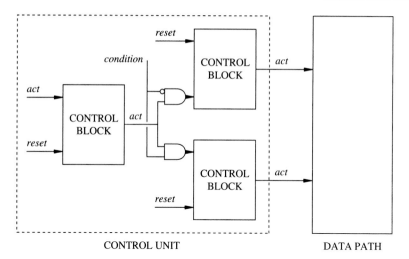

FIGURE 4.18
Example of interconnecting a hierarchical control structure.

Example 4.7.6. Consider now the complete differential equation integrator, as described in Example 1.5.3. Assume that the data path is required to have at most two multipliers and two ALUs, as shown in Figure 1.14. We show next a specification of the control unit, as synthesized by a computer program. The control unit is specified as a synchronous logic network (see Section 3.3.2), where the keyword "@D" denotes a delay-type register.

```
.model control;
.inputs CLK reset c ;
.outputs
        s1 s2 s3 s4 s5 s6
        s1s3 s2s3 s2s4 s1s2s3 s1s3s4 s2s3s4
        s1s5s6 s2s4s5s6 s2s3s4s6 s3s4s5s6 ;

## state machine

s1_in = reset + s6 ;                    # read inputs
s2_in = (s1 + c s5) reset' ;            # first step of loop
s3_in = (s2) reset' ;                   # second step
s4_in = (s3) reset' ;                   # third step
s5_in = (s4) reset' ;                   # fourth step
s6_in = (c' s5) reset' ;                # write outputs

s1 = @D(s1_in, CLK) ;
s2 = @D(s2_in, CLK) ;
s3 = @D(s3_in, CLK) ;
s4 = @D(s4_in, CLK) ;
s5 = @D(s5_in, CLK) ;
s6 = @D(s6_in, CLK) ;
```

```
## signals needed for muxing data

s1s3 = s1 + s3 ;
s2s3 = s2 + s3 ;
s2s4 = s2 + s4 ;
s1s2s3 = s1 + s2 + s3 ;
s1s3s4 = s1 + s3 + s4 ;
s1s5s6 = s1 + s5 + s6 ;
s2s3s4 = s2 + s3 + s4 ;
s2s3s4s6 = s2 + s3 + s4 + s6 ;
s2s4s5s6 = s2 + s4 + s5 + s6 ;
s3s4s5s6 = s3 + s4 + s5 + s6 ;
.endmodel control ;
```

4.7.5   Control Synthesis for Unbounded-Latency Sequencing Graphs*

Unbounded-latency sequencing graphs contain unbounded-delay operations that provide *completion* signals to notify the end of execution. We shall assume that the *completion* signal is raised during the last cycle of execution of an operation, so that no control step is wasted in detecting a completion and starting the successor operations. Similarly, the control unit of the implementation of an unbounded-latency graph is assumed to provide its own *completion* signal to denote the end of execution of all operations. This *completion* signal is used when composing control blocks to form a hierarchical controller.

We mention here three approaches for synthesizing control units for unbounded-latency graphs. The first one is the *clustering* method. It clusters the graph into bounded-latency subgraphs, as previously shown in Example 4.3.9. The number of clusters depends on the number of unbounded-delay operations. The method is efficient (in terms of control-unit area) when this number is small. Control implementations can be in terms of microcode-based or hard-wired styles. The second approach, called *adaptive* control synthesis, is reminiscent of some control synthesis techniques for self-timed circuits. It leads to a hard-wired implementation and it is efficient when the number of unbounded-delay operations is high. The third method is based on a particular scheduling technique called *relative scheduling* and uses an interconnection of *relative control units*, mentioned briefly in Section 5.3.4 and described in reference [10].

Let us consider the clustering method first. Bounded-latency subgraphs can be extracted by using the following procedure. Consider the unbounded-delay vertices in the graph one at a time, in a sequence consistent with the partial order represented by the graph itself. Let $S \subset V$ be the subset of vertices that are neither unbounded-delay vertices nor their successors. Then the subgraph induced by S can be made polar by adding a sink vertex representing a *No-Operation* and edges from the vertices in S with no successors to the sink. This subgraph can be scheduled and its control unit can be generated with a microcoded or hard-wired style. The vertices S can then be deleted from the graph and the unbounded-delay vertex under consideration replaced by a *No-Operation* that is now the source vertex of the subgraph induced by the remaining vertices.

A synchronizer is added to the control unit in correspondence to the unbounded-delay vertex previously under consideration. The synchronizer is a control primitive that can be implemented by a simple finite-state machine. (See Section 3.3.3.) The synchronizer takes as input the *completion* signal of the controller of the subgraph just extracted and the *completion* signal of the unbounded-delay operation itself. The synchronizer issues an *activation* signal to the controller of the subsequent operations. The synchronizer memorizes the arrival of both *completion* signals into two independent states. The *activation* signal is asserted either in coincidence of both *completion* signals or when one *completion* signal is received and the finite-state machine is in the state that memorizes the arrival of the other one at some previous time step.

> **Example 4.7.7.** Consider the graph of Figure 4.8 (a). Let S be equal to $\{v_1, v_2\}$. The subgraph induced by S is shown in the upper part of Figure 4.8 (c), where the added edge (v_2, v_a) and the replacement of v_a by a *No-Operation* sink vertex makes the subgraph abide the sequencing graph definition. The subgraph can then be scheduled and its control unit built. After deleting $\{v_1, v_2, v_a\}$ the remaining cluster has only vertex v_3, and its schedule and control block can be easily synthesized. The overall hard-wired control implementation is described by a state transition diagram in Figure 4.19. The shaded area on the upper left controls operations of the cluster $\{v_1, v_2\}$. The shaded area on the right is a synchronizer circuit that operates as follows. Its reset state is s_a. A transition $s_a \rightarrow s_b$ is caused by the completion of the controller of the first cluster, while a transition $s_a \rightarrow s_c$ is caused by the completion of the unbounded-delay operation. When the synchronizer is in s_b (or in s_c) and the completion of the unbounded-delay operation (or of the controller of the first cluster) is detected, the synchronizer issues the activation signal and goes to state s_d. If the two *completion* signals are simultaneous, there is a state transition $s_a \rightarrow s_d$ and the activation is asserted.

The *adaptive* control synthesis scheme provides a method for constructing a hard-wired control unit in a regular and modular fashion [10]. The controller is the interconnection of basic primitives, called *atomic finite-state machines*. There is an atomic finite-state machine in correspondence to each vertex of the sequencing graph, regardless of the boundedness of the corresponding delay. *No-Operations* and operations that do not require a control step for execution (e.g., some combinational

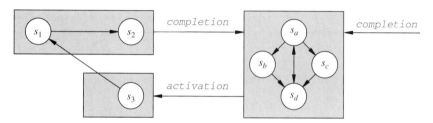

FIGURE 4.19
State transition diagram for a sequencing graph with two clusters and a synchronizer.

operations) have simpler, degenerate finite-state machines. The overall interconnection of the atomic finite-state machines is a finite-state machine itself that can be implemented as a single control unit or distributed.

The rationale of adaptive control is that any operation can start executing as soon as all its immediate predecessors have completed. In the case of concurrent hardware synthesis, as in our case, we must memorize the completion of an operation, because the completion of two (or more) immediate predecessors may not be simultaneous. Therefore, two states are associated with each operation. Thus each atomic finite-state machine has two states, corresponding to one storage bit. The first state is called *ready*, and it denotes that either the operation is ready for execution or it is executing. The second state is called *wait*, and it denotes that the operation has finished execution.

Each atomic finite-state machine communicates with the corresponding operation in the data path via the *activation* and *completion* signals. It communicates with the other atomic finite-state machines by means of two other signals: an *enable* that notifies that the corresponding operation can start execution and a *done* signal that is raised when the operation has completed its execution. The state transitions in the atomic finite-state machine are defined as follows. A transition from the *ready* state to the *wait* state is triggered by the *completion* signal of the corresponding operation. The reverse transition is in response to the *reset* signal or to the *completion* signal of the control block, which is the *done* signal of the sink operation. An atomic finite-state machine is shown in Figure 4.20 and the transitions are summarized in Table 4.1.

The atomic finite-state machines are interconnected with the same topology as the sequencing graph. In particular, the *enable* signal for an atomic finite-state machine is the conjunction of the *done* signals of its immediate predecessors. The *activation* signal is asserted when the atomic finite-state machine is in the *ready* state and the

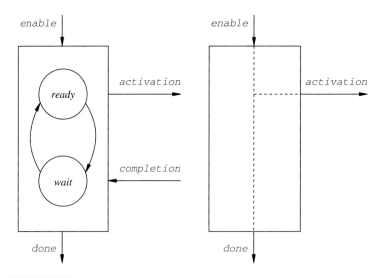

FIGURE 4.20
Atomic and degenerate finite-state machine.

TABLE 4.1

State transitions and outputs from an atomic finite-state machine.

$ready \rightarrow wait$:	$completion \cdot (done_n + reset)'$
$wait \rightarrow ready$:	$done_n + reset$
$activation$:	$ready \cdot enable$
$done$:	$wait + ready \cdot completion$

enable signal is detected. The *done* signal is true when the atomic finite-state machine is in the *wait* state or when it is in the *ready* state and the *completion* signal is detected.

Degenerate finite-state machines are associated with operations that do not require a control state. Such machines have no state information, and their *done* signal is equivalent to their *enable*, computed as in the previous case. The *activation* signal is equivalent to *enable* and the *completion* signal is not used.

> **Example 4.7.8.** An example of a sequencing graph and its adaptive control scheme is shown in Figure 4.21. For the sake of the example, we assume that the adders take one cycle to execute, and therefore their *completion* signal is raised in the cycle they execute. In this example, the multiplier uses a multiplication algorithm based on repeated additions of partial products with a Booth encoding scheme. It is a data-dependent operation, and

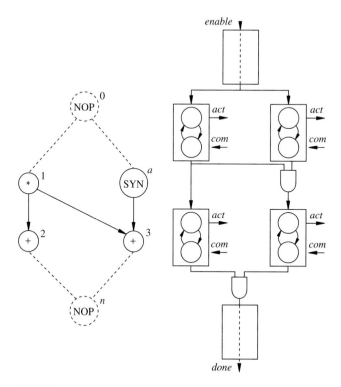

FIGURE 4.21
A sequencing graph and its adaptive control implementation.

it raises its own *completion* signal. The source and sink operations are *No-Operations*, and their atomic finite-state machines are degenerate.

The control unit corresponding to a sequencing graph has two interface signals, in addition to the *reset*, namely the *enable* of the atomic finite-state machine of the source vertex ($enable_0$) and the *done* of the atomic finite-state machine of the sink vertex ($done_n$). Synthesis of hierarchical controllers can be achieved by using the techniques presented in Section 4.7.4. In particular, model calls can be achieved by connecting the *activation* signal of the calling control block to the $enable_0$ entry point and the $done_n$ signal to the *completion* entry. This scheme can be extended to handle branching constructs by controlling the $enable_0$ signals of the branches by the conditional clause. Extension to iterative constructs, including data-dependent loops, is also possible with this scheme. The loop exit clause controls the repeated call to the loop body. Therefore, the $enable_0$ signal of the loop body is asserted when the iteration can be started and it remains so as long as the exit condition is not met. Two important considerations must be highlighted. First, the adaptive scheme provides automatically for repeated execution of the operations in a sequencing graph, because the completion of the sink operation resets all atomic finite-state machines. Therefore, the loop body is automatically restarted. Second, the loop exit condition must be evaluated at the loop boundary to avoid premature exit from a loop. Therefore, the loop exit condition must be temporarily stored.

The control synthesis of arbitrarily nested sequencing graphs can be achieved with no execution delay penalty in transferring control across the hierarchy. This is a desirable feature, because the control unit should never increase the circuit latency determined by the data path. A requirement in satisfying this goal is that the *completion* signals $done_n$ are asserted during the last execution step of the controller for each graph in the hierarchy.

The actual implementation of the adaptive scheme requires other considerations, such as the clocking choice for the data path and control. When the data path and control operate on the same phase of the clock, look-ahead techniques may be required to ensure that the $done_n$ signal is raised on time. Consider, for example, a sequencing graph at some level of the hierarchy whose last operation is a data-dependent branch. The two branches are a *No-Operation* and an operation requiring a finite number of steps. Then, the branch clause should be used directly to determine the *completion* signal, because the selection of a *No-Operation* must correspond to an immediate transfer of execution to the control block at the higher level of the hierarchy. We refer the interested reader to reference [10] for the details of an adaptive scheme that incorporates look-ahead techniques and that guarantees no performance penalty in conjunction with control transfers and unbounded-delay operations.

4.8 SYNTHESIS OF PIPELINED CIRCUITS

Pipelining is a common technique to enhance the circuit performance. In a pipelined implementation, the circuit is partitioned into a linear array of *stages*, each executing concurrently and feeding its results to the following stage. Pipelining does not affect the latency of the computation, but it may increase the input/output data rate, called *throughput*.

Pipelining has been applied to instruction set and to signal/image processors. Pipeline design is more challenging in the former case, because it must support efficient execution of different instruction streams. Conversely, pipelined DSP design may be simpler, because often the processor is dedicated to an application.

At the time of this writing (1993), synthesis techniques for pipelined circuits are still in their infancy. Few techniques for synthesizing pipelined data paths have been proposed, under some limiting assumptions such as constant data rates. Unfortunately, efficient instruction set processor design requires handling variable data rates as well as a variety of other issues, such as stage bypasses, hazard analysis and support for controlling the pipeline by allowing stalling and flushing. As a result, present synthesis techniques are not yet applicable to the design of competitive instruction set processors, although they have been applied successfully to some DSP designs.

Some *ad hoc* representation paradigms have been developed for pipelined circuits. For example, pipelined circuits can be specified by modeling each stage independently as well as the stage interfaces and the synchronization mechanisms. This corresponds to providing a mixed structural/behavioral circuit model that may preclude architectural optimization of the circuit as a whole. In particular, the number of stages is prescribed.

We consider here a simplified model for pipelined circuits that supports architectural optimization. Circuits are modeled by pipelined sequencing graphs where the source vertex is fired at the *throughput* rate. The inverse of this rate, i.e., the time separation between two successive firings of the source vertex, normalized to the *cycle-time*, is called a *data introduction interval* and is represented by δ_0. The data introduction interval is smaller than the latency. In the limiting case that the data introduction interval equals the latency, the circuit no more performs operations in a pipelined fashion but just restarts after the last operation. With this model, the number of stages of a pipelined circuit is the rounded-up quotient of the latency to the data introduction interval.

> **Example 4.8.1.** Consider the sequencing graph of Figure 4.2. Assume that the operations have unit execution delays and that the number of resources is not constrained, so that a latency of 4 cycles can be achieved.
>
> For $\delta_0 = 2$, input and output data are requested and made available at every other cycle. Figure 4.22 shows the sequencing graph partitioned into two stages.
>
> The repetitive execution of the operations of Figure 4.2 in a non-pipelined mode corresponds to $\delta_0 = 4$. Note that halving δ_0 corresponds to doubling the throughput for the same latency.

The design objectives of pipelined circuits are four: *area, latency, cycle-time, throughput*. Architectural exploration and optimization relate to determining the trade-off points in the corresponding four-dimensional space. Most approaches consider the sequences of problems derived by choosing particular values of the data introduction interval and of the cycle-time (hence determining implicitly the throughput) and a search for the optimal *(area, latency)* trade-off points. In the case of resource-dominated circuits, the trade-off curve can be computed as the solution to minimum-

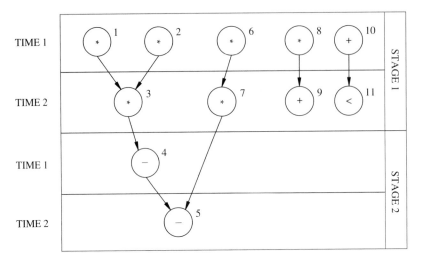

FIGURE 4.22
Pipelined sequencing graph partitioned into two stages. (The source and sink vertices have been omitted.)

latency scheduling problems, under constraints on the resource usage and the data introduction interval. This approach is justified by the fact that the choices of data introduction interval and cycle-time are limited in number due to environmental constraints.

> **Example 4.8.2.** Consider again the sequencing graph of Figure 4.2. The assumptions of Example 4.4.1 still hold (including the bound on area of 20 units) and the circuit is considered here as resource dominated. Assume a cycle-time of 40 nsec, so that all resources have unit execution delay.
>
> Assume first $\delta_0 = 1$. Hence the throughput is 25 MHz. All operations must have dedicated resources, yielding an area of 36 units, violating the bound. A latency of 4 cycles can be achieved by the schedule of Figure 4.22.
>
> Consider then $\delta_0 = 2$. Hence the throughput is 12.5 MHz. We can construct a schedule with latency of 4 cycles, requiring three multipliers and three ALUs (area = 19), as will be shown in detail in Example 5.6.4. (Recall that we assign 1 unit of area to the control unit, as done in previous examples.) Implementations with larger resource usage are not Pareto points; implementations with smaller resource usage are incompatible with the data introduction interval constraint.
>
> For $\delta_0 = 4$ the circuit is not pipelined. The area/latency trade-off points are those shown in Figures 1.15 and 4.9. Points of the design evaluation space are shown in Figure 4.23.

The scheduling and binding problems are more complex in the case of pipelined circuits, because each operation may be executing concurrently with its predecessors and successors. Lower bounds on the resource usage are also induced by the constraints on the data introduction interval. Algorithms for scheduling pipelined circuits will be described in Section 5.6 and considerations for binding in Section 6.7.

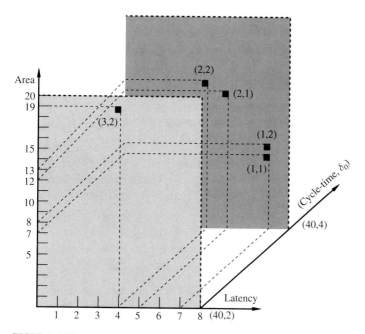

FIGURE 4.23
Design evaluation space: some area/latency trade-off points for values of the pair (*cycle-time, data introduction interval*). Note that (1,2) is not a Pareto point.

Data-path synthesis for pipelined circuits can be achieved by means of the same techniques used for non-pipelined circuits. Synthesis of control units for pipelined data paths with fixed data rates (hence involving operations with data-independent delays) can be achieved by extending the techniques shown in Sections 4.7.1 and 4.7.3.

> **Example 4.8.3.** Consider the example shown in Figure 4.22. For $\delta_0 = 2$ there are two control steps. Assume resources are dedicated. The controller will activate the resources scheduled in each step of both stages concurrently. Namely, the first control state (or microword) will activate operations $\{v_1, v_2, v_6, v_8, v_{10}, v_4\}$, the second state operations $\{v_3, v_7, v_9, v_{11}, v_5\}$.

In the general case, when considering pipelined circuits with variable data rates and/or data-dependent delays (that may require stalling the pipeline), synthesis and even estimation of the control unit is much more complex. Architectural exploration of pipelined circuits becomes even harder, when general circuits (not necessarily resource dominated) are considered. Synthesis and optimization of pipelined circuits with advanced features is still the subject of ongoing research.

4.9 PERSPECTIVES

Architectural synthesis provides a means of constructing the macroscopic structure of a circuit, starting from a behavioral model, while exploring the trade-off among the figures of merit. The importance of architectural exploration goes beyond its application in CAD synthesis systems, because it supports a way of reasoning about multiple objectives in digital design. Thus it is also an important means of analysis in the design conceptualization phase.

Several CAD synthesis systems are available today that incorporate the ideas presented in this chapter. Architectural synthesis has shown to be beneficial not only in producing optimized designs in a short time but also in raising the modeling abstraction level of digital designers to that of hardware languages.

Dedicated architectural synthesis tools for specific circuit applications have also been developed, for example, in the area of digital image processing. Whereas specific architectural synthesis tools can yield higher quality solutions in restrictive domains, as compared to general purpose architectural synthesis tools, it is not obvious if such tools should be developed for all major classes of circuits. The debate between supporters of general purpose HDLs and synthesis tools and those advocating specific ones is still open today.

Few problems in architectural synthesis have not yet found satisfactory solutions. Examples are synthesis of pipelined circuits with advanced features and synthesis of memory subsystems, including automatic partitioning into memory hierarchies. Most architectural synthesis techniques have considered single-chip implementations. Nevertheless architectural synthesis is very appealing for large-system design, where the implementation may require several chips, possibly embedding existing components. Modeling multiple-module circuits and their environmental constraints, as well as synthesis from these models, is still the subject of ongoing research.

4.10 REFERENCES

Early design systems that performed architectural synthesis are MIMOLA [19], SILC [2], MAC PITTS [15] and the YORKTOWN SILICON COMPILER SYSTEM [8]. A survey of recent systems is proposed by some edited books [6–8, 18] or tutorials [11]. Data-path and control synthesis have been the object of intensive investigation. Most data-path compilers have dealt with the complete synthesis of the data path, including the physical layout [8]. Agerwala's survey [1] is one of several on microcode optimization techniques.

1. T. Agerwala, "Microprogram Optimization: A Survey," *IEEE Transactions on Computers*, Vol. C-25, No. 10, pp. 962–973, October 1976.
2. T. Blackman, J. Fox and C. Rosebrugh, "The Silc Silicon Compiler: Language and Features," *DAC, Proceedings of the Design Automation Conference*, pp. 232–237, June 1985.
3. G. Borriello and R. Katz, *Synthesis and Optimizations of Interface Transducer Logic, ICCAD, Proceedings of the International Conference on Computer Aided Design*, pp. 274–277, 1987.
4. G. Borriello, "Combining Events and Data-Flow Graphs in Behavioral Synthesis," *ICCAD, Proceedings of the International Conference on Computer Aided Design*, pp. 56–59, 1988.
5. R. Camposano and W. Rosenstiel, "Synthesizing Circuits from Behavioral Descriptions," *IEEE Transactions on CAD/ICAS*, Vol. 8, No. 2, pp. 171–180, February 1989.
6. R. Camposano and W. Wolf, Editors, *High-Level VLSI Synthesis*, Kluwer Academic Publishers, Boston, MA, 1991.
7. G. De Micheli, P. Antognetti and A. Sangiovanni-Vincentelli, Editors, *Design Systems for VLSI Circuits: Logic Synthesis and Silicon Compilation*, M. Nijhoff, Dordrecht, The Netherlands, 1987.
8. D. Gajski, Editor, *Silicon Compilation*, Addison-Wesley, Reading, MA, 1987.
9. D. Knapp, "Synthesis from Partial Structure," in D. Edwards, Editor, *Design Methodologies for VLSI and Computer Architecture*, North Holland, Amsterdam, pp. 35–51, 1989.

10. D. Ku and G. De Micheli, *High-Level Synthesis of ASICS under Timing and Synchronization Constraints*, Kluwer Academic Publishers, Boston, MA, 1992.

11. M. J. McFarland, A. Parker and R. Camposano, "The High-Level Synthesis of Digital Systems," *IEEE Proceedings*, Vol. 78, No. 2, pp. 301–318, February 1990.

12. M. J. McFarland, "Using Bottom-Up Design Techniques in the Synthesis of Digital Hardware from Abstract Behavioral Descriptions," *DAC, Proceedings of the Design Automation Conference*, pp. 474–480, 1986.

13. M. J. McFarland, "Reevaluating the Design Space for Register-Transfer Hardware Synthesis," *DAC, Proceedings of the Design Automation Conference*, pp. 262–265, 1987.

14. B. Preas and M. Lorenzetti, *Physical Design Automation of VLSI Systems*, Benjamin Cummings, Menlo Park, CA, 1988.

15. J. Southard, "MacPitts, An Approach to Silicon Compilation," *IEEE Transactions on Computers*, Vol. 16, No. 12, pp. 59–70, December 1983.

16. D. Thomas, E. Lagnese, R. Walker, J. Nestor, J. Rajan and R. Blackburn, *Algorithmic and Register Transfer Level Synthesis: The System Architect's Workbench*, Kluwer Academic Publishers, Boston, MA, 1990.

17. C. Tseng and D. Siewiorek, "Automated Synthesis of Data Paths in Digital Systems," *IEEE Transactions on CAD/ICAS*, Vol. CAD-5, pp. 379–395, July 1986.

18. R. Walker and R. Camposano, *A Survey of High-Level Synthesis Systems*, Kluwer Academic Publishers, Boston, MA, 1991.

19. G. Zimmermann, "The MIMOLA Design System: Detailed Description of the Software System," *DAC, Proceedings of the Design Automation Conference*, pp. 56–63, 1979.

4.11   PROBLEMS

1. Consider the sequencing graph of Figure 4.2. Assume you have a resource that performs the operations $(+, -, <, *)$ in one cycle and that occupies 1 unit of area. Determine the area/latency trade-off points. Consider the circuit as resource dominated with no overhead.

2. Consider the sequencing graph of Figure 4.2. Consider a partial binding of operations $\{v_1, v_2\}$ and $\{v_3, v_6\}$ to two resources. Draw the sequencing graphs with task serialization edges corresponding to this binding that resolve the potential conflicts. Rank the sequencing graphs in terms of minimum latency. Assume all operations have unit delay.

3. Consider the loop of the differential equation integrator. Assume a word length of 8 bits in the data path. Consider the binding shown in Figure 4.5. Perform connectivity synthesis and show a diagram with the detailed interconnection among the building blocks of the data path. Estimate the area, including resources, multiplexers and registers. (Assume that the areas of the multiplier and ALU are 5 and 1, respectively, that of a multiplexer is 0.1 unit per input and that of each register is 0.2 unit.)

4. Consider the full example of the differential equation integrator. Assume a word length of 8 bits in the data path. Consider an implementation with one multiplier and one ALU. Complete the connectivity synthesis sketched in Figure 4.12 by showing all connections in detail. Construct the state diagram of a finite-state machine controller for the circuit.

5. Consider a data path modeled by the following assignments:

    a = i + j; b = k + l; c = a + b; d = b + n; e = c + m; f = c + e;

    The only available resource type is a two-input adder. Its delay is 50 nsec. Ignore overflow problems and assume that the input carry is unused. Apply operation chaining when possible.

    Consider the circuit as resource dominated. Determine a minimum-latency schedule for 100-nsec cycle-time, with the reduction of resource usage as a secondary goal. Determine then a minimum-latency schedule with two resources. Draw the corresponding data path showing the resources, registers and multiplexers.

Assuming that multiplexers have 10-nsec delay, reschedule the operations to meet the cycle-time constraint.

6. Consider the following set of scheduled operations:

```
t1:   o1 o6 o10
t2:   o2 o9
t3:   o1 o8
t4:   o3 o5
t5:   o4
t6:   o2 o8
t7:   o7 o9
t8:   o1 o10
t9:   o7 o8
t10:  o9
t11:  o5
```

Draw the operation conflict graph and outline a minimum vertex coloring.

Derive the personality matrix of a microcode ROM for a corresponding control unit. Use an encoding scheme for the *activation* signals of the operations that preserves operation concurrency while reducing its width. Minimize the number of decoders. Report the encoding of the signals and the ROM.

7. Consider the full differential equation integrator. Assume that the ALU executes in one cycle and that the multiplier uses an iterative algorithm with data-dependent delay. A *completion* signal is provided by the multiplier. Assume dedicated resources. Draw a diagram showing an implementation of the control unit using the adaptive scheme, and derive a set of logic equations specifying the controller.

8. Consider the full differential equation integrator. Assume that the ALU executes in one cycle and that the multiplier uses an iterative algorithm with data-dependent delay. A *completion* signal is provided by the multiplier. Assume one multiplier and one ALU are used. Draw the state diagram of the finite-state machine implementing the control unit.

CHAPTER
5

SCHEDULING ALGORITHMS

There's a time for all things.

Shakespeare. The Comedy of Errors.

5.1 INTRODUCTION

Scheduling is a very important problem in architectural synthesis. Whereas a sequencing graph prescribes only dependencies among the operations, the scheduling of a sequencing graph determines the precise start time of each task. The start times must satisfy the original dependencies of the sequencing graph, which limit the amount of parallelism of the operations, because any pair of operations related by a sequency dependency (or by a chain of dependencies) may not execute concurrently.

Scheduling determines the concurrency of the resulting implementation, and therefore it affects its performance. By the same token, the maximum number of concurrent operations of any given type at any step of the schedule is a lower bound on the number of required hardware resources of that type. Therefore the choice of a schedule affects also the area of the implementation.

The number of resources (of any given type) may be bounded from above to satisfy some design requirement. For example, a circuit with a prescribed size may have at most one floating point multiplier/divider. When resource constraints are imposed, the number of operations of a given type whose execution can overlap in time is limited by the number of resources of that type. A spectrum of solutions may be obtained by scheduling a sequencing graph with different resource constraints. Tight bounds on the number of resources correlate to serialized implementations. As a limiting case, a scheduled sequencing graph may be such that all operations are executed in a linear sequence. This is indeed the case when only one resource is available to implement all operations.

Area/latency trade-off points can be derived as solutions to constrained scheduling problems for desired values of the cycle-time. As mentioned in Section 4.4, the area evaluation is just a weighted sum of the resource usage for resource-dominated circuits. When considering other circuits, an additional component must be taken into account, corresponding to steering logic, registers, wiring and control area.

In this chapter we model the scheduling problem and we describe the major algorithms for scheduling. We consider first sequencing graphs that are not hierarchical, hence not modeling branching or iterative constructs, and representative of the model data flow. Such graphs represent non-pipelined circuit models. We address unconstrained, timing-constrained and resource-constrained problems by presenting some algorithms to solve them (Sections 5.3 and 5.4). We consider then the extensions to the full hierarchical model in Section 5.5 as well as scheduling techniques for pipelined circuits (Section 5.6).

5.2 A MODEL FOR THE SCHEDULING PROBLEMS

We use in the following sections a non-hierarchical sequencing graph model, as introduced in Section 3.3.4. We recall that the sequencing graph is a polar directed acyclic graph $G_s(V, E)$, where the vertex set $V = \{v_i;\ i = 0, 1, \ldots, n\}$ is in one-to-one correspondence with the set of operations and the edge set $E = \{(v_i, v_j);\ i, j = 0, 1, \ldots, n\}$ represents dependencies. We recall also that $n = n_{ops} + 1$ and that we denote the source vertex by v_0 and the sink by v_n; both are *No-Operations*. Let $D = \{d_i;\ i = 0, 1, \ldots, n\}$ be the set of operation *execution delays*; the execution delays of the source and sink vertices are both zero, i.e., $d_0 = d_n = 0$. We assume in this chapter that the delays are data independent and known. Extensions are described in Section 5.3.4. To be concrete, we assume a synchronous mono-phase circuit implementation. Therefore the execution delays are integers representing clock cycles. We denote by $T = \{t_i;\ i = 0, 1, \ldots, n\}$ the *start time* for the operations, i.e., the cycles in which the operations start. We use often the vector notation **t** to represent all start times in a compact form. The *latency* of the schedule is the number of cycles to execute the entire schedule, or equivalently the difference in start time of the sink and source vertices: $\lambda = t_n - t_0$. Without loss of generality, in our examples we assume $t_0 = 1$, i.e., the first operations start execution in the first cycle.

The sequencing graph requires that the start time of an operation is at least as large as the start time of each of its direct predecessor plus its execution delay, i.e., that the following relations hold:

$$t_i \geq t_j + d_j \quad \forall\ i, j\ \ :\ (v_j, v_i) \in E \qquad (5.1)$$

An unconstrained schedule is a set of values of the start times T that satisfies the above relation. An unconstrained minimum-latency schedule is one such that its latency λ attains a minimum value. Note that in this case the latency of the schedule equals the weight of the longest path from source to sink, defined by the execution delays associated with the vertices. More formally, the *unconstrained minimum-latency scheduling problem* can be defined as follows.

> **Definition 5.2.1.** Given a set of operations V with integer delays D and a partial order on the operations E, find an integer labeling of the operations $\varphi : V \rightarrow Z^+$ such that $t_i = \varphi(v_i)$, $t_i \geq t_j + d_j \ \forall \ i, j \ : \ (v_j, v_i) \in E$ and t_n is minimum.

We analyze the unconstrained scheduling problem and its extensions in Section 5.3.

Let us consider now the hardware resources that implement the operations. We assume that there are n_{res} resource types, and we denote by function $\mathcal{T} : V \rightarrow \{1, 2, \ldots, n_{res}\}$ the unique resource type that implements an operation. A *resource-constrained scheduling problem* is one where the number of resources of any given type is bounded from above by a set of integers $\{a_k; k = 1, 2, \ldots, n_{res}\}$. Therefore the operations are scheduled in such a way that the number of operations of any given type executing in any schedule step does not exceed the bound. The *minimum-latency resource-constrained scheduling problem* can be defined more formally as follows.

> **Definition 5.2.2.** Given a set of operations V with integer delays D, a partial order on the operations E and upper bounds $\{a_k; k = 1, 2, \ldots, n_{res}\}$, find an integer labeling of the operations $\varphi : V \rightarrow Z^+$ such that $t_i = \varphi(v_i)$, $t_i \geq t_j + d_j \ \forall \ i, j \ s.t. \ (v_j, v_i) \in E$, $|\{v_i : \mathcal{T}(v_i) = k \text{ and } t_i \leq l < t_i + d_i\}| \leq a_k$ for each operation type $k = 1, 2, \ldots, n_{res}$ and schedule step $l = 1, 2, \ldots, t_n$, and t_n is minimum.

If all resources are of a given type (e.g., ALUs), then the problem reduces to the classical multiprocessor scheduling problem. The minimum-latency multiprocessor scheduling problem is intractable.

5.3 SCHEDULING WITHOUT RESOURCE CONSTRAINTS

Before considering the algorithms for unconstrained scheduling, we would like to comment on the relevance of the problem. Unconstrained scheduling is applied when dedicated resources are used. Practical cases leading to dedicated resources are those when operations differ in their types or when their cost is marginal when compared to that of steering logic, registers, wiring and control.

Unconstrained scheduling is also used when resource binding is done prior to scheduling, and resource conflicts are solved by serializing the operations that share the same resource. In this case, the area cost of an implementation is defined before and independently from the scheduling step.

Eventually unconstrained scheduling can be used to derive bounds on latency for constrained problems. A lower bound on latency can be computed by unconstrained scheduling, because the minimum latency of a schedule under some resource constraint is obviously at least as large as the latency computed with unlimited resources.

5.3.1   Unconstrained Scheduling: The ASAP Scheduling Algorithm

The unconstrained minimum-latency scheduling problem can be solved in polynomial time by topologically sorting the vertices of the sequencing graph. This approach is called in jargon *as soon as possible* (ASAP) scheduling, because the start time for each operation is the least one allowed by the dependencies.

We denote by $\mathbf{t}^S$ the start times computed by the ASAP Algorithm 5.3.1, i.e., by a vector whose entries are $\{t_i^S; i = 0, 1, \ldots, n\}$.

5.3.2   Latency-Constrained Scheduling: The ALAP Scheduling Algorithm

We consider now the case in which a schedule must satisfy an upper bound on the latency, denoted by $\bar{\lambda}$. This problem may be solved by executing the ASAP scheduling algorithm and verifying that $(t_n^S - t_0^S) \leq \bar{\lambda}$.

If a schedule exists that satisfies the latency bound $\bar{\lambda}$, it is possible then to explore the range of values of the start times of the operations that meet the bound. The ASAP scheduling algorithm yields the minimum values of the start times. A complementary algorithm, the *as late as possible* (ALAP) scheduling Algorithm 5.3.2, provides the corresponding maximum values.

We denote by $\mathbf{t}^L$ the start times computed by the ALAP algorithm.

The ALAP scheduling algorithm is also used for unconstrained scheduling. In this case, the latency bound $\bar{\lambda}$ is chosen to be the length of the schedule computed by the ASAP algorithm, i.e., $\bar{\lambda} = t_n^S - t_0^S$.

An important quantity used by some scheduling algorithms is the *mobility* (or *slack*) of an operation, corresponding to the difference of the start times computed by the ALAP and ASAP algorithms. Namely $\mu_i = t_i^L - t_i^S; \; i = 0, 1, \ldots, n$.

```
ASAP ( G_s(V, E)) {
        Schedule v_0 by setting t_0^S = 1;
        repeat {
                Select a vertex v_i whose predecessors are all scheduled;
                Schedule v_i by setting t_i^S =    max      t_j^S + d_j;
                                                j:(v_j, v_i)∈E
        }
        until (v_n is scheduled) ;
        return (t^S);
}
```

ALGORITHM 5.3.1

$ALAP(\ G_s(V, E),\ \overline{\lambda})\ \{$

    Schedule v_n by setting $t_n^L = \overline{\lambda} + 1$;

    repeat {

        Select vertex v_i whose successors are all scheduled;

        Schedule v_i by setting $t_i^L = \min\limits_{j:(v_i, v_j)\in E} t_j^L - d_i$;

    }

    until (v_0 is scheduled) ;

    return $(\mathbf{t}^L)$;

$\}$

ALGORITHM 5.3.2

Zero mobility implies that an operation can be started only at one given time step in order to meet the overall latency constraint. When the mobility is larger than zero, it measures the span of the time interval in which it may be started.

Example 5.3.1. Consider the sequencing graph of Figure 3.12, reported here again for convenience in Figure 5.1. Assume all operations have unit execution delay. The ASAP algorithm would set first $t_0^S = 1$. Then, the vertices whose predecessors have been scheduled are $\{v_1, v_2, v_6, v_8, v_{10}\}$. Their start time is set to $t_0^S + d_0 = 1 + 0 = 1$. And so on. Note that the start time of the sink $t_n^S = 5$, and thus latency is $\lambda = 5 - 1 = 4$. An example ASAP schedule is shown in Figure 5.2 (corresponding to Figure 4.3).

    Consider now the ALAP algorithm with $\overline{\lambda} = 4$. The algorithm would set first $t_n^L = 5$. Then, the vertices whose successors have been scheduled are $\{v_5, v_9, v_{11}\}$. Their start time is set to $t_n^L - 1 = 4$. And so on. An example ALAP schedule is given in Figure 5.3.

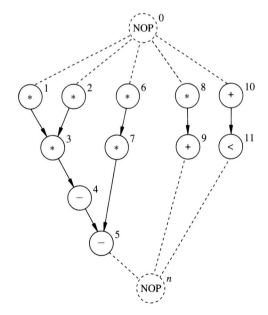

FIGURE 5.1
Sequencing graph.

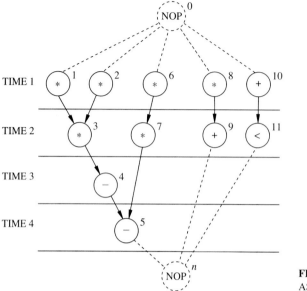

FIGURE 5.2
ASAP schedule.

By comparing the two schedules, it is possible to deduce that the mobility of operations $\{v_i; i = 1, 2, 3, 4, 5\}$ is zero, i.e., they are on a critical path. The mobility of operations v_6 and v_7 is 1, while the mobility of the remaining ones is 2.

The ASAP and ALAP algorithms can be implemented in slightly different ways, according to the data-structure used for storing the sequencing graph. Their computational complexity varies accordingly. At best, it matches that of topological sort, i.e., $O(|V| + |E|)$.

5.3.3 Scheduling Under Timing Constraints

The scheduling problem under latency constraints can be generalized to the case in which *deadlines* need to be met by other operations, or even further by considering *release times* for the operations. The release times and deadlines are nothing but *absolute* constraints on the start time of the operations.

A further generalization is considering *relative* timing constraints that bind the time separation between operations pairs, regardless of their absolute value. Note that absolute timing constraints can be seen as constraints relative to the source operation.

Relative constraints are very useful in hardware modeling, because the absolute start times are not known *a priori*. Minimum timing constraints between any two operations can be used to ensure that an operation follows another by at least a prescribed number of time steps, regardless of the existence of a dependency between them. It is often also important to limit the maximum distance in time between two operations by means of maximum timing constraints. The combination of maximum and minimum timing constraints permits us to specify the exact distance in time between two operations and, as a special case, their simultaneity.

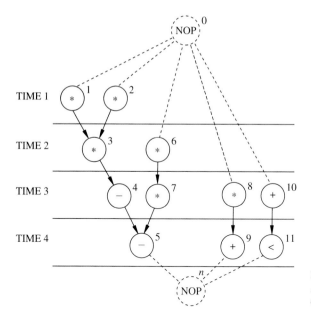

FIGURE 5.3
ALAP schedule under latency
constraints of four steps.

Example 5.3.2. Consider a circuit that reads data from a bus, performs a computation and writes the result back onto the bus. The bus interface prescribes that the data are written exactly three cycles after it is read. This specification can be interpreted by setting a minimum and a maximum timing constraint of three cycles between the read and the write operations.

Consider now another circuit with two independent streams of operations that are constrained to communicate simultaneously to the external circuits by providing two pieces of data at two interfaces. The cycle in which the data are made available is irrelevant although the simultaneity of the operations is important. This requirement can be captured by setting a minimum and a maximum timing constraint of zero cycles between the two write operations.

We define formally the relative timing constraints as follows.

Definition 5.3.1. Relative timing constraints are positive integers specified for some operation pair $v_i, v_j; i, j \in \{0, 1, \ldots, n\}$.

- A **minimum** timing constraint $l_{ij} \geq 0$ requires: $t_j \geq t_i + l_{ij}$.
- A **maximum** timing constraint $u_{ij} \geq 0$ requires: $t_j \leq t_i + u_{ij}$.

A schedule under relative timing constraints is a set of start times for the operations satisfying the requirements stated in Definition 5.2.1 and in addition:

$$t_j \geq t_i + l_{ij} \quad \forall \text{ specified } l_{ij} \tag{5.2}$$

$$t_j \leq t_i + u_{ij} \quad \forall \text{ specified } u_{ij} \tag{5.3}$$

A consistent modeling of minimum and maximum timing constraints can be done by means of a constraint graph $G_c(V_c, E_c)$, that is, an edge-weighted directed graph derived from the sequencing graph as follows. The constraint graph $G_c(V_c, E_c)$ has the same vertex set as $G_s(V, E)$ and its edge set includes the edge set E. Such edges are weighted by the delay of the operation corresponding to their tail. The weight on the edge (v_i, v_j) is denoted by w_{ij}. Additional edges are related to the timing constraints. For every minimum timing constraint l_{ij}, we add a *forward* edge (v_i, v_j) in the constraint graph with weight equal to the minimum value, i.e., $w_{ij} = l_{ij} \geq 0$. For every maximum timing constraint u_{ij}, we add a *backward* edge (v_j, v_i) in the constraint graph with weight equal to the opposite of the maximum value, i.e., $w_{ji} = -u_{ij} \leq 0$, because $t_j \leq t_i + u_{ij}$ implies $t_i \geq t_j - u_{ij}$. Note that the overall latency constraint can be modeled as a maximum timing constraint $\bar{\lambda} = u_{0,n}$ between the source and sink vertices.

> **Example 5.3.3.** Consider the example in Figure 5.4. We assume one cycle for addition and two for multiplication. A minimum timing constraint requires operation v_4 to execute at least $l_{04} = 4$ cycles after operation v_0 has started. A maximum timing constraint requires operation v_2 to execute at most $u_{12} = 3$ cycles after operation v_1 has started. Note that the constraint graph has a backward edge with negative weight (e.g., -3).

The presence of maximum timing constraints may prevent the existence of a consistent schedule, as in the case of the latency constraint. In particular, the requirement of an upper bound on the time distance between the start time of two operations may be inconsistent with the time required to execute the first operation, plus possibly the time required by any sequence of operations in between. Similarly, minimum timing constraints may also conflict with maximum timing constraints.

A criterion to determine the existence of a schedule is to consider in turn each maximum timing constraint u_{ij}. The longest weighted path in the constraint graph between v_i and v_j (that determines the minimum separation in time between operations v_i and v_j) must be less than or equal to the maximum timing constraint u_{ij}. As a consequence, any cycle in the constraint graph including edge (v_j, v_i) must have negative or zero weight. Therefore, a necessary condition for the existence of the

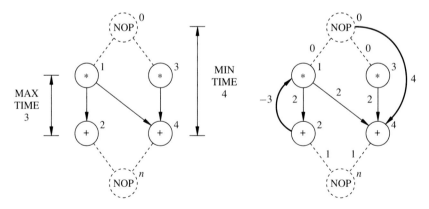

FIGURE 5.4
Example of a constraint graph, with minimum and maximum relative timing constraints.

schedule is that the constraint graph does not have positive cycles. The condition is also sufficient, as stated by Theorem 2.4.1 in Section 2.4.1.

The existence of a schedule under timing constraints can be checked using the Bellman-Ford algorithm, described in Section 2.4.1. It is often the case that the number of maximum timing constraints is small when compared to the number of edges in the constraint graph. Then, relaxation-based algorithms like Liao-Wong's can be more efficient. When a schedule exists, the weight of the longest path in the constraint graph from the source to a vertex is also the minimum start time, as shown in Section 2.4.1. Thus the Bellman-Ford or the Liao-Wong algorithm provides also the schedule.

Example 5.3.4. A schedule for the constraint graph of the Example 5.3.3, satisfying timing constraints, is given by the following table:

Vertex	Start time
v_0	1
v_1	1
v_2	3
v_3	1
v_4	5
v_n	6

Note that operation v_4 is delayed by two cycles when compared to the unconstrained case. The difference in start times of operations v_2, v_1 is two cycles, less than the required upper bound. The latency is $\lambda = t_n - t_0 = 6 - 1 = 5$.

5.3.4   Relative Scheduling*

We extend scheduling to the case of operations with *unbounded delays*. We assume that these operations issue completion signals when execution is finished. We associate also a start signal to the source vertex, which is also its completion signal (because the source is a *No-Operation*), that represents a means of activating the operations modeled by the graph. The scheduling problem can still be modeled by a sequencing graph $G_s(V, E)$, where a subset of the vertices has unspecified execution delay. Such vertices, as well as the source vertex, provide a frame of reference for determining the start time of the operations.

Definition 5.3.2. The **anchors** of a constraint graph $G(V, E)$ consist of the source vertex v_0 and of all vertices with unbounded delay. Anchors are denoted by $A \subseteq V$.

The start time of the operations cannot be determined on an absolute scale in this case. Nevertheless, the start times of the operations can be computed as a function of the completion signals of the anchors and of the schedule of the operations *relative* to the anchors.

Example 5.3.5. Consider the sequencing graph of Figure 4.8 (a), reported again in Figure 5.5. There are three operations with known delays v_1, v_2, v_3 and one synchronization

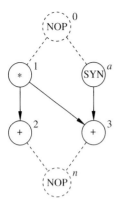

FIGURE 5.5
Example of a sequencing graph, with a synchronization operation with unknown delay.

operation, denoted by a. We assume again one cycle for addition and two for multiplication. The start times of the operations depend on the start time of the source vertex t_0 and the completion signal of the synchronization operation. Such a signal arrives at time $t_a + d_a$, where t_a is the time at which the synchronizer is activated and d_a is the unknown synchronization delay. The start times of v_1 and v_2 can be computed with reference to t_0. Namely, v_1 can start at t_0 and v_2 can start at $t_0 + 2$. The third operation can start no earlier than the synchronization signal at time $t_a + d_a$ and no earlier than $t_0 + 2$, i.e., its start time is $\max\{t_0 + 2; t_a + d_a\}$.

We summarize now a method, known as relative scheduling [10, 11], to schedule the sequencing graph relative to the anchors. We require the following definitions.

Definition 5.3.3. A **defining path** $\rho(a, v_i)$ from anchor a to vertex $v_i \in V$ is a path in $G_s(V, E)$ with one and only one unbounded weight d_a.

Definition 5.3.4. The **relevant anchor set** of a vertex $v_i \in V$ is the subset of anchors $R(v_i) \subseteq A$ such that $a \in R(v_i)$ if there exists a defining path $\rho(a, v_i)$.

The defining path denotes the dependency of the operation v_i on a given anchor a. Note that when considering one path only and when anchors are cascaded along the path, only the last one affects the start time of the operation at the head of the path. This motivates the restriction to at most one unbounded weight on the path. In addition, not all relevant anchors may affect the start time of a vertex. Some relevant anchors may still be *redundant* for the computation of the start time of a vertex. Let us denote by $|\rho(a, v_i)|$ the weight of the longest defining path, excluding d_a.

Definition 5.3.5. An anchor a is **redundant** for vertex v_i when there is another relevant anchor $b \in R(v_i)$ such that $|\rho(a, v_i)| = |\rho(a, b)| + |\rho(b, v_i)|$.

For any given vertex $v_i \in V$, the *irredundant relevant anchor set* $IR(v_i) \subseteq R(v_i)$ represents the smallest subset of anchors that affects the start time of that vertex.

Example 5.3.6. Consider again the sequencing graph of Figure 5.5. The relevant anchor sets are as follows: $R(v_1) = \{v_0\}$; $R(v_2) = \{v_0\}$; $R(v_3) = \{v_0, v_a\}$. They correspond to the irredundant anchor sets.

For the sake of the example, assume that operation v_1 executes in zero cycles. Then the source vertex is redundant for v_3, because operation v_3 can never be activated sooner than the completion time of v_a. Hence $IR(v_3) = \{v_a\}$.

The start times of the operations are defined on the basis of partial schedules relative to the *completion* time of each anchor in their irredundant relevant anchor sets as follows. Let t_i^a be the schedule of operation v_i with respect to anchor a, computed on the polar subgraph induced by a and its successors, assuming that a is the source of the subgraph and that all anchors have zero execution delay. Then:

$$t_i = \max_{a \in IR(v_i)} \{t_a + d_a + t_i^a\} \tag{5.4}$$

Note that if there are no operations with unbounded delays, then the start times of all operations will be specified in terms of time offsets from the source vertex, which reduces to the traditional scheduling formulation.

A *relative schedule* is a collection of schedules with respect to each anchor, or equivalently a set of offsets with respect to the irredundant relevant anchors for each vertex. Thus relative scheduling consists of the computation of the offset values t_i^a for all irredundant relevant anchors $a \in R(v_i)$ of each vertex v_i; $i = 0, 1, \ldots, n$. From a practical point of view, the start times of Equation 5.4 cannot be computed. However, the offset values and the anchor completion signals are sufficient to construct a control circuit that activates the operations at the appropriate times, as shown by the following example.

Example 5.3.7. Consider the graph of Figure 5.5. The irredundant relevant anchor sets and the offset with respect to the anchors v_0 and a are reported in the following table:

| Vertex | Irredundant relevant anchor set | Offsets | |
v_i	$IR(v_i)$	t_0	t_a
a	$\{v_0\}$	0	-
v_1	$\{v_0\}$	0	-
v_2	$\{v_0\}$	2	-
v_3	$\{v_0, a\}$	2	0

A control unit, corresponding to a relative schedule, can be thought of as a set of *relative control blocks*, each one related to an anchor of the sequencing graph, and a set of synchronizers. Each relative control unit can be implemented by microcode or be hard wired. Each block implements the schedule relative to an anchor and it is started by the corresponding *completion* signal. Each operation, with two or more elements in its irredundant relevant anchor set, has a synchronizer circuit that activates the operation in response to the conjunction of the arrival (past or present) of the relative activation signals coming from the relative control blocks.

A control unit for the graph of Figure 5.5 is shown in Figure 5.6. For the sake of the example, the first relative control unit is shown in a microcode-based implementation, while the second is trivial. Other styles of implementing control units based on relative scheduling are described in reference [11].

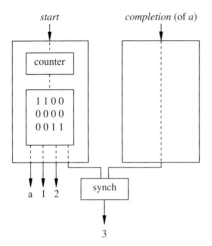

FIGURE 5.6
Example of a control unit, consisting of two relative control blocks and a synchronizer. The second relative control block is trivial.

RELATIVE SCHEDULING UNDER TIMING CONSTRAINTS.* Scheduling under timing constraints is more complex in the presence of operations with unbounded delays. The constraint graph formulation still applies although the weights on the edges whose tails are anchors are unknown. The definitions of anchor sets and defining paths are then extended to the constraint graph in a straightforward way.

Also in this case a schedule may or may not exist under the timing constraint. It is important to be able to assess the existence of a schedule for any value of the unbounded delays, because these values are not known when the schedule is computed. For this reason, we introduce the following definitions.

> **Definition 5.3.6.** A constraint graph is **feasible** if all timing constraints are satisfied when the execution delays of the anchors are zero.

Feasibility is a necessary condition for the existence of a relative schedule, and it can be verified by the Bellman-Ford or Liao-Wong algorithm. Unfortunately it is not sufficient to guarantee the satisfaction of all constraints for all possible delay values of the anchors.

> **Definition 5.3.7.** A constraint graph is **well-posed** if it can be satisfied for all values of the execution delays of the anchors.

Note that well-posedness implies feasibility.

> **Example 5.3.8.** The graph of Figure 5.7 (a) is ill-posed. Indeed, the maximum timing constraint bounds from above the time separation between operations v_i and v_j, while the time separation between these two operations depends on the unknown delay of anchor a. Therefore the constraints may be satisfied only for some values of d_a, if any at all.
>
> The graph of Figure 5.7 (b) shows two operations in two independent streams. A maximum timing constraint bounds again from above by the time separation between operations v_i and v_j. The constraint graph is ill-posed, because there may be delay values of anchor a_2 that would violate the constraint.

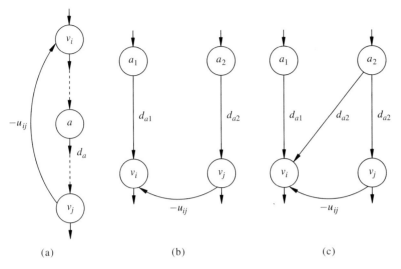

FIGURE 5.7
Examples of ill-posed timing constraints (a, b) and a well-posed constraint (c).

Let us assume now that the graph of Figure 5.7 (c) admits a schedule for $d_{a_2} = 0$. Then it is possible to verify that the graph has a schedule for any value of d_{a_2}. It is interesting to note that this graph can be derived from the second one by adding a sequencing dependency from a_2 to v_i.

Relative schedules can be defined only for well-posed graphs. Some feasible graphs that are not well-posed can be transformed into well-posed ones by serializing some operations. This transformation can be applied under the assumptions of the following theorem.

Theorem 5.3.1. A feasible constraint graph $G_c(V_c, E_c)$ is well-posed or it can be made well-posed if and only if no cycles with unbounded weight exist in $G_c(V_c, E_c)$.

Proof. The proof is reported in reference [10].

Example 5.3.9. Refer again to the three graphs of Figure 5.7. The first graph has an unbounded cycle and cannot be made well-posed. The second graph can be made well-posed by adding the edge (a_2, v_i).

Ku developed polynomial-time algorithms for detecting well-posedness and for making a graph well-posed (when possible) as part of a comprehensive theory of relative scheduling [10, 11].

The minimum-latency schedule of an operation with respect to an anchor can be computed as the weighted longest path from the anchor to the corresponding vertex when all unbounded delays are set to zero. Therefore the offsets with respect to the anchors can be computed by repeated applications of the Bellman-Ford or the Liao-Wong algorithm or by an iterative incremental algorithm presented in reference [10].

Once the offsets have been computed, the start times of the operations can be derived by Equation 5.4. We refer the interested reader to [10, 11] for further details.

5.4 SCHEDULING WITH RESOURCE CONSTRAINTS

Scheduling under resource constraints is an important and difficult problem. Resource constraints are motivated by the fact that the resource usage determines the circuit area of resource-dominated circuits and represents a significant component of the overall area for most other circuits.

The solution of scheduling problems under resource constraints provides a means for computing the (area/latency) trade-off points. In practice two difficulties arise. First, the resource-constrained scheduling problem is intractable, and only approximate solutions can be found for problems of reasonable size. Second, the area-performance trade-off points are affected by other factors when considering non-resource-dominated circuits.

We consider first modeling and exact solution methods for the resource-constrained scheduling problem. We describe then heuristic algorithms.

5.4.1 The Integer Linear Programming Model

A formal model of the scheduling problem under resource constraints can be achieved by using binary decision variables with two indices: $X = \{x_{il}; i = 0, 1, \ldots, n; l = 1, 2, \ldots, \bar{\lambda} + 1\}$. (The range of the indexes is justified by the fact that we consider also the source and sink operations and that we start the schedule on cycle 1. We use the set notation for the variables in X, rather than a matrix notation, because the expressions based on scalar variables are easier to understand.) The number $\bar{\lambda}$ represents an upper bound on the latency, because the schedule latency is unknown. The bound can be computed by using a fast heuristic scheduling algorithm, such as a list scheduling algorithm (Section 5.4.3).

The indices of the binary variables relate to the operations and schedule steps respectively. In particular, a binary variable, x_{il}, is 1 only when operation v_i starts in step l of the schedule, i.e., $l = t_i$. Equivalently, we can write $x_{il} = \delta_{t_i,l}$ by using the Kronecker delta notation.

We denote the summations over all operations as $\sum_i$ (instead of $\sum_{i=0}^{n}$) and those over all schedule steps as $\sum_l$ (instead of $\sum_{l=1}^{\bar{\lambda}+1}$) for the sake of simplicity. Note that upper and lower bounds on the start times can be computed by the ASAP and ALAP algorithms on the corresponding unconstrained problem. Thus x_{il} is necessarily zero for $l < t_i^S$ or $l > t_i^L$ for any operation v_i; $i = 0, 1, \ldots, n$. Therefore the summations over the time steps l with argument x_{il} can be restricted to $\sum_{l=t_i^S}^{t_i^L}$ for any $i; i = 0, 1, \ldots, n$.

We consider now constraints on the binary variables X to model the resource-constrained scheduling problem. First, the start time of each operation is unique:

$$\sum_l x_{il} = 1, \quad i = 0, 1, \ldots, n \tag{5.5}$$

Therefore the start time of any operation $v_i \in V$ can be stated in terms of x_{il} as $t_i = \sum_l l \cdot x_{il}$.

Second, the sequencing relations represented by $G_s(V, E)$ must be satisfied. Therefore, $t_i \geq t_j + d_j \ \forall i, j : (v_j, v_i) \in E$ implies:

$$\sum_l l \cdot x_{il} \geq \sum_l l \cdot x_{jl} + d_j, \quad i, j = 0, 1, \ldots, n \quad : (v_j, v_i) \in E \qquad (5.6)$$

Third, the resource bounds must be met at every schedule time step. An operation v_i is executing at time step l when $\sum_{m=l-d_i+1}^{l} x_{im} = 1$. The number of all operations executing at step l of type k must be lower than or equal to the upper bound a_k. Namely:

$$\sum_{i:T(v_i)=k} \sum_{m=l-d_i+1}^{l} x_{im} \leq a_k, \quad k = 1, 2, \ldots, n_{res}, \quad l = 1, 2, \ldots, \bar{\lambda} + 1 \qquad (5.7)$$

Let us denote by $\mathbf{t}$ the vector whose entries are the start times. Then, the *minimum-latency* scheduling problem *under resource constraints* can be stated as follows:

$$\text{minimize } \mathbf{c}^T \mathbf{t} \text{ such that}$$

$$\sum_l x_{il} = 1, \quad i = 0, 1, \ldots, n \qquad (5.8)$$

$$\sum_l l \cdot x_{il} - \sum_l l \cdot x_{jl} - d_j \geq 0, \quad i, j = 0, 1, \ldots, n \quad : (v_j, v_i) \in E \qquad (5.9)$$

$$\sum_{i:T(v_i)=k} \sum_{m=l-d_i+1}^{l} x_{im} \leq a_k, \quad k = 1, 2, \ldots, n_{res}, \quad l = 1, 2, \ldots, \bar{\lambda} + 1 \qquad (5.10)$$

$$x_{il} \in \{0, 1\}, \quad i = 0, 1, \ldots, n, \quad l = 1, 2, \ldots, \bar{\lambda} + 1 \qquad (5.11)$$

Note that it suffices to require the variables in X to be non-negative integers to satisfy 5.11. Hence the problem can be formulated as an ILP and not necessarily as a ZOLP. (See Section 2.5.3.)

The choice of vector $\mathbf{c} \in \mathbf{Z}^n$ relates to slightly different optimization goals. Namely $\mathbf{c} = [0, \ldots, 0, 1]^T$ corresponds to minimizing the latency of the schedule, because $\mathbf{c}^T \mathbf{t} = t_n$ and an optimal solution implies $t_0 = 1$. Therefore the objective function in terms of the binary variables is $\sum_l l \cdot x_{nl}$. The selection of $\mathbf{c} = \mathbf{1} = [1, \ldots, 1]^T$ corresponds to finding the earliest start times of all operations. This is equivalent to minimizing $\sum_i \sum_l l \cdot x_{il}$.

This model can be enhanced to support relative timing constraints by adding the corresponding constraint inequality in terms of the variables X. For example, a maximum timing constraint u_{ij} on the start times of operations v_i, v_j can be expressed as $\sum_l l \cdot x_{jl} \leq \sum_l l \cdot x_{il} + u_{ij}$.

Example 5.4.1. Let us consider again the sequencing graph of Figure 5.1. We assume that there are two types of resources: a multiplier and an ALU that performs addition/subtraction and comparison. Both resources execute in one cycle. We also assume that the upper bounds on the number of both resources is 2; i.e., $a_1 = 2$ and $a_2 = 2$.

By using a heuristic (list scheduling) algorithm we find an upper bound on the latency of $\bar{\lambda} = 4$ steps. We defer a detailed description of such an algorithm to Example 5.4.7. By applying the ASAP and ALAP algorithms on the corresponding unconstrained model, we can derive bounds on the start times. Incidentally, by noticing that the latency computed by the unconstrained ASAP algorithm matches the upper bound computed by the heuristic algorithm, we realize that the heuristic solution is an optimum. We pretend now not to pay attention to this fact, for the sake of showing the ILP constraint formulation.

Let us consider the constraint sets one at a time. First, all operations must start only once:

$$x_{0,1} = 1$$

$$x_{1,1} = 1$$

$$x_{2,1} = 1$$

$$x_{3,2} = 1$$

$$x_{4,3} = 1$$

$$x_{5,4} = 1$$

$$x_{6,1} + x_{6,2} = 1$$

$$x_{7,2} + x_{7,3} = 1$$

$$x_{8,1} + x_{8,2} + x_{8,3} = 1$$

$$x_{9,2} + x_{9,3} + x_{9,4} = 1$$

$$x_{10,1} + x_{10,2} + x_{10,3} = 1$$

$$x_{11,2} + x_{11,3} + x_{11,4} = 1$$

$$x_{n,5} = 1$$

We consider then the constraints based on sequencing. We report here the non-trivial constraints only, i.e., those involving more than one possible start time for at least one operation:

$$2x_{7,2} + 3x_{7,3} - x_{6,1} - 2x_{6,2} - 1 \geq 0$$

$$2x_{9,2} + 3x_{9,3} + 4x_{9,4} - x_{8,1} - 2x_{8,2} - 3x_{8,3} - 1 \geq 0$$

$$2x_{11,2} + 3x_{11,3} + 4x_{11,4} - x_{10,1} - 2x_{10,2} - 3x_{10,3} - 1 \geq 0$$

$$4x_{5,4} - 2x_{7,2} - 3x_{7,3} - 1 \geq 0$$

$$5x_{n,5} - 2x_{9,2} - 3x_{9,3} - 4x_{9,4} - 1 \geq 0$$

$$5x_{n,5} - 2x_{11,2} - 3x_{11,3} - 4x_{11,4} - 1 \geq 0$$

Finally we consider the resource constraints:

$$x_{1,1} + x_{2,1} + x_{6,1} + x_{8,1} \leq 2$$

$$x_{3,2} + x_{6,2} + x_{7,2} + x_{8,2} \leq 2$$

$$x_{7,3} + x_{8,3} \leq 2$$

$$x_{10,1} \leq 2$$

$$x_{9,2} + x_{10,2} + x_{11,2} \leq 2$$

$$x_{4,3} + x_{9,3} + x_{10,3} + x_{11,3} \leq 2$$

$$x_{5,4} + x_{9,4} + x_{11,4} \leq 2$$

Any set of start times satisfying these constraints provides us with a feasible solution. In this particular case, the lack of mobility of the sink ($x_{n,5} = 1$) shows that any feasible solution is optimum for $\mathbf{c} = [0, \ldots, 0, 1]^T$.

On the other hand, when considering $\mathbf{c} = [1, \ldots, 1]^T$, an optimum solution minimizes $\sum_i \sum_l l \cdot x_{il}$, or equivalently:

$$x_{6,1} + 2x_{6,2} + 2x_{7,2} + 3x_{7,3} + x_{8,1} + 2x_{8,2} + 3x_{8,3} + 2x_{9,2} + 3x_{9,3} +$$

$$+ 4x_{9,4} + x_{10,1} + 2x_{10,2} + 3x_{10,3} + 2x_{11,2} + 3x_{11,3} + 4x_{11,4}$$

A solution is shown in Figure 5.8.

We consider next the *minimum-resource* scheduling problem *under latency constraints*, which is the dual of the previously considered problem.

The optimization goal is a weighted sum of the resource usage represented by **a**. Hence the objective function can be expressed by $\mathbf{c}^T \mathbf{a}$, where $\mathbf{c} \in \mathbf{R}^{n_{res}}$ is a vector whose entries are the individual resource (area) costs. Inequality constraints 5.8, 5.9 and 5.10 still hold; in the last inequality the resource usage $a_k; k = 1, 2, \ldots, n_{res}$, are now unknown auxiliary (slack) variables. The latency constraint is expressed by $\sum_l l x_{nl} \leq \bar{\lambda} + 1$.

Example 5.4.2. Let us consider again the sequencing graph of Figure 5.1. We assume again two types of resources: a multiplier and an ALU, both executing in one cycle.

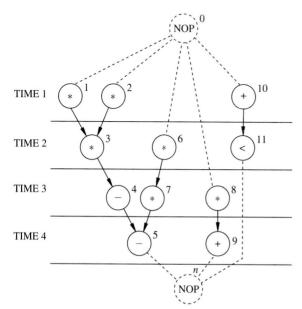

FIGURE 5.8
Optimum schedule under resource constraints.

Let us assume that the multiplier costs five units of area and the ALU one unit. Hence $\mathbf{c} = [5, 1]^T$. We assume that an upper bound on the latency is $\bar{\lambda} = 4$.

The uniqueness constraints on the start time of the operations and the sequencing dependency constraints are the same as those shown in Example 5.4.1. The resource constraints are in terms of the unknown variables a_1 and a_2:

$$x_{1,1} + x_{2,1} + x_{6,1} + x_{8,1} - a_1 \leq 0$$

$$x_{3,2} + x_{6,2} + x_{7,2} + x_{8,2} - a_1 \leq 0$$

$$x_{7,3} + x_{8,3} - a_1 \leq 0$$

$$x_{10,1} - a_2 \leq 0$$

$$x_{9,2} + x_{10,2} + x_{11,2} - a_2 \leq 0$$

$$x_{4,3} + x_{9,3} + x_{10,3} + x_{11,3} - a_2 \leq 0$$

$$x_{5,4} + x_{9,4} + x_{11,4} - a_2 \leq 0$$

The objective function to minimize is $\mathbf{c}^T \mathbf{a} = 5 \cdot a_1 + 1 \cdot a_2$. A solution is shown in Figure 5.8, with cost equal to 12 units.

ILP SOLUTION. The ILP formulation of the constrained scheduling problems is attractive for three major reasons. First it provides an exact solution to the scheduling problems. Second, general purpose software packages can be used to solve the ILP. Last, additional constraints and problem extensions (e.g., scheduling pipelined circuits) can be easily incorporated. The disadvantage of the ILP formulation is the computational complexity of the problem. The number of variables, the number of inequalities and their tightness affect the ability of computer programs to find a solution.

Gebotys *et al.* analyzed the ILP model of the scheduling and binding problems in detail. They developed a set of tighter constraints for these problems that reduce the feasible solution region and as a consequence the computing time of an optimum solution [4]. Moreover, the addition of these constraints have the desirable property that the computed solution is often integer (and in particular binary valued) even when the integer constraints on the decision variables are relaxed. This allows us to use linear program (LP) solvers, instead of ILP solvers, to compute the exact solution in shorter time in some cases or bounds on the exact solution in the general case. These bounds can be used as a starting point for computing an optimum solution by a subsequent application of an ILP solver.

Practical implementations of ILP schedulers have been shown to be efficient for medium scale examples but to fail to solve problems with several hundreds of variables or constraints.

5.4.2   Multiprocessor Scheduling and Hu's Algorithm

Resource-constrained scheduling problems have been the subject of intensive investigation due to their relevance to the field of operations research and to the challenge posed by their intrinsic difficulty. When we assume that all operations can be solved by the same type of resource, the problem is often referred to as a *precedence-constrained multiprocessor* scheduling problem, which is a well-known intractable problem. The problem remains difficult to solve even when all operations have unit execution delay. With both these simplifications, the resource-constrained minimum-latency scheduling problem can be restated as follows:

$$\text{minimize } \mathbf{c}^T \mathbf{t} \text{ such that}$$

$$\sum_l x_{il} = 1, \quad i = 0, 1, \dots, n \tag{5.12}$$

$$\sum_l l \cdot x_{il} - \sum_l l \cdot x_{jl} \geq 1, \quad i, j = 0, 1, \dots, n : (v_j, v_i) \in E \tag{5.13}$$

$$\sum_i x_{il} \leq a, \quad l = 1, 2, \dots, \bar{\lambda} + 1 \tag{5.14}$$

$$x_{il} \in \{0, 1\}, \quad i = 0, 1, \dots, n, \quad l = 1, 2, \dots, \bar{\lambda} + 1 \tag{5.15}$$

where $\mathbf{c}$ is a suitable vector. The latency-constrained minimum-latency problem can be restated in a similar way.

Despite the simplifications, these two scheduling problems are still relevant in architectural synthesis. For example, all operations in a DSP may be implemented by ALUs having the same execution delay.

We compute first a lower bound on the number of resources required to schedule a graph with a latency constraint under these assumptions. The bound was derived first by Hu [8]. We use the sequencing graph model where the source vertex is ignored, because it is irrelevant as far as this problem is concerned.

Definition 5.4.1. A **labeling** of a sequencing graph consists of marking each vertex with the weight of its longest path to the sink, measured in terms of edges.

Let us denote the labels by $\{\alpha_i; i = 1, 2, \dots, n\}$ and let $\alpha = \max_{1 \leq i \leq n} \alpha_i$. Let $p(j)$ be the number of vertices with label equal to j, i.e., $p(j) = |\{v_i \in V : \alpha_i = j\}|$. It is obvious that the latency is greater than or equal to the weight of the longest path, i.e., $\lambda \geq \alpha$.

Example 5.4.3. Let us consider the graph of Figure 5.1, where we assume that all operations can be executed by a general purpose ALU with a unit execution delay. A labeled sequencing graph is shown in Figure 5.9. In this example, $\alpha = 4$ and obviously $\lambda \geq 4$. Also $p(0) = 1, p(1) = 3, p(2) = 4, p(3) = 2, p(4) = 2$.

Theorem 5.4.1. A lower bound on the number of resources to complete a schedule with latency λ is

$$\bar{a} = \max_\gamma \left\lceil \frac{\sum_{j=1}^{\gamma} p(\alpha + 1 - j)}{\gamma + \lambda - \alpha} \right\rceil \tag{5.16}$$

where γ is a positive integer.

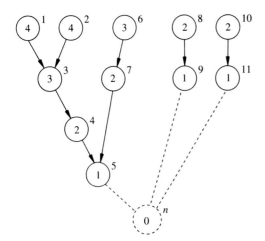

FIGURE 5.9
Labeled scheduled sequencing graph. Labels are shown inside the vertices.

Proof. Let

$$\gamma^* = \arg\max \left\lceil \frac{\sum_{j=1}^{\gamma} p(\alpha + 1 - j)}{\gamma + \lambda - \alpha} \right\rceil \tag{5.17}$$

For the sake of contradiction, let us assume that a schedule exists with a resources that satisfy the latency bound λ, where:

$$a < \left\lceil \frac{\sum_{j=1}^{\gamma^*} p(\alpha + 1 - j)}{\gamma^* + \lambda - \alpha} \right\rceil \tag{5.18}$$

The vertices scheduled up to step l cannot exceed $a \cdot l$. At schedule step $l = \gamma^* + \lambda - \alpha$ the scheduled vertices are at most:

$$a \cdot (\gamma^* + \lambda - \alpha) < \sum_{j=1}^{\gamma^*} p(\alpha + 1 - j) \tag{5.19}$$

This implies that at least a vertex with label $\alpha_i = p(\alpha + 1 - \gamma^*)$ has not been scheduled yet. Therefore, to schedule the remaining portion of the graph we need at least $\alpha + 1 - \gamma^*$ steps. Thus, the schedule length is at least:

$$\gamma^* + \lambda - \alpha + \alpha + 1 - \gamma^* = \lambda + 1 \tag{5.20}$$

which contradicts our hypothesis of satisfying a latency bound of λ. Therefore, at least $\bar{a}$ resources are required.

Example 5.4.4. Consider again the problem of Example 5.4.3, where $\alpha = 4$. Let us compute a resource lower bound to achieve a latency $\lambda = 4$ cycles. Then:

$$\bar{a} = \left\lceil \max \left\{ \frac{p(4)}{1}, \frac{p(4) + p(3)}{2}, \frac{p(4) + p(3) + p(2)}{3}, \frac{p(4) + p(3) + p(2) + p(1)}{4}, \right. \right.$$

$$\left. \left. \frac{p(4) + p(3) + p(2) + p(1) + p(0)}{5} \right\} \right\rceil$$

$$= \lceil \max\{2, 2, 8/3, 11/4, 12/5\} \rceil$$

$$= 3$$

```
HU( G_s(V, E), a ) {
        Label the vertices;
        l = 1;
        repeat {
                U = unscheduled vertices in V without predecessors
                        or whose predecessors have been scheduled;
                Select S ⊆ U vertices, such that |S| ≤ a and labels in S are maximal;
                Schedule the S operations at step l by setting t_i = l ∀i : v_i ∈ S;
                l = l + 1;
        }
        until (v_n is scheduled);
        return (t);
}
```

ALGORITHM 5.4.1

Hence, 3 resources are required to schedule the operations in 4 steps.

This theorem can be applied to derive resource bounds even when operations take more than one cycle, by modeling them as sequences of single-cycle operations.

In the remaining part of this section, we assume that the sequencing graph $G_s(V, E)$ (after the source vertex has been deleted) is a tree. Under this additional simplification, the problem can be solved exactly in polynomial time. This new assumption makes the problem less relevant to hardware synthesis, because sequencing graphs are not trees in general. Indeed, such a model applies well to the assembly line problems, where products are constructed by assembling subcomponents and the composition rules form a tree. We present here the algorithm proposed by Hu [8] for two major reasons. First, Hu's algorithm is one of the few exact polynomial-time algorithms for resource-constrained scheduling. Second, some heuristic algorithms for solving the general scheduling problem are based on Hu's ideas.

Hu's Algorithm 5.4.1 applies a greedy strategy. At each step, it schedules as many operations as possible among those whose predecessors have been scheduled. In particular, the selection is based on the labels. Vertices with the largest labels are chosen first. Let a denote the upper bound on the resource usage. The computational complexity of the algorithm is $O(n)$.

> **Example 5.4.5.** Let us consider the labeled graph of Figure 5.9, with a resource bound $a = 3$. Then, the first iteration of Hu's algorithm would select $U = \{v_1, v_2, v_6, v_8, v_{10}\}$ and schedule operations $\{v_1, v_2, v_6\}$ at the first time step, because their labels $\{\alpha_1 = 4, \alpha_2 = 4, \alpha_6 = 3\}$ are not smaller than any other label of unscheduled vertices in U. At the second iteration $U = \{v_3, v_7, v_8, v_{10}\}$ and $\{v_3, v_7, v_8\}$ are scheduled at the second time step. Operations $\{v_4, v_9, v_{10}\}$ are scheduled at the third step, $\{v_5, v_{11}\}$ at the fourth and $\{v_n\}$ at the last.

Whereas the idea in Hu's algorithm is very simple and intuitive, its power lies in the fact that the computed solution is an optimum one. We analyze here the algorithm in a formal way. We show first that the algorithm can always achieve a latency λ with $\bar{a}$ resources, where $\bar{a}$ is defined as in Theorem 5.4.1. Note that $\bar{a}$ is a function of λ. Since the number of resources used by the algorithm is equal to the lower bound for the problem (Theorem 5.4.2), the algorithm achieves an optimum schedule with minimum resource usage under latency constraints. As a consequence, the algorithm achieves also the minimum-latency schedule under resource constraints, as shown by Corollary 5.4.1.

Let us consider the operations scheduled by the algorithm. The algorithm schedules a operations at each step, starting from the first one until a *critical* step, after which less than a resources can be scheduled due to the precedence constraints. We denote by c the critical step. Then $c + 1$ is the first step in which less than a operations are scheduled. The vertices scheduled up to the critical step are $a \cdot c$ and those scheduled up to step $c + 1$ are $a \cdot (c + \delta)$ where $0 < \delta < 1$.

We denote by γ' the largest integer such that all vertices with labels larger than or equal to $\alpha + 1 - \gamma'$ have been scheduled up to the critical step and the following one, i.e., up to step $c + 1$.

At any schedule step following the critical step, all unscheduled vertices with the largest label will be less than a in number. Then $\alpha - \gamma'$ schedule steps are used by the algorithm to schedule the remaining operations after step $c + 1$.

> **Example 5.4.6.** Consider the schedule computed by Hu's algorithm in Example 5.4.5. The critical step is the third one, i.e., $c = 3$, because thereafter fewer than $a = 3$ resources are assigned to each step. When considering the steps up to $c + 1 = 4$, all vertices with labels larger than or equal to 2 have been scheduled.
>
> Hence $\gamma' = 3$. Thus $\alpha - \gamma' = 4 - 3 = 1$ step is needed to complete the schedule after step $c + 1 = 4$. Another way of saying this is that only vertices with label 0 are left to be scheduled.

The number of steps to reach the critical point is derived analytically by the following theorem.

> **Theorem 5.4.2.** Hu's algorithm achieves latency λ with as many resources as:
>
> $$\bar{a} = \max_{\gamma} \left\lceil \frac{\sum_{j=1}^{\gamma} p(\alpha + 1 - j)}{\gamma + \lambda - \alpha} \right\rceil \tag{5.21}$$
>
> where γ is a positive integer.
>
> **Proof.** The resource constraint implies that:
>
> $$\bar{a} \geq \left\lceil \frac{\sum_{j=1}^{\gamma'} p(\alpha + 1 - j)}{\gamma' + \lambda - \alpha} \right\rceil \tag{5.22}$$
>
> and therefore:
>
> $$\bar{a} \cdot (\gamma' + \lambda - \alpha) \geq \sum_{j=1}^{\gamma'} p(\alpha + 1 - j) \tag{5.23}$$

Since $\sum_{j=1}^{\gamma'} p(\alpha + 1 - j) \geq \overline{a} \cdot (c + \delta)$ by definition of γ', then

$$\gamma' + \lambda - \alpha \geq c + \delta \qquad (5.24)$$

which implies that $c \leq \lambda - \alpha + \gamma' - 1 + (1 - \delta)$. By noting that all quantities are integers, except δ, which satisfies $0 \leq \delta \leq 1$,

$$c \leq \lambda - \alpha + \gamma' - 1 \qquad (5.25)$$

$$c + 1 \leq \lambda - \alpha + \gamma' \qquad (5.26)$$

By recalling that the $\alpha - \gamma'$ schedule steps are used by the algorithm to schedule the remaining operations after step $c + 1$, the total number of steps used by the algorithm is:

$$\lambda - \alpha + \gamma' \;\; + \;\; \alpha - \gamma' \;\; = \;\; \lambda \qquad (5.27)$$

We turn now to the minimum-latency scheduling problem under resource constraints. We show that Hu's algorithm gives the minimum-length schedule for a given resource constraint.

Corollary 5.4.1. Let λ be the latency of the schedule computed by Hu's algorithm with a resources. Then, any schedule with a resources has latency larger than or equal to λ.

Proof. Let us assume first that $\lambda = \alpha$. Then the computed number of steps is minimum, because it is equal to the weight of the longest path. Let us assume that $\lambda > \alpha$. Select the smallest integer $\lambda' > \alpha$ such that

$$\max_{\gamma} \left\lceil \frac{\sum_{j=1}^{\gamma} p(\alpha + 1 - j)}{\gamma + \lambda' - \alpha} \right\rceil \leq a < \max_{\gamma} \left\lceil \frac{\sum_{j=1}^{\gamma} p(\alpha + 1 - j)}{\gamma + \lambda' - \alpha - 1} \right\rceil \qquad (5.28)$$

Then by Theorem 5.4.2 we know that Hu's algorithm can schedule all operations with latency $\lambda = \lambda'$. By Theorem 5.4.1 we know that no schedule exists with a resources and latency $\lambda' - 1$. Then the algorithm schedules all operations with a minimum latency.

It is important to remark that a key factor in achieving the optimality in Hu's algorithm is to have single paths from each vertex to the sink with monotonically unit-wise decreasing labels. For generic sequencing graphs (i.e., not restricted to be trees) and multiple-cycle operations (which can be modeled by a sequence of unit-delay operations), the lower bound of Theorem 5.4.1 still holds. However, it is easy to verify that the algorithm does not meet the bound for some cases.

5.4.3 Heuristic Scheduling Algorithms: List Scheduling

Practical problems in hardware scheduling are modeled by generic sequencing graphs, with (possibly) multiple-cycle operations with different types. With this model, the *minimum-latency* resource-constrained scheduling problem and the *minimum-resource*

```
LIST_L( G_s(V, E), a ) {
    l = 1;
    repeat {
        for each resource type k = 1, 2, ..., n_res {
            Determine candidate operations U_{l,k};
            Determine unfinished operations T_{l,k};
            Select S_k ⊆ U_{l,k} vertices, such that |S_k| + |T_{l,k}| ≤ a_k;
            Schedule the S_k operations at step l by setting t_i = l  ∀i : v_i ∈ S_k;
        }
        l = l + 1;
    }
    until (v_n is scheduled);
    return (t);
}
```

ALGORITHM 5.4.2

latency-constrained problem are known to be intractable. Therefore, heuristic algorithms have been researched and used. We consider in this section a family of algorithms called *list scheduling* algorithms.

We consider first the problem of minimizing latency under resource constraints, represented by vector **a**. Algorithm 5.4.2 is an extension of Hu's algorithm to handle multiple operation types and multiple-cycle execution delays.

The candidate operations $U_{l,k}$ are those operations of type k whose predecessors have already been scheduled early enough, so that the corresponding operations are completed at step l. Namely: $U_{l,k} = \{v_i \in V : \mathcal{T}(v_i) = k \text{ and } t_j + d_j \leq l \;\; \forall j : (v_j, v_i) \in E\}$ for any resource type $k = 1, 2, \ldots, n_{res}$. The unfinished operations $T_{l,k}$ are those operations of type k that started at earlier cycles and whose execution is not finished at step l. Namely: $T_{l,k} = \{v_i \in V : \mathcal{T}(v_i) = k \text{ and } t_i + d_i > l\}$. Obviously, when the execution delays are 1, the set of unfinished operations is empty.

The computational complexity of the algorithm is $O(n)$. It constructs a schedule that satisfies the resource constraints by construction. However, the computed schedule may not have minimum latency.

The list scheduling algorithms are classified according to the selection step. A *priority list* of the operations is used in choosing among the operations, based on some heuristic urgency measure.

A common priority list is to label the vertices with weights of their longest path to the sink and to rank them in decreasing order. The most urgent operations are scheduled first. Note that when the operations have unit delay and when there is only one resource type, the algorithm is the same as Hu's and it yields an optimum solution for tree-structured sequencing graphs.

Scheduling under resource and relative timing constraints can be handled by list scheduling [12]. In particular, minimum timing constraints can be dealt with by delaying the selection of operations in the candidate set. The priority list is modified to reflect the proximity of an unscheduled operation to a deadline related to a maximum timing constraint. Schedules constructed by the algorithms satisfy the required constraints by construction. Needless to say, the heuristic nature of list scheduling may prevent finding a solution that may exist.

Example 5.4.7. Let us consider the sequencing graph of Figure 5.1. Assume that all operations have unit delay. Assume that $a_1 = 2$ multipliers and $a_2 = 2$ ALUs are available. The priority function is given by the labeling shown in Figure 5.9.

At the first step for $k = 1$, $U_{1,1} = \{v_1, v_2, v_6, v_8\}$. The selected operations are $\{v_1, v_2\}$ because their label is maximal. For $k = 2$, $U_{1,2} = \{v_{10}\}$, which is selected and scheduled.

At the second step for $k = 1$, $U_{2,1} = \{v_3, v_6, v_8\}$. The selected operations are $\{v_3, v_6\}$ because their label is maximal. For $k = 2$, $U_{2,2} = \{v_{11}\}$, which is selected and scheduled.

At the third step for $k = 1$, $U_{3,1} = \{v_7, v_8\}$, which are selected and scheduled. For $k = 2$, $U_{3,2} = \{v_a\}$, which is selected and scheduled.

At the fourth step $\{v_5, v_9\}$ are selected and scheduled.

The schedule is shown in Figure 5.8.

Example 5.4.8. Let us consider the sequencing graph of Figure 5.1. Assume we have $a_1 = 3$ multipliers and $a_2 = 1$ ALU. Let us assume that the execution delays of the multiplier and the ALU are 2 and 1, respectively.

A list schedule, where the priority function is based on the weight of the longest path to the sink vertex, is the following:

Operation		
Multiply	**ALU**	**Start time**
$\{v_1, v_2, v_6\}$	v_{10}	1
–	v_{11}	2
$\{v_3, v_7, v_8\}$	–	3
–	–	4
–	v_4	5
–	v_5	6
–	v_9	7

The scheduled sequencing graph is shown in Figure 5.10. It is possible to verify that the latency of this schedule is minimum.

Assume now that another priority function is used that assigns operations v_2, v_6, v_8 to the first step. Then the schedule would require at least 8 steps, and it would not be minimum.

List scheduling can also be applied to minimize the resource usage under latency constraints. At the beginning, one resource per type is assumed, i.e., **a** is a vector with all entries set to 1. For this problem, the *slack* of an operation is used to rank the operations, where the slack is the difference between the latest possible start time (computed by an ALAP schedule) and the index of the schedule step under consideration. The lower the slack, the higher the urgency in the list is. Operations with zero slack are always scheduled; otherwise the latency bound would be violated. Scheduling such operations may require additional resources, i.e., updating **a**. The remaining operations are scheduled only if they do not require additional resources (see Algorithm 5.5.1).

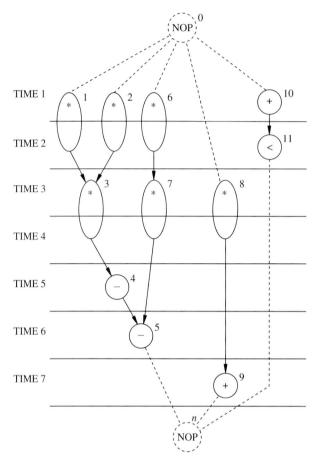

FIGURE 5.10
Optimum schedule under resource constraints.

Note that the algorithm exits prematurely when $t_0^L < 0$, i.e., when the ALAP algorithm detects no feasible solution with dedicated resources. Hence the latency bound is too tight for the problem.

Example 5.4.9. Let us consider the sequencing graph of Figure 5.1. Assume that all operations have unit delay and a latency of 4 cycles is required. The corresponding ALAP schedule is shown in Figure 5.3.

Let $\mathbf{a} = [1, 1]^T$ in the beginning.

At the first step for $k = 1$, $U_{1,1} = \{v_1, v_2, v_6, v_8\}$. There are two operations with zero slack, namely $\{v_1, v_2\}$, which are scheduled. Thus vector $\mathbf{a} = [2, 1]^T$. For $k = 2$, $U_{1,2} = \{v_{10}\}$, which is selected and scheduled.

At the second step for $k = 1$, $U_{2,1} = \{v_3, v_6, v_8\}$. There are two operations with zero slack, namely $\{v_3, v_6\}$, which are scheduled. For $k = 2$, $U_2 = \{v_{11}\}$, which is selected and scheduled.

```
LIST_R( G(V, E), λ̄ ) {
    a = 1;
    Compute the latest possible start times tᴸ by ALAP ( G(V, E), λ̄);
    if ( t₀ᴸ < 0 )
        return (∅);
    l = 1;
    repeat {
        for each resource type k = 1, 2, ..., n_res {
            Determine candidate operations U_lk;
            Compute the slacks {s_i = t_iᴸ - l ∀v_i ∈ U_lk};
            Schedule the candidate operations with zero slack and update a;
            Schedule the candidate operations requiring no additional resources;
        }
        l = l + 1;
    }
    until (v_n is scheduled);
    return (t, a);
}
```

ALGORITHM 5.4.3

At the third step for $k = 1$, $U_{3,1} = \{v_7, v_8\}$, which are selected. For $k = 2$, $U_{3,2} = \{v_a\}$, which is selected and scheduled.

At the fourth step $U_{4,2} = \{v_5, v_9\}$. Both operations have zero slack. They are selected and scheduled. Vector **a** is updated to $= [2, 2]^T$. Hence two resources of each type are required. The schedule is shown in Figure 5.8. In this case, it matches the one computed in Example 5.4.7. In general, schedules computed with different list scheduling algorithms may differ.

Overall list scheduling algorithms have been widely used in synthesis systems, because the low computational complexity of the algorithm makes it applicable to large graphs. Solutions have been shown not to differ much in latency from the optimum ones, for those (small) examples whose optimum solutions are known.

5.4.4 Heuristic Scheduling Algorithms: Force-directed Scheduling*

Force-directed scheduling algorithms were proposed by Paulin and Knight [15] as heuristic approaches to solve both the resource-constrained and the latency-constrained scheduling problems. Paulin called *force-directed list scheduling* the algorithm for the former problem (because it is an extension of list scheduling) and *force-directed scheduling* the algorithm for the latter. Before describing both algorithms, we explain the underlying concepts.

The *time frame* of an operation is the time interval where it can be scheduled. Time frames are denoted by $\{[t_i^S, t_i^L]; i = 0, 1, \ldots, n\}$. The earliest and latest start times in a frame can be computed by the ASAP and ALAP algorithms. Thus the width of the time frame of an operation is equal to its mobility plus 1. The *operation probability* is a function that is zero outside the corresponding time frame and is equal to the reciprocal of the frame width inside it. We denote the probability of the operations at time l by $\{p_i(l); i = 0, 1, \ldots, n\}$. The significance of operation probability is as follows. Operations whose time frame is one unit wide are bound to start in one specific time step. For the remaining operations, the larger the width, the lower the probability that the operation is scheduled in any given step inside the corresponding time frame.

The *type distribution* is the sum of the probabilities of the operations implementable by a specific resource type in the set $\{1, 2, \ldots, n_{res}\}$ at any time step of interest.[1] We denote the type distribution at time l by $\{q_k(l); k = 1, 2, \ldots, n_{res}\}$. A *distribution graph* is a plot of an operation-type distribution over the schedule steps.

The distribution graphs show the likelihood that a resource is used at each schedule step. A uniform plot in a distribution graph means that a type is evenly scattered in the schedule and it relates to a good measure of utilization of that resource.

Example 5.4.10. Consider the sequencing graph of Figure 5.1 with two unit-delay resources (a multiplier and an ALU). We consider the time frames related to a latency bound of 4 steps, which can be derived from the ASAP and ALAP schedules of Figures 5.2 and 5.3, respectively, and the mobilities computed in Example 5.3.1.

Operation v_1 has zero mobility. Hence $p_1(1) = 1, p_1(2) = p_1(3) = p_1(4) = 0$. Similar considerations apply to operation v_2. Operation v_6 has mobility 1. Its time frame is $[1, 2]$. Hence $p_6(1) = p_6(2) = 0.5$ and $p_6(3) = p_6(4) = 0$. Operation v_8 has mobility 2. Its time frame is $[1, 3]$. Hence $p_8(1) = p_8(2) = p_8(3) = 0.3$ and $p_8(4) = 0$. Thus the type distribution for the multiplier ($k = 1$) at step 1 is $q_1(1) = 1 + 1 + 0.5 + 0.3 = 2.8$.

Figure 5.11 shows the distribution plots for both the multiplier and the ALU.

In force-directed scheduling the selection of a candidate operation to be scheduled in a given time step is done using the concept of *force*. Forces attract (repel) operations into (from) specific schedule steps. The concept of force has a direct mechanical analogy. The force exerted by an elastic spring is proportional to the displacement of its end points. The elastic constant is the proportionality factor. Paulin and Knight [15] envisioned forces relating operations to control steps. The assignment

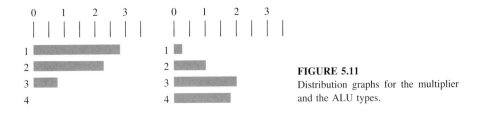

FIGURE 5.11
Distribution graphs for the multiplier and the ALU types.

----

[1]The type distribution is not bound to be between zero and unity unless it is normalized. We do not consider a normalization for compatibility with the work of Paulin and Knight, who referred to it as a distribution graph.

of an operation to a control step corresponds to changing its probability. Indeed such a probability is 1 in that step and 0 elsewhere once the assignment has been done. The variation in probability is analogous to the displacement of a spring. The value of the type distribution given by the distribution graph at that step is analogous to the elastic constant.

The value of the mechanical analogy is limited as far as gaining intuition into the problem. An important point to remember is that forces are related to the concurrency of operations of a given type. The larger the force, the larger the concurrency. Thus measuring the forces is useful as a heuristic to assign operations to time steps.

For the sake of simplicity, we assume in this section that operations have unit delays. The formalism can be extended to multi-cycle operations as well. The assignment of an operation to a step is chosen while considering all forces relating it to the schedule steps in its time frame. Forces can be categorized into two classes. The former is the set of forces relating an operation to the different possible control steps where it can be scheduled and called *self-forces*. The latter is related to the operation dependencies and called *predecessor/successor forces*.

Let us consider self-forces first. To be more specific, let us consider operation v_i of type $k = \mathcal{T}(v_i)$ when scheduled in step l. The force relating that operation to a step $m \in [t_i^S, t_i^L]$ is equal to the type distribution $q_k(m)$ times the variation in probability $1 - p_i(m)$. The self-force is the sum of the forces relating that operation to all schedule steps in its time frame. Namely:

$$\text{self-force}(i, l) = \sum_{m=t_i^S}^{t_i^L} q_k(m)(\delta_{lm} - p_i(m)) \tag{5.29}$$

where δ_{lm} denotes a Kronecker delta function.

This expression can be rewritten as:

$$\text{self-force}(i, l) = q_k(l) - \frac{1}{\mu_i + 1} \sum_{m=t_i^S}^{t_i^L} q_k(m) \tag{5.30}$$

by noticing that: $p_i(m) = \frac{1}{\mu_i + 1}, \quad \forall m \in [t_i^S, t_i^L]$.

> **Example 5.4.11.** Consider the operation v_6 in Figure 5.1. Its type is multiply (i.e., $k = 1$). It can be scheduled in the first two schedule steps, and its probability is $p_6 = 0.5$ in those steps and zero elsewhere. The distribution graph of Figure 5.11 yields the type probability, that is, $q_1(1) = 2.8$ and $q_1(2) = 2.3$. When the operation is assigned to step 1, its probability variations are $1 - 0.5$ for step 1 and $0 - 0.5$ for step 2. Therefore the self-force is $2.8 * (1 - 0.5) + 2.3 * (0 - 0.5) = 0.25$. Note that the force is positive, because the concurrency at step 1 of the multiplication is higher than at step 2. Similarly, when the operation is assigned to step 2, its self-force is $2.8 * (0 - 0.5) + 2.3 * (1 - 0.5) = -0.25$.

It is important to note that assigning an operation to a specific step may reduce the time frame of other operations. Therefore, the effects implied by one assignment must be taken into account by considering the *predecessor/successor* forces, which are the forces on other operations linked by dependency relations.

> **Example 5.4.12.** The assignment of operation v_6 to step 2 implies the assignment of operation v_7 to step 3. Therefore the force of v_7 related to step 3, i.e., $q_1(2)(0 - p_7(2)) + q_1(3)(1 - p_7(3)) = 2.3 * (0 - 0.5) + 0.8 * (1 - 0.5) = -0.75$ is the successor force of v_6. The total force on v_6 at step 2 is the sum of its self-force and successor force, i.e., $-0.25 - 0.75 = -1$.
>
> The total forces on v_6 at step 1 and 2 are 0.25 and -1, respectively. Scheduling v_6 at step 1 would thus increase the concurrency as compared to scheduling v_6 at step 2.

In general, the predecessor/successor forces are computed by evaluating the variation on the self-forces of the predecessors/successors due the restriction of their time frames. Let $[t_i^S, t_i^L]$ be the initial time frame and $[\tilde{t}_i^S, \tilde{t}_i^L]$ be the reduced one. Then, from Equation 5.30, the force variation is:

$$\text{ps-force}(i, l) = \frac{1}{\tilde{\mu}_i + 1} \sum_{m=\tilde{t}_i^S}^{\tilde{t}_i^L} q_k(m) - \frac{1}{\mu_i + 1} \sum_{m=t_i^S}^{t_i^L} q_k(m) \tag{5.31}$$

> **Example 5.4.13.** The assignment of operation v_8 to step 2 implies the assignment of operation v_9 to step 3 or 4. Therefore the variation on the force of v_9 is $1/2(q_2(3) + q_2(4)) - 1/3(q_2(2) + q_2(3) + q_2(4)) = 0.5(2 + 1.6) - 0.3(1 + 2 + 1.6) = 0.3$.

The total force on an operation related to a schedule step is computed by adding to its self-force the predecessor/successor forces of all its predecessors and successors (whose time frame is affected).

We consider now the scheduling algorithms. The *force-directed list scheduling* algorithm addresses the minimum-latency scheduling problem under resource constraints. The outer structure of the algorithm is the same as the LIST_L algorithm. A specific selection procedure is used to select the $S_k \subseteq U_{l,k}$ operations of type $k = 1, 2, \ldots, n_{res}$ at each schedule step l. Namely the selected candidates are determined by reducing iteratively the candidate set $U_{l,k}$ by deferring those operations with the least force until the resource bound is met. The rationale is to maximize the local concurrency (by selecting operations with large forces) while satisfying the resource bound.

Note that force calculation requires the computation of the time frames, or their update, at each outer iteration of the algorithm, and when all candidate operations are critical (i.e., their mobility is zero) and one (or more) must be deferred, thus causing a latency increase. The computational complexity of the algorithm is quadratic in the number of operations, because the predecessor/successor force computation requires considering operation pairs.

The *force-directed scheduling* Algorithm 5.4.3 instead addresses the minimum-resource scheduling problem under latency constraints. The algorithm considers the operations one at a time for scheduling, as opposed to the strategy of considering each schedule step at a time as done by list scheduling. At each iteration, the time frames,

FDS($G(V, E), \bar{\lambda}$) {

    repeat {

        Compute the time-frames;

        Compute the operation and type probabilities;

        Compute the self-forces, predecessor/successor forces and total forces;

        Schedule the operation with least force and update its time-frame;

    }

    until (all operations are scheduled);

    return (**t**);

}

ALGORITHM 5.4.4

probabilities and forces are computed. The operation with the least force is scheduled in the corresponding step. The rationale is to minimize the local concurrency (related to the resource usage) while meeting the latency bound that is satisfied by construction because operations are always scheduled in their time frames.

The computational complexity of the algorithm is cubical in the number of operations, because there are as many iterations as operations and each requires the calculation of the forces. The algorithm may be enhanced by an efficient force computation, leading to an overall quadratic complexity in the number of operations. The algorithm can also be applied to more complex graph models. We refer the interested reader to reference [15] for further details.

The force-directed scheduling algorithm has been implemented in different design systems. The results have been shown to be superior to list scheduling. However, its run times tend to be long for large graphs, limiting its practical use for large designs.

5.4.5 Other Heuristic Scheduling Algorithms*

Several other heuristic scheduling algorithms have been proposed. Some are derived from software design techniques, such as *trace scheduling* [3] and *percolation scheduling* [16]. The latter has been used in different forms in various synthesis systems, even though some of its implementations are more restrictive in scope and power than the original algorithm. We comment here briefly on scheduling algorithms using a transformational approach, where an initial schedule is improved upon in a stepwise fashion. These methods have to be contrasted to the list and force-directed scheduling algorithms, which are constructive.

In percolation scheduling, operations are assigned initially to some time steps and the schedule is iteratively refined by moving operations from one step to another. For example, an initial schedule may place the operations in a linear order, i.e., requiring full serialization. Operations can be rescheduled to share control steps when this is allowed by sequencing, timing and resource constraints. Then unused schedule steps may be removed. Rules may be used to define the valid transformations.

The scheduling problem can obviously be solved by general iterative methods, like simulated annealing [2]. Also in this case, an initial scheduling is refined by moving operations from one step to another. Moves are generated randomly and accepted when the cost function (latency or resource count) decreases. Uphill moves (i.e., causing a cost increase) are also accepted, but with a probability that depends on some parameters of the algorithm and that decreases monotonically with the execution of the algorithm. This reduces the probability of finding a local minimum solution, and hence the likelihood of finding an exact solution with this approach is often higher than with other heuristic algorithms. Experiments have shown that this approach yields results that are close in quality to those computed by list scheduling, with possibly much higher computing time.

5.5 SCHEDULING ALGORITHMS FOR EXTENDED SEQUENCING MODELS*

When considering hierarchical sequencing graphs, scheduling and latency computation can be performed bottom up, as mentioned in Section 4.3.3. The computed start times are relative to those of the source vertices in the corresponding graph entities.

Whereas unconstrained hierarchical scheduling is straightforward, timing and resource-constrained scheduling is not. Satisfaction of relative timing constraints is hard to verify when the constraints are applied to operations in different graph entities of the hierarchy [11]. Similar considerations apply to resource constraints. Obvious exceptions are cases when the model is a tree-structured hierarchy of submodels that can be flattened explicitly or implicitly.

The difficulties of hierarchical scheduling can be avoided under the restrictive and simplifying assumption that no resource can be shared across different graph entities in the hierarchy, and that timing and resource constraints apply within each graph entity. In this case, each graph entity can be scheduled independently. Otherwise, either the entities are scheduled concurrently or each scheduled sequencing graph entity poses some restriction on the time-step assignment and resource usage of the others.

A very relevant problem is scheduling graph models expressing conditional constructs, which are very common in behavioral circuit models. Operations (or a group of operations) may represent alternatives of a branching choice. Alternative operations can share resources even when their execution time frames overlap. This factor may affect scheduling under resource constraints.

5.5.1 Scheduling Graphs with Alternative Paths*

We assume in this section that the sequencing graph contains alternative paths related to branching constructs. This extended sequencing graph model can be obtained by expanding the branching vertices, i.e., by replacing them by the corresponding graph entities. The mutual exclusive execution semantics of the graphs representing the branching bodies gives rise to alternative paths in the graph.

An exact formulation of the scheduling problem can be achieved by modifying the ILP model of Section 5.4.1. In particular, the resource constraints represented by inequality 5.7 must express the fact that operations in alternative branches can have overlapping execution intervals without affecting the resource usage.

We consider here the special case in which the graph can be partitioned into n_c groups of mutually exclusive operations and let $C : V \rightarrow \{1, 2, \ldots, n_c\}$ denote the group to which an operation belongs. Then, inequality 5.7 can be restated as follows:

$$\sum_{i:T(v_i)=k \text{ and } C(v_i)=c} \sum_{m=l-d_i+1}^{l} x_{im} \leq a_k \quad k = 1, 2, \ldots, n_{res},$$

$$l = 1, 2, \ldots, \overline{\lambda} + 1, \quad c = 1, 2, \ldots, n_c \quad (5.32)$$

Obviously more complex constraints can be constructed for more general cases.

> **Example 5.5.1.** Consider the graph shown in Figure 5.1, where all operations have unit delay. Assume that the path (v_0, v_8, v_9, v_n) is mutually exclusive with the remaining operations. The constraints related to the uniqueness of the start times and to the sequencing constraints are the same as those of Example 5.4.1.
>
> The non-trivial resource constraints are the following:
>
> $$x_{1,1} + x_{2,1} + x_{6,1} \leq a_1$$
>
> $$x_{3,2} + x_{6,2} + x_{7,2} \leq a_1$$
>
> $$x_{10,2} + x_{11,2} \leq a_2$$
>
> $$x_{4,3} + x_{10,3} + x_{11,3} \leq a_2$$
>
> $$x_{5,4} + x_{11,4} \leq a_2$$

The list scheduling and the force-directed scheduling algorithms can support mutually exclusive operations by modifying the way in which the resource usage is computed at each step. For example, Wakabayashi and Yoshimura [19] extended list scheduling by adding a condition vector to each vertex, whose entries encode the branching conditions taken to reach the corresponding operation. Condition vectors and branching probabilities are used to determine the urgency of the operations for candidate selection.

In the case of force-directed scheduling, attention needs to be paid to the computation of the type distribution. When two mutually exclusive operations of the same type can occur in a time step, their probability is not added together to compute the type distribution, because it is impossible that they are both executed at that time step. Instead, the maximum of their values contributes to the type distribution [15].

Camposano proposed a specialized scheduling algorithm for graphs where all paths from source to sink represent alternative flows of operations, called *as fast as possible*, or AFAP [1]. In addition, operation chaining may also be considered. The method is a two-step procedure. First, all paths are scheduled independently. Then, the scheduled paths are merged.

When scheduling each path, resource and timing constraints are taken into account. Resource constraints may limit the amount of operation chaining in any single step, because each resource can execute once in any given time step. Camposano proposed a path scheduling algorithm based on the search for the intersection of the constraints that delimit the time-step boundaries. Scheduling each path corresponds to determining the *cuts* in each path, where a cut is a subset of adjacent operations in a path such that any can define a schedule step boundary.

Once the paths are scheduled, they are merged together. Another graph is derived, where the cuts are represented by vertices and their intersection by edges. A clique in the graph corresponds to a subset of operations that can be started at some time step. Therefore a minimum clique partitioning of the graph provides a minimum latency solution. The computational complexity of the approach is related to solving the clique partitioning problem, which is intractable, and to the fact that the number of paths may grow exponentially with the number of vertices in the sequencing graph. However, an implementation of this algorithm with an exact solution of the clique partitioning problem has given good results [1]. The path-based scheduling formulation fits well with processor synthesis problems, where a large amount of alternative paths is related to executing different (alternative) instructions.

Example 5.5.2. Consider the example of Figure 5.12 (a), where all paths are alternative execution streams. The numbers inside the circles denote the propagation delay. The required cycle-time is 60 nsec.

There are four alternative paths: (v_1, v_3, v_5, v_6); (v_1, v_3, v_5, v_7); (v_2, v_4, v_5, v_6); (v_2, v_4, v_5, v_7). By analyzing the first path, it is clear that a cut is required. It can be done after operation v_1 or after v_3 or after v_5. We indicate this cut by $c_1 = \{1, 3, 5\}$. A similar analysis of the second paths suggests that a cut is required after v_3 or v_5, i.e., $c_2 = \{3, 5\}$. The other cuts are $c_3 = \{2\}$, $c_4 = \{2\}$ and $c_5 = \{4, 5\}$. Note that the two last cuts are related to the last path. Cuts are shown in Figure 5.12 (b). Let us consider now the intersection graph corresponding to the cuts and shown in Figure 5.12 (c). The clique partition indicates that operations v_2 and v_5 are the last before starting a new time step. This yields a schedule, as shown in Figure 5.12 (d).

5.6   SCHEDULING PIPELINED CIRCUITS*

We consider in this section extending scheduling algorithms to cope with pipelined circuits. In general, the specification of a pipelined circuit consists of a sequencing graph model and a corresponding data rate. For hierarchical models, the submodels can be pipelined or not.

We present here specific subproblems and solutions. First we consider non-pipelined sequencing graphs whose operations can be bound to pipelined resources (i.e., pipelined submodels). Second, we consider non-hierarchical pipelined models with non-pipelined resources. These two problems are referred to in the literature as

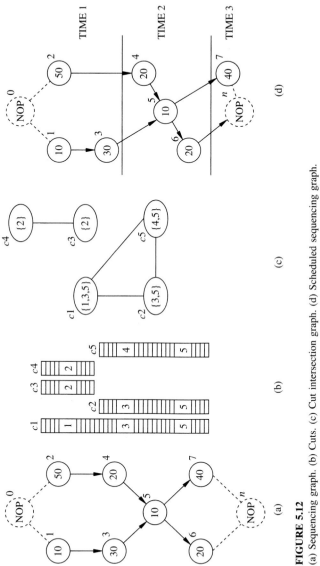

FIGURE 5.12

(a) Sequencing graph. (b) Cuts. (c) Cut intersection graph. (d) Scheduled sequencing graph.

structural pipelining and *functional pipelining*, respectively. Last, we consider the use of pipelining to speed up the execution of loops. This technique is also called *loop winding* and *loop folding*.

5.6.1   Scheduling with Pipelined Resources*

Pipelined resources consume and produce data at time intervals that are smaller than the execution delays. The intervals, called *data introduction intervals*, of the resources are assumed here to be constant. In the limiting case of non-pipelined resources, the data introduction interval is equal to the execution delay.

> **Example 5.6.1.** A typical example is provided by parallel multipliers that have an execution delay of two cycles. One cycle is spent in the Wallace-tree reduction of the partial products into two summands. The second cycle is spent in the final addition and rounding. Since the partial-product reduction and the final addition can overlap execution, data can be produced and consumed at every cycle, i.e., the data introduction interval is 1.

Pipelined resources can be shared, even when the corresponding operations overlap execution. In this case, necessary requirements are that no data dependency exists between the operations and that the operations do not start in the same time step. The former requirement is due to the unavailability of data, the latter to avoid overlapping of data.

> **Example 5.6.2.** Consider operations v_1, v_2 and v_3 in the sequencing graph of Figure 5.1. Assume that a pipelined multiplier resource is available, with execution delay of 2 and a data introduction interval of 1. Then operations v_1 and v_2 can overlap execution by starting at steps 1 and 2 of the schedule. However, operations v_1 and v_3 cannot overlap execution, because the result of v_1 (and of v_2) are inputs to v_3.

The list scheduling algorithm can be extended to handle pipelined resources by allowing the scheduling of overlapping operations with different start times and no data dependencies. Assume for the sake of simplicity that the data introduction intervals of the resources are 1. Then, the resource-constrained list scheduling algorithm LIST_L can be used while assuming that the set of unfinished operations is always empty, because all resources are ready to accept data at each cycle.

> **Example 5.6.3.** Let us consider the sequencing graph of Figure 5.1, with $a_1 = 3$ multipliers and $a_2 = 1$ ALU as resources. Let us assume that the execution delays of the multiplier and the ALU are 2 and 1, respectively, and that the multiplier is pipelined with a unit data introduction interval. Note that the last assumption differentiates this example from Example 5.4.8.
>
> Multiplications $\{v_1, v_2, v_6\}$ are scheduled at the first step. Multiplication v_8 can start at the second step, but $\{v_3, v_7\}$ must be delayed until the third step due to data dependencies.

A schedule is summarized by the following table:

| Operation | | Start time |
Multiply	ALU	
$\{v_1, v_2, v_6\}$	v_{10}	1
v_8	v_{11}	2
$\{v_3, v_7\}$	–	3
–	v_9	4
–	v_4	5
–	v_5	6

Note that the latency is 6 cycles, one cycle less than in Example 5.4.8. The scheduled sequencing graph is shown in Figure 5.13.

Other scheduling algorithms can be extended to cope with pipelined resources. We refer the interested reader to reference [9] for an enhanced ILP model and to reference [15] for the extensions to the force-directed scheduling algorithm.

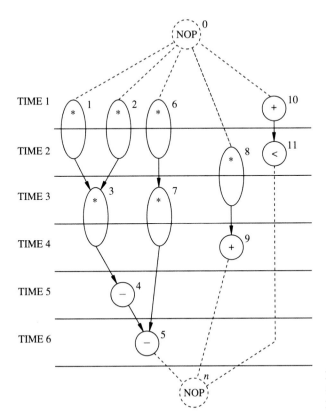

FIGURE 5.13
Scheduled sequencing graph with
pipelined resources.

5.6.2   Functional Pipelining*

We consider here scheduling sequencing graphs under a global *data introduction interval* (δ_0) constraint, representing the time interval between two consecutive executions of the source vertex. We assume that δ_0 is constant and that it is smaller than the latency. We assume also that resources are not pipelined and that the graph model is not hierarchical for the sake of simplicity.

The number of required resources in pipelined implementations depends on δ_0. The smaller δ_0, the larger is the number of operations executing concurrently, and consequently the larger is the resource requirement. Conversely, an upper bound on the resource usage implies a lower bound on δ_0. For example, assume unit execution delays. Let us denote by n_k the number of operations of type $k, k = 1, 2, \ldots, n_{res}$. Then, for maximum-rate pipelining $(\delta_0 = 1)$ and for any resource type, the resource usage a_k equals n_k, because all operations are concurrent. For larger values of δ_0, a lower bound on the resource usage is $\bar{a}_k = \lceil n_k/\delta_0 \rceil$ for any type k. This result can be understood by thinking that we can serialize at most δ_0 operations of each type when we assign them to a single resource. This relation was introduced first by Park and Parker [14].

When considering scheduling pipelined circuits, it is convenient to preserve the notation used for non-pipelined circuits. Thus we represent the start times of the operations $\{v_i, i = 0, 1, \ldots, n\}$ in the interval $\{1, 2, \ldots, \bar{\lambda} + 1\}$. Nevertheless, the pipelined nature of the circuit requires simultaneous activation of the operations with start times: $l + p\delta_0$; $\forall l, p \in Z : 1 \leq l + p\delta_0 \leq \bar{\lambda} + 1$. Therefore, a control unit for a pipelined circuit with this model will have δ_0 control steps.

The ILP formulation can be extended to cope with functional pipelining. In particular, inequality 5.7 representing the resource bounds is modified to reflect the increased concurrency of the operations:

$$\sum_{p=0}^{\lceil \bar{\lambda}/\delta_0 \rceil - 1} \sum_{i:\mathcal{T}(v_i)=k} \sum_{m=l-d_i+1+p\delta_0}^{l+p\delta_0} x_{im} \leq a_k, \quad k = 1, 2, \ldots, n_{res}, \quad l = 1, 2, \ldots, \delta_0 \quad (5.33)$$

Example 5.6.4.   Consider the sequencing graph of Figure 5.1, where all resources have unit execution delay. Assume $\delta_0 = 2$.

Then a lower bound on the resource usage is $\bar{a}_1 = \lceil 6/2 \rceil = 3$ multipliers and $\bar{a}_2 = \lceil 5/2 \rceil = 3$ ALUs. Assume then that a schedule with $\mathbf{a} = [3, 3]^T$ resources is sought for.

The uniqueness constraints on the start time of the operations and the sequencing dependency constraints are the same as those shown in Example 5.4.1. The resource constraints can be derived by inequality 5.33, which can be simplified for unit-delay resources to:

$$\sum_{p=0}^{\lceil \bar{\lambda}/\delta_0 \rceil - 1} \sum_{i:\mathcal{T}(v_i)=k} x_{i,l+p\delta_0} \leq a_k, \quad k = 1, 2, \ldots, n_{res}, \quad l = 1, 2, \ldots, \delta_0$$

Assuming a latency bound of 4, the constraint for this example is:

$$\sum_{p=0}^{1} \sum_{i:\mathcal{T}(v_i)=k} x_{i,l+2p} \le a_k, \quad k = 1, 2, \quad l = 1, 2$$

Thus yielding:

$$x_{1,1} + x_{2,1} + x_{6,1} + x_{8,1} + x_{7,3} + x_{8,3} \le 3$$

$$x_{3,2} + x_{6,2} + x_{7,2} + x_{8,2} \le 3$$

$$x_{10,1} + x_{4,3} + x_{9,3} + x_{10,3} + x_{11,3} \le 3$$

$$x_{9,2} + x_{10,2} + x_{11,2} + x_{5,4} + x_{9,4} + x_{11,4} \le 3$$

A schedule is shown in Figure 5.14.

The heuristic scheduling algorithms can be extended to support functional pipelining. For example, the list scheduling algorithm can be used to solve the resource-constrained scheduling problem for a given δ_0. Inequality 5.33 can be used to check whether the resource bound is violated at any given step and therefore to determine the schedulable candidates. An example of such a list scheduling algorithm was implemented in program SEHWA [14] (with operation chaining), where the priority function is based on the sum of the operation propagation delays from the candidate to the sink vertex. The force-directed scheduling algorithm can also support functional pipelining. The type distribution at a given step must be computed differently to take into account the actual operation concurrency in the pipelined implementation [15].

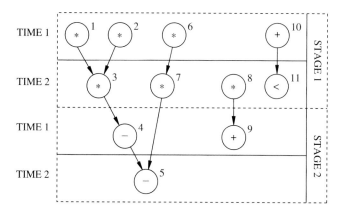

FIGURE 5.14
Scheduled sequencing graph with $\delta_0 = 2$ and three resources per type.

5.6.3 Loop Folding*

Loop folding is an optimization technique to reduce the execution delay of a loop. It was first introduced as a method for optimizing software compilation. It can be applied to hardware optimization as well.

Let us consider a loop with a fixed number of iterations, say n_l. The execution delay of the loop is equal to $n_l \cdot \lambda_l$, where λ_l is the latency of the loop body. If it is possible to pipeline to the loop body itself with local data introduction interval $\delta_l < \lambda_l$, then the overall loop execution delay is approximately $n_l \cdot \delta_l < n_l \cdot \lambda_l$. More precisely, some overhead has to be accounted for in starting the pipeline, and thus the loop execution delay is $(n_l + \lceil \lambda_l / \delta_l \rceil - 1) \cdot \delta_l$.

> **Example 5.6.5.** Consider the sequencing graph of Figure 5.15 (a). Assume that all operations take one unit of time. Then the loop body executes one iteration in 4 steps, i.e., its latency is 4. Figure 5.15 (b) shows a folded loop implementation with $\delta_l = 2$. Each iteration requires $\delta_l = 2$ control steps. Starting the pipeline requires $\lceil \lambda_l / \delta_l \rceil - 1 = 1$ iteration. To execute $n_l = 10$ iterations, 22 time steps are needed in the folded implementation, while 40 steps are needed in the original one.

Therefore loop folding can reduce the latency of a scheduled sequencing graph that contains loops, possibly nested loops. When considering the hierarchical nature of sequencing graphs, loop folding can be viewed as pipelining the graph entity of the loop body with the goal of reducing the execution delay of the corresponding iteration vertex at the higher level in the hierarchy. Note that this vertex (representing the loop

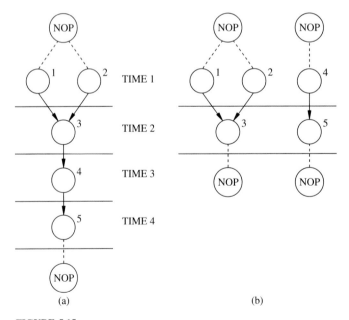

(a)   (b)

FIGURE 5.15
(a) Sequencing graph of a loop body. (b) Sequencing graph after loop folding.

as a whole) is not a pipelined operation, because it can consume new data only when the loop exit condition is met.

5.7   PERSPECTIVES

Scheduling is a fascinating problem, because many creative solutions can be researched to address several related problems. Examples are scheduling under timing constraints, data-rate constraints, and scheduling with unbounded-delay operations. Specialized algorithms have been proposed to address each of these issues.

In practice, most scheduling problems are computationally hard because of the constraints on the resource usage. Recent interest in exact solution methods, based on the ILP formulation, has been successful in providing us with schedules with global minimum-latency or minimum-resource usage. Unfortunately, the potentially explosive computational cost of solving ILPs has precluded the use of these methods for large-scale graph models.

Heuristic algorithms, and in particular list scheduling algorithms, have been found adequate for most practical applications. Even though the optimality of the solution is not guaranteed, list schedules offer a good compromise between computing time and latency (or resource usage). Many flavors of list scheduling algorithms exist, with support for conditional constructs, timing and data introduction interval constraints and other features.

It is important to remember that the value of a solution to a scheduling problem for architectural synthesis depends on the underlying assumption of modeling the operations by execution delays and area costs. While this model fits well with resource-dominated circuits, the actual area and performance of other circuit implementations may differ from those estimated by scheduling. Thus both the area and the cycle-time have to account for the components contributed by registers, steering logic circuits, wires and control circuits. For this reason, the value of the scheduling algorithms must always be considered in the context of the other tasks of architectural synthesis and optimization.

5.8   REFERENCES

Most of the early work on scheduling was done in the software domain (e.g., list, trace and percolation scheduling [3, 16]) or in operation research (e.g., Hu's algorithm [8]). The list scheduling technique was borrowed from the former domain. It has been used in several systems, including the EMERALD/FACET program [18], the MAHA program [13] and the SYSTEM ARCHITECT'S WORKBENCH [17].

A formal model of the scheduling problem for architectural synthesis was presented first by Hafer and Parker [7]. Hwang *et al.* [9] reported about an efficient ILP scheduling implementation in the ALPS program. Recently, Gebotys and Elmasry [4] enhanced the ILP formulation by showing extensions and tighter bounds that increase the efficiency of ILP solution methods and extend the applicability of this approach.

One of the most used techniques for scheduling is the force-directed scheduling method, which was introduced by Paulin and Knight [15]. Many variations and extension of this technique have been proposed and used.

Recently, some specialized techniques for scheduling have been presented. They include the AFAP scheduling method, used in the HIS system [1], relative scheduling implemented in the HEBE system [10] and the list-based scheduling algorithm for the functional pipeline of the SEHWA program [14]. Loop folding was introduced by Girczyc [5] as a set of transformations at the behavioral model and called *loop winding*. It was later applied at the register-transfer level in the ATOMIC scheduler of the CATHEDRAL-II system by Goossens *et al.* [6].

1. R. Camposano, "Path-based Scheduling for Synthesis," *IEEE Transactions on CAD/ICAS*, Vol. CAD-10, No. 1, pp. 85–93, January 1990.

2. S. Devadas and R. Newton, "Algorithms for Hardware Allocation in Data-Path Synthesis," *IEEE Transactions on CAD/ICAS*, Vol. CAD-8, No. 7, pp. 768–781, July 1989.

3. J. Fisher, "Trace Scheduling: A Technique for Global Microcode Compaction," *IEEE Transactions on Computers*, Vol. C-30, No. 7, pp. 478–490, July 1981.

4. C. Gebotys and M. Elmasry, *Optimal VLSI Architectural Synthesis*, Kluwer Academic Publishers, Boston, MA, 1992.

5. E. Girczyc, "Loop Winding—A Data Flow Approach to Functional Pipelining," *ISCAS, Proceedings of the International Symposium on Circuits and Systems*, pp. 382–385, 1987.

6. G. Goossens, J. Vandewalle and H. De Man, "Loop Optimization in Register-Transfer Level Scheduling for DSP Systems," *DAC, Proceedings of the Design Automation Conference*, pp. 826–831, 1989.

7. L. Hafer and A. Parker, "Automated Synthesis of Digital Hardware," *IEEE Transactions on Computers*, Vol. C-31, No. 2, February 1982.

8. T. C. Hu, "Parallel Sequencing and Assembly Line Problems," *Operations Research*, No. 9, pp. 841–848, 1961.

9. C-T. Hwang, J-H. Lee and Y-C. Hsu, "A Formal Approach to the Scheduling Problem in High-Level Synthesis," *IEEE Transactions on CAD/ICAS*, Vol. CAD-10, No. 4, pp. 464–475, April 1991.

10. D. Ku and G. DeMicheli, "Relative Scheduling under Timing Constraints: Algorithms for High-Level Synthesis of Digital Circuits," *IEEE Transactions on CAD/ICAS*, pp. 696–718, June 1992.

11. D. Ku and G. DeMicheli, *High-Level Synthesis of ASICS under Timing and Synchronization Constraints*, Kluwer Academic Publishers, Boston, MA, 1992.

12. J. Nestor and D. Thomas, "Behavioral Synthesis with Interfaces," *ICCAD, Proceedings of the International Conference on Computer Aided Design*, pp. 112–115, 1986.

13. A. Parker, J. Pizarro and M. Mlinar, "MAHA: A Program for Data-Path Synthesis," *DAC, Proceedings of the Design Automation Conference*, pp. 461–466, 1986.

14. N. Park and A. Parker, "Sehwa: A Software Package for Synthesis of Pipelines from Behavioral Specifications," *IEEE Transactions on CAD/ICAS*, Vol. CAD-7, No. 3, pp. 356–370, March 1988.

15. P. Paulin and J. Knight, "Force-Directed Scheduling for the Behavioral Synthesis of ASIC's," *IEEE Transactions on CAD/ICAS*, Vol. CAD-8, No. 6, pp. 661–679, July 1989.

16. R. Potasman, J. Lis, A. Nicolau and D. Gajski, "Percolation Based Scheduling," *DAC, Proceedings of the Design Automation Conference*, pp. 444–449, 1990.

17. D. Thomas, E. Lagnese, R. Walker, J. Nestor, J. Rajan and R. Blackburn, *Algorithmic and Register Transfer Level Synthesis: The System Architect's Workbench*, Kluwer Academic Publishers, Boston, MA, 1990

18. C. Tseng and D. Siewiorek, "Automated Synthesis of Data Paths in Digital Systems," *IEEE Transactions on CAD/ICAS*, Vol. CAD-5, pp. 379–395, July 1986.

19. K. Wakabayashi and T. Yoshimura, "A Resource Sharing and Control Synthesis Method for Conditional Branches," *ICCAD, Proceedings of the International Conference on Computer Aided Design*, pp. 62–65, 1989.

5.9 PROBLEMS

1. Consider the graph of Figure 5.1. Assume the execution delays of the multiplier and of the ALU are 2 and 1 cycle, respectively. Schedule the graph using the ASAP algorithm. Assuming a latency bound of $\bar{\lambda} = 8$ cycles, schedule the graph using the ALAP algorithm. Determine the mobility of the operations.

2. Consider the graph of Figure 5.1. Assume the execution delays of the multiplier and of the ALU are 2 and 1 cycle, respectively. Consider the following timing constraints between

the start times of the operations:

- Operation v_6 starts at least 1 cycle after v_4 starts.
- Operations v_5 and v_9 start simultaneously.
- Operation v_9 starts at most 2 cycles after operation v_{10} starts.

Construct a constraint graph. Schedule the operations by assuming that $t_0 = 1$.

3. Consider operations $v_i; i = 1, 2, \ldots, 7$, with the corresponding execution delays $D = \{0, 1, 3, 1, 1, 1, 4\}$. Assume the dependencies $E = \{(v_1, v_2), (v_2, v_3), (v_5, v_6)\}$ and the following constraints:

- Operation v_1 has release time 1.
- Operation v_5 has release time 4.
- Operation v_7 has release time 8.
- Operation v_4 starts at least (no sooner than) 4 cycles after operation v_2 starts.
- Operation v_4 starts at most 1 cycle after operation v_3 starts.
- Operation v_7 starts at most 2 cycles after operation v_6 starts.
- Operation v_5 starts exactly 2 cycles after operation v_1 starts.

Construct a constraint graph. Schedule the operations by assuming that $t_0 = 1$ with the Liao-Wong algorithm. Show all steps.

4. Let $\widetilde{G}_c(V, E)$ be the subgraph of $G_c(V_c, E_c)$ after having removed all edges corresponding to maximum timing constraints. Assume $\widetilde{G}_c(V, E)$ is acyclic and assume $G_c(V_c, E_c)$ is feasible. Let the anchor set $A(v_i) \subseteq A$ of a vertex $v_i \in V$ be the subset of A such that $a \in A(v_i)$ if there is a path in $\widetilde{G}_c(V, E)$ from a to v_i with an edge weighted by d_a.

   Prove that a feasible maximum timing constraint u_{ij} is well-posed (i.e., satisfiable for any value of the anchor delays) if and only if $A(v_j) \subseteq A(v_i)$.

5. Using the definition of Problem 4, show that in a well-posed constraint graph the anchor sets on a cycle are identical.

6. Using the definition of Problem 4, show that in a well-posed constraint graph the weight of any cycle is bounded.

7. Under the definitions and assumptions of Problem 4, show that a constraint graph is well-posed if and only if $R(v_i) \subseteq A(v_i) \; \forall v_i \in V$.

8. Consider the sequencing graph of Figure 5.1. Assume that all operations have unit delay. Formulate the ILP constraint inequalities with a latency bound $\bar{\lambda} = 5$. Write the objective function that models latency minimization.

9. Determine the type distributions for the *scheduled* sequencing graph of Figure 5.8.

10. Explain how you would extend the force-directed scheduling algorithm to multi-cycle operations. Describe the extensions to the computation of the time frames, operation and type probabilities.

11. Write the pseudo-code of a list scheduler for the minimum-latency problem under resource and relative timing constraints.

12. Consider the problem of unconstrained scheduling with chaining. Assume that the propagation delay of the resources is bounded by the cycle-time. Show how to extend the ASAP scheduling algorithm to incorporate chaining.

13. Consider the ILP formulation for the minimum-latency scheduling problem under resource constraints and with alternative operations. Assume that the groups of alternative operations do not form a partition of the set V. Determine an inequality constraint representing resource usage bounds.

14. Consider the ILP formulation for scheduling with pipelined resources. Derive a set of inequality constraints for the case of multi-cycle resources. Apply the inequalities to Example 5.6.3.

15. Determine the type distributions for the pipelined *scheduled* sequencing graph of Figure 5.14.

16. Consider the sequencing graph of Figure 5.1. Assume that the multiplier and the ALU have execution delays of 2 and 1 cycle, respectively. Formulate the ILP constraint inequalities for a pipeline schedule with $\delta_0 = 1$.

CHAPTER
6

RESOURCE
SHARING
AND BINDING

In der Beschränkung zeigt sich erst der Meister.
The master distinguishes himself in coping with the limitations.

J. Goethe. Epigrammatisch: Natur und Kunst.

6.1 INTRODUCTION

Resource sharing is the assignment of a resource to more than one operation. The primary goal of resource sharing is to reduce the area of a circuit, by allowing multiple non-concurrent operations to share the same hardware operator. Resource sharing is often mandatory to meet specified upper bounds on the circuit area (or resource usage).

Resource binding is the explicit definition of a mapping between the operations and the resources. A binding may imply that some resources are shared. Resource binding (or partial binding) may be part of the original circuit specifications, and thus some sharing may be defined explicitly in the hardware model. Resource usage constraints may infer implicitly some resource sharing, even though they do not imply a particular binding.

Resource binding can be applied to sequencing graphs that are scheduled or unscheduled. Specific strategies for architectural synthesis, as described in Section 4.5, use either approach. In the former case, the schedule provides some limitations to the extent to which resource sharing can be applied. For example, concurrent operations cannot share a resource. In the latter case, resource binding may affect the circuit latency because operations with a shared resource cannot execute concurrently, and therefore some operations may have to be postponed.

When considering resource-dominated circuits that have been scheduled under resource constraints, the area is already determined by the resource usage. Thus binding and sharing serve just the purpose of refining the structural information so that connectivity synthesis can be performed. On the other hand, for general circuits, the overall area and performance depend on the resource usage as well as on registers, wiring, steering logic and control circuits. Then binding and sharing affect the figure of merit of the circuit. Obviously, when considering sequencing graphs that are not scheduled, sharing and binding affect both area and performance.

Different binding and sharing problems can be defined according to the types of operations and resources being considered. The simplest case is when operations can be matched to resources with the same type. A slight generalization being considered in this chapter is when operations with different types can be implemented (covered) by one resource of appropriate type. Examples are operations with types *addition, subtraction* and *comparison* being implemented by a resource with type ALU. These two cases are modeled by using a single-valued function $T : V \rightarrow \{1, 2, \ldots, n_{res}\}$ that associates with any operation the resource type that can implement it. A further generalization is the case when any operation can be implemented by more than one resource type, possibly with different area and performance. An example is given by an addition operation that can be implemented by different types of adders. The choice of such a resource type, called a *resource-type* (or *module*) *selection problem*, is described in Section 6.6.

In this chapter we consider primarily sharing functional resources, called briefly resources, for non-pipelined circuit models. We address also the problems of sharing registers and of using memory arrays efficiently, as well as sharing interface resources such as busses. First we present resource sharing models and algorithms for resource-dominated circuits in Section 6.2. Then we extend the analysis to general circuits and comment on the impact of sharing on area and performance in Section 6.3. We then revisit the relations between scheduling and binding and we present methods for concurrent scheduling and binding in Section 6.4 and for binding unscheduled models in Section 6.5. We conclude the chapter by considering other extensions, such as the module selection problem and resource sharing and binding for pipelined circuits as well as relations between sharing and testability.

6.2 SHARING AND BINDING FOR RESOURCE-DOMINATED CIRCUITS

We consider sharing and binding problems for resource-dominated circuits modeled by scheduled sequencing graphs $G_s(V, E)$. In this chapter, we drop from consideration the source and sink vertices v_0, v_n that are *No-Operations*. Hence the operation set of interest is $\{v_i, i = 1, 2, \ldots, n_{ops}\}$.

Two (or more) operations may be bound to the same resource if they are not concurrent and they can be implemented by resources of the same type. When these conditions are met, the operations are said to be *compatible*. Two operations are not concurrent when either one starts after the other has finished execution or when they are alternative, i.e., they are mutually exclusive choices of a branching decision. Therefore, an analysis of the sequencing graph is sufficient to determine the compatibility of two or more operations for sharing. We postpone this analysis to the following two sections and we concentrate now on the compatibility issue.

Definition 6.2.1. The **resource compatibility graph** $G_+(V, E)$ is a graph whose vertex set $V = \{v_i, i = 1, 2, \ldots, n_{ops}\}$ is in one-to-one correspondence with the operations and whose edge set $E = \{\{v_i, v_j\} \ i, j = 1, 2, \ldots, n_{ops}\}$ denotes the compatible operation pairs.

The resource compatibility graph has at least as many disjoint components as the resource types. A group of mutually compatible operations corresponds to a subset of vertices that are all mutually connected by edges, i.e., to a clique. Therefore a *maximal* set of mutually compatible operations is represented by a *maximal clique* in the compatibility graph.

An optimum resource sharing is one that minimizes the number of required resource instances. Since we can associate a resource instance to each clique, the problem is equivalent to partitioning the graph into a minimum number of cliques. Such a number is the *clique cover number* of $G_+(V, E)$, denoted by $\kappa(G_+(V, E))$.

Example 6.2.1. Let us consider the scheduled sequencing graph of Figure 5.8, which we repeat in Figure 6.1 for convenience. We assume again that there are two resource types: a multiplier and an ALU, both with 1 unit execution delay. The compatibility graph is shown in Figure 6.2. Examples of compatible operations are $\{v_1, v_3\}$ and $\{v_4, v_5\}$ among others. Examples of cliques are the subgraphs induced by $\{v_1, v_3, v_7\}$, $\{v_2, v_6, v_8\}$ and $\{v_4, v_5, v_{10}, v_{11}\}$. These cliques, in addition to $\{v_9\}$, cover the graph. Since they are disjoint, they form a clique partition. The clique cover number κ is then equal to 4, corresponding to two multipliers and two ALUs. Edges of the cliques are emboldened in Figure 6.2.

An alternative way of looking at the problem is to consider the *conflicts* between operation pairs. Two operations have a conflict when they are not compatible. Conflicts can be represented by *conflict graphs*.

Definition 6.2.2. The **resource conflict graph** $G_-(V, E)$ is a graph whose vertex set $V = \{v_i, i = 1, 2, \ldots, n_{ops}\}$ is in one-to-one correspondence with the operations and whose edge set $E = \{\{v_i, v_j\} \ i, j = 1, 2, \ldots, n_{ops}\}$ denotes the conflicting operation pairs.

It is obvious that the conflict graph is the *complement* of the compatibility graph. A set of mutually compatible operations corresponds to a subset of vertices that are not connected by edges, also called the *independent set* of $G_-(V, E)$. A proper vertex coloring of the conflict graph provides a solution to the sharing problem: each color

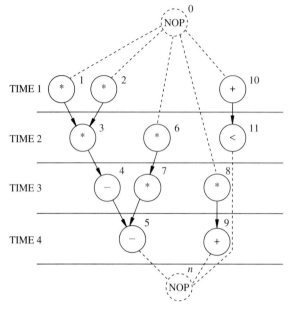

FIGURE 6.1
Scheduled sequencing graph.

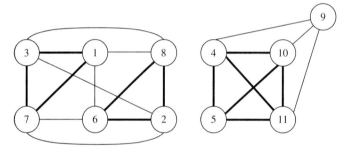

FIGURE 6.2
Compatibility graph.

corresponds to a resource instance. An optimum resource sharing corresponds to a vertex coloring with a minimum number of colors. Such a number is the *chromatic number* of $G_-(V, E)$ and is denoted by $\chi(G_-(V, E))$. Note that $\chi(G_-(V, E))$ is equal to $\kappa(G_+(V, E))$.

Since operations with different types are always conflicting, it is convenient to consider the conflict graphs for each type independently. Such graphs are the complements of the corresponding compatibility subgraphs for operations of that type. The overall conflict graph can be obtained by adding edges joining any vertex pair with different types to the union of all partial conflict graphs.

> **Example 6.2.2.** Consider again the scheduled sequencing graph of Figure 6.1. We show in Figure 6.3 the conflict graphs for the multiplier and ALU types. Examples of independent sets are $\{v_1, v_3, v_7\}$ and $\{v_4, v_5, v_{10}, v_{11}\}$, among others. Each graph can be colored

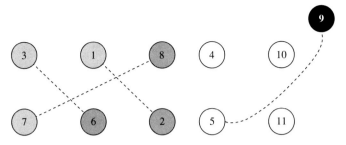

FIGURE 6.3
Conflict graphs for the multiplier and ALU types.

with two colors, yielding an overall resource requirement of 4. Different shadings of the vertices show a minimum coloring in Figure 6.3.

The *clique partitioning* and *vertex coloring* problems have been studied extensively. (See Sections 2.4.4 and 2.4.3.) Both problems are intractable for general graphs, and exact and heuristic solution methods have been proposed. According to the specific circuit type under consideration, the compatibility graph may be sparser than the conflict graph (or *vice versa*). In this case, clique partitioning (or vertex coloring) may be easier to solve.

In some particular cases, it is possible to exploit the structure of the sequencing graph to derive compatibility and conflict graphs with special properties that make the partitioning and coloring tractable. This will be considered in the following section.

6.2.1 Resource Sharing in Non-Hierarchical Sequencing Graphs

We consider here non-hierarchical sequencing graphs, i.e., we exclude for now alternative paths, and each path from source to sink represents a parallel stream of operations. We recall that the resource type implementing an operation is denoted by $T(v_i), i = 1, 2, \ldots, n_{ops}$, the start times by $T = \{t_i; \ i = 1, 2, \ldots, n_{ops}\}$ and the execution delays by $D = \{d_i; \ i = 1, 2, \ldots, n_{ops}\}$. Recall also that an operation v_i with execution delay d_i starts executing at t_i and completes at $t_i + d_i - 1$. Data-dependent delays are not considered here, because the sequencing graph is assumed to be scheduled. Resource sharing for unbounded-latency sequencing graphs is described in Section 6.5.

Two operations are compatible if they can be implemented by resources of the same type and if they are not concurrent. Therefore, the compatibility graph $G_+(V, E)$ is described by the following set of edges: $E = \{\{v_i, v_j\} \mid T(v_i) = T(v_j)$ and $((t_i + d_i \leq t_j)$ or $(t_j + d_j \leq t_i)), \ i, j = 1, 2, \ldots, n_{ops}\}$. Such a graph can be constructed by traversing the sequencing graph in $O(|V|^2)$ time. This graph is a *comparability* graph because it has a transitive orientation property. Indeed, a corresponding directed graph could be derived by assigning an orientation to the edges consistent with the relations

$\{((t_i + d_i \leq t_j) \text{ or } (t_j + d_j \leq t_i)), \ i, j = 1, 2, \ldots, n_{ops}\}$, which are *transitive*. (See Section 2.2.3.)

The search for a minimum clique partition of a comparability graph can be achieved in polynomial time, as mentioned in Section 2.4.4.

> **Example 6.2.3.** Consider again the scheduled sequencing graph of Figure 6.1, where all operations have unit execution delay. Let us consider operation v_1, with $t_1 = 1$. Now $\mathcal{T}(v_1) = multiplier$. Then, all operations whose type is a multiplier and whose start time is larger than or equal to 2 are compatible with v_1. (Obviously, no operation can be compatible by having a start time less than or equal to zero.) Such operations are $\{v_3, v_6, v_7, v_8\}$. The corresponding vertices are incident to edges that stem from v_1. The compatibility graph can be constructed by visiting each operation and checking for others with the same type with non-overlapping execution intervals.
>
> Note that a directed graph could be constructed having the compatibility graph as the underlying graph. The orientation is determined by comparing the start times. In this case, it would have the edges $\{(v_1, v_3), (v_1, v_6), (v_1, v_7), (v_1, v_8)\}$, among others. Note also that the relations $\{(v_1, v_3), (v_3, v_7)\}$ imply $\{(v_1, v_7)\}$, because ordering is a transitive relation. Hence the compatibility graph is a comparability graph. The transitive orientation of the compatibility graph is shown in Figure 6.4.

Let us consider now the conflict graphs for each resource type. The execution intervals for each operation are $\{[t_i, t_i + d_i - 1]; \ i = 1, 2, \ldots, n_{ops}\}$ and the edges of the conflict graphs denote intersections among intervals; hence they are *interval graphs*. The search for a minimum coloring of an interval graph can be achieved in polynomial time. A few algorithms can be used, including the *LEFT_EDGE* algorithm described in Section 2.4.3. Usually, resource sharing and binding is achieved by considering the conflict graphs for each type, because resources are assumed to have a single type. Thus, the overall conflict graph is of limited interest, even though it can be derived from the conflict graphs of each resource type in a straightforward way.

> **Example 6.2.4.** Consider again the scheduled sequencing graph of Figure 6.1, where all operations have unit execution delay. The set of intervals corresponding to the conflict

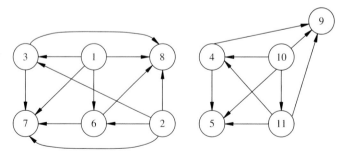

FIGURE 6.4
Transitive orientation of the compatibility graph.

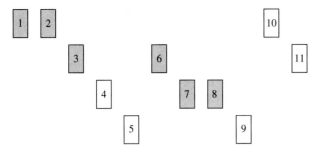

FIGURE 6.5
Intervals corresponding to the conflict graph.

graphs is shown in Figure 6.5. Overlapping intervals correspond to edges in the conflict graph for each type. When considering the multiplier, the conflict edges are $\{v_1, v_2\}$, $\{v_3, v_6\}$ and $\{v_7, v_8\}$. When considering the ALU, the conflict edge is $\{v_5, v_9\}$.

It is instructive to show that the binding problem can be formulated with an ILP model. We consider a similar framework to the one presented for scheduling in Section 5.4.1. For the sake of simplicity, we assume first that all operations and resources have the same type. We use a set of binary decision variables with two indices, $B = \{b_{ir}; i = 1, 2, \ldots, n_{ops}; r = 1, 2, \ldots, a\}$, and a set of binary decision constants with two indices, $X = \{x_{il}; i = 1, 2, \ldots, n_{ops}; l = 1, 2, \ldots, \lambda + 1\}$, where $a \leq n_{ops}$ is an upper bound on the number of resources to be used. We use the set notation for the variables in B, rather than a matrix notation, because we do not make use of matrix operations. The binary variable, b_{ir}, is 1 only when operation v_i is bound to resource r, i.e., $\beta(v_i) = (1, r)$. The binary constant, x_{il}, is 1 only when operation v_i starts in step l of the schedule, i.e., $l = t_i$, as defined in Section 5.4.1. These values are known constants, because we consider scheduled sequencing graphs.

Searching for a binding compatible with a given schedule (represented by X) and a resource bound a is equivalent to searching for a set of values of B satisfying the following constraints:

$$\sum_{r=1}^{a} b_{ir} = 1, \quad i = 1, 2, \ldots, n_{ops} \tag{6.1}$$

$$\sum_{i=1}^{n_{ops}} b_{ir} \sum_{m=l-d_i+1}^{l} x_{im} \leq 1, \quad l = 1, 2, \ldots, \lambda + 1, \quad r = 1, 2, \ldots, a \tag{6.2}$$

$$b_{ir} \in \{0, 1\}, \quad i = 1, 2, \ldots, n_{ops}, \quad r = 1, 2, \ldots, a \tag{6.3}$$

Constraint 6.1 states that each operation v_i should be assigned to one and only one resource. Constraint 6.2 states that at most one operation can be executing, among those assigned to resource r, at any time step. Note that it suffices to require the variables in B to be non-negative integers to satisfy 6.3. Hence the problem can be formulated as an ILP and not necessarily as a ZOLP.

This model can be easily extended to include multiple operation types. Nevertheless, the disjointness of the types makes it easier to formulate and solve as many independent binding problems as each type.

Example 6.2.5. Consider again the scheduled sequencing graph of Figure 6.1. The operations have two types (labeled 1 for the multiplier and 2 for the ALU) and unit execution delays. Therefore a feasible binding satisfies the constraints:

$$\sum_{r=1}^{a_1} b_{ir} = 1, \quad \forall i : T(v_i) = 1$$

$$\sum_{i:T(v_i)=1} b_{ir} \, x_{il} \leq 1, \quad l = 1, 2, \ldots, \lambda + 1, \quad r = 1, 2, \ldots, a_1$$

$$\sum_{r=1}^{a_2} b_{ir} = 1, \quad \forall i : T(v_i) = 2$$

$$\sum_{i:T(v_i)=2} b_{ir} \, x_{il} \leq 1, \quad l = 1, 2, \ldots, \lambda + 1, \quad r = 1, 2, \ldots, a_2$$

The constants in the set X are all zero, except for $x_{1,1}, x_{2,1}, x_{3,2}, x_{4,3}, x_{5,4}, x_{6,2}, x_{7,3}, x_{8,3}, x_{9,4}, x_{10,1}, x_{11,2}$, which are 1.

Then, an implementation with $a_1 = 1$ multiplier would correspond to finding a solution to:

$$b_{i1} = 1, \quad \forall i \in \{1, 2, 3, 6, 7, 8\}$$

$$\sum_{i \in \{1,2,3,6,7,8\}} b_{i1} \, x_{il} \leq 1, \quad l = 1, 2, \ldots, 5$$

Such a solution does not exist, because the second constraint would imply $b_{1,1} + b_{2,1} \leq 1$, which contradicts the first one.

An implementation with $a_1 = 2$ multipliers would correspond to finding a solution to:

$$b_{i1} + b_{i2} = 1, \quad \forall i \in \{1, 2, 3, 6, 7, 8\}$$

$$\sum_{i \in \{1,2,3,6,7,8\}} b_{i1} \, x_{il} \leq 1, \quad l = 1, 2, \ldots, 5$$

$$\sum_{i \in \{1,2,3,6,7,8\}} b_{i2} \, x_{il} \leq 1, \quad l = 1, 2, \ldots, 5$$

which admits the solution $b_{1,1} = 1, b_{2,2} = 1, b_{3,1} = 1, b_{6,2} = 1, b_{7,1} = 1, b_{8,2} = 1$, all other elements of B with first subscript $i \in \{1, 2, 3, 6, 7, 8\}$ being zero. The tabulation of the binding is the following, where the binding of the ALUs can be computed in a similar way. The bound sequencing graph is shown in Figure 6.6.

$\beta(v_1)$	$(1, 1)$
$\beta(v_2)$	$(1, 2)$
$\beta(v_3)$	$(1, 1)$
$\beta(v_4)$	$(2, 1)$
$\beta(v_5)$	$(2, 1)$
$\beta(v_6)$	$(1, 2)$
$\beta(v_7)$	$(1, 1)$
$\beta(v_8)$	$(1, 2)$
$\beta(v_9)$	$(2, 2)$
$\beta(v_{10})$	$(2, 1)$
$\beta(v_{11})$	$(2, 1)$

Note that in the particular case of a non-hierarchical graph, solving the ILP problem may be far less efficient than coloring the corresponding interval graph.

6.2.2   Resource Sharing in Hierarchical Sequencing Graphs

Let us now consider hierarchical sequencing graphs. A simplistic approach to resource sharing is to perform it independently within each sequencing graph entity. Such an approach is overly restrictive, because it would not allow sharing resources in different entities. Therefore we consider here resource sharing across the hierarchy levels.

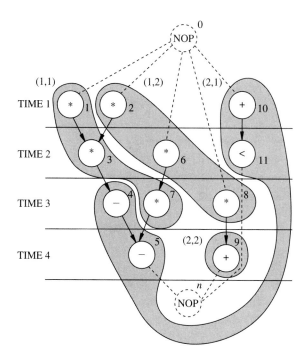

FIGURE 6.6
Scheduled and bound sequencing graph.

Let us first restrict our attention to sequencing graphs where the hierarchy is induced by *model calls*. When two link vertices corresponding to different called models are not concurrent, any operation pair implementable by resources with the same type and in the different called models is compatible. Conversely, concurrency of the called models does not necessarily imply conflicts of operation pairs in the models themselves.

> **Example 6.2.6.** Consider a model *a* consisting of two operations: an addition followed by a multiplication. Consider also a model *b* consisting of two operations: a multiplication followed by an addition. Assume that the addition has 1-unit delay and the multiplication 2-unit delay. When a model *m*1 has a call to model *a* followed by a call to model *b*, *a* and *b* are not concurrent and the corresponding additions and multiplications are compatible.
>
> Consider another model *m*2 with two calls to *a* and *b* that overlap in time, say with start times $t_a = 1$ and $t_b = 3$. Then we cannot say *a priori* that the operations of *a* and *b* are conflicting. Indeed the multiplications are not compatible while the additions are! Both situations are shown in Figure 6.7.

Therefore the appropriate way of computing the compatibility of operations across different levels of the hierarchy is to flatten the hierarchy. Such an expansion can be done explicitly, by replacing the link vertices by the graphs of the corresponding models, or implicitly, by computing the execution intervals of each operation with respect to the source operation of the root model in the hierarchy.

To determine the properties of the compatibility and conflict graphs, we need to distinguish the cases when models are called once or more than once. In both cases, model calls make the sequencing graph representation modular. In the latter case, model calls express also the sharing of the application-specific resource corresponding to the model.

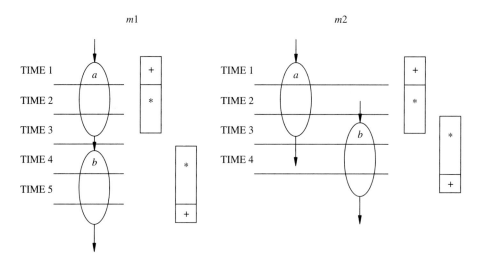

FIGURE 6.7
Hierarchical conflicts and compatibility.

When all models are called only once, the hierarchy is only a structured representation of the data-flow information. Thus compatibility and conflict graphs have the special properties described in the previous section.

Let us consider now multiple calls to a model. We question the compatibility or conflict of the operations in the called model with those in the calling one, and the properties of the corresponding graphs.

> **Example 6.2.7.** Consider again model a consisting of two operations: an addition followed by a multiplication. By assuming that the addition has 1-unit delay and the multiplication 2-unit delay, model a has an overall delay of 3 units. Consider then model $m3$ with two calls to model a that are not concurrent scheduled at times 1 and 5, respectively. Assume also that model $m3$ has three other multiplication operations. We question the sharing of the multipliers across the hierarchy.
>
> A sequencing graph fragment (related to $m3$), the execution intervals and the conflict graph for the multiplier type are shown in Figure 6.8. Note that the double call to a results in two non-contiguous execution intervals for the multiplication in a. As a result, the conflict graph is not an intersection among intervals and therefore not an interval graph. It is not even a chordal graph, as shown in Figure 6.8 (c).

Whereas the computation of the compatibility and conflict graphs is still straightforward, the resulting graphs may no longer have special properties. Therefore their clique partitioning and vertex coloring are now intractable problems. Heuristic algorithms may be used, as described in Section 2.4.3.

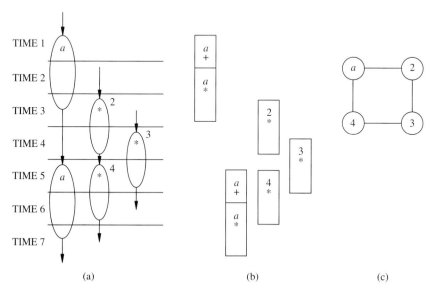

FIGURE 6.8
Hierarchical conflicts. (a) Sequencing graph fragment. (b) Execution intervals. (c) Non-chordal conflict graph.

The compatibility of the operations across the hierarchy can be computed in a similar way in the presence of *iterative* constructs that can be unrolled. Note that a resource bound to one operation in a loop corresponds to a resource bound to multiple instances of that operation when the loop is unrolled. Moreover, that resource may be bound to other operations outside the loop model. Note also that a single model call inside a loop body becomes a multiple call when the loop body is unrolled.

Let us consider now *branching* constructs. When considering operation pairs in two alternative branching bodies, their compatibility corresponds to having the same type. The computation of the compatibility and conflict graphs can still be done by traversing the hierarchy and using Definitions 6.2.1 and 6.2.2. The resulting compatibility and conflict graphs may not have any special property, as shown by the following example.

> **Example 6.2.8.** Consider the sequencing graph of Figure 6.9 (a). We assume that all operations take 2 time units to execute and that the start times are the following: $t_a = 1$; $t_b = 3$; $t_c = t_d = 2$. The intervals are shown in Figure 6.9 (b) and the conflict graph in Figure 6.9 (c). Note that the alternative nature of operations c and d makes them compatible and prevents a chord $\{v_c, v_d\}$ to be present in the conflict graph. Hence the conflict graph is not an interval graph.

6.2.3 Register Sharing

We consider in this section those registers that hold the values of the variables. Recall that each variable has a *lifetime* that is the interval from its *birth* to its *death*, where the former is the time at which the value is generated as an output of an operation and the latter is the latest time at which the variable is referenced as an input to another operation. We assume that those variables with multiple assignments within one model are aliased, so that each variable has a single lifetime interval in the frame of reference corresponding to the sequencing graph entity where it is used. Note that the lifetimes can be data dependent, for example, due to branching and iterative constructs.

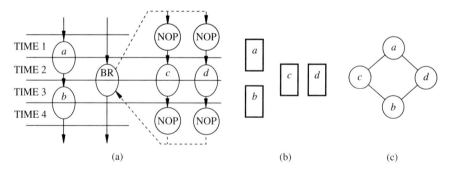

FIGURE 6.9
Conditional execution. (a) Sequencing graph fragment. (b) Execution intervals. (c) Non-chordal conflict graph.

Whereas an implementation that associates a register with each variable suffices, it is obviously inefficient. Indeed, variables that are alive in different intervals or under alternative conditions can share the same register. Such variables are called *compatible*.

The *register compatibility* and *conflict* graphs are defined analogously to the resource compatibility and conflict graphs. The problem of minimizing the number of registers can be cast in a minimum clique partitioning problem of the compatibility graph or into a minimum coloring problem of the conflict graph. We consider now how these graphs are generated and their properties.

Let us consider first non-hierarchical sequencing graphs. In this model, a conflict between two variables corresponds to a lifetime overlap. Since in this model the variable lifetimes are intervals, the conflict graph is an interval graph and its complement is a comparability graph. Therefore, optimum register sharing can be computed in polynomial time, for example, by optimum coloring using the *LEFT_EDGE* algorithm [6].

Example 6.2.9. Consider just a portion of the sequencing graph of Figure 6.1, shown in Figure 6.10 (a). There are six intermediate variables, named $\{z_i; i = 1, 2, \ldots, 6\}$, that must be stored in registers. The lifetime of three pairs of these variables is conflicting, as shown by the conflict graph in Figure 6.10 (c). These variables can be implemented by two registers, which is the chromatic number of the conflict graph.

Let us now consider sequencing models of iterative bodies. In this case, some variables are alive across the iteration boundary, for example, the loop-counter variable. The cyclicity of the lifetimes is modeled accurately by circular-arc graphs that represent the intersection of arcs on a circle.

Example 6.2.10. We consider the full differential equation integrator. There are 7 intermediate variables $\{z_i, i = 1, 2, \ldots, 7\}$, 3 loop variables (x, y, u) and 3 loop invariants $(a, 3, dx)$. We consider here the intermediate and loop variables and their assignment to registers. The data-flow graph is shown in Figure 6.11 (a) along with the explicit

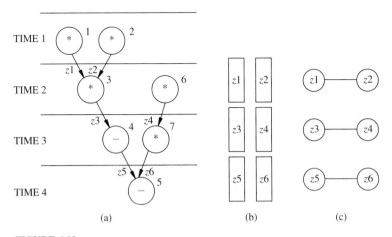

(a) Sequencing graph fragment. (b) Variable intervals. (c) Conflict graph.

FIGURE 6.10
(a) Sequencing graph fragment. (b) Variable intervals. (c) Conflict graph.

annotation of the variables. The variable lifetimes are shown in Figure 6.11 (b) and are represented as arcs on a circle in Figure 6.12. The corresponding circular-arc conflict graph is shown in Figure 6.13. Five registers suffice to store the 10 intermediate and loop variables.

The register sharing problem can then be cast as a minimum coloring of a circular-arc graph that unfortunately is intractable. Stok [15] has shown that this problem can be transformed into a multi-commodity flow problem and then solved by a primal algorithm.

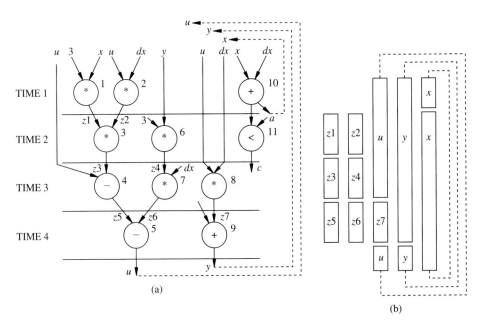

(a)

(b)

FIGURE 6.11
(a) Sequencing graph. (b) Variable lifetimes.

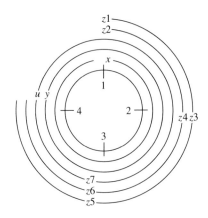

FIGURE 6.12
Variable lifetimes as arcs on a circle.

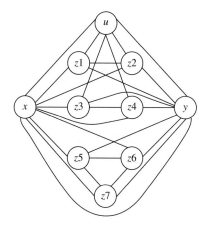

FIGURE 6.13
Circular-arc conflict graph.

The register sharing problem can be extended to cope with hierarchical models. The register compatibility and conflict graphs can be derived by traversing the hierarchy and comparing the variable lifetimes. The properties of these graphs depend on the structure of the hierarchy, as in the case of the resource compatibility and conflict graphs. In particular, interval conflict graphs can be derived from hierarchical models with only single model calls by considering the variable lifetimes with reference to the start time of the sequencing graph entity in the root model of the hierarchy. In the general case, register compatibility and conflict graphs may not have any special property, and therefore the corresponding optimum register sharing problem is intractable. Springer and Thomas [14] have shown that polynomial-time colorable conflict graphs can be achieved by enforcing some restrictions on the model calls and on the branch types.

The register sharing problem can be formulated by ILP constraints, *mutatis mutandis*, similarly to the resource sharing problem.

6.2.4   Multi-Port Memory Binding

We consider now the problem of using multi-port memory arrays to store the values of the variables. Let us assume a memory with a ports for either *read* or *write* requiring one cycle per access. Such a memory array can be a general purpose register file common to RISC architectures. We assume the memory to be large enough to hold all data. We consider in this section non-hierarchical sequencing graphs; extensions are straightforward and similar to those presented in the previous sections.

A first problem corresponds to computing the minimum number of ports a required to access as many variables as needed. If each variable accesses the memory always through the same port, then the problem reduces to binding variables to ports. Thus the considerations for functional resource binding presented in Section 6.2.1 can be applied to the ports, which can be seen as interface resources.

On the other hand, if the variables can access the memory array through any port, the minimum number of ports is equal to $\max\limits_{1 \leq l \leq \lambda+1} \sum\limits_{i=1}^{n_{var}} x_{il}$, where n_{var} is the total number of variables and X is a set of binary constants determined by scheduling, where x_{il} is 1 when the access of variable i, $i = 1, 2, \ldots, n_{var}$, is at step l, $l = 1, 2, \ldots, \lambda + 1$. In other words, the maximum number of concurrent accesses is the minimum number of ports.

Balakrishnan *et al.* [2] considered the dual problem. They assumed a fixed number of ports and maximized the number of variables to be stored in the multi-port memory array, subject to the port limitation. They formulated the problem as follows. Let $\mathbf{b} \in \{0, 1\}^{n_{var}}$ be a binary-valued vector whose entries denote whether the corresponding variable is stored in the array or not. Let $\mathbf{1}$ be a n_{var}-dimensional vector of all 1s. Then the desired variables are indicated by a solution to:

$$\max \ \mathbf{1}^T \mathbf{b} \quad \text{such that}$$

$$\sum_{i=1}^{n_{var}} b_i \, x_{il} \ \leq \ a, \qquad l = 1, 2, \ldots, \lambda + 1 \qquad (6.4)$$

This problem can be extended easily to handle separate *read* and *write* ports and to the interconnect minimization problem [2].

Example 6.2.11. This example is borrowed from reference [2]. Consider the following scheduled sequence of operations, which requires the storage of variables $\{z_i, i = 1, 2, \ldots, 15\}$:

> *Time-step 1:* $z_3 = z_1 + z_2$; $z_{12} = z_1$
>
> *Time-step 2:* $z_5 = z_3 + z_4$; $z_7 = z_3 * z_6$; $z_{13} = z_3$
>
> *Time-step 3:* $z_8 = z_3 + z_5$; $z_9 = z_1 + z_7$; $z_{11} = z_{10}/z_5$
>
> *Time-step 4:* $z_{14} = z_{11} \wedge z_8$; $z_{15} = z_{12} \vee z_9$
>
> *Time-step 5:* $z_1 = z_{14}$; $z_2 = z_{15}$

Let us consider a memory array with a ports. Then, the problem can be represented by maximizing $\sum_{i=1}^{15} b_i$ under the following set of constraints:

$$b_1 + b_2 + b_3 + b_{12} \leq a$$

$$b_3 + b_4 + b_5 + b_6 + b_7 + b_{13} \leq a$$

$$b_1 + b_3 + b_5 + b_7 + b_8 + b_9 + b_{10} + b_{11} \leq a$$

$$b_8 + b_9 + b_{11} + b_{12} + b_{14} + b_{15} \leq a$$

$$b_1 + b_2 + b_{14} + b_{15} \leq a$$

This problem with a one-port memory (i.e., $a = 1$) yields a solution where only $\{b_2, b_4, b_8\}$ are non-zero, i.e., where only variables z_2, z_4 and z_8 are stored in the memory. Increasing the number of ports helps in storing more variables into the array. For example, with a two-port memory (i.e., $a = 2$), variables $z_2, z_4, z_5, z_{10}, z_{12}$ and z_{14} can be stored and with a three-port memory (i.e., $a = 3$), variables $z_1, z_2, z_4, z_6, z_8, z_{10}, z_{12}, z_{13}$ and z_{14} can be stored.

6.2.5   Bus Sharing and Binding

Busses act as transfer resources that feed data to functional resources. The operation of writing a specific bus can be modeled explicitly as a vertex in the sequencing graph model. In this case, the compatible (or conflicting) data transfers may be represented by compatibility (or conflict) graphs, as in the case of functional resources. Alternatively, busses may not be explicitly described in the sequencing graph model. Their (optimum) usage can be derived by exploiting the timing of the data transfers. Since busses have no memory, we consider only the transfers of data within each schedule step (or across two adjacent schedule steps, when we assume that the bus transfer is interleaved with the computation).

Two problems then arise: first, to find the minimum number of busses to accommodate all (or part of) the data transfers and, second, to find the maximum number of data transfers that can be done through a given number of busses. These problems are analogous to the multi-port binding problem and can be modeled by ILP constraints.

> **Example 6.2.12.** Consider the sequencing graph fragment of Figure 6.10 (a). Let us assume that the variables of interest are $\{z_i, i = 1, 2, \ldots, 6\}$ and that busses can transfer the information across two adjacent steps. Let the timing scheduled data transfers be modeled by constants $X = \{x_{il}, i = 1, 2, \ldots, 6; l = 1, 2, \ldots, 5\}$. The values of the elements of X are all 0s, except for $x_{11}, x_{21}, x_{32}, x_{42}, x_{53}, x_{63}$, which are 1s. Then Equation 6.4 yields:
>
> $$b_1 + b_2 \ < a$$
>
> $$b_3 + b_4 \ < a$$
>
> $$b_5 + b_6 \ < a$$
>
> Let us consider first the case of $a = 1$ bus. Then at most three variables can be transferred on the bus, for example $\{z_1, z_3, z_5\}$. With $a = 2$ busses all variables can be transferred.

6.3   SHARING AND BINDING FOR GENERAL CIRCUITS*

We consider now general circuit models where area and performance do not depend solely on the resource usage. We extend our considerations to include the influence of steering logic and wiring on the binding choices that optimize area and performance. In particular, we consider multiplexers whose area and propagation delays depend on the number of inputs and wires whose lengths can be derived from statistical models (see Section 4.4). We neglect the control unit when considering scheduled models, because its contribution to the overall area and performance is affected marginally by resource binding.

As mentioned in Section 4.5, the scheduling and binding problems are intertwined for general circuit models. We present in this section the binding problem for scheduled sequencing graphs and we defer the considerations about concurrent scheduling and binding to Section 6.4.

We consider first functional resource binding, and we revisit the unconstrained minimum-area binding problem in Section 6.3.1. We then consider performance issues in Section 6.3.2. Eventually, we comment on register, multi-port memory arrays and bus binding in Section 6.3.3.

6.3.1 Unconstrained Minimum-Area Binding*

We consider the unconstrained minimum-area binding problem for scheduled graphs. Whereas compatibility and conflict graphs, as defined in Section 6.2, still model the latitude at which sharing can be performed, the area cost function depends on several factors (e.g., resource count, steering logic and wiring). In some limiting cases, resource sharing may affect adversely the circuit area.

> **Example 6.3.1.** Assume here, for the sake of simplicity, that the circuit area depends on the resources and multiplexers only.
>
> Consider a circuit with n 1-bit add operations. Assume that the area of a 1-bit adder resource is $area_{add}$ and the area of a multiplexer is $area_{mux}$, which is a function of the number of inputs, namely $area_{mux} = area_{mux}^{\Delta}(i-1)$, where $area_{mux}^{\Delta}$ is a constant that relates the increase in size of the multiplexer to the number of its inputs i. Note that a one-input mux has zero cost, because it is redundant.
>
> Then the total area of a binding with a resources is $a(area_{add} + area_{mux}) \approx a(area_{add} - area_{mux}^{\Delta}) + n \cdot area_{mux}^{\Delta}$, which may be an increasing or decreasing function of a according to the value of the relation $area_{add} > area_{mux}^{\Delta}$.

In general, the optimum sharing problem can be cast as a non-linear optimization problem under linear constraints, expressed by inequalities 6.1 and 6.2 in the case of non-hierarchical graphs. The area cost function may include dependence on the components of interest. While such an optimization problem can be solved by standard nonlinear program solvers, this approach is practical for small-scale circuits only.

The minimum-area binding problem can be modeled by a *weighted compatibility graph*. Different types of weighted compatibility graphs can be defined. For example, weights may be associated with cliques and model the cost of assigning the subgraph to a single resource, the cost being comprehensive of the related steering logic and wiring cost components. Thus, optimum binding can be modeled as a weighted clique partitioning problem. Note that an optimum solution to this problem may involve more cliques than the clique-cover number of the graph, because non-maximal cliques may be chosen.

With this formulation, two difficulties must be overcome. First is solving the weighted clique partitioning problem, which is intractable. Heuristic algorithms are commonly used for this purpose. Second is assigning weights to vertices and cliques that model correctly the problem objective. Heuristic methods have been proposed to assign weights to the compatibility graph, representative of the desired factors to be taken into account in sharing.

Springer and Thomas proposed a weighted compatibility graph model based on the use of a set inclusion cost function [14]. In this model each vertex of the compatibility graph is given a set of weights, called a *cost set*. The cost set of any clique is the union of the cost sets of the corresponding vertices, and its weight is the sum of the elements in its cost set. The overall cost is the sum of all weights associated with the cliques of a partition.

The cost-set model is very general and it allows us to compute weights on the cliques from the attributes of the vertices. Note that the clique weight is not necessarily the sum of the weight on the vertices. Thus, this model can be useful to represent the steering logic and wiring area, which depend on the grouping of operations and their binding to a shared resource. This model applies very well to other more general binding problems. We refer the interested reader to reference [14] for details.

> **Example 6.3.2.** We model the total area as a weighted sum of resource usage and multiplexing cost. We assume that the cost of multiplexing a signals is $area_{mux} = area_{mux}^\Delta (a - 1)$ and is expressed as $area_M^0 + \sum_{i=1}^{a} area_M^i$, where $area_{mux}^\Delta = area_M^i = - area_M^0$, $1 \leq i \leq a$. The purpose of the model is to spread the multiplexer cost over the operations.
>
> Consider four operations of the same type (e.g., addition), namely v_1, v_2, v_3 and v_4, and their compatibility graph, shown in Figure 6.14. We assume that the cost set of each vertex v_i; $i = 1, 2, 3, 4$ consists of the triple $\{area_+, area_M^0, area_M^0\}$. The area of the corresponding adder resource is $area_+$; the other elements relate to the offset and incremental costs of the multiplexers.
>
> Let us assume first the use of dedicated resources. Then the overall cost is $4 \cdot (area_+ + area_M^0) + \sum_{i=1}^{4} area_M^i = 4 \ area_+$.
>
> Let us assume instead that v_1, v_2 and v_3 share a resource. Then the cost is $\{area_+ + area_M^0 + \sum_{i=1}^{3} area_M^i\} + \{area_+, area_M^0, area_M^4\} = 2 \ area_+ + area_{mux}^\Delta (3 - 1)$.

A different approach for resource sharing was proposed by Tseng and Siewiorek [18], who devised a heuristic clique partitioning algorithm that constructs cliques from compatible pairs of operations. Thus they used edge weights as representative of the level of desirability for sharing. Even though edge weights may be related to the cost of steering logic and wiring, the overall area cost of the circuit cannot be inferred from the weights. The edge-weighted compatibility graph is denoted here by $G_+(V, E, W)$.

Tseng and Siewiorek's algorithm considers repeatedly the subgraphs induced by the vertices that are end-points of edges with the same weights for decreasing values of the weights. The algorithm performs unweighted clique partitioning of these subgraphs by iterating the following steps until the subgraph is empty.

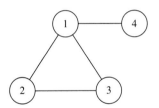

FIGURE 6.14
Compatibility graph.

```
TSENG( G₊(V, E, W) ) {
    while (E ≠ ∅) do {
        lw = max  w;                                          /* largest edge weight */
              w∈W
        E' = {{vᵢ, vⱼ} ∈ E such that wᵢⱼ = lw};
        G'₊(V', E', W') = subgraph of G₊(V, E, W) induced by E';
        while (E' ≠ ∅) do {
            Select {vᵢ, vⱼ} ∈ E' such that vᵢ and vⱼ have the most neighbors in common;
            C = {vᵢ, vⱼ};
            Delete edges {vₗ, vᵢ} if {vₗ, vⱼ} ∉ E' ∀vₗ ∈ V';
            Delete vertex vⱼ from V';
            while (one vertex adjacent to vᵢ in G'₊(V', E', W')) do {
                Select vₖ such that {vᵢ, vₖ} ∈ E' and vᵢ and vₖ have the
                    most neighbors in common;
                C = C ∪ {vₖ};
                Delete edges {vₗ, vᵢ} if {vₗ, vₖ} ∉ E' ∀vₗ ∈ V';
                Delete vertex vₖ from V';
            }
            Save clique C in the clique list;
        }
        Delete the vertices in the clique list from V;
    }
}
```

ALGORITHM 6.3.1

The two vertices in the subgraph with most common neighbors (i.e., adjacent vertices) are merged and the edges from any vertex, say v_l, to the merged pair are deleted unless v_l was incident to both vertices before merging. The two merged vertices form the seed of a clique. The seed is then expanded by selecting iteratively other vertices having the most neighbors in common with the clique seed. The selected vertices are merged with the clique seed and the graph is updated accordingly. When no more vertices can be added to the clique, the clique is saved on a list and the process is repeated. (See Algorithm 6.3.1.)

Example 6.3.3. Consider the compatibility graph of Figure 6.2 and let us focus on the subgraph related to the multiplication type, i.e., induced by vertices $\{v_1, v_2, v_3, v_6, v_7, v_8\}$. Such a subgraph is shown in Figure 6.15 (a). Assume first all edges have unit weight. Thus, the first step of the second loop of Tseng and Siewiorek's algorithm would select vertices $\{v_1, v_3\}$ as clique seed, because they have two common neighbors (i.e., v_8 and v_7). Then, edge $\{v_1, v_6\}$ is removed, because edge $\{v_3, v_6\}$ is not in the graph. Vertex v_3 is deleted (with all edges incident to it). The corresponding graph is shown in Figure 6.15 (b). Another vertex is then sought in the remaining graph. Vertex v_7 is selected, because it is adjacent to v_1, and added to the clique seed. Then, edge $\{v_1, v_8\}$ is deleted as well as v_7. A clique is then identified as $\{v_1, v_3, v_7\}$. A new vertex pair is then searched for

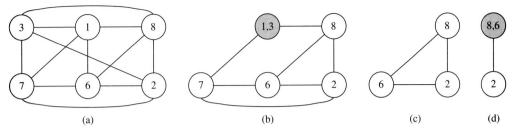

FIGURE 6.15
(a) Compatibility graph for the multiplier type. (b) Reduced compatibility graph with clique seed. (c) Fragment of compatibility graph after one clique has been removed. (d) Reduced fragment with clique seed.

in the remaining graph, shown in Figure 6.15 (c). Let us say $\{v_6, v_8\}$ is chosen as the seed. At the next step, v_2 is added to form the second clique $\{v_6, v_8, v_2\}$.

Assume now that edges $\{v_1, v_3\}$, $\{v_1, v_6\}$, $\{v_1, v_7\}$, $\{v_3, v_7\}$, $\{v_7, v_6\}$ have weight 2 and the remaining ones have weight 1. Then the subgraph induced by $\{v_1, v_3, v_6, v_7\}$ is considered. Assuming that vertices $\{v_1, v_3\}$ are selected as the clique seed, because they have one common neighbor (i.e., v_7), clique $\{v_1, v_3, v_7\}$ is identified. Since there is no edge left in E', the algorithm would look for cliques in the graph obtained from the original one by removing $\{v_1, v_3, v_7\}$. The second clique $\{v_6, v_8, v_2\}$ is thus found.

6.3.2 Performance-Constrained and Performance-Directed Binding*

Whereas binding does not affect the performance of resource-dominated circuits, in the general case it changes the path delay between register boundaries. Since we consider here scheduled graphs, we must ensure that such path delays are bounded from above by the cycle-time. Otherwise, the computed schedule is invalid.

For the sake of simplicity, consider combinational resources with propagation delays $\{\widehat{d}_i; i = 1, 2, \ldots, n_{ops}\}$ bounded from above by the cycle-time ϕ. Operations can be chained together. Let us call a *path* a maximal set of vertices that induce a path in the sequencing graph and that are scheduled in the same time step. (In the limiting case where chaining is not applied, each path degenerates to a single vertex.) The path delay is:

$$\sum_{i \in path} \widehat{d}_i + multiplexer\_delay(B) + wiring\_delay(B) \qquad (6.5)$$

where B denotes the binding.

Then, minimum-area resource sharing under timing constraints can be solved by a non-linear program by adding the constraint

$$\sum_{i \in path} d_i + multiplexer\_delay(B) + wiring\_delay(B) < \phi \quad \forall path \qquad (6.6)$$

to those represented by inequalities 6.1 and 6.2. Note that both the objective function and the constraints are now non-linear.

The dependency of the cycle-time on the binding allows us to define a performance-oriented binding problem, where the goal is to find a sharing that minimizes the cycle-time. Such a problem can be formalized as one where the maximum path delay is minimized subject to the constraints 6.1 and 6.2.

In this problem, steering logic and wiring delay play conflicting roles. Indeed, if wiring delays were negligible, then fully dedicated resource implementations would optimize the cycle-time, because no multiplexing would be required. Unfortunately, the choice of dedicated resources correlates to more subcircuits, a larger physical layout and longer wiring delays. Conversely, if steering logic delays were negligible, then full sharing may reduce the delay because of the reduction in area. The optimum solution that can be achieved by solving a general non-linear problem reflects a trade-off between wiring and multiplexing delays.

6.3.3   Considerations for Other Binding Problems*

Let us consider register sharing first, as described in Section 6.2.3. Sharing of a register involves steering data to and from a register. Therefore the area and delay of the steering logic, e.g., multiplexers, should be taken into account when considering optimum register sharing. The considerations on functional resource sharing reported in the previous sections can be extended, *mutatis mutandis*, to register sharing.

In the multi-port memory binding problem, it is also important to consider the steering logic and wiring costs. Wires connect the memory ports to the functional resources that have as inputs and/or outputs those variables stored in the array. A particular optimization problem is related to maximizing the assignment of the same port to all those variables having the same source and/or destination. This correlates to reducing both the wiring and steering logic cost. Balakrishnan *et al.* considered this problem and solved it by an integer linear program formulation. We refer the interested reader to [2] for further details.

Finally we comment here on sharing of busses. Busses provide distributed multiplexing and wiring, and therefore their area and delay cost captures the steering logic and wiring delays. The cost of the bus drivers (in terms or area and delay) and the physical length of the bus affect the overall design objectives. The problems of optimizing the bus usage, described in Section 6.2.5, can be extended to cope with these factors. In particular, analogous to the multi-port memory binding problem, it is convenient to maximize the assignment to the same bus of those variables having the same source and/or destination. This correlates to reducing both the bus length and the number of bus drivers.

6.4   CONCURRENT BINDING AND SCHEDULING

We comment here on a general framework that can be used to unify the binding and scheduling problems. For the sake of simplicity, we consider operations with one type and non-hierarchical graphs.

We use two sets of binary decision variables with two indices: $B = \{b_{ir}; i = 1, 2, \ldots, n_{ops}; r = 1, 2, \ldots, \bar{a}\}$ and $X = \{x_{il}; i = 0, 1, \ldots, n; l = 1, 2, \ldots, \bar{\lambda} + 1\}$, where $\bar{a}$ and $\bar{\lambda}$ are upper bounds on resource usage and latency, respectively. (Note that we consider variables to represent the start times of the source and sink operations, but we do not consider their binding, because they are *No-Operations*. Recall also that $n = n_{ops} + 1$.)

The binary variable b_{ir} is 1 when operation v_i is bound to resource r, i.e., $\beta(v_i) = (1, r)$. The binary variable x_{il} is 1 when operation v_i starts in step l of the schedule.

We can write the set of constraints in terms of B and X as follows:

$$\sum_l x_{il} = 1, \quad i = 0, 1, \ldots, n \tag{6.7}$$

$$\sum_l l \cdot x_{il} - \sum_l l \cdot x_{jl} - d_j \geq 0, \quad i, j = 0, 1, \ldots, n : (v_j, v_i) \in E \tag{6.8}$$

$$\sum_{r=1}^{\bar{a}} b_{ir} = 1, \quad i = 1, 2, \ldots, n_{ops} \tag{6.9}$$

$$\sum_{i=1}^{n_{ops}} b_{ir} \sum_{m=l-d_i+1}^{l} x_{im} \leq 1, \quad l = 1, 2, \ldots, \bar{\lambda} + 1, \quad r = 1, 2, \ldots, \bar{a} \tag{6.10}$$

$$x_{il} \in \{0, 1\}, \quad i = 0, 1, \ldots, n, \quad l = 1, 2, \ldots, \bar{\lambda} + 1 \tag{6.11}$$

$$b_{ir} \in \{0, 1\}, \quad i = 1, 2, \ldots, n_{ops}, \quad r = 1, 2, \ldots, \bar{a} \tag{6.12}$$

where constraint 6.7 states that each operation has to be started once, 6.8 states that the sequencing constraints must be satisfied, 6.9 states that each operation has to be bound to one and only one resource and 6.10 states that operations bound to the same resource must not be concurrent. The bounds on the summations in 6.7 and 6.8 have been discussed in Section 5.4.1. Note that it suffices to require the variables in X and in B to be non-negative integers to ensure that their value is binary valued.

Consider the search for area/latency trade-off points for a given cycle-time. Latency is determined by the variables X, namely as $\lambda = \sum_l l x_{nl} - \sum_l l x_{0l}$. The area is determined as a function of B.

Let us review briefly the case of resource-dominated circuits. The execution delays D do not depend on the binding B and are constant for a given cycle-time. The area depends on the resource usage a. Hence the latency (or area) minimization problems under area (or latency) constraints can be modeled by an ILP [3]. In addition, since:

$$\sum_{i=1}^{n_{ops}} \sum_{m=l-d_i+1}^{l} x_{im} \leq a, \quad l = 1, 2, \ldots, \bar{\lambda} + 1 \tag{6.13}$$

implies that constraint 6.10 can be satisfied by some binding with a resources, then the overall problem can be solved by determining first the values of X satisfying constraints 6.7, 6.8 and 6.13 and then by determining the values of B satisfying constraints 6.9 and 6.10. This corresponds to performing scheduling before binding.

Let us turn now our attention to general (non-resource-dominated) circuits. For a given cycle-time, the execution delays D depend non-linearly on the binding B. The area cost function is a non-linear function of the binding B (and of the schedule X when the control-unit cost is considered). Thus optimizing area, subject to latency constraints, corresponds to optimizing a non-linear function with non-linear constraints. Optimizing latency, under area constraints, corresponds to optimizing a linear function with non-linear constraints. Both problems can be solved by standard non-linear program solvers, but the approach has been shown to be practical for small-scale circuits only.

Recent research efforts have tried to model the optimization of general circuits with linear (or piecewise linear) cost functions and/or constraints. Obviously some restrictive assumptions are needed. If the execution delays can be assumed to be constant (by computing them as the rounded-up quotient of the propagation delays plus some margin to the cycle-time), then all constraints are linear. By restricting the area computation to resources, registers and steering logic with piecewise linear area models (excluding wiring and control cost), the combined scheduling and binding problem can be modeled by an ILP. Gebotys showed that this approach is practical for medium-scale circuits by using additional constraints to reduce the feasible solution region as well as the computation time [3].

Most synthesis programs perform scheduling before binding or *vice versa*, sometimes iterating the two tasks. This can be seen as searching for the values of X and B by means of a relaxation technique. In general, scheduling and binding are performed with an educated guess on the execution delays D. The maximum path delay is evaluated after binding, by taking into account all factors of interest, and is compared to the cycle-time. If the cycle-time constraint is not met, scheduling and binding are repeated after increasing the execution delays. Otherwise, the iterative process may stop or another iteration tried with shorter execution delays when the cycle-time constraint is satisfied by a wide margin. Similar considerations apply to area constraints, when specified. Note that in the case of resource-dominated circuits, the execution delays and area do not depend on the binding and therefore one iteration is sufficient.

Some architectural-level synthesis systems iterate scheduling and binding along with an evaluation of the design objectives, and possibly a reassessment of the design constraints. This latter approach has been named the *knobs and gauges* approach [11].

6.5 RESOURCE SHARING AND BINDING FOR NON-SCHEDULED SEQUENCING GRAPHS

We consider in this section the problems related to applying resource sharing and binding before scheduling a sequencing graph model. Before delving into the details, we comment on the relevance of this approach for different classes of circuits.

Consider first resource-dominated circuits with operations having data-independent delays. The area and latency do not depend on binding, but just on the resource usage and on the schedule. Hence the area/latency trade-off is best computed by scheduling and preventive binding restricts the design space. Thus, applying binding before scheduling has limited value.

When considering resource-dominated circuits with unbounded-delay operations, latency cannot be determined. Such circuits may have to satisfy some relative timing constraints (see Section 5.3.3). The corresponding scheduling algorithms may require that binding is performed prior to scheduling [5].

It is convenient to perform binding before scheduling for general (non-resource-dominated) circuits, so that the steering logic and wiring area and delay can be derived from the binding. Thus the overall area can be estimated with accuracy. Moreover the execution delays to be used in scheduling can incorporate steering logic and wiring delays.

6.5.1 Sharing and Binding for Non-Scheduled Models

We consider first the two limiting cases in applying resource sharing to non-scheduled sequencing graphs, namely, minimum-area and minimum-latency sharing and binding. Then we comment on the intermediate solutions.

Minimum-area binding is achieved by using one resource per type. This problem can be modeled by defining two operations *weakly compatible* when they can be implemented by resources of the same type and by partitioning into cliques the corresponding weak-compatibility graph, which is obviously trivial. For each type, all operations that are not alternative need to be serialized with respect to each other to remove possible conflicts due to concurrency. This task is called *conflict resolution*, and it can be done by adding appropriate edges to the sequencing graph.

There are many possible ways of resolving the conflicts. When all operation delays are data independent and minimizing the circuit latency is a secondary objective, conflict resolution is achieved by computing a minimum-latency schedule under a single resource constraint per type. Otherwise, any operation serialization consistent with the partial order implied by the sequencing graph can be chosen. Thus conflict resolution can be done by sorting topologically the operations.

> **Example 6.5.1.** Consider the sequencing graph of Figure 4.8, reported again for convenience in Figure 6.16 (a), with an unbounded-delay operation. Assume one resource is used. Then the two additions must be serialized, as shown in Figure 6.16 (b) or Figure 6.16 (c).

Let us consider now the case of resource sharing and binding that does not require additional operation serialization. Thus, when all operation delays are data independent, a subsequent scheduling can yield minimum latency. Otherwise, the binding still guarantees the existence of a minimum-latency circuit implementation for every possible value of the data-dependent delays. We assume here that operation chaining is not used in scheduling.

A conservative approach is used, with a restricted notion of compatibility. Two operations are said to be *strongly compatible* when they can be implemented by resources of the same type and they are either alternative or serialized with respect to

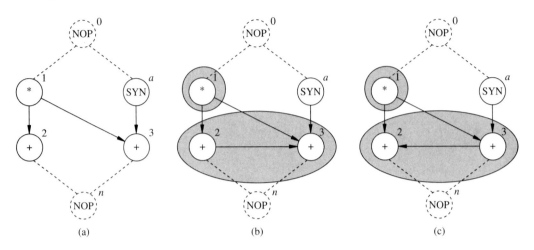

FIGURE 6.16
(a) Sequencing graph. (b) Conflict resolution. (c) Other conflict resolution.

each other, i.e., when the corresponding vertices are the head and tail of a directed path. Clearly, strong compatibility implies compatibility. In addition, no resource conflict arises in sharing strongly compatible operations, because they are either already constrained to be serialized with respect to each other or alternative. Therefore the binding does not induce any serialization constraint on the schedule, which must just satisfy the sequencing dependencies.

The strong-compatibility graph of a non-hierarchical sequencing graph is the underlying undirected graph of its transitive closure (excluding the source and sink vertices). Hence it is a comparability graph, and its partitioning into cliques can be done in polynomial time. In the case of hierarchical graphs, the strong-compatibility graph can be computed by expanding the hierarchy in a straightforward way. In general, such graphs may not have any special property.

> **Example 6.5.2.** Consider the example shown in Figure 6.17 (a). The strong-compatibility graph is shown in Figure 6.17 (b) and a binding in Figure 6.17 (c).

There is a spectrum of intermediate binding solutions that trade off resource usage for operation serialization and hence latency. Area/delay exploration in this context is meaningful especially for non-resource-dominated circuits. Indeed the knowledge of a binding allows us to estimate accurately the overall area and delays. Unfortunately, no efficient algorithm is known to compute minimum-area (or minimum-latency) binding under latency (area) constraints, aside from enumerative techniques.

The consideration of all different configurations in the design space is potentially computationally expensive, because the size of the design space grows with a greater-than-polynomial rate with the number of operations and resources, as shown in the next section. The complete search is justified in the case of those circuits where the number of operations of any given type is small and therefore the size of the design space size is within reach. For circuit models with many possible bindings, heuristic methods have

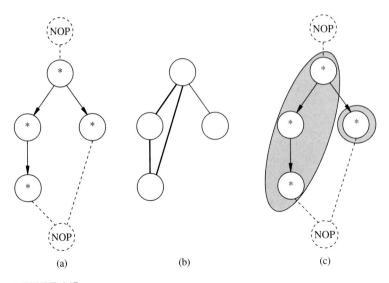

FIGURE 6.17
(a) Sequencing graph. (b) Compatibility graph. (c) Bound sequencing graph.

been proposed to prune the design space by selecting specific bindings that are likely to satisfy the design goals. Once such bindings are scheduled, an accurate evaluation of the objectives can be performed and the trade-off curve determined. Unfortunately, the heuristic nature of pruning is such that bindings corresponding to some trade-off point may not be considered. We refer the interested reader to reference [5] for details.

6.5.2 Size of the Design Space*

Different measures can be given to the size of the design space, which is the set of all feasible implementations. For resource-dominated circuits, the design space can be characterized by the resource usage, because the circuit objectives (area, latency) depend on that. Characterizing the design space for general circuits is more complex. A possible measure is the number of feasible bindings.

We consider now the enumeration of the possible bindings for a sequencing graph and a set of resource instances. We perform first the analysis by considering each resource type one at a time. We assume that there are n operations of a given type and that exactly $a \leq n$ resource instances have to be used.

Let us consider first the number of ways $m(n, a)$ in which the n operations can be partitioned into a blocks corresponding to the a resource instances. Let us assume first that the resource instances are labeled (i.e., ordered) and the operations are not. We compute this number by means of a recursive formula:

$$
m(\widehat{n}, \widehat{a}) = \begin{cases} 1 & \text{when } \widehat{n} = \widehat{a} & (6.14) \\ 1 & \text{when } \widehat{a} = 1 & (6.15) \\ \sum_{i=\widehat{a}-1}^{\widehat{n}-1} m(i, \widehat{a} - 1) & \text{otherwise} & (6.16) \end{cases}
$$

The formula can be explained as follows. When the number of operations $n = \widehat{n}$ matches the number of resources $a = \widehat{a}$, the only possible partition is to assign one operation to each resource. When only one resource is available ($a = \widehat{a} = 1$), the only partition is the one that assigns all operations to that resource. In the remaining cases, the number of partitions is computed by assuming that the first resource can be assigned to a number of operations ranging from 1 to $\widehat{n} - \widehat{a} + 1$ operations. (Note that by assigning more than $\widehat{n} - \widehat{a} + 1$ operations to the first resource, at least one of the remaining $\widehat{a} - 1$ resources will remain unutilized.) Then the number of cases is computed recursively by summing the number of partitions of i operations to $\widehat{a} - 1$ resources, where i ranges from $\widehat{a} - 1$ (where $\widehat{n} - \widehat{a} + 1$ are assigned to the first resource) to $\widehat{n} - 1$ (where one operation is assigned to the first resource).

Example 6.5.3. We consider the ways in which $n = 4$ operations can be assigned to $a = 3$ resources, called A, B and C. Assume that two operations are assigned to A. Then the two remaining operations are assigned to B and C, one each. Suppose now that one operation is assigned to A. Then in this case we have to assign the three remaining operations to B and C. Assume now that two operations are assigned to B. Then the remaining operation is assigned to C. Else, one operation is assigned to B and two to C. Thus, there are three possible cases.

Let us use the recursive formula now. The total number of possible assignments is $m(4, 3) = \sum_{i=2}^{3} m(i, 2) = m(2, 2) + m(3, 2)$. Now $m(2, 2) = 1$ (corresponding to assigning two operations to A, and consequently one to B and to C). Similarly, $m(3, 2) = \sum_{i=1}^{2} m(i, 1) = m(1, 1) + m(2, 1)$, where $m(1, 1) = 1$ (corresponding to assigning one operation to A, one to B and consequently two to C), and $m(2, 1) = 1$ (corresponding to assigning one operation to A, two to B and consequently one to C). Thus $m(4, 3) = 3$.

Consider now any one of the $m(n, a)$ possible partitions of n operations to a labeled resources and let us consider the individual operations. Let m_j; $j = 1, 2, \ldots, a$ be the number of elements in each partition block. Then there are $\binom{n}{m_1}$ ways of selecting m_1 operations out of n, $\binom{n - m_1}{m_2}$ ways of selecting m_2 operations out of the remaining $n - m_1$, and so on. The number of possible assignments of the n individual operations to the a labeled resources is:

$$\binom{n}{m_1} \cdot \binom{n - m_1}{m_2} \cdots \binom{n - \sum_{j=1}^{a-1} m_j}{m_a} \tag{6.17}$$

The total number of bindings for n operations and a resources is then:

$$p(n, a) = \sum_{i=1}^{m(n,a)} \left[\frac{\binom{n}{m_1^i} \cdot \binom{n - m_1^i}{m_2^i} \cdots \binom{n - \sum_{j=1}^{a-1} m_j^i}{m_a^i}}{a!} \right] \tag{6.18}$$

where the superscript on variable m denotes the particular partition under consideration. The denominator $a!$ in Equation 6.18 represents the fact that the labeling of the resources is irrelevant, since resources of the same type are interchangeable.

When the exact number of resource instances is not given, the overall number of bindings can be computed by considering the possible cases. For example, for n operations of a given type, this number is $\sum_{i=1}^{n} p(n, i)$, because the number of resources can be any ranging from 1 to n. This computation gives the number of bindings for a given operation type. The grand total number of bindings is the product over the different operation types.

Example 6.5.4. The following table lists some examples of partitions and the number of bindings for different values of n operations and a resources.

n	a	Partitions	$m(n,a)$	$p(n,a)$
1	1	(1)	1	1
2	1	(2)	1	1
2	2	(1,1)	1	1
3	1	(3)	1	1
3	2	(1,2) (2,1)	2	3
3	3	(1,1,1)	1	1
4	1	(4)	1	1
4	2	(1,3) (2,2) (3,1)	3	7
4	3	(1,1,2) (1,2,1) (2,1,1)	3	6
4	4	(1,1,1,1)	1	1
5	1	(5)	1	1
5	2	(1,4) (2,3) (3,2) (4,1)	4	15
5	3	(1,1,3) (1,3,1) (3,1,1)		
		(1,2,2) (2,1,2) (2,2,1)	6	25
5	4	(1,1,1,2) (1,1,2,1) (1,2,1,1) (2,1,1,1)	4	10
5	5	(1,1,1,1,1)	1	1

In the particular case of the previous example, i.e., four operations to be bound to three resources, there are $p(4, 3) = 6$ possible cases. Let us call the operations v_1, v_2, v_3 and v_4. Then the cases correspond to $\{\{v_1, v_2\}; \{v_3\}; \{v_4\}\}$, $\{\{v_1, v_3\}; \{v_2\}; \{v_4\}\}$, $\{\{v_1, v_4\}; \{v_3\}; \{v_2\}\}$, $\{\{v_2, v_3\}; \{v_1\}; \{v_4\}\}$, $\{\{v_2, v_4\}; \{v_3\}; \{v_1\}\}$ and $\{\{v_3, v_2\}; \{v_1\}; \{v_4\}\}$.

Example 6.5.5. Consider the unscheduled sequencing graph of Figure 4.2. Assume that we use two ALUs. Then there are $p(5, 2) = 15$ possible bindings for the corresponding operations to the two ALUs. We leave as an exercise for the reader (see Problem 4) to compute how many possible bindings exist of the six multiplications to two multipliers.

6.6 THE MODULE SELECTION PROBLEM

The *resource-type selection* problem, called also *module selection*, is a generalization of the binding problem, where we assume that more than one *resource type* can match the functional requirements of an operation type. In other words $\mathcal{T} : V \rightarrow \{1, 2, \ldots, n_{res}\}$ is a multi-valued function, i.e., a relation. In the particular case that no resource sharing is allowed, $\mathcal{T}$ is a one-to-many mapping.

We say that two resource types are *compatible* if they can perform the same operation. An example is given by the addition operation and the resource types corresponding to *ripple-carry* and *carry look-ahead* adders. Both resource types fulfill the required functionality, but with different area and propagation delay parameters. Another example is provided by different multiplier implementations, e.g., *fully parallel, serial-parallel* and *fully serial*. In this case the resource types differ in area, cycle-time and execution delay.

For the sake of simplicity and illustration, we consider here the module selection problem for resource-dominated circuits. In practice this problem often arises in conjunction with the selection of arithmetic units in DSPs. We also assume that the cycle-time is constant and that compatible resource types differ in area and execution delays. Thus the resource types are characterized by the pairs $\{(area\_type_k, delay\_type_k);$ $k = 1, 2, \ldots, n_{res}\}$.

The compatible resource types offer a spectrum of implementations, ranging from larger and faster implementations to smaller and slower ones.

> **Example 6.6.1.** We consider some resource types compatible with multiplication: a fully parallel, a serial-parallel and a fully parallel 32-bit multiplier. The first requires 1024 cycles, the second 32 and the third 1. We assume that the cycle-time is the same, because it is dictated by other considerations. The area of the first multiplier is proportional to 1, the second to 32 and the third to 1024.
>
> When considering only these resource types, $n_{res} = 3$ and the resource types are characterized by the pairs $\{(1, 1024), (32, 32), (1024, 1)\}$.

The module selection problem is the search for the best resource type for each operation. Resource sharing and module selection are deeply related, because all operations sharing the same resource must have the same resource type. In addition, scheduling and module selection are related, because the execution delays of the operations depend on the chosen module.

The most general formulation is an extension of the combined scheduling and binding formulation, presented in Section 6.4. For the sake of simplicity, let us consider non-hierarchical graphs and only one type of operation.

Different formulations are possible. (See Problem 7.) To be compatible with the notation of Section 6.4, given n_{res} resource types, we define $area_r = area\_type_{(r \bmod n_{res})}$ and $delay_r = delay\_type_{(r \bmod n_{res})}$; $r = 1, 2, \ldots, \bar{a}\}$. Thus we can pretend to have only one resource type whose instances have different area/delay parameters. Then, the scheduling, binding and module selection problems can be formalized by the constraint inequalities 6.7, 6.8, 6.9 and 6.10, in addition to the execution delay equation for each bound operation:

$$d_j = \sum_{r=1}^{\bar{a}} b_{jr} \cdot delay_r \; ; \; j = 1, 2, \ldots, n_{ops} \tag{6.19}$$

Several heuristic methods have been proposed for the module selection problem. When a minimum-latency schedule is sought, the fastest resource types are used in scheduling. Resource sharing and module selection are then applied to minimize the weighted resource usage corresponding to the area. Thus, slower and smaller resources can be assigned to non-critical operations.

Example 6.6.2. Consider the sequencing graph of Figure 4.2. Assume first that two multipliers are available, with area and delay equal to $(5, 1)$ and $(2, 2)$, respectively, and one ALU with area and delay equal to $(1, 1)$. Assume first that a minimum-latency implementation is sought.

Then, operations v_1, v_2 and v_3 must be assigned to the fastest multiplier, and two instances are required because v_1 is concurrent with v_2. The smaller multiplier may be assigned to v_8 and to either v_6 or v_7. However, it is not convenient to use it because it precludes some sharing. Indeed, if only the fastest type is used, only two instances suffice, leading to an area cost of 10 (excluding the ALU cost). Using an additional smaller multiplier would increase the cost to 12.

Assume next that the schedule has to satisfy a latency bound of five cycles. Then operations v_1, v_2, v_3 and v_7 can share the fastest multiplier, while v_6 and v_8 can share the smallest, as shown in Figure 6.18. The area cost is now 7 excluding the ALU cost and 9 including it.

We comment now on module selection for scheduled sequencing graphs and we concentrate on the relations between sharing and module selection in the search for minimum-area implementations.

As an example, we consider circuit models using addition and comparison as operations. We assume we have adder and ALU resource types, where the ALU can perform both addition and comparison but is larger than the adder. A circuit model with additions and without comparisons would make use of adders only. A

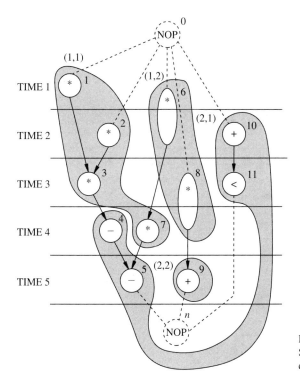

FIGURE 6.18
Scheduled and bound graph with two different multiplier types.

circuit model with additions and comparisons would make use of ALUs. It would also make use of adders when some additions cannot share the ALU resources because of conflicts. A module selection would then be finding the least-area cost implementation by choosing among types. Again, module selection is intertwined with the resource sharing. Springer and Thomas [14] modeled this problem by means of a weighted clique partitioning problem, with the cost set inclusion model described in Section 6.3.

> **Example 6.6.3.** Consider again the compatibility graph of Figure 6.14, derived after scheduling. Assume that operations v_1, v_2 and v_3 are additions and that v_4 is a comparison. We have two resources available: an adder with area $area_+$ and an ALU with area $area_{ALU}$. Let us define $area_- = area_{ALU} - area_+$. Then the cost set associated with vertices v_1, v_2 and v_3 is $\{area_+\}$, while the cost set associated with vertex v_4 is $\{area_+, area_-\}$.
>
> Let us assume that operations v_1, v_2 and v_3 share an adder. The cost set associated with the clique $\{v_1, v_2, v_3\}$ is $\{area_+\}$. The cost of a clique partition is $area_+ + (area_+ + area_-) = area_+ + area_{ALU}$.
>
> Let us assume now that operations v_1, v_2 share an adder and operations v_3, v_4 share an ALU. The cost set associated with clique $\{v_1, v_2\}$ is $\{area_+\}$. The cost set associated with clique $\{v_3, v_4\}$ is $\{area_+, area_-\}$. The cost of a clique partition is again $area_+ + (area_+ + area_-) = area_+ + area_{ALU}$.
>
> Finally, let us assume that the operations have dedicated resources. Then the total cost is $4\ area_+ + area_- = 3\ area_+ + area_{ALU}$.

6.7 RESOURCE SHARING AND BINDING FOR PIPELINED CIRCUITS

We comment in this section on resource-dominated circuits only. Resource sharing in pipelined implementations is limited by the pipeline throughput, or equivalently by the data introduction interval δ_0. Indeed, by increasing the throughput (or decreasing δ_0), we increase the concurrency of the operations and therefore their conflicts, as mentioned in Section 5.6.

To be more specific, let us consider a scheduled sequencing graph. For the sake of simplicity, let us assume that all operations have unit execution delay. Then operations with start time $l + p\delta_0$; $\forall l, p \in Z : 1 \leq l + p\delta_0 \leq \lambda + 1$ are concurrent. Compatibility and conflict graphs are defined accordingly.

> **Example 6.7.1.** Consider the pipelined scheduled sequencing graph of Figure 5.14, reported again in Figure 6.19, with $\delta_0 = 2$. It is convenient to fold the graph as shown in Figure 6.20, to highlight the concurrency of the operations. The corresponding compatibility graph is shown in Figure 6.21, where the emboldened edges denote the clique partition. Three resource instances of each type are required.

When considering hierarchical sequencing graphs, special attention has to be paid to branching constructs. When pipelining circuit models with branching, the execution of a branch body may be initiated while an alternative body is still executing. Thus, operations of alternative constructs may no longer be compatible. In particular,

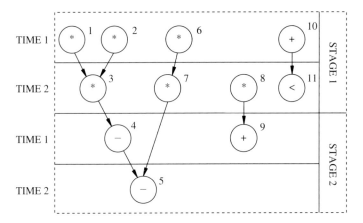

FIGURE 6.19
Scheduled sequencing graph with $\delta_0 = 2$.

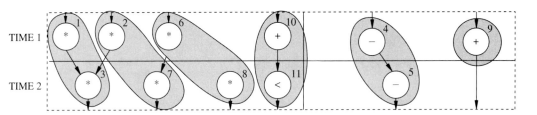

FIGURE 6.20
Folded scheduled sequencing graph with $\delta_0 = 2$ and binding annotation.

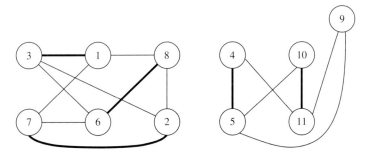

FIGURE 6.21
Compatibility graph for $\delta_0 = 2$.

they are not compatible when two pairs have twisted dependencies, and they are called in jargon *twisted pairs*. Only one pair can share a resource.

Example 6.7.2. Consider the fragment of the sequencing graph shown in Figure 6.22. The two paths are alternative. Nevertheless binding $\{v_1, v_4\}$ to a resource and $\{v_2, v_3\}$ to another resource is disallowed. When two successive data forms require the execution

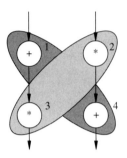

FIGURE 6.22
Example of twisted pair of operations.

of different paths, there would be simultaneous access to the same resource and hence a deadlock.

Two approaches have been proposed to cope with this problem. Park and Parker [13] suggested limiting sharing to alternative operations in the same schedule step, thus avoiding twisted pairs. Hwang *et al.* [7] did not accept this limitation and developed an algorithm to detect and avoid the twisted pairs.

6.8 SHARING AND STRUCTURAL TESTABILITY*

Testability is a broad term that relates to the ease of testing manufactured circuits [9]. We consider here testability of the data path only, which we assume to be an interconnection of combinational resources and registers, with data steered by multiplexers. We are interested in testing the combinational logic blocks (consisting of resources and steering logic) between register pairs. We defer consideration on synthesis for testability of combinational circuits to Sections 7.2.4 and 8.5. We concentrate here on *structural testability* issues, i.e., on circuit testability as related to the macroscopic circuit structure in terms of registers and combinational blocks.

Architectural synthesis for testability is the subject of ongoing research and only few results are available at present. We shall restrict our comments to some issues related to resource and register sharing of scheduled sequencing graphs, as suggested by Papachristou *et al.* [12] and Avra [1]. To be more concrete, Papachristou *et al.* and Avra dealt with testability analysis for circuits with a *built-in self-test* (BIST) mode [9]. With this testing strategy, the data path must be reconfigured in the testing mode so that each combinational logic block can accept inputs from one or more registers (which are also reconfigured to generate pseudo-random patterns) and feed its output to another register (that compresses the output data into a signature [9]).

Problems arise when a register stores both an input and an output of a combinational logic block. In this case the register is called *self-adjacent*, and it must be replaced by a complex circuit (providing concurrently for pattern generation and signature compression). Thus some strategies propose to avoid self-adjacent registers completely [12], while others try to minimize their occurrences [1]. The rationale depends on the actual areas of the subcircuits being involved.

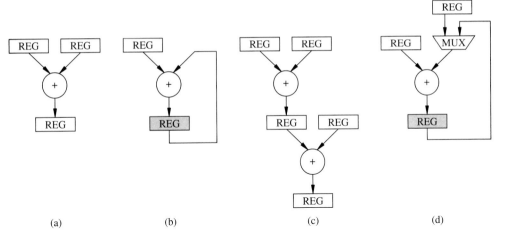

FIGURE 6.23
(a) Self-testable resource. (b) Partially testable resource, due to a self-adjacent register (shaded). (c) Testable resource pair. (d) Partially testable combinational logic block, due to a self-adjacent register (shaded).

There are two situations leading to self-adjacent registers: first, sharing a register that holds both an operand and the result of an operation and, second, binding a shared resource to two operations with data dependency and in consecutive schedule steps.

> **Example 6.8.1.**   Consider the circuit of Figure 6.23 (a) and an implementation in which the right operand and the result share a register [Figure 6.23 (b)]. This register is self-adjacent.
>
> Consider now two cascaded operations in adjacent control steps [Figure 6.23 (c)] and a shared resource implementation [Figure 6.23 (d)]. The register storing the result is self-adjacent, because it is simultaneously the source and destination of data traversing the combinational logic block consisting of the adder and the multiplexer.

Papachristou *et al.* [12] proposed both an exact and a heuristic resource sharing method that avoids self-adjacent registers. Both algorithms prohibit data-dependent consecutively-scheduled operations of the same type to share resources. Avra [1] investigated a register sharing method based on coloring a conflict graph, where appropriate testability conflict edges model those assignments of variables to registers that would become self-adjacent. Avra's approach tolerates self-adjacent registers due to resource sharing, but it attempts to minimize their count. Recently, Papachristou *et al.* [12] as well as Lee *et al.* [8] investigated testability-oriented scheduling techniques such that the corresponding scheduled data path can be made testable. We refer the interested reader to the specialized literature for details.

6.9   PERSPECTIVES

Resource sharing is an important problem of data-path synthesis that allows us to leverage on the multiple use of hardware resources for implementing the operations.

Different types of resources can be shared, such as functional resources, registers, busses, ports, etc. The complexity of the binding and sharing problems depends on modeling. For resource-dominated circuits, where optimum binding corresponds to a minimum-resource usage, sharing can be solved in polynomial time for circuits modeled by non-hierarchical graphs. For more general circuits and graph models, the optimum sharing problem is computationally intractable and strongly tied to scheduling.

Resource sharing can be extended to cope with different operation and resource types and generic relations that represent the possible assignments of resources to operations. A family of problems that go under the name of module selection have been proposed and partially investigated. When operations can be implemented by more than one resource type, with different areas and execution delays, the module selection problem affects the schedule as well as the possible sharing.

Scheduling, binding and module selection can be formulated as a single problem. The advantage of dealing simultaneously with all decisions about a circuit implementation is offset by the computational cost of searching for a solution under a large set of constraints. Thus most architectural synthesis systems decompose these problems as a sequence of subproblems to be solved independently. This approach is justified by the fact that most subproblems are still intractable and are solved by heuristic algorithms when considering circuit models that are detailed enough to represent the features of practical circuits.

6.10 REFERENCES

Early work on resource sharing was reported by Hafer and Parker [4] and by Thomas *et al.* [16]. A clique partitioning algorithm for resource sharing was presented first by Tseng and Siewiorek [18]. Later, the clique partitioning and the coloring problems were analyzed in detail by Springer and Thomas [14] and by Stok [15], who considered the special properties of the chordal, comparability and interval graphs.

The register and bus sharing problems were also approached by Hafer and Parker [4] and by Tseng and Siewiorek [18]. Balakrishnan *et al.* [2] reported on the multi-port memory binding problem. Kurdahi and Parker [6] showed how the *LEFT_EDGE* algorithm could be used for register sharing. An ILP formulation for concurrent scheduling and binding was recently presented by Gebotys and Elmasry [3], who reported on the successful solution of ILP problems for circuits of interesting size. McFarland [10] commented first on the importance of considering better area/delay models for the design space evaluation.

Thomas and Leive [17] considered the first module selection problem, and opened the door to further research on the subject. Resource sharing for pipelined circuits was pioneered by Park and Parker [13]. Hwang *et al.* [7] suggested scheduling methods and ways to avoid the twisted pairs. Papachristou *et al.* [12] analyzed the structural testability problem in connection with resource sharing. Avra [1] extended the analysis to register sharing. Both references [1] and [12] proposed algorithms for architectural synthesis of structurally testable designs.

1. L. Avra, "Allocation and Assignment in High-Level Synthesis of Self-Testable Data-paths," *ITC, Proceedings of the International Testing Conference*, pp. 463–472, 1988.
2. M. Balakrishnan, A. Majumdar, D. Banerji, J. Linders and J. Majithia, "Allocation of Multiport Memories in Data Path Synthesis," *IEEE Transactions on CAD/ICAS*, Vol. CAD-7, No. 4, pp. 536–540, April 1988.
3. C. Gebotys and M. Elmasry, *Optimal VLSI Architectural Synthesis*, Kluwer Academic Publishers, Boston, MA, 1992.
4. L. Hafer and A. Parker, "Automated Synthesis of Digital Hardware," *IEEE Transactions on Computers*, Vol. C-31, No. 2, February 1982.

5. D. Ku and G. DeMicheli, *High-Level Synthesis of ASICS under Timing and Synchronization Constraints*, Kluwer Academic Publishers, Boston, MA, 1992.

6. F. Kurdahi and A. Parker, "REAL: A Program for Register Allocation," *DAC, Proceedings of the Design Automation Conference*, pp. 210–215, 1987.

7. K. Hwang, A. Casavant, M. Dragomirecky and M. d'Abreu, "Constrained Conditional Resource Sharing in Pipeline Synthesis," *ICCAD, Proceedings of the International Conference on Computer Aided Design*, pp. 52–55, 1988.

8. T. Lee, W. Wolf and N. Jha, "Behavioral Synthesis for Easy Testability in Data-Path Scheduling," *ICCAD, Proceedings of the International Conference on Computer Aided Design*, pp. 616–619, 1992.

9. E. McCluskey, *Logic Design Principles*, Prentice-Hall, Englewood Cliffs, NJ, 1986.

10. M. J. McFarland, "Reevaluating the Design Space for Register Transfer Hardware Synthesis," *ICCAD, Proceedings of the International Conference on Computer Aided Design*, pp. 184–187, 1987.

11. B. Pangrle and D. Gajski, "Design Tools for Intelligent Silicon Compilation," *IEEE Transactions on CAD/ICAS*, Vol. CAD-6, No. 6, pp. 1098–1112, November 1987.

12. C. Papachristou, S. Chiu and H. Harmanani, "A Data-Path Synthesis Method for Self-Testable Designs," *DAC, Proceedings of the Design Automation Conference*, pp. 378–384, 1991.

13. N. Park and A. Parker, "Sehwa: A Software Package for Synthesis of Pipelines from Behavioral Specifications," *IEEE Transactions on CAD/ICAS*, Vol. CAD-7, No. 3, pp. 356–370, March 1988.

14. D. Springer and D. Thomas, "Exploiting the Special Structure of Conflict and Compatibility Graphs in High-Level Synthesis," *ICCAD, Proceedings of the International Conference on Computer Aided Design*, pp. 254–259, 1990.

15. L. Stok, "Architecture Synthesis and Optimization of Digital Systems," Ph.D. Dissertation, Eindhoven University, The Netherlands, 1991.

16. D. Thomas, C. Hitchcock, T. Kowalski, J. Rajan and R. Walker, "Automatic Data-Path Synthesis," *IEEE Computer*, No. 12, pp. 59–70, December 1983.

17. D. Thomas and G. Leive, "Automating Technology Relative Logic Synthesis and Module Selection," *IEEE Transactions on CAD/ICAS*, Vol. CAD-2, No. 2, pp. 94–105, April 1983.

18. C. J. Tseng and D. Siewiorek, "Automated Synthesis of Data Paths in Digital Systems," *IEEE Transactions on CAD/ICAS*, Vol. CAD-5, No. 3, pp. 379–395, July 1986.

6.11 PROBLEMS

1. Draw the compatibility and conflict graphs for the scheduled sequencing graph of Figure 5.2. Determine an optimum clique partition of the former and a coloring of the latter. How many resources are required for each type?

2. Assume only one resource type. Determine a property of the conflict graphs related to the non-hierarchical sequencing graphs when all operations have unit execution delay.

3. Consider the register assignment problem of Example 1.2.10. Assume that the intermediate variables are labeled as in Example 1.2.10 and that the inner loop of the differential equation integrator is scheduled as in Figure 4.3. Determine the circular-arc conflict graph of the intermediate and loop variables and a minimum coloring.

4. Consider the unscheduled sequencing graph of Figure 4.2. Assume that we use two ALUs and two multipliers. Compute how many possible bindings exist of the six multiplications and the overall number of bindings when using two resources per type.

5. Consider non-hierarchical sequencing graphs with one resource type and unbounded delays. Propose a method to determine a bound on the minimum number of resources needed, so that the overall execution delay for any values of the unbounded delays is the same as when all resources are dedicated. Extend the analysis to hierarchical graphs and to multiple resource types.

6. Formulate the concurrent scheduling and binding problems in the case of multiple resource types by extending inequalities 1.7, 1.8, 1.9 and 1.10. Apply the formulation to the se-

quencing graph of Figure 4.2, while assuming that multipliers have two-unit delay and adders one-unit delay. Determine a minimum-latency solution with two resources per type.

7. Formulate the concurrent scheduling, binding and module selection problems by an ILP model by rewriting inequalities 6.7, 6.8, 6.9, 6.10 and 6.19 in terms of n_{res} explicit resource types that can implement an operation type. Assume a single operation type.

8. Draw the compatibility and conflict graphs for the scheduled sequencing graph of Figure 5.2 assuming that it is pipelined with $\delta_0 = 2$. Determine an optimum clique partition of the former and a coloring of the latter. How many resources are required for each type?

PART

III

LOGIC-LEVEL
SYNTHESIS
AND
OPTIMIZATION

This part deals with synthesis and optimization of circuits at the logic level, which are represented by abstract models such as those described in Section 3.3. In particular, we distinguish between sequential and combinational circuits, whose behavior can be captured by finite-state machine state diagrams or by Boolean functions and relations. Often, sequential and combinational circuits are represented conveniently by mixed structural/behavioral models, such as *logic networks*. We assume that the abstract models either are derived from HDL models by compilation or are the result of synthesis from architectural-level models.

The goal of logic-level synthesis and optimization is to determine the microscopic structure of a circuit, i.e., its gate-level representation. This is accomplished through the application of several techniques. For example, a common design flow for sequential circuit synthesis is first to optimize the corresponding finite-state machine representation by state minimization and encoding, second to minimize the related combinational component and last to bind it to cells of a given library. Alternative flows do exist, because the circuit may be specified in terms of a network, where states are already encoded implicitly.

267

Since sequential synthesis and optimization techniques rely on those for combinational circuits, we present first methods for optimizing combinational circuits. We consider the optimization of circuits in two-level forms in Chapter 7 and that for multiple-level forms in Chapter 8. We describe then sequential circuit synthesis in Chapter 9 and conclude this part with library binding in Chapter 10.

CHAPTER
7

TWO-LEVEL
COMBINATIONAL
LOGIC
OPTIMIZATION

C'est la possibilité permanente du non être ... qui conditionne nos questions sur l'être. ....
Quelle que soit cette réponse, elle pourra se formuler ainsi: "L'être est cela et, en dehors de
cela, rien."
The permanent possibility of non-being ... conditions our questions about being.... Whatever
being is, it will allow this formulation: "Being is that and outside of that, nothing."

J.P. Sartre. L'être et le néant.

7.1 INTRODUCTION

This chapter deals with the optimization of combinational logic circuits, modeled by
two-level *sum of products* expression forms, or equivalently by tabular forms such
as implicant tables. The translations between these two formats is straightforward
in either direction. Therefore we assume that circuit representations in *sum of products* forms are available. Transformations of two-level forms into multiple-level forms
and *vice versa* will be described in the next chapter.

Two-level logic optimization is important in three respects. First, it provides a means for optimizing the implementation of circuits that are direct translations of two-level tabular forms. Thus two-level logic optimization has a direct impact on macro-cell design styles using programmable-logic arrays (PLAs). Second, two-level logic optimization allows us to reduce the information needed to express any logic function that represents a component of a multiple-level representation. Specifically, a module of a logic network may be associated with a logic function in two-level form, whose optimization benefits the overall multiple-level representation. Thus, two-level optimization is of key importance to multiple-level logic design. Last but not least, two-level logic optimization is a formal way of processing the representation of systems that can be described by logic functions. Hence its importance goes beyond logic circuit design.

The bulk of this chapter is devoted to exact and heuristic minimization methods for two-level forms. The exact algorithms find their roots in early work on logic design by Quine [21] and McCluskey [15]. Several heuristic logic minimization algorithms have been proposed, mostly based on *iterative improvement* strategies. Examples of heuristic minimizers are programs MINI, PRESTO and ESPRESSO, the last being considered today as the standard tool for two-level logic optimization. The underlying principles of exact and heuristic logic optimization are described first, followed by the algorithmic details of the procedures for manipulating Boolean functions. This allows the reader to appreciate the overall architecture of the Boolean optimizers before delving into the details.

The last part of this chapter is devoted to extensions of two-level logic optimization to symbolic functions and to the optimization of Boolean relations. Even though these methods find their application in sequential and multiple-level logic design, they are presented here because they extend the methods presented in the earlier parts of this chapter.

7.2 LOGIC OPTIMIZATION PRINCIPLES

We present in this section the major ideas of both exact and heuristic two-level logic minimization, while the algorithmic details are deferred to the following sections. The objective of two-level logic minimization is to reduce the size of a Boolean function in either *sum of products* or *product of sums* form. Since either form can be derived from the other by using De Morgan's law, which preserves the number of terms and literals, we can concentrate on the minimization of one form, in particular *sum of products* without loss of generality.

The detailed goals of logic optimization may vary slightly, according to the implementation styles. Let us consider first the case of PLA implementations of logic functions [12]. A PLA is a rectangular macro-cell consisting of an array of transistors aligned to form rows in correspondence with product terms and columns in correspondence of inputs and outputs. The input and output columns partition the array into two subarrays, called *input* and *output planes*, respectively. A symbolic diagram and a transistor-level diagram of a PLA are shown in Figures 7.1 (b) and (c), respectively, implementing the function represented by the two-level tabular representation of

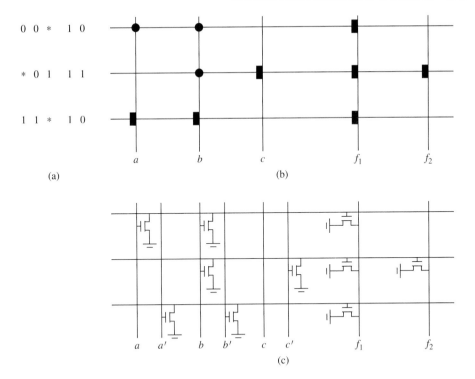

FIGURE 7.1

(a) Tabular representation of a three-input, two-output function: $f_1 = a'b' + b'c + ab$; $f_2 = b'c$. (b) Symbolic diagram of a PLA. In the input plane a circle corresponds to a 0 and a rectangle to a 1. In the output plane a rectangle corresponds to a 1. (c) Transistor-level schematic of a PLA core (excluding pull-ups, drivers and output inverters). In MOS technology both *sums* and *products* are implemented as NORs. Thus the outputs of the core need to be complemented and the transistors of the input plane are gated by the complement of the signals related to the products.

Figure 7.1 (a). Each row of a PLA is in one-to-one correspondence with a product term of the *sum of products* representation. Each transistor in the input plane is in one-to-one correspondence with a literal of the *sum of products* form. Each transistor in the output plane relates to the dependence of a scalar output on a product term. Therefore, the primary goal of logic minimization is the reduction of terms and a secondary one the reduction of literals.

Other optimization goals are relevant when two-level forms model functions to be implemented differently from PLAs. For example, two-level logic representations of single-output functions may be implemented by complex gates [12], whose size relates to the number of literals in a factored form, as shown in Section 8.2.1. Thus the minimization of the number of literals is the primary objective.

Logic minimization of single-output and multiple-output functions follows the same principle, but the latter task is obviously more complex. Note that the disjoint minimization of the scalar components of a multiple-output function may lead to suboptimal results, because the optimization cannot exploit product term sharing. An important result in logic optimization, which we shall show later, is the equivalence of multiple-output binary-valued functions with single-output multiple-valued functions. For this reason we concentrate first on techniques for optimizing single-output functions. We then extend the techniques to multiple-valued functions, and by doing so we present the solution to the multiple-output minimization problem. This approach not only is justified by didactic purposes, but it also parallels the operation of existing minimizers.

7.2.1 Definitions

We consider in this chapter incompletely specified Boolean functions, because completely specified functions are the special case of functions with no *don't care* conditions. A Boolean function $\mathbf{f} : \mathsf{B}^n \to \{0, 1, *\}^m$ can be represented in several ways, as shown in Section 2.5.2. For each output $i = 1, 2, \ldots, m$ we define the corresponding *on set*, *off set* and *dc set* as the subsets of B^n whose image under the ith component of $\mathbf{f}$ (i.e., under f_i) is $1, 0$ and $*$, respectively. Thus an incompletely specified function can be seen as a triple of completely specified functions.

We represent functions as lists of implicants. The concept of multiple-output implicant is general in nature, because it combines an input pattern with an implied value of the function. In this chapter, we restrict the notion of an implicant to the implication of a TRUE or *don't care* value of a function. Therefore the output part of a multiple-output implicant is binary valued, with the following meaning: a 1 implies a TRUE or *don't care* value of the function; a 0 does not imply a value of the function.

> **Definition 7.2.1.** A **multiple-output implicant** of a Boolean function $\mathbf{f} : \mathsf{B}^n \to \{0, 1, *\}^m$ is a pair of row vectors of dimensions n and m called the **input part** and **output part**, respectively. The input part has entries in the set $\{0, 1, *\}$ and represents a product of literals. The output part has entries in the set $\{0, 1\}$. For each output component, a 1 implies a TRUE or *don't care* value of the function in correspondence with the input part.

Multiple-output minterms are implicants under special restrictions.

> **Definition 7.2.2.** A **multiple-output minterm** of a Boolean function $\mathbf{f} : \mathsf{B}^n \to \{0, 1, *\}^m$ is a multiple-output implicant whose input part has elements in the set $\{0, 1\}$ (i.e., a zero-dimensional cube involving the product of exactly n literals) and that implies the TRUE value of one and only one output of the function.

A multiple-output implicant corresponds to a subset of minterms of the function (and of its *don't care* set). Therefore we can define implicant *containment* (or *cover*) and *size* in a conceptual way by relating implicants to sets and to their containment and cardinality. Similarly, we can define containment and intersection among sets of implicants. From an operative point of view, there are efficient algorithms for computing containment and intersection, which will be described in Section 7.3.

> **Example 7.2.1.** Consider the three-input, two-output function $\mathbf{f} = \begin{bmatrix} f_1 \\ f_2 \end{bmatrix}$, where $f_1 = a'b'c' + a'b'c + ab'c + abc + abc'$; $f_2 = a'b'c + ab'c$, whose implicants and the minterms

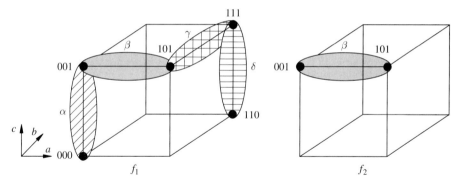

FIGURE 7.2
Minterms and implicants of the function.

are shown in Figure 7.2. A multiple-output implicant of **f** is $\beta = *01\ \ 11$, which implies both outputs to be TRUE whenever the second input is FALSE and the third one is TRUE. Its input part corresponds to the cube $b'c$. The implicant corresponds to the following four minterms: 001 10; 101 10; 001 01; 101 01.

Consider now a second multiple-output implicant, $\gamma = 1*1\ \ 10$, corresponding to the minterms 101 10 and 111 10.

Implicant $\beta = *01\ \ 11$ covers minterm 101 10 but not minterm 111 10. The intersection of β with γ yields 101 10.

Definition 7.2.3. A **cover** of a Boolean function is a set (list) of implicants that covers its minterms.

We denote by F a cover of a function **f**. The *on set*, *off set* and *dc set* of a function **f** can be modeled by covers, where implicants and minterms are related to the completely specified functions that specify them. The covers of the *on set*, *off set* and *dc set* of a function **f** are denoted by F^{ON}, F^{OFF} and F^{DC}, respectively.

A cover F of a function **f** satisfies the bounds $F^{ON} \subseteq F \subseteq F^{ON} \cup F^{DC}$. The upper bound is always satisfied by the definition of implicant. The lower bound is directly related to the concept of cover. When dealing with completely specified functions, F and F^{ON} are interchangeable.

The *size*, or *cardinality*, of a cover is the number of its implicants. There may be several covers for a Boolean function. *Don't care* conditions can be effectively used to reduce the size of a cover of an incompletely specified function.

Definition 7.2.4. A **minimum cover** is a cover of minimum cardinality.

Note that according to the previous discussion, a minimum-cost cover may differ from a minimum cover according to the implementation style. We shall consider the cost of the cover as its cardinality, for the sake of simplicity and uniformity.

The objective of exact two-level logic minimization is to determine a minimum cover of a function. Methods for determining a minimum cover are described in the next section. It is often useful to determine local minimum covers, called *minimal covers*, because their computation can be achieved with smaller memory and computing-time requirements. The goal of heuristic logic minimization is to determine minimal covers. Such covers are often close in cardinality to minimum covers and can provide good solutions to practical problems.

It is customary [4] to define local minimality with respect to containment.

> **Definition 7.2.5.** A **minimal**, or **irredundant**, cover of a function is a cover that is not a proper superset of any cover of the same function.

In other words, the removal of any implicant from a minimal cover does not cover the function. Equivalently, no implicant is contained in any subset of implicants of the cover.

A weaker minimality property is the minimality with respect to single-implicant containment.

> **Definition 7.2.6.** A cover is **minimal with respect to single-implicant containment**, if no implicant is contained in any other implicant of the cover.

An irredundant cover is also minimal with respect to single-implicant containment, but the converse is not true. The reason is that an implicant may be redundant because it is contained by a subset of implicants of the cover, but not by a single implicant. Therefore irredundancy is a stronger property.

> **Example 7.2.2.** Consider again the three-input, two-output function $\mathbf{f} = \begin{bmatrix} f_1 \\ f_2 \end{bmatrix}$, where
> $f_1 = a'b'c' + a'b'c + ab'c + abc + abc'$; $f_2 = a'b'c + ab'c$. A minimum cover of cardinality 3 is given by:
>
> | | |
> |---|---|
> | 00* | 10 |
> | *01 | 11 |
> | 11* | 10 |
>
> In expression form, it reads $f_1 = a'b' + b'c + ab$; $f_2 = b'c$.
> An irredundant cover of cardinality 4 is given by:
>
> | | |
> |---|---|
> | 00* | 10 |
> | *01 | 01 |
> | 1*1 | 10 |
> | 11* | 10 |
>
> A redundant cover of cardinality 5 that is minimal with respect to single-implicant containment is given by:
>
> | | |
> |---|---|
> | 00* | 10 |
> | *01 | 01 |
> | *01 | 10 |
> | 1*1 | 10 |
> | 11* | 10 |

Note that the third implicant is contained in the union of the first and fourth implicants and thus the cover is redundant, even though it is minimal with respect to single-implicant containment. The covers can be visualized in Figure 7.3.

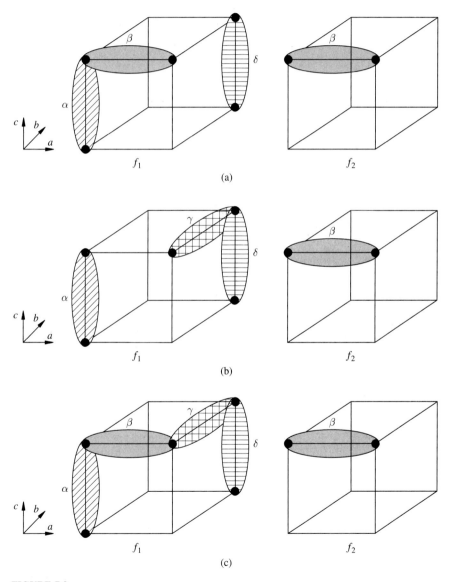

FIGURE 7.3
(a) Minimum cover. (b) Irredundant cover. (c) Minimal cover with respect to single-implicant containment.

Another property of implicants and covers is primality.

Definition 7.2.7. An implicant is **prime** if it is not contained by any implicant of the function. A cover is prime if all its implicants are prime.

Note that the definition of primality is related to all possible implicants of the function, and not just to those of the cover under consideration. For single-output functions, a prime implicant corresponds to a product of literals where no literal can be dropped while preserving the implication property. In geometric terms, a prime implicant corresponds to a maximal-dimension cube, where maximal means of largest size without intersecting the *off set*. Moreover, for multiple-output functions, a prime implicant implies a maximal number of outputs. Prime implicants are often called *primes* for brevity.

Example 7.2.3. Consider again the function used in the previous examples. The following table collects all primes:

$$
\begin{array}{ll}
00* & 10 \\
*01 & 11 \\
1*1 & 10 \\
11* & 10
\end{array}
$$

By definition, and as shown in Figure 7.3, no prime is contained in another prime of the function. The table above represents a prime cover, but it is not minimal. Indeed the third prime is contained in the union of the second and fourth ones. By definition of prime implicant, the cover is still minimal with respect to single-implicant containment. By deleting the third prime, we obtain a minimum cover, which is necessarily irredundant.

Suppose now that the function is incompletely specified and that *don't care* conditions for both outputs are expressed by an implicant with input part 100. Then, the prime implicants would be:

$$
\begin{array}{ll}
0 & 10 \\
*01 & 11 \\
1** & 10
\end{array}
$$

which represent a minimum cover.

Some prime implicants have the special property that must be contained in any cover of the function. They are called *essential prime implicants*.

Definition 7.2.8. A prime implicant is **essential** if there exists one minterm of the function covered by only that implicant among all prime implicants of the function.

Example 7.2.4. Consider again the previous function and its primes:

$$
\begin{array}{ll}
00* & 10 \\
*01 & 11 \\
1*1 & 10 \\
11* & 10
\end{array}
$$

All primes but the third are essential. Indeed, while considering the first output, the first prime is necessary to cover minterm 000 10 and the fourth for minterm 110 10. The second output is implied by only one implicant, the second one, which therefore is essential.

7.2.2 Exact Logic Minimization

Exact logic minimization addresses the problem of computing a minimum cover. It is considered a classic problem in switching theory and it was addressed first by Quine [21] and McCluskey [15].

The solution of the problem hinges on Quine's theorem, which delimits the search space for an optimum solution.

> **Theorem 7.2.1.** There is a minimum cover that is prime.

> **Proof.** Consider a minimum cover that is not prime. Each non-prime implicant can be replaced by a prime implicant that contains it. Thus the resulting set of implicants is a cover and has the same cardinality as the original cover. Hence there is a minimum cover that is prime.

The theorem allows us to limit the search for a minimum cover to those covers which consist entirely of prime implicants. Note that the theorem can be generalized to handle broader definitions of minimum covers, where the cost of an implicant is always no greater than the cost of an implicant it contains. For example, the theorem applies to the case of literal minimization for single-output functions.

McCluskey formulated the search for a minimum cover as a covering problem by means of a *prime implicant table*. We explain his formulation by considering completely specified single-output functions.

A prime implicant table is a binary-valued matrix $\mathbf{A}$ whose columns are in one-to-one correspondence with the prime implicants of the function f and whose rows are in one-to-one correspondence with its minterms.[1] An entry $a_{ij} \in \mathbf{A}$ is 1 if and only if the jth prime covers the ith minterm. A minimum cover is a minimum set of columns which covers all rows, or equivalently a minimum set of primes covering all minterms. Therefore the covering problem can be viewed as the problem of finding a binary vector $\mathbf{x}$ representing a set of primes with minimum cardinality $|\mathbf{x}|$ such that:

$$\mathbf{Ax} \geq \mathbf{1} \tag{7.1}$$

where the dimensions of matrix $\mathbf{A}$ and of vectors $\mathbf{x}$ and $\mathbf{1}$ match the number of minterms and primes.

Note that matrix $\mathbf{A}$ can be seen as the incidence matrix of a hypergraph, whose vertices correspond to the minterms and whose edges correspond to the prime implicants. Hence the covering problem corresponds to an *edge cover* of the hypergraph.

---

[1] Some authors [e.g., 15] use a table that is the transpose of the one used here. All algorithms still apply, *mutatis mutandis*.

Example 7.2.5. Consider the function $f = a'b'c' + a'b'c + ab'c + abc + abc'$, whose primes are:

$$
\begin{array}{c|cc}
\alpha & 00* & 1 \\
\beta & *01 & 1 \\
\gamma & 1*1 & 1 \\
\delta & 11* & 1
\end{array}
$$

where the Greek letters are identifiers. The following is a prime implicant table for f:

	α	β	γ	δ
000	1	0	0	0
001	1	1	0	0
101	0	1	1	0
111	0	0	1	1
110	0	0	0	1

Vector $\mathbf{x} = [1101]^T$ represents a cover, because $\mathbf{Ax} \geq \mathbf{1}$. The vector selects implicants $\{\alpha, \beta, \delta\}$. The function and all its primes are represented in Figure 7.4 (a). The minimum cover corresponding to $\mathbf{x} = [1101]^T$ is shown in Figure 7.4 (b).

Note that a hypergraph can be identified by a set of vertices corresponding to the minterms and by a set of edges corresponding to primes, as shown in Figure 7.4 (c). In this particular example, each edge is incident to two vertices and the hypergraph is a graph. This is not true in general, as shown, for example, by Figure 7.5 (b).

Exact minimization can be solved by computing the prime implicant table first and then solving the resulting covering problem. Note that the covering problem is unate (see Section 2.5.3), because the covering clauses can all be expressed as a disjunction of implicants. The difficulty of the approach lies in the intractability of the covering problem and in the size of the implicant table. Indeed, an n-input, single-output Boolean function can have as many as $3^n/n$ prime implicants and 2^n minterms. Therefore, an exponential algorithm on a problem of exponential size is likely to require prohibitive memory size and computing time. Never-

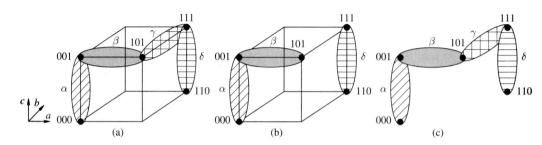

FIGURE 7.4
(a) Prime implicants of $f = a'b'c' + a'b'c + ab'c + abc + abc'$. (b) Minimum cover of f. (c) Hypergraph representation of the covering problem.

theless, many instances of two-level minimization problems can be solved exactly today by smart algorithms that exploit the nature of the problem and efficient data structures.

The implicant table can be reduced by techniques similar to those used for the generic covering problem described in Section 2.5.3. In particular, each essential column corresponds to an essential prime, which must be part of any solution. Row and column dominance can be used to reduce the size of matrix **A**. Extraction of the essentials and removal of dominant rows and dominated columns can be iterated to yield a *reduced prime implicant table*. If this table happens to be void, a minimum cover has been constructed by setting aside the essential primes. Otherwise, the reduced table is said to be *cyclic* and to model the so-called *cyclic core* of the problem. The original method proposed by McCluskey then resorts to branching, i.e., selecting the different column (prime) combinations and evaluating the corresponding cost. Even though the selection of a column (prime) may lead to simplification based on the essential and dominance rules, the process is exponential in the size of the reduced implicant table in the worst case.

> **Example 7.2.6.** Consider the prime implicant table of Example 7.2.5. Implicants $\{\alpha, \delta\}$ are essentials. They are part of any cover. Therefore, the corresponding columns can be deleted from the tables, as well as the rows incident to them. This leads to a reduced prime implicant table:
>
> | | β | γ |
> |---|---|---|
> | 101 | 1 | 1 |
>
> In this case, the table states that either implicant β or γ can be used to complete the cover. Thus the table is not cyclic, and no branching is required.

Another classic approach, often referred to as Petrick's method [15], consists of writing down the covering clauses of the (reduced) implicant table in a *product of sums* form. Each clause (or equivalently each product term) corresponds to a minterm and it represents the disjunction of the primes that cover it. The *product of sums* form is then transformed into a *sum of products* form by carrying out the products of the sums. The corresponding *sum of products* expression is satisfied when any of its product terms is TRUE. In this case, product terms represent the primes that have been selected. The cost of a cover then relates to the number of literals in the product. As a result, a minimum cover is identified by any product term of the *sum of products* form with the fewest literals.

> **Example 7.2.7.** Let us apply Petrick's method to the full prime implicant table of the previous examples. The clause that states the covering of the first minterm is α; the clause related to the second minterm is $\alpha + \beta$, etc. The *product of sums* expression is then:
> $$(\alpha)(\alpha + \beta)(\beta + \gamma)(\gamma + \delta)(\delta) = 1$$
> By carrying out the products, we get the corresponding *sum of products* form:
> $$\alpha\beta\delta + \alpha\gamma\delta = 1$$

which shows that there are two minimum covers of cardinality 3. The first one is the sum of implicants $\{\alpha, \beta, \delta\}$ and the second of $\{\alpha, \gamma, \delta\}$.

Note that Petrick's method could have been applied to the reduced prime implicant table, yielding:

$$\beta + \gamma = 1$$

Thus either β or γ, in conjunction with the essential primes $\{\alpha, \delta\}$, provides a minimum cover.

Whereas the formation of the *product of sums* form and the product term selection in the *sum of products* form are straightforward, the transformation of the *product of sums* form into a *sum of products* form is not as easy as it seems from a conceptual standpoint. Indeed, carrying out the products involves an exponential number of operations. This limits the applicability of Petrick's method to small-sized tables.

The Quine-McCluskey algorithm can be extended to multiple-output functions by computing the multiple-output prime implicants and the corresponding table. Extensions to cope with incompletely specified functions are also straightforward. Details and examples of the method are reported in standard textbooks [14, 16].

RECENT RESULTS IN EXACT MINIMIZATION. Despite the exact two-level minimization algorithms that have been known for many years, several recent developments have drastically improved the Quine-McCluskey procedure and made it applicable to functions modeling realistic design problems. Recall that the difficulties in exact two-level logic minimization are due to the size of the covering table and to solving the covering problem. Thus most recent approaches have addressed both representing efficiently the cyclic core and developing algorithms for unate covering.

Rudell and Sangiovanni improved the Quine-McCluskey algorithm and implemented it in the program ESPRESSO-EXACT [23]. While the principles of exact minimization in ESPRESSO-EXACT are the same as those of the Quine-McCluskey procedure, the underlying algorithms are significantly different. The program has been shown to be effective in minimizing exactly 114 out of 134 benchmark functions of a standard set [23].

The description of the ESPRESSO-EXACT algorithm of this section applies to both single-output and multiple-output functions, which are treated as multiple-valued-input, single-output functions. We shall show in Section 7.3 that the same algorithms can be used in both cases for the generation of prime implicants and detection of essential primes. For the time being, we assume we have a set of prime implicants of a single-output function.

The major improvements of the ESPRESSO-EXACT algorithm over the Quine-McCluskey algorithm consist of the construction of a smaller reduced prime implicant table and of the use of an efficient branch-and-bound algorithm for covering. Their covering algorithm was presented already in Section 2.5.2, because of its wide applicability to covering problems.

ESPRESSO-EXACT partitions the prime implicants into three sets: *essentials, partially redundant* and *totally redundant*. The essentials have the usual meaning, the totally redundant primes are those covered by the essentials and the *don't care* set, and the partially redundant set includes the remaining ones. The last set is the only one relevant to the covering phase, and it corresponds to the columns of the reduced implicant table.

The rows of the reduced implicant table correspond to sets of minterms, rather than to single minterms as in the case of Quine-McCluskey's algorithm. Namely each row corresponds to all minterms which are covered by the same subset of prime implicants. A method for generating the reduced implicant table is described in Section 7.4.3.

Example 7.2.8. Consider the function defined by the following minterm table:

$$
\begin{array}{ll}
0000 & 1 \\
0010 & 1 \\
0100 & 1 \\
0110 & 1 \\
1000 & 1 \\
1010 & 1 \\
0101 & 1 \\
0111 & 1 \\
1001 & 1 \\
1011 & 1 \\
1101 & 1 \\
\end{array}
$$

The corresponding prime implicants are:

$$
\begin{array}{c|cc}
\alpha & 0**0 & 1 \\
\beta & *0*0 & 1 \\
\gamma & 01** & 1 \\
\delta & 10** & 1 \\
\epsilon & 1*01 & 1 \\
\zeta & *101 & 1 \\
\end{array}
$$

Prime implicants γ and δ are essential, because they cover minterms 0111 1 and 1011 1, respectively, which are not covered by any other prime. The remaining primes are partially redundant. The reduced prime implicant table is then:

	α	β	ϵ	ζ
0000,0010	1	1	0	0
1101	0	0	1	1

Therefore the matrix is reducible. Partitioning yields two subproblems. Any cover is the union of the first subcover (containing either α or β), the second subcover (containing either ϵ or ζ) and the essentials (γ and δ). Hence there are four different covers with minimum cardinality. The minimum cover $\{\alpha, \gamma, \delta, \epsilon\}$ is shown in Figure 7.5 (b). An alternative minimum cover $\{\beta, \gamma, \delta, \epsilon\}$ is shown instead in Figure 7.5 (c).

Dagenais, *et al.* suggested another method for exact minimization [8] which is based on the Quine-McCluskey method but avoids the explicit creation of the prime

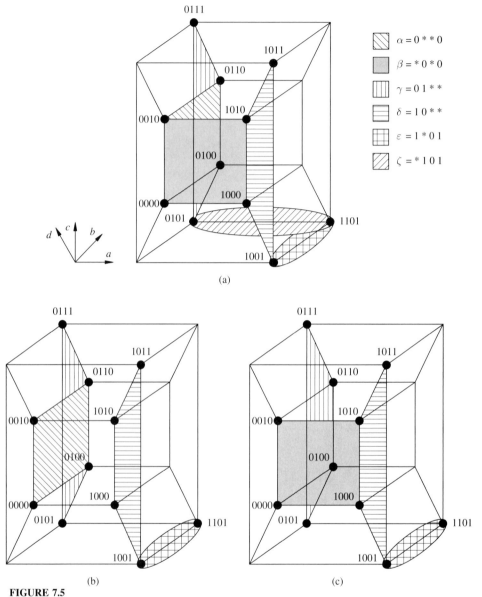

FIGURE 7.5
(a) Prime implicants of the function. (b) A minimum cover. (c) Alternative minimum cover.

implicant table. The program implementation of their algorithm, called MCBOOLE, first generates the prime implicants and stores them into two lists, called *retained* and *undecided* implicants. Storing implicants in lists is more efficient than using tables. Initially all implicants are in the undecided list. Then, by exploiting essential and dominance properties, some implicants are moved from the undecided to the retained list. Each implicant in the undecided list has a corresponding *uncovered part*, i.e., the portion that is not covered by the implicants in the retained list or by the *don't care* conditions. A minimum cover is found when all uncovered parts are void.

Branching is necessary to solve the cyclic core by selecting a subset of the elements in the undecided list and adding them to the retained list. Even though branching is inherently a trial-and-error process with worst-case exponential complexity, it is implemented efficiently in MCBOOLE by means of a graph that stores the relations among primes, which is constructed when the primes are generated. MCBOOLE has been shown to be successful, even though only 86/134 functions of the standard benchmark set could be minimized exactly. We refer the interested reader to reference [8] for further details.

Coudert and Madre [6] devised a way of representing implicitly (i.e., without enumerating them) all prime implicants of a function using binary decision diagrams. This avoids listing all prime implicants, which for some benchmark functions can exceed a trillion. A major problem in exact minimization is weeding out the prime implicants by using the dominance property to determine the cyclic core of the problem. Often, a small fraction of the primes are part of the cyclic core. Coudert *et al.* [7] proposed also a new exact minimization procedure that takes advantage of a transformation of the covering problem and that makes the computational complexity of the algorithm independent of the number of minterms and prime implicants of the function. With their method, all current benchmark examples in the standard set could be minimized exactly.

An alternative approach was recently proposed by McGeer *et al.* [17] which departs from the Quine-McCluskey framework. Instead of computing all prime implicants, their algorithm derives the covering problem implicitly and then generates only those prime implicants involved in the covering problem. This avoids completely the problem of computing (implicitly or explicitly) the set of all primes, which may be very large. The algorithm relies on the concept of *signature cubes* that identify sets of primes. Namely, a signature cube identifies uniquely the set of primes covering each minterm: it is the largest cube of the intersection of the corresponding primes. The set of maximal signature cubes, i.e., those not singly contained within any other, defines a *minimum canonical cover* of the logic function being minimized and represents the prime implicant table implicitly. The minimum canonical cover of a logic function is unique and irredundant. Thus exact minimization consists then in first determining such a minimum canonical cover, computing the primes related to each signature cube and solving the corresponding covering problem. This method has been implemented in program ESPRESSO-SIGNATURE, which has been shown to be nearly twice as fast as ESPRESSO-EXACT and able to solve, to date, all but three benchmark problems of the standard set. We refer the reader to reference [17] for details.

7.2.3 Heuristic Logic Minimization

We describe here the major principles of heuristic logic minimization. Details of algorithms for heuristic minimization are deferred to Section 7.4.

Heuristic minimization is motivated by the need of reducing the size of two-level forms with limited computing time and storage. Some minimizers, like ESPRESSO,[2] yield prime and minimal covers whose cardinality is often close to minimum. Hence such minimizers are used for most practical applications.

Many approaches to heuristic minimization have been proposed, some of which are summarized in reference [4]. Early heuristic minimizers computed all primes and used heuristic algorithms to cover the prime implicant table. Most recent heuristic logic minimizers do not compute all prime implicants, thus avoiding the potential bottleneck of storing a large set. They compute instead a prime cover starting from the initial specification of the function. This cover is then manipulated by modifying and/or deleting implicants until a cover with a suitable minimality property is found. Therefore these heuristic minimizers use an *iterative improvement* strategy.

Heuristic logic minimization can be viewed as applying a set of operators to the logic cover, which is initially provided to the minimizer along with the *don't care* set. The cover is declared to be the final outcome of minimization when the available operators cannot reduce the size of the cover any further.

The most common operators in heuristic minimization are the following:

- **Expand** makes a cover prime and minimal with respect to single-implicant containment. Implicants are processed one at a time. Each non-prime implicant is expanded to a prime, i.e., it is replaced by a prime implicant that contains it. Then, all other implicants covered by the expanded implicant are deleted.

- **Reduce** transforms a cover into a non-prime cover with the same cardinality. Implicants are processed one at a time. The reduce operator attempts to replace each implicant with another that is contained in it, subject to the condition that the reduced implicants, along with the remaining ones, still cover the function.

- **Reshape** modifies the cover while preserving the cardinality. Implicants are processed in pairs. One implicant is expanded while the other is reduced subject to the condition that the reshaped implicants, along with the remaining ones, still cover the function.

- **Irredundant** makes a cover irredundant. A minimal subset of implicants is selected such that no single implicant in that subset is covered by the remaining ones in that subset.

A characterization of heuristic minimizers can be done on the basis of the operators they use and the order in which they are applied. A simple heuristic minimizer can be implemented by applying the *expand* operator once. For example, this was the strategy used by the original PRESTO program. Primality and minimality with re-

---

[2] We refer to ESPRESSO as a heuristic minimizer and to ESPRESSO-EXACT as an exact minimizer for consistency with the literature. More precisely, the program referenced and described here is ESPRESSO-IIc. The ESPRESSO program being distributed implements both the heuristic and exact methods. It performs by default heuristic minimization and exact minimization with the option *-Dexact*.

spect to single-implicant containment can be achieved in this way. Unfortunately, the cardinality of the final cover may be much larger than the minimum cardinality.

To avoid low-quality solutions, program MINI iterates the application of operators *expand, reduce* and *reshape*. The prime cover computed by *expand* is manipulated by the other two operators so that another expansion is likely to reduce its size. The algorithm terminates when the iteration of these three operators does not reduce further the cover cardinality. Note that this approach does not ensure the irredundancy of the final cover. A similar approach is used by programs POP and PRESTO-L, both of which extend the capabilities of the original PRESTO program.

Program ESPRESSO uses the *irredundant* operator to ensure irredundancy of a cover. The main algorithm of the ESPRESSO program first computes a prime and irredundant cover by applying the *expand* and *irredundant* operators. Then, it improves the cover, when possible, by computing a sequence of prime and irredundant covers of decreasing cardinality. This is achieved by iteratively applying *reduce, expand* and *irredundant* as well as by switching heuristics inside these operators.

Example 7.2.9. Consider the function of Example 7.2.8, its minterm list and the labels given to the primes.

Suppose that the *expand* operator is applied first. Assume that the initial cover is the list of minterms and that they are expanded in the order they are listed. Minterms covered by the expanded ones are dropped. Thus expanding minterm 0000 1 to prime $\alpha = 0 * *0$ 1 allows us to drop minterms $\{0010 \quad 1; 0100 \quad 1; 0110 \quad 1\}$ from the initial cover. Note that minterm 0000 1 could have been expanded differently, i.e., replaced by another prime. Heuristic rules determine the direction of an expansion. For example, assume that the expansion of all minterms but the last yield $\{\alpha, \beta, \gamma, \delta\}$. Then the expansion of 1101 1 can yield either $\epsilon = 1 * 01$ 1 or $\zeta = *101$ 1 according to the direction of the expansion. If we assume that ϵ is chosen, then *expand* returns the cover $\{\alpha, \beta, \gamma, \delta, \epsilon\}$ with cardinality 5. The cover is prime, redundant and minimal w.r.t. single-implicant containment. Note that the cover differs from the list of all primes.

Let us assume that *reduce* is applied to the primes in alphabetical order. Prime α can be reduced to an empty implicant, because all its minterms are covered by other implicants of the cover. Prime $\beta = *0 * 0$ 1 can be reduced to $\tilde{\beta} = 00 * 0$ 1, because part of it is covered by δ. Similarly $\epsilon = 1 * 01$ 1 can be reduced to $\tilde{\epsilon} = 1101$ 1. The overall result of *reduce* is the cover $\{\tilde{\beta}, \gamma, \delta, \tilde{\epsilon}\}$ with cardinality 4.

An example of *reshape* is the following. The implicant pair $\{\tilde{\beta}, \delta\}$ can be changed into the pair $\{\beta, \tilde{\delta}\}$, where $\tilde{\delta} = 10 * 1$ 1. Thus the cover is now $\{\beta, \gamma, \tilde{\delta}, \tilde{\epsilon}\}$.

Another expansion would lead to the cover $\{\beta, \gamma, \delta, \epsilon\}$, which is prime and minimum. The cover is shown in Figure 7.5 (c). Note that neither a minimum cover nor an irredundant cover is guaranteed by iterating the *expand, reduce* and *reshape* operators. Irredundancy is guaranteed, though, by invoking *irredundant* after *expand*, which would immediately detect that either α or β should be dropped from the cover.

It is important to realize that heuristic minimizers also differ widely in the way in which the minimization operators are implemented, because all operators use heuristics. For example, the order in which the implicants are considered during an expansion affects the size of the resulting cover, and therefore heuristic rules are used for ordering the implicants. As a result, two implementations of the *expand* operator would yield both prime and minimal covers (w.r.t. single-implicant containment) but with possibly different cardinalities. Some algorithms implementing the operators have above-polynomial complexity. The execution times and storage requirements depend on the heuristics, which are the key factor in making heuristic minimization viable for large covers. We shall describe the heuristic algorithms in detail in Section 7.4.

RECENT DEVELOPMENTS IN HEURISTIC MINIMIZATION. The discovery of the properties of signature cubes, as described in Section 7.2.2, has led us to rethink how heuristic minimization can be implemented [17]. Recall that a minimum canonical cover is a set of maximal signature cubes that uniquely identify the set of primes covering each minterm. Thus a minimum canonical cover can be computed and used as a starting point for heuristic minimization. The application of the *expand* operator leads to a prime cover that in this case can be shown to be also irredundant [17]. This allows us to skip the application of the *irredundant* operator after *expand*. In addition, the *reduce* operator can be seen as a return to the minimum canonical cover. Thus, heuristic minimization can be seen as a sequence of expansions, with different expansion orders guided by heuristics of an initial minimum canonical cover. The minimum canonical cover can be computed by an algorithm described in reference [17].

These new and revolutionary ideas [17] have been proposed at the time of this writing and have not yet been validated by extensive experimentation and comparisons against program ESPRESSO.

7.2.4 Testability Properties

When a circuit has been manufactured, it is important to test it to detect possible malfunctions. Malfunctions are represented by *fault models*. Typical fault models used in logic design include the *single* and *multiple stuck-at* faults. A stuck-at-1 fault means that the malfunction can be modeled as if an input to a logic gate were permanently stuck at a logic TRUE value. A stuck-at-0 fault is similarly defined. A single- (or multiple-) fault model means we consider a possible malfunction related to a single fault (or to multiple faults). Even though more elaborate fault models do exist, stuck-at models are the most common models used in testing.

In general, the *testability* of a circuit measures the ease of testing it [16]. It is desirable to detect as many faults as possible. Therefore testability is a measure of circuit quality as well as area and performance.

We are interested in relating the minimality of a two-level cover to the testability of its implementation. Thus, we associate with the cover an equivalent circuit in terms of AND and OR gates and a specific meaning to testability. Namely a *fully testable* circuit for a given fault model (e.g., single/multiple stuck-at faults) is a circuit that has test patterns that can detect all faults. Testability of a fault relates to the capability of propagating a test pattern from the primary inputs to a test point of the circuit and to the possibility of observing the different responses of a correct and faulty circuit at their primary outputs.

The following theorem relates cover minimality to testability of its implementation.

Theorem 7.2.2. A necessary and sufficient condition for full testability for single stuck-at faults of an AND-OR implementation of a two-level cover is that the cover is prime and irredundant.

This theorem is a special case of Theorem 8.5.1 and the proof is reported in Section 8.5 [2].

Example 7.2.10. Consider the function f_1 of Example 7.2.1 and a redundant cover $a'b' + b'c + ac + ab$, whose AND-OR implementation is shown in Figure 7.6. If any input to the gate labeled A_2 in the figure is stuck at 0, no input pattern can detect that fault, because the faulty circuit implements function $a'b' + ac + ab$, which is equivalent to the behavior of the non-faulty implementation.

Consider now the irredundant cover $a'b' + b'c + ab$ and its implementation obtained by removing the gate labeled A_3. It is easy to verify that test patterns exist to detect all stuck-at faults.

In the case of single-output circuits, full testability for single stuck-at faults implies full testability for multiple faults [26]. Therefore, a cover that is prime and irredundant yields a circuit that is fully testable for multiple faults when it has a single output. In the case of multiple outputs, a prime and irredundant cover yields a fully testable circuit for multiple faults when there is no product term sharing or when each scalar output is represented by a prime and irredundant cover itself. While these conditions are sufficient, the definition of necessary conditions is still an open problem.

For this reason, primality and irredundancy are very important goals. They can be achieved by using exact or (some) heuristic minimizers. Note that a minimum prime cover (irredundant by definition) would not yield better testability properties than a non-minimum irredundant prime cover.

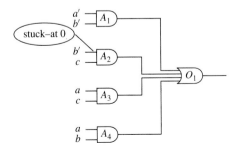

FIGURE 7.6
AND-OR implementation of redundant cover: $a'b' + b'c + ac + ab$.

7.3   OPERATIONS ON TWO-LEVEL LOGIC COVERS

The operators used in exact and heuristic logic minimization can be implemented in different ways. Since most logic minimizers (e.g., MINI, ESPRESSO, ESPRESSO-EXACT, etc.) use tables to represent covers, we describe in this section how operations on two-level forms can be implemented efficiently with tabular models. We shall present the operators in detail in Section 7.4, after having described the implementation of the fundamental operations on logic covers.

To avoid confusion, it is important to remember the context in which operations are performed. When we are considering a function **f**, its cover F is a suitable set of its implicants. We often consider sets of cubes and operations among cube sets, without reference to a specific function. Nevertheless, a set of cubes F defines a completely specified function **f**. The cubes are implicants of such a function **f**. The set F is a cover of **f**. For this reason, we often interchange the names "cube" and "implicant," as well as "set (list) of cubes" and "cover." The precise meaning is obvious from the context.

In the following subsections we first consider a binary encoding of implicants that enhances the efficiency of their manipulation. We then consider list-oriented and recursive algorithms for implicant manipulation and their extension to multiple-valued functions.

7.3.1   The Positional-Cube Notation

The *positional-cube* notation is a binary encoding of implicants. Let us first consider single-output functions. Thus a cover can be represented by just the input part of the implicants, with the output part being implicitly 1. The symbols used in the input part are $\{0, 1, *\}$. The positional-cube notation encodes each symbol by 2-bit *fields* as follows:

∅	00
0	10
1	01
*	11

don't Care

where the symbol ∅ means none of the allowed symbols. As a result, the presence of ∅ means that the implicant is void and should be deleted. Whereas this notation doubles the number of columns in an implicant table, it makes the implicant manipulation easier. For example, the intersection of two implicants in the positional-cube notation is their bitwise product.

Example 7.3.1. Consider the function $f = a'd' + a'b + ab' + ac'd$, whose minterms and primes are shown in Figure 7.5 (b). The corresponding implicant table in the positional-cube notation is:

$$
\begin{array}{cccc}
a & b & c & d \\
10 & 11 & 11 & 10 \\
10 & 01 & 11 & 11 \\
01 & 10 & 11 & 11 \\
01 & 11 & 10 & 01 \\
\end{array}
$$

The intersection of the first two implicants is 10 01 11 10, i.e., $a'bd'$. The intersection of the first and the third implicants is 00 10 11 10, i.e., it is void.

Multiple-output functions can be represented by appending the output part of the implicant to the encoded input part. This yields a binary-valued row vector, with as many fields as the number of inputs n plus 1. The first n fields have 2 bits; the last has as many bits as the number of outputs m. Therefore the positional-cube notation for each implicant of a binary-valued function of n inputs and m outputs requires $2n + m$ bits.

Any multiple-output function $\mathbf{y} = \mathbf{f}(\mathbf{x})$ can be written implicitly as $\chi(\mathbf{x}, \mathbf{y}) = 1$. This implicit representation of $\mathbf{f}$ is called the *characteristic equation* of $\mathbf{f}$. The positional-cube notation of an implicant of $\mathbf{f}$ can be seen as the input part of an implicant of its characteristic function $\chi(\mathbf{x}, \mathbf{y})$.

Example 7.3.2. The positional-cube representation of the two-input, three-output function $f_1 = a'b' + ab$; $f_2 = ab$; $f_3 = ab' + a'b$ is:

$$
\begin{array}{ccc}
10 & 10 & 100 \\
10 & 01 & 001 \\
01 & 10 & 001 \\
01 & 01 & 110 \\
\end{array}
$$

7.3.2 Functions with Multiple-Valued Inputs

A *multiple-valued function* is a function whose input and output variables can take two or more values. Multiple-valued functions and their optimization have been studied for several reasons [22], originally to provide support for designing logic circuits with more than two logic values, such as three- and four-valued PLAs and ROMs. Recently, techniques for manipulating multiple-valued logic functions have been used to design binary-valued circuits that implement some encoding of the multiple values. Examples are the design of PLAs with decoders [24] and of finite-state machines (see Sections 7.5 and 9.2.2). Overall multiple-valued logic representations provide a unified framework for optimizing single/multiple-output binary/multiple-valued functions, as shown in the following subsections.

We consider a class of multiple-valued functions that have n multiple-valued inputs and one binary-valued output. Each input $i = 1, 2, \ldots, n$ can take values in a corresponding interval $\{0, 1, \ldots, p_i - 1\}$. Thus $f : \{0, 1, \ldots, p_1 - 1\} \times \{0, 1, \ldots, p_2 - 1\} \times \ldots \times \{0, 1, \ldots, p_n - 1\} \rightarrow \{0, 1, *\}$. We call these functions *mvi-functions*. (Similarly, we refer to binary-valued functions as *bv-functions*.) The notion of *on set*, *off set* and *dc set* can be generalized to mvi-functions.

Let x be a p-valued variable. A *literal* over x is a subset of values $S \subseteq \{0, 1, \ldots, p - 1\}$, and it is denoted by x^S. Hence the literal over x corresponding to value j is denoted by $x^{\{j\}}$. An *empty* literal corresponds to an empty set (i.e., $S = \emptyset$) and a *full* literal to $S = \{0, 1, \ldots, p - 1\}$, which is equivalent to a *don't care* condition on that variable. The *complement* of a literal is the complementary set.

Mvi-functions can be represented in different ways, including *sum of products* and *product of sums* of literals. A literal is TRUE when the variable takes a value in the corresponding set; otherwise it is FALSE. The concepts of minterm, implicant, prime and essential can be extended in a straightforward way.

> **Example 7.3.3.** Consider an mvi-function with two inputs x and y. The first input variable is binary valued and the second is ternary valued. Thus $n = 2$, $p_1 = 1$ and $p_2 = 2$.
>
> Assume the function is TRUE under two conditions: (i) when the first input is 1 and the second input is either 0 or 1 and (ii) when the second input is 2. Thus a *sum of products* representation of the function is $x^{\{1\}}y^{\{0,1\}} + x^{\{0,1\}}y^{\{2\}}$. Literal $x^{\{0,1\}}$ is full and can be dropped from consideration because it represents a *don't care* condition on x.
>
> An implicant of this function is $x^{\{1\}}y^{\{0,1\}}$ and a minterm $x^{\{1\}}y^{\{0\}}$. Prime implicants of this function are $x^{\{1\}}$ and $y^{\{2\}}$.

When considering any multiple-output function, it is always possible to represent it as the corresponding single-output characteristic function. In particular, when considering a multiple-output binary-valued function, it is possible to interpret the output part of its implicants (with m bits) as an encoding of an m-valued input of a corresponding mvi-representation.

Indeed, the positional-cube notation can be extended to multiple-valued representation. Whereas a binary-valued literal can take two values (0 or 1) and be represented by a 2-bit field, an m-valued literal is represented by an m-bit field. Value i, $i = 0, 1, \ldots, m - 1$, corresponds to bit i of the field. For bv-functions a *don't care* condition on a variable means either 0 or 1 and therefore the encoding 11. For mvi-functions, a *don't care* on a variable corresponds to any value i, $i = 0, 1, \ldots, m - 1$, and therefore to a field of 1s. In both cases, the corresponding literal is full. In general, an implicant depends on a variable when the corresponding literal is not full. An implicant with an empty literal is void.

Given a (multiple-valued) literal or product of literals, the corresponding positional cube is denoted by $C(\cdot)$. For example, the positional cube for literal a is denoted by $C(a)$. The positional cube representing the Boolean space of interest (i.e., with appropriate dimensions) is denoted by U.

> **Example 7.3.4.** The positional-cube representation of the function $x^{\{1\}}y^{\{0,1\}} + x^{\{0,1\}}y^{\{2\}}$ introduced in Example 7.3.3 is:
>
> $$\begin{array}{cc} 01 & 110 \\ 11 & 001 \end{array}$$
>
> The positional cube corresponding to $x^{\{1\}}$ is $C(x^{\{1\}}) = 01\ 111$.

Conversely, the positional-cube representation of the two-input, three-output function in Example 7.3.2 can be seen as the specification of an mvi-function whose *sum of products* representation is:

$$a^{\{0\}}b^{\{0\}}z^{\{0\}} + a^{\{0\}}b^{\{1\}}z^{\{2\}} + a^{\{1\}}b^{\{0\}}z^{\{2\}} + a^{\{1\}}b^{\{1\}}z^{\{0,1\}}$$

where z is a three-valued output variable.

As a result of the analogy in the representations of binary-valued multiple-output functions with mvi-functions, minimization of mvi-functions subsumes minimization of bv-functions as a special case. In the sequel we shall describe the logic minimization algorithms by using the positional-cube notation and the mvi-function paradigm.

7.3.3 List-Oriented Manipulation

We consider implicants in the positional-cube representation and we review some of their properties. The *size* of a literal is the cardinality of its defining set S. Equivalently, the size of the related field is the number of 1s in the field. The size of an implicant is the product of the sizes of its literals (or of the corresponding fields). A minterm is an implicant of unit size. Note that in the case of binary-valued functions, the size of an implicant corresponds to the number of *don't cares* in its input part times the number of implied outputs. It is straightforward to realize that the definitions match.

We consider implicants as sets and we apply set operators (e.g., $\cap, \cup, \subseteq$) to them. Nevertheless, bitwise operations can be performed on positional cubes and their fields. We denote bitwise product, sum and complement by $\cdot, +$ and $'$, respectively. Note that a bitwise operation may have a different meaning than the corresponding set operations. For example, the complement of an implicant is not represented by the bitwise complement of its positional cube in the general case.

Let us consider first pairwise operations between implicants. The *intersection* of two implicants is the largest cube contained in both. It is computed by the bitwise product. The *supercube* of two implicants is the smallest cube containing both. It is computed by the bitwise sum. The *distance* between two implicants is the number of empty fields in their intersection. When the distance is zero, the two implicants intersect; otherwise they are disjoint. An implicant covers another one when the bits of the former cover (are greater than or equal to) those of the latter.

Example 7.3.5. Consider again the mvi-function represented by:

$$
\begin{array}{ccc}
10 & 10 & 100 \\
10 & 01 & 001 \\
01 & 10 & 001 \\
01 & 01 & 110 \\
\end{array}
$$

The first three implicants have size 1; the fourth has size 2. None of the implicants intersect: hence it is a disjoint cover. The supercube of the first two implicants is 10 11 101. It covers the first two implicants but not the last two.

The *sharp* operation, when applied to two implicants, returns a set of implicants covering all and only those minterms covered by the first one and not by the second one. The sharp operator is denoted by #. Namely, let $\alpha = a_1 a_2 ... a_n$ and $\beta = b_1 b_2 ... b_n$, where $a_i, b_i, i = 1, 2, \ldots, n$, represent their fields. An operative definition of the sharp operation is the following:

$$\alpha \# \beta = \begin{cases} a_1 \cdot b_1' & a_2 & \ldots & a_n \\ a_1 & a_2 \cdot b_2' & \ldots & a_n \\ \ldots & \ldots & \ldots & \ldots \\ a_1 & & \ldots & a_n \cdot b_n' \end{cases} \tag{7.2}$$

All void implicants (i.e., with an empty literal or equivalently a field equal to a string of 0s) are dropped. Note that the operative definition, which is an algorithm by itself, really implements the sharp operation. Indeed all and only the minterms covered by α and not by β are covered by the result. When considering the implicants of $\alpha \# \beta$ as defined above, they are all contained by α, because all fields but one match, this last one being contained. Moreover, all implicants of $\alpha \# \beta$ do not intersect β, because at least one field doesn't.

A similar operation, called *disjoint sharp* and denoted by ⊕, yields the sharp of two implicants in terms of mutually disjoint implicants. An operative definition of the disjoint sharp is the following:

$$\alpha \oplus \beta = \begin{cases} a_1 \cdot b_1' & a_2 & \ldots & a_n \\ a_1 \cdot b_1 & a_2 \cdot b_2' & \ldots & a_n \\ \ldots & \ldots & \ldots & \ldots \\ a_1 \cdot b_1 & a_2 \cdot b_2 & \ldots & a_n \cdot b_n' \end{cases} \tag{7.3}$$

The previous argument about the correctness of the result still applies. In addition, all implicants in $\alpha \oplus \beta$ are disjoint, because all implicant pairs have at least a field with void intersection. Namely, field i of implicant i (i.e., $a_i \cdot b_i'$) has void intersection with field i of other implicants $j > i$ (i.e., $a_i \cdot b_i$).

Example 7.3.6. Let us consider the two-dimensional space denoted by $U = 11\ 11$. Consider cube ab with $C(ab) = 01\ 01$. Let us compute the complement of cube ab by subtracting $C(ab)$ from U with the sharp operation:

$$11\ 11\ \#\ 01\ 01 = \begin{matrix} 10\ 11 \\ 11\ 10 \end{matrix} \tag{7.4}$$

The expression form for the complement is $a' + b'$.

Let us use now the disjoint sharp operator:

$$11\ 11\ \oplus\ 01\ 01 = \begin{matrix} 10\ 11 \\ 01\ 10 \end{matrix} \tag{7.5}$$

The expression form for the complement is now $a' + ab'$, which is a disjoint cover.

The *consensus* operator between implicants is defined as follows:

$$\mathcal{CONSENSUS}(\alpha, \beta) = \begin{cases} a_1 + b_1 & a_2 \cdot b_2 & \ldots & a_n \cdot b_n \\ a_1 \cdot b_1 & a_2 + b_2 & \ldots & a_n \cdot b_n \\ \ldots & \ldots & \ldots & \ldots \\ a_1 \cdot b_1 & a_2 \cdot b_2 & \ldots & a_n + b_n \end{cases} \tag{7.6}$$

The consensus operation among implicants is different in nature than the consensus of a function with respect to a variable (see Section 2.5.1). For this reason, we keep a separate notation. The consensus is void when two implicants have distance larger than or equal to 2. When they have distance 1, the consensus yields a single implicant.

Example 7.3.7. Consider the following implicants:

α	01	10	01
β	01	11	10
γ	10	01	01

The first two implicants have distance 1: $\mathcal{CONSENSUS}(\alpha, \beta) = 01\ 10\ 11 = C(ab')$. The distance between the other pairs is 2: their consensus is void. The implicants and $\mathcal{CONSENSUS}(\alpha, \beta)$ are shown in Figure 7.7.

Let us consider the *cofactor* computation in terms of implicants. The cofactor of an implicant α w.r.t. an implicant β is void when α does not intersect β. Otherwise it is given by the formula:

$$\alpha_\beta = a_1 + b'_1 \quad a_2 + b'_2 \quad \ldots \quad a_n + b'_n \tag{7.7}$$

Likewise, the cofactor of a set w.r.t. an implicant is the set of the corresponding cofactors.

To understand the formula, let us consider the familiar case in which we take the cofactor of a function, represented by an implicant, w.r.t. a binary-valued variable.

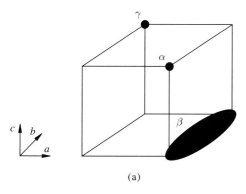
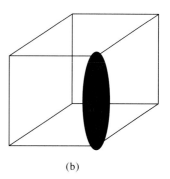

(a)   (b)

FIGURE 7.7
(a) Three cubes. (b) The $\mathcal{CONSENSUS}(\alpha, \beta)$.

A single variable (or complemented variable) is represented by the positional cube with 01 (or 10) in the corresponding field, all other entries being 1s. If the variable appears in the function with the opposite polarity, the two implicants are disjoint and the cofactor is empty. Otherwise, the bitwise sum with the complemented positional cube leaves all fields unaltered, except the one corresponding to the variable which is set to 11 (i.e., *don't care* on that variable). In the general binary-valued case, the cofactor w.r.t. an implicant is the same as the cofactor w.r.t. all variables it represents. The generalization to multiple-valued representations is straightforward.

> **Example 7.3.8.** Consider the function $f = a'b' + ab$, represented by the following cover in the positional-cube notation:
>
> $$\begin{array}{cc} a & b \\ 10 & 10 \\ 01 & 01 \\ a & b \end{array}$$
>
> The cofactor w.r.t. variable a is equivalent to the cofactor w.r.t. $C(a) = 01\ 11$. The cofactor of the first implicant in the cover is void. The cofactor of the second implicant is 11 01. Equivalently $f_a = b$.

Let us consider now operations on sets of implicants, i.e., on covers of logic functions. The union of two sets, i.e., the sum of two functions, can be obtained by merging the sets (and removing duplicated and void implicants). The intersection of two sets, i.e., the product of two functions, is given by all pairwise implicant intersections that are not void. The complexity of this computation is the product of the cardinalities of the two sets.

The sharp (and the disjoint sharp) operation can be applied to sets of implicants. The (disjoint) sharp of an implicant and a set is computed by iteratively applying the sharp operator between the implicant and the elements of the set (in jargon *sharping off*). The (disjoint) sharp of two sets of implicants is the union of the results of sharping off the second set from each element of the first one.

The (disjoint) sharp operation gives one way of computing the complement of a function. The complement is the result of applying the (disjoint) sharp operator between implicant U representing the Boolean space and the union of the *on set* and *dc set*, namely, $R = U \# (F^{ON} \cup F^{DC})$.

Note that a double complementation of a cover of a function yields again a cover of the same function. It is possible to show that the use of the sharp operator yields a cover in terms of all prime implicants of the function [11]. Thus the sharp operator allows us to compute the set of all primes of a function f as $P(f) = U \# (U \# (F^{ON} \cup F^{DC}))$. Despite the usefulness of the sharp and disjoint sharp operators for complementation and prime computation, today other algorithms are preferred for these tasks for efficiency reasons.

7.3.4   The Unate Recursive Paradigm

The recursive paradigm for operating on lists of implicants exploits a corollary of the orthonormal expansion Theorem 2.5.2. Let f and g be two mvi-functions expressed by an expansion with respect to the set of values that a variable, say x, can take. Recall that the literal corresponding to the ith value (denoted by $x^{\{i\}}$) is TRUE when x takes value i.

> **Corollary 7.3.1.** Let f and g be two mvi-functions. Let $\odot$ be an arbitrary binary operator representing a Boolean function of two arguments. Let x be a p-valued variable in the support of f and g. Then:
>
> $$f \odot g = \sum_{k=0}^{p-1} x^{\{k\}} \cdot (f_{x^{\{k\}}} \odot g_{x^{\{k\}}}) \qquad (7.8)$$

The corollary reduces to Corollary 2.5.1 for bv-functions.

The meaning of this corollary is that an operation on a set of implicants can be done on a variable-by-variable basis. An operation can be carried out by merging the results of the operation applied to the cofactors with respect to a chosen variable. Thus operations on sets of implicants can be computed recursively. At every level of the recursion a simpler problem needs to be solved.

Whereas the recursive approach to handling logic functions has been known for a long time, it was not until recently that its full power was exploited. The key lies in the fact that operations on unate functions are simpler to be solved. While most logic functions are not unate, the recursive decomposition may lead to cofactors that are unate and whose processing is very efficient. The so-called *unate recursive paradigm* proposed by Brayton, *et al.* [4] consists of exploiting the properties of unate functions within a recursive expansion.

We shall review next the properties of unate functions in conjunction with Boolean operation on implicants represented in the positional-cube notation. Then we shall consider the fundamental operations and how they are solved with the unate recursive paradigm. The techniques presented in the rest of this section are those implemented by program ESPRESSO. They are also described in detail in reference [4] for bv-functions and in reference [23] for mvi-functions.

UNATE FUNCTIONS AND UNATE COVERS. We recall that a bv-function f is positive (negative) unate in variable x if $f_x \supseteq f_{x'}$ ($f_x \subseteq f_{x'}$). We extend first the unateness concept to a cover F of a function f. A cover F is positive (negative) unate in a variable x if all its implicants have 11 or 01 (10) in the corresponding field. If a cover F of a function f is unate in x, so is the function f [4].

The unateness property is extended to mvi-functions as follows. Let us consider the set of values that a p-valued variable, say x, can take and let us impose an order on the values. Let $j, k : j > k$ be any two values that the variable can take. A function f is positive (negative) unate in variable x if $f_{x^{\{j\}}} \supseteq f_{x^{\{k\}}}$ ($f_{x^{\{j\}}} \subseteq f_{x^{\{k\}}}$) $\forall j, k \in (0, 1, \ldots, p-1) : j > k$. This unateness property of mvi-functions, called also *strong unateness*, is the counterpart of the unateness property for bv-functions.

A weaker unateness property is often very useful. A function f is *weakly unate* in variable x if there exists a value j such that, for all other values k, $f_{x^{\{k\}}} \supseteq f_{x^{\{j\}}}$. In other words, when a function is weakly unate in a variable, there exists a value j

such that changing the value of x from j to any other value causes the function to change, if it does, from 0 to 1 [23].

A cover F of an mvi-function is weakly unate in a variable x if the subset of all implicants that depend on variable x (i.e., that do not have a full literal in the corresponding field) has a column of 0s in the field corresponding to x. If a cover F of f is weakly unate in x, so is the function f [23].

A function or a cover is (strongly/weakly) unate if it is (strongly/weakly) unate in all variables.

Example 7.3.9. The following cover is weakly unate:

$$11 \quad 101$$
$$01 \quad 111$$
$$01 \quad 001$$

Let a and b be the variables associated with the two fields. Consider the matrix defined by the fields related to a, excluding the *don't care* field 11. Such a matrix has a column of 0s and thus the function is weakly unate in a.

Similarly, consider the matrix defined by the fields related to b, excluding the *don't care* field 111. Such a matrix has also a column of 0s and thus the function is weakly unate in b.

The following properties of weakly unate functions are important for determining if the function represented by a cover is a tautology [23]. In this case we say that the cover is a tautology.

Theorem 7.3.1. Let F be a weakly unate cover in variable x and G the subset of F that does not depend on x (i.e., where the field corresponding to x is full). Then F is a tautology if and only if G is a tautology.

Proof. It is obvious that if a subset of F is a tautology, then F is also a tautology.

Let us assume that F is a tautology and let j be a column of F, where all entries are 0 except for the implicants that do not depend on x. For the sake of contradiction, assume that G is not a tautology. Hence, there exists a minterm μ of its complement with a 1 in column j. Since F is weakly unate in x, no implicant of F has a 1 in column j. Hence μ belongs to the complement of F and F is not a tautology.

Theorem 7.3.2. A weakly unate cover F is a tautology if and only if one of its implicants is the universal cube U.

Proof. Apply repeatedly Theorem 7.3.1 on all variables. The cover is a tautology if and only if its subset G, consisting of the implicants that do not depend an all variables, is a tautology. Only the universal cube U can be in G, and hence G is a tautology if and only if F contains U.

TAUTOLOGY. Tautology plays an important role in all algorithms for logic optimization. Despite the intractability of the problem, the question if a function is a tautology can be answered efficiently using the recursive paradigm.

For binary-valued functions, a function f is a tautology if and only if its cofactors with respect to any variable and its complement are both a tautology. For mvi-functions, a function f is a tautology if and only if all its generalized cofactors with respect to the values of any variable are a tautology. Both statements are immediate consequences of Corollaries 2.5.1 and 7.3.1, respectively, by replacing function g by 1 and operator $\odot$ by $\overline{\oplus}$. (Thus $f \odot g$ corresponds to $f \overline{\oplus} 1$.)

Therefore, the tautology question can be answered by expanding a function about a variable and checking recursively whether its cofactors are a tautology. Note that when a function is binary valued and unate in a variable, only one cofactor needs to be checked for tautology, because one cofactor being a tautology implies that the other cofactor is a tautology as well. As an example, if a function f is positive unate in x, then $f_x \supseteq f_{x'}$ and therefore f_x is a tautology when $f_{x'}$ is so. The same argument applies to negative unate functions *mutatis mutandis* and can be extended to mvi-functions which are unate in a variable.

There are six rules for terminating or simplifying the recursive procedure: two rely on weak unateness.

- A cover is a tautology when it has a row of all 1s (tautology cube).

- A cover is not a tautology when it has a column of 0s (a variable that never takes a value).

- A cover is a tautology when it depends on one variable only and there is no column of 0s in that field.

- A cover is not a tautology when it is weakly unate and there is not a row of all 1s (by Theorem 7.3.2).

- The problem can be simplified when the cover is weakly unate in some variable. By Theorem 7.3.1, the tautology check can be done on the subset of the rows that do not depend on the weakly unate variable.

- When a cover can be written as the union of two subcovers that depend on disoint subsets of variables, the tautology question reduces to the checking tautology of both subcovers.

When a tautology cannot be decided upon these rules, the function is expanded about a variable. The choice of the splitting variable is very important. Brayton *et al.* proposed first the unate heuristic for bv-functions [4]. The heuristic consists of selecting a variable that is likely to create unate subproblems. Hence a binate variable is chosen, in particular the one that has the most implicants dependent on it. In this way, the size of the cofactors is minimized. In case of a tie, the heuristic chooses the variable minimizing the difference between the number of occurrences with positive polarity and the number of occurrences with negative polarity. The rationale is to keep the recursion tree as balanced as possible.

Example 7.3.10. Consider the function $f = ab + ac + ab'c' + a'$. We question if it is a tautology. Let us represent it first by a cover F:

$$
\begin{array}{ccc}
01 & 01 & 11 \\
01 & 11 & 01 \\
01 & 10 & 10 \\
10 & 11 & 11
\end{array}
$$

All variables are binate. The first variable, a, affects most implicants. Let us consider it as first splitting variable. We take then the cofactor w.r.t. $C(a') = 10\ 11\ 11$ and w.r.t. $C(a) = 01\ 11\ 11$. The first cofactor has only one row, namely $11\ 11\ 11$, which is a tautology. The second cofactor is:

$$
\begin{array}{c|cc}
11 & 01 & 11 \\
11 & 11 & 01 \\
11 & 10 & 10
\end{array}
$$

Let us concentrate on the right part of the array. All variables are still binate. Let us select the second variable for splitting, i.e., let us compute the cofactors w.r.t. $C(b') = 11\ 10\ 11$ and to $C(b) = 11\ 01\ 11$. The first cofactor is:

$$
\begin{array}{cc|c}
11 & 11 & 01 \\
11 & 11 & 10
\end{array}
$$

which is a tautology. The second cofactor is $11\ 11\ 11$, which also is a tautology. Hence f is a tautology.

Example 7.3.11. Let us consider now function $f = ab + ac + a'$, whose cover is:

$$
\begin{array}{ccc}
\overset{a}{01} & 01 & 11 \\
01 & 11 & 01 \\
10 & 11 & 11
\end{array}
$$

A representation of the minterms and implicant is shown in Figure 7.8 (a). Assume that the first variable is chosen for splitting. We take then the cofactor again w.r.t. $10\ 11\ 11$ and w.r.t. $01\ 11\ 11$. The first cofactor has $11\ 11\ 11$ in the first row and yields a tautology. The second cofactor is:

$$
\begin{array}{c|cc}
11 & 01 & 11 \\
11 & 11 & 01
\end{array}
$$

The cover is weakly unate and there is not a row of 1s. Hence it is not a tautology and f is not a tautology.

Rudell and Sangiovanni extended the heuristics for variable selection to mvi-functions [23]. The splitting variable is the one with the most values. Ties are broken by choosing the variable with the most 0s in the corresponding field when summed over all implicants. In the case of further ties, the selected variable has the fewest 0s in the column with the largest number of 0s.

The recursive expansion of multiple-valued functions poses an additional problem. An expansion in p values leads to p children in the decomposition tree. Program

$$b+c+b'c'$$

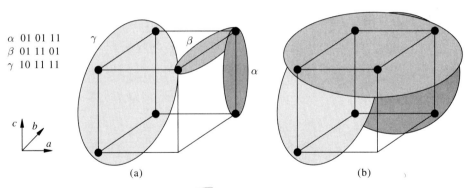

α 01 01 11
β 01 11 01
γ 10 11 11

(a)   (b)

FIGURE 7.8
(a) Representation of function $f = ab + ac + a'$. (b) The set of prime implicants of f.

ESPRESSO implements the p-valued expansion by means of a sequence of binary expansions. This choice leads to heuristic decisions on how to transform the p-valued expansion into a sequence of binary expansions. We refer the interested reader to reference [23] for further details.

CONTAINMENT. The question of whether a cover F contains (i.e., covers) an implicant α can be reduced to a tautology problem by using the following result [4].

> **Theorem 7.3.3.** A cover F contains an implicant α if and only if F_α is a tautology.

> **Proof.** Note first that $F \supseteq \alpha \iff F \cap \alpha = \alpha$. Suppose: $F \cap \alpha = \alpha$ holds. Let us take the cofactor of both sides w.r.t. α. Then $(F \cap \alpha)_\alpha = \alpha_\alpha = \text{TRUE}$. Since $(F \cap \alpha)_\alpha = F_\alpha \cap \alpha_\alpha$, F_α is a tautology. Conversely, let F_α be a tautology. Then $F_\alpha \cap \alpha = \alpha$, which implies $F \cap \alpha = \alpha$.

Therefore any algorithm for verifying tautology can be used to check for containment.

> **Example 7.3.12.** Let us consider again function $f = ab + ac + a'$, whose cover is shown in Figure 7.8 (a). Let us test whether the product bc is covered. The corresponding positional cube is $C(bc) = 11\ 01\ 01$. By taking the cofactor, we obtain:
>
> $$\begin{matrix} \overset{\alpha}{01} & 11 & 11 \\ 01 & 11 & 11 \\ 10 & 11 & 11 \end{matrix}$$
>
> which is a tautology, because the list depends on one variable and there is no column of 0s. Hence cube bc is covered by f.

COMPLEMENTATION. The complementation of a function implemented by using the unate recursive paradigm has been shown to be very efficient. We describe first the original complementation algorithm of program ESPRESSO for bv-functions [4] and then its extension to mvi-functions [23]. The original algorithm was designed for single-output bv-functions; complementation of multiple-output bv-functions was achieved by repeated invocation of the algorithm for each output.

Recursive complementation exploits Corollary 2.5.1. In particular, given a function f and a variable x in its support set, the following identity holds:

$$f' = x \cdot f'_x + x' \cdot f'_{x'} \tag{7.9}$$

Therefore the complement of a function f can be computed recursively by using the (complemented) covers of the cofactors of f. We denote the cover of f' by F' and those of f_x and $f_{x'}$ by F'_x and $F'_{x'}$ respectively. The recursive procedure terminates, or can be simplified, when the cover of F satisfies one of the following conditions:

- The cover F is void. Hence its complement is the universal cube.
- The cover F has a row of 1s. Hence F is a tautology and its complement is void.
- The cover F consists of one implicant. Hence the complement is computed by De Morgan's law.
- All implicants of F depend on a single variable, and there is not a column of 0s. Hence the function is a tautology, and its complement is void.
- The cover has a column of 0s, say in position j. Let α be a cube of all 1s, except in position j. Then $F = \alpha \cap F_\alpha$ and $F' = \alpha' \cup F'_\alpha$. Hence recursively compute the complement of F_α and of α and return their union.

Two issues are important in the implementation of the algorithm: the construction of the cover F' of f' from its components at each step of the recursion and the selection of the splitting variable.

The construction of the cover F' has to be done with care to keep its size as small as possible, namely, $F' = (C(x) \cap F'_x) \cup (C(x') \cap F'_{x'})$. A simple and effective optimization step is to compare each cube of F'_x to each case of $F'_{x'}$ for pairwise containment. When this happens, the covered cube is removed from the set and added to a new list G. Then the merging algorithm returns: $F' = (C(x) \cap \widetilde{F'_x}) \cup (C(x') \cap \widetilde{F'_{x'}}) \cup G$, where $\widetilde{F'_x}$ and $\widetilde{F'_{x'}}$ are the modified covers of the cofactors. Other more sophisticated techniques can be used, such as those described in reference [23].

The choice of the splitting variable is based upon the unate recursive paradigm. Indeed, the computation of the complement is simpler for unate functions, as shown by the following theorem.

Theorem 7.3.4. Let f be a positive unate function in x. Then:

$$f' = f'_x + x' \cdot f'_{x'} \tag{7.10}$$

Let f be a negative unate function in x. Then:

$$f' = x \cdot f'_x + f'_{x'} \tag{7.11}$$

Proof. From the expansion theorem (2.5.1) and the positive unateness condition it follows that $f = x \cdot f_x + x' \cdot f_{x'} = x \cdot f_x + f_{x'}$. By complementing the last expression, the thesis is proven. For negative unate functions the proof is obtained by exchanging x with x'.

This result generalizes to unate covers. Hence the computation of a cover of the complement of a unate function is simpler than in the case of a binate function. Since functions are in general binate, the selection of the splitting variable is done to improve the likelihood of generating unate cofactors. The variable is then chosen among the binate ones, as in the case of the tautology algorithm.

Example 7.3.13. Let us consider again function $f = ab + ac + a'$, whose cover is shown in Figure 7.8 (a) and reported here for convenience:

$$01 \ 01 \ 11$$
$$01 \ 11 \ 01$$
$$10 \ 11 \ 11$$

The only binate variable is a, which is chosen for splitting. Thus we compute the complement as $F' = (C(a') \cap F'_{a'}) \cup (C(a) \cap F'_{a})$. The computation of F_a yields a tautology (as shown in Example 7.3.11), hence $F'_{a'}$ is void. The computation of F_a yields:

$$\begin{array}{c|cc} 11 & 01 & 11 \\ 11 & 11 & 01 \end{array}$$

which is positive unate in both the second and third variables. Let us select the second variable, i.e., b. We compute now $F'_a = F'_{ab} \cup (C(b') \cap F'_{ab'})$. Now F_{ab} is a tautology and its complement is void. Then $F_{ab'} = 11 \ 11 \ 01$ and its complement is $11 \ 11 \ 10$, which intersected with $C(b') = 11 \ 10 \ 11$ yields $F'_a = 11 \ 10 \ 10$. The complement of f is $F' = C(a) \cap F'_a = 01 \ 10 \ 10$, because $C(a) = 01 \ 11 \ 11$ and $F'_{a'}$ is empty. An expression form for the complement is $f' = ab'c'$.

Let us consider now the case of mvi-functions. A multi-way (non-binary) recursive complementation algorithm could be devised to exploit directly Corollary 7.3.1. Nevertheless the implementation of such an algorithm may be inefficient, because of the multi-way recursion tree and of the complexity of merging multiple components of the complement at each level of the recursion. Rudell and Sangiovanni [23] suggested a two-way expansion of a multiple-valued variable x about two orthonormal literals x^A and x^B (i.e., such that $A \cup B = S$, where S is the set of all values for x and $A \cap B = \emptyset$). Let $\alpha = C(x^A)$ and $\beta = C(x^B)$. Then:

$$F' = (\alpha \cap F'_\alpha) \ \cup \ (\beta \cap F'_\beta) \tag{7.12}$$

The size of x^A and x^B is kept as close as possible, and a sequence of two-way expansions on variable x is performed until all possible values of x are exhausted.

Example 7.3.14. Consider the mvi-function represented by the cover:

$$F = \begin{array}{c} 10 \ 011 \\ 01 \ 101 \end{array}$$

Let us expand it about the second variable, which has three values, i.e., $S = \{0, 1, 2\}$. Consider orthonormal literals x^A and x^B, where $A = \{0\}$ and $B = \{1, 2\}$. Thus $\alpha = C(x^A) = 11 \ 100$ and $\beta = C(x^B) = 11 \ 011$.

The cofactor w.r.t. α is $F_\alpha = 01\ 111$ and the cofactor w.r.t. β is:

$$F_\beta = \begin{matrix} 10\ 111 \\ 01\ 101 \end{matrix}$$

We expand then F_β further by splitting B as $C \cup D$, where $C = \{1\}$ and $D = \{2\}$. Thus $\gamma = C(x^C) = 11\ 010$ and $\delta = C(x^D) = 11\ 001$. Then $F_\gamma = 10\ 111$ and $F_\delta = 11\ 111$. Hence:

$$F = (\alpha \cap F_\alpha) \cup (\gamma \cap F_\gamma) \cup (\delta \cap F_\delta) = \begin{matrix} 01\ 100 \\ 10\ 010 \\ 11\ 001 \end{matrix}$$

Hence the complement is:

$$\begin{matrix} 10\ 100 \\ 01\ 010 \\ 00\ 001 \end{matrix}$$

where the last term is void and can be dropped.

COMPUTATION OF ALL PRIME IMPLICANTS. The computation of all primes is relevant to exact minimization. It can be achieved by recursive methods for both binary-valued and mvi-functions. Let us consider again an expansion of a variable x about two orthonormal literals x^A and x^B. Let $\alpha = C(x^A)$, $\beta = C(x^B)$, $f^A = x^A \cdot f_{x^A}$ and $f^B = x^B \cdot f_{x^B}$. Then $f = f^A + f^B$, where f^A and f^B do not intersect. Hence the distance between any implicant of f^A and any implicant of f^B is larger than or equal to 1. As a result, there are three possibilities for a prime implicant of f:

- It is a prime of f^A.
- It is a prime of f^B.
- The prime is the consensus of two implicants: one in f^A and the other in f^B. The reason is that the prime is contained in $f^A + f^B$ and it consists of the union of two subsets of minterms, one in f^A and the other in f^B, that can be put in one-to-one correspondence and have pairwise distance 1.

We consider now sets of implicants and the cover notation. Let $P(F)$ be the set of primes of function f whose cover is F. Let F^A be a cover of f^A and F^B be a cover of f^B. Then $P(F^A) = \alpha \cap P(F_\alpha)$ and $P(F^B) = \beta \cap P(F_\beta)$.

Therefore the primes can be computed by the recursive formula:

$$P(F) = \mathcal{SCC}(\ (\alpha \cap P(F_\alpha)) \ \cup \ (\beta \cap P(F_\beta))$$
$$\cup \ \mathcal{CONSENSUS}((\alpha \cap P(F_\alpha)), (\beta \cap P(F_\beta)))\ ) \qquad (7.13)$$

The formula represents the union of three components according to the afore-mentioned argument. Since this union may yield some non-prime cubes (because some cubes of the first two components may be contained in their consensus), the $\mathcal{SCC}$ operator is applied to make the set minimal set w.r.t. single-implicant containment. The $\mathcal{SCC}$ operator checks for pairwise covering of implicants and removes the covered ones. The recursion terminates when f is represented by a single-implicant cover F, because $P(F) = F$.

The choice of the splitting variable plays again a role in this computation. When the cover F of a bv-function f is unate (or the cover of an mvi-function f is strongly unate), $P(F) = SCC(F)$. The reason is that each prime of a (strongly) unate function is essential [4] and therefore any cover of a unate function contains all primes. The non-prime implicants can be eliminated by applying SCC. As a result, the evaluation of the primes is very fast. Again, the choice of the splitting variable is such that unate cofactors can be found in early levels of the recursion.

Example 7.3.15. Let us consider again function $f = ab + ac + a'$, whose cover is shown in Example 7.8 (a). The only binate variable is a, which is chosen for splitting. Hence:

$$P(F) = SCC((C(a) \cap P(f_a)) \ \cup \ (C(a') \cap P(f_{a'}))$$

$$\cup \ CONSENSUS(C(a) \cap P(f_a), C(a') \cap P(f_{a'})))$$

$$= SCC(P_1 \cup P_2 \cup P_3)$$

where we represent the primes of a cover as the primes of the corresponding function, for the sake of simplicity. Note first that $f_{a'}$ is a tautology and thus $P(f_{a'}) = U$. Therefore $P_2 = C(a') \cap P(f_{a'}) = $ 10 11 11.

Consider then f_a, which is a unate function. A unate cover of f_a, minimal w.r.t. single-implicant containment, represents all primes of f_a. Thus $P(f_a)$:

$$11 \ 01 \ 11$$
$$11 \ 11 \ 01$$

Since $C(a) = $ 01 11 11, $P_1 = C(a) \cap P(f_a)$:

$$01 \ 01 \ 11$$
$$01 \ 11 \ 01$$

The last term, $P_3 = CONSENSUS(P_1, P_2)$, is:

$$11 \ 01 \ 11$$
$$11 \ 11 \ 01$$

The prime implicants of f are $P(f) = SCC(P_1 \cup P_2 \cup P_3)$:

$$10 \ 11 \ 11$$
$$11 \ 01 \ 11$$
$$11 \ 11 \ 01$$

or, equivalently, $\{a', b, c\}$. Note that the implicants of P_1 are contained in those of P_3 and are eliminated by SCC. The primes are shown in Figure 7.8 (b).

7.4 ALGORITHMS FOR LOGIC MINIMIZATION

We describe here the most important operators for two-level logic optimization and how they are implemented in detail. Our goal is to minimize a two-level cover F of a bv- (or mvi-) function f, which may be incompletely specified. Recall that we represent by F^{ON}, F^{OFF} and F^{DC} the *on set*, *off set* and *dc set* of the function f, respectively. We assume that at least two of these sets are specified as inputs to the minimization problem. The third set can be derived by complementing the union of the other two, if needed. In the particular case of completely specified functions, F^{DC} is empty and F and F^{ON} are used interchangeably.

We assume in this section that covers are represented in the positional-cube notation. Most procedures in this section are based on Rudell and Sangiovanni's description of program ESPRESSO [23].

7.4.1 Expand

The *expand* operator is used by most heuristic minimizers. Its goal is to increase the size of each implicant of a given cover F, so that implicants of smaller size can be covered and deleted. Maximally expanded implicants are primes of the function. As a result, the *expand* operator makes a cover prime and minimal with respect to single-implicant containment.

The expansion of an implicant is done by raising one (or more) of its 0s to 1. This corresponds to increasing its size (by a factor of 2 per raise), and therefore to covering more minterms. The fundamental question in the expansion process is whether the expanded cube is still *valid*, i.e., it is still an implicant of the function f. There are two major approaches for this test.

- *Checking for an intersection of the expanded implicant with F^{OFF}*. The intersection can be computed very efficiently, but the *off set* F^{OFF} is required. Sometimes F^{OFF} can be very large, and in some cases it may overflow the memory size and cause the minimizer to fail. Nevertheless this is the approach used by most minimizers, such as MINI and ESPRESSO.

- *Checking for the covering of the expanded implicant by $F^{ON} \cup F^{DC}$*. In this approach, the complement F^{OFF} is not needed. This method was used by program PRESTO, where the covering test was solved recursively. The containment check can also be reduced to a tautology problem, as described in Section 7.3.4.

The *expand* procedure sorts the implicants, one at a time and expands the 0 entries to 1 subject to the validity of the expanded implicant. The computational efficiency and the quality of the result depend on two important factors: (i) the order in which the implicants are selected and (ii) the order in which the 0 entries are raised to 1. Heuristic rules are used in both cases.

Programs MINI and ESPRESSO use the same heuristic for ordering the implicants.[3] The rationale is to expand first those cubes that are unlikely to be covered by other

---

[3] An earlier version of ESPRESSO used to sort the implicants for decreasing size [4].

cubes. The technique is the following. A vector is computed whose entries are the column sums of the matrix representing F. Each cube is assigned a weight that is the inner product of the cube itself and the previously computed vector. A low weight of a cube correlates to the cube having few 1s in the densely populated columns. Therefore the implicants are sorted for ascending weights. Needless to say, if an implicant is known to be prime (because of a previous expansion), there is no need to expand it.

Example 7.4.1. Consider the function $f = a'b'c' + ab'c' + a'bc' + a'b'c$ with abc' as a *don't care* condition. Then the set F^{ON}, in the positional-cube notation, is:

$$
\begin{array}{ccc}
10 & 10 & 10 \\
01 & 10 & 10 \\
10 & 01 & 10 \\
10 & 10 & 01
\end{array}
$$

The set F^{DC} is:

$$
\begin{array}{ccc}
01 & 01 & 10
\end{array}
$$

The *off set* F^{OFF} is then computed by complementation:

$$
\begin{array}{ccc}
01 & 11 & 01 \\
11 & 01 & 01
\end{array}
$$

Let us determine first the order of expansion. Let the columns be labeled from 1 to 6. The column count vector is $[313131]^T$ and the weights of the implicants in F are $(9, 7, 7, 7)$. Hence, the second implicant is processed first, i.e., 01 10 10.

In this example, we assume that entries are raised from left to right for the sake of simplicity. The raising of the 0 in column 1 yields 11 10 10, which does not intersect the *off set* and therefore is valid. The raising of the 0 in column 4 yields 11 11 10, which is also valid, but the raising of the 0 in column 6 would intersect the *off set* and it is rejected. Hence the expanded implicant is 11 11 10, which covers the first and third implicants of the original cover $F = F^{ON}$.

Thus F can be updated to:

$$
\begin{array}{ccc}
11 & 11 & 10 \\
10 & 10 & 01
\end{array}
$$

The last implicant is processed next. A raise of the literal in column 2 or 4 would make the implicant intersect the *off set*. On the contrary, a raise in column 5 is valid, yielding 10 10 11. The expanded cover is then:

$$
\begin{array}{ccc}
11 & 11 & 10 \\
10 & 10 & 11
\end{array}
$$

which is prime and minimal with respect to single-implicant containment. The different steps of the expansion are shown in Figure 7.9.

The heuristics for selecting the entries to expand are complex. It is obvious that any order in raising the 0s to 1s will lead to a prime. Many different primes can be

α 10 10 10
β 01 10 10
γ 10 01 10
δ 10 10 01

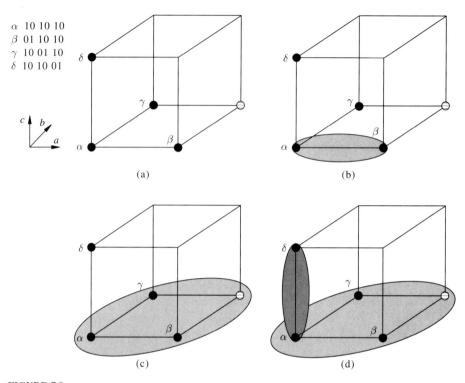

FIGURE 7.9
(a) Representation of function $f = a'b'c + ab'c' + a'bc' + a'b'c$. (b) A first expansion of β. (c) A second expansion of β. (d) Prime cover.

derived by expanding the cube in different directions. Each prime may cover different subsets of F. The major goal is to cover the largest subset of implicants of F.

We survey here the basic principles used in ESPRESSO as reported in reference [23]. We refer the interested reader to references [13], [4] and [23] for details. In ESPRESSO, the selection of an entry for the raise and the validity check are merged into a single step.

Let us describe first informally the rationale behind the heuristic expansion. Our goal is to make the implicant prime, while covering a maximal number of other implicants. A quick analysis of the position of the implicant in the Boolean space with respect to the *off set* allows us to determine some feasible and infeasible directions for expansion, but not all. Thus, the next step is to search for other cubes in the cover that would be covered by a specific feasible expansion. When this search is exhausted, we look for expansion directions that allow the expanded implicant to overlap a maximal number of other cubes. In the end, we just make the implicant prime by making it as large as possible.

To be more specific, we introduce some definitions. Let α be the cube to be expanded. Let *free* be a set denoting the candidate entries that can be raised to 1. Initially it consists of all entries that are 0. As the procedure progresses, elements are removed from $free$ and the process terminates when $free$ is empty. The *over-expanded* cube is the cube whose entries in $free$ are raised. For any implicant β of F, the smallest cube containing α and β is their supercube. Cube β is *feasibly covered* if the supercube is an implicant of f, i.e., if it does not intersect the *off set*.

There are four major steps in the expand procedure of program ESPRESSO.

1. *Determine the essential parts.*
   - Determine which entries can never be raised and remove them from $free$. (Search for any cube in F^{OFF} that has distance 1 from α.)
   - Determine which parts can always be raised, raise them and remove them from $free$. (Search for any column that has only 0s in F^{OFF}.)

2. *Detection of feasibly covered cubes.*
   If there is an implicant β whose supercube with α is feasible, repeat the following steps.
   - Raise the appropriate entry of α and remove it from $free$.
   - Remove from $free$ entries that can never be raised or that can always be raised and update α (as before).

3. *Expansion guided by the overexpanded cube.*
   While the overexpanded cube of α covers some other cubes of F, repeat the following steps.
   - Raise a single entry of α as to overlap a maximum number of those cubes.
   - Remove from $free$ entries that can never be raised or that can always be raised and update α (as before).

4. *Find the largest prime implicant.*
   Formulate a covering problem and solve it by a heuristic method. In the covering problem, the goal is to find a maximal subset of $free$ to be raised, while ensuring no intersection of the expanded implicant with F^{OFF}. It corresponds to finding the largest prime covering α.

Example 7.4.2. Let us consider again the function of Example 7.4.1 as shown in Figure 7.9 (a). Assume that the implicants are labeled $\{\alpha, \beta, \gamma, \delta\}$. Implicant $\beta = 01\ 10\ 10$ is selected first for expansion. The $free$ set includes the following columns: $\{1, 4, 6\}$.

Let us consider first entries that can never be raised. Cube $01\ 11\ 01$ of the *off set* has distance 1 from the selected implicant and it differs only in its third field. Hence column 6 cannot be raised and can be removed from $free$. It is not possible to determine entries that can always be raised by looking at the *off set*.

We consider then the other cubes of the original cover. The supercube of β and α is valid, and hence the first entry of β can be raised to yield $\widehat{\beta} = 11\ 10\ 10$. Similarly, the supercube of β and γ is valid, and hence the fourth entry of β can be raised to yield $\widehat{\beta} = 11\ 11\ 10$. The supercube of β and δ is not valid. The expanded implicant $\widetilde{\beta}$ is then $11\ 11\ 10$, its $free$ set is now empty, and no further expansion is possible. Implicants $\{\alpha, \gamma\}$ can be deleted from F, because they are covered by $\widetilde{\beta}$.

Consider now implicant $\delta = 10\ 10\ 01$, whose $free$ set is $\{2, 4, 5\}$. Since δ has distance 1 from both elements of F^{OFF}, the elements in columns 2 and 4 cannot be raised and they are dropped from $free$. Column 5 of F^{OFF} has only 0s. Hence, the 0 in column 5 can be raised, yielding $\widetilde{\delta} = 10\ 10\ 11$. The final cover $\{\widetilde{\beta}, \widetilde{\delta}\}$ is the same as that computed in Example 7.4.1 and shown in Figure 7.9 (d).

7.4.2   Reduce*

The *reduce* operator is used by most heuristic minimizers. Its goal is to decrease the size of each implicant of a given cover F so that a successive expansion may lead to another cover of smaller cardinality. A reduced implicant is *valid* when, along with the remaining implicants, it still covers the function. The reduced cover has the same cardinality as the original one. Reduced covers are not prime (unless no implicant can be reduced). Note that a redundant implicant would be reduced to a void one that can be discarded if it is reduced before the implicants that cover it. Since this depends on the order in which implicants are processed, the *reduce* operator does not guarantee the irredundancy of the cover.

Let us consider first how we can compute a maximally reduced cube. For the sake of explanation, we describe first the implicant reduction with an informal argument, while considering a completely specified function whose cover is F. To reduce an implicant α of F, we must remove from α those minterms that are covered by $F - \{\alpha\}$. Whereas this could in principle be computed by the intersection of α with the complement of $F - \{\alpha\}$, we must ensure that the result yields a single implicant. Since the complement of $F - \{\alpha\}$ is a set of cubes, we can replace it by its supercube (e.g., the smallest single cube containing all cubes) whose intersection with α yields the desired result.

> **Example 7.4.3.**   Let us consider the function of Example 7.4.1. Let F be:
>
> $$\begin{array}{c|ccc} \alpha & 11 & 11 & 10 \\ \beta & 10 & 10 & 11 \end{array}$$
>
> Let us try to reduce α. The complement of β is:
>
> $$\begin{array}{ccc} 01 & 11 & 11 \\ 11 & 01 & 11 \end{array}$$
>
> whose intersection with α yields:
>
> $$\begin{array}{ccc} 01 & 11 & 10 \\ 11 & 01 & 10 \end{array}$$
>
> Thus α should be replaced by two cubes, which is not what we want. By taking the supercube of the complement of β, 11 11 11, and intersecting it with α we obtain again α, which shows us that α cannot be reduced.

The aforementioned procedure can be formalized by the following theorem [23].

> **Theorem 7.4.1.**   Let $\alpha \in F$ be an implicant and $Q = F \cup F^{DC} - \{\alpha\}$. Then, the maximally reduced cube is $\widetilde{\alpha} = \alpha \cap supercube(Q'_\alpha)$.

Proof. The maximally reduced cube can be computed by deleting from α those minterms that belong to $Q = F \cup F^{DC} - \{\alpha\}$ under the restriction that the result must be a single cube. Now $\alpha \,\#\, Q = \alpha \cap Q'$ can yield a set of cubes. Hence the maximally reduced cube is $\widetilde{\alpha} = \alpha \cap supercube(Q')$, which is equivalent to $\widetilde{\alpha} = \alpha \cap supercube(Q'_\alpha)$, because Q' can be expanded as $(\alpha \cap Q'_\alpha) \cup (\alpha' \cap Q'_{\alpha'})$ and the second term has void intersection with α. The cube is maximally reduced, because the supercube is the smallest cube containing Q'.

The *reduce* operator entails then two steps: sorting the implicants and replacing each implicant by the maximally reduced one. Heuristic rules are used to sort the implicants. Programs MINI and ESPRESSO sort the implicants by weighting them first, as for the *expand* operator. The implicants are sorted in descending order of weight to first process those that are large and overlap many other implicants.

The computation of the maximally reduced cube can be done in several ways. In particular, ESPRESSO uses the unate recursive paradigm to compute the supercube of Q'_α. Indeed, it is easy to show that the following expansion about any p-valued variable x holds:

$$supercube(F') = supercube(\cup_{k=0}^{p-1} \; C(x^{\{k\}}) \cap supercube(F'_{x^{\{k\}}})) \qquad (7.14)$$

At the end of the recursion, we compute the supercube of the complement of a cube, which is:

- Empty when the cube depends on no variables.
- The cube's complement when it depends on one variable.
- The universe when it depends on two or more variables.

The unate heuristics leads to determining unate subproblems, for which the supercube can be more easily computed. We refer the interested reader to [23] for details.

Example 7.4.4. Let us consider again the function of Example 7.4.1. Assume it has been expanded as:

$$
\begin{array}{ccc}
11 & 11 & 10 \\
10 & 10 & 11
\end{array}
$$

Assume that the implicants are labeled as $\{\alpha, \beta\}$. The vector representing the column counts is $[212121]^T$ and the weights are 8 and 7, respectively. Hence implicant α is selected first for reduction. This implicant cannot be reduced, as shown by Example 7.4.3.

Let us consider now implicant β. The maximally reduced cube is $\widetilde{\beta} = \beta \cap supercube(Q'_\beta)$, where $Q = F \cup F^{DC} - \{\beta\}$, namely 11 11 10. Hence $Q' = 11\ 11\ 01$, $Q'_\beta = Q'$ and $supercube(Q'_\beta) = Q'$. Then $\widetilde{\beta} = \beta \cap Q' = 10\ 10\ 01$ and the reduced cover is:

$$
\begin{array}{ccc}
11 & 11 & 10 \\
10 & 10 & 01
\end{array}
$$

7.4.3   Irredundant*

An irredundant cover is minimal with respect to implicant containment. Thus its cardinality is a local minimum. The *irredundant* operator makes a cover irredundant, while trying to achieve a cover with the least cardinality. In other words, the *irredundant* operator makes the cover irredundant by deleting a maximal number of redundant implicants.

The *irredundant* operator was used first in heuristic logic minimization by program ESPRESSO [4]. This description follows the outline of reference [23]. We assume that the cover under consideration is prime, as constructed by the *expand* operator.

The cover F is split first into three sets:

- The *relatively essential* set E^r, which contains those implicants that cover some minterms of the function not covered by other implicants of F.

- The *totally redundant* set R^t, which contains those implicants that are covered by the relatively essential set.

- The *partially redundant* set R^p, which contains the remaining implicants.

Note that this classification is similar to the one used in Section 7.2.2 for exact minimization. The difference is that the cover F does not necessarily contain all prime implicants of the function. Hence, the set E^r contains the essential primes, but possibly other implicants.

The computation of the relatively essential set is based on checking whether an implicant $\alpha \in F$ is not covered by $F \cup F^{DC} - \{\alpha\}$. Similarly, the computation of the totally redundant set is based on checking whether an implicant $\alpha \in F$ is covered by $E^r \cup F^{DC}$. Then $R^p = F - \{E^r \cup R^t\}$. The major task of *irredundant* is to find a subset of R^p that together with E^r covers F^{ON}.

> **Example 7.4.5.**   Consider the following cover, shown in Figure 7.10:
>
> | | | | |
> |---|---|---|---|
> | α | 10 | 10 | 11 |
> | β | 11 | 10 | 01 |
> | γ | 01 | 11 | 01 |
> | δ | 01 | 01 | 11 |
> | ϵ | 11 | 01 | 10 |
>
> It is easy to verify that $E^r = \{\alpha, \epsilon\}$, $R^t = \emptyset$ and $R^p = \{\beta, \gamma, \delta\}$. Our problem is to find a minimal subset of R^p that, together with E^r, covers $F = F^{ON}$. Incidentally, note that in this example the cover F is not the set of all primes, and therefore E^r is larger than the set of the essentials.

Before describing the irredundant procedure, we describe informally a simple algorithm and its pitfall. It is conceivable to select one element $\alpha \in R^p$ at a time and verify whether it is contained by $H = E^r \cup R^p \cup F^{DC} - \{\alpha\}$. In the affirmative case, the cube can be deleted. By iterating this test, an irredundant cover can be constructed. Unfortunately the number of redundant cubes that can be deleted may be far less than

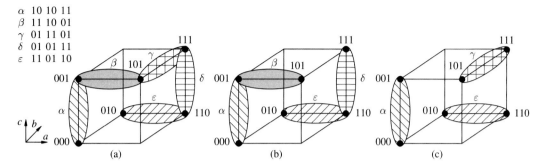

FIGURE 7.10
(a) Redundant cover of a logic function. (b) Irredundant cover. (c) Minimum irredundant cover.

the maximum. The simple algorithm gives a suboptimal result because it considers the partially redundant cubes in a sequential fashion. Indeed, the removal of a maximum number of redundant implicants requires considering the mutual covering relations among implicants.

> **Example 7.4.6.** Consider the redundant cover of Example 7.4.5. Assume that implicant γ is selected for examination by the simple algorithm. Since γ is covered by $\{\beta, \delta\}$, it can be deleted from the cover. No other implicant can be deleted from the cover, and thus cover $\{\alpha, \beta, \delta, \epsilon\}$ is irredundant. Nevertheless another irredundant cover, namely $\{\alpha, \gamma, \epsilon\}$, has lower cardinality. To detect such a cover, we must consider the interplay of the covering relations among implicants β, γ and δ. The two irredundant covers are shown in Figures 7.10 (b) and (c), respectively.

We describe now the irredundant procedure used in program ESPRESSO. The algorithm uses a modification of the tautology check. The key idea is the following. Rather than checking the tautology, we determine the set of cubes that, when removed, prevents the function from being a tautology. Recall that tautology of the cofactor can be used to determine containment. Hence we can determine the set of cubes that, when removed, prevents the containment.

Consider cube $\alpha \in R^p$ and test whether it is contained by $H = E^r \cup R^p \cup F^{DC} - \{\alpha\}$, which is equivalent to testing H_α for tautology. The test yields a positive answer by definition of a partially redundant prime. The test for tautology can be seen as the conjunction of the tests at each leaf of the decomposition tree induced by the tautology algorithm. Each leaf represents a set of conditions for tautology or, equivalently, for covering. Namely, each leaf is related to those cubes that cover a portion of the cube α under consideration.

When the tautology algorithm reaches a leaf, we mark the corresponding cubes of H_α that cause the tautology. If a leaf is related to any cube in $E^r \cup F^{DC}$, we drop the leaf from consideration, because the corresponding portion of cube α is covered by either a relatively essential prime or part of the *don't care* set. Otherwise we form a binary row vector whose entries are 1 in correspondence to those cubes of R^p, whose joint removal uncovers that portion of α. By repeating these steps for every leaf and every cube in R^p and by piling the vectors, we obtain a covering matrix $\mathbf{A}$.

The overall significance of $\mathbf{A}$ is the following. Each column corresponds to an element of R^p and each row to leaves of the tautology decomposition tree. Each row represents a subset of cubes of R^p, whose joint removal uncovers (a portion of) α. Therefore the problem of selecting an irredundant subset of R^p is equivalent to finding a minimum subset of columns such that all rows have at least a 1. This unate covering problem can be solved exactly or approximately by a heuristic algorithm. ESPRESSO solves it by a greedy algorithm. (Note that an exact solution would not necessarily yield a minimum cover, because not all primes were considered.)

Example 7.4.7. Let us consider again the cover of Example 7.4.5. Consider first implicant β. Then H is given by:

$$
\begin{array}{ccc}
10 & 10 & 11 \\
01 & 11 & 01 \\
01 & 01 & 11 \\
11 & 01 & 10
\end{array}
$$

and H_β is:

$$
\begin{array}{ccc}
10 & 11 & 11 \\
01 & 11 & 11
\end{array}
$$

We then test H_β for tautology. Here H_β provides already a split about the first variable. It is a leaf of the decomposition tree of the tautology algorithm, because H_β depends on one variable and there is no column of 0s. The two cubes in H_β are related to α and to γ. Since α is relatively essential and must be part of the cover, the relevant information is that the removal of γ uncovers β. The corresponding row vector is [110], where the first entry is trivially 1 because β covers itself.

Next we consider implicant γ. Then H is given by:

$$
\begin{array}{ccc}
10 & 10 & 11 \\
11 & 10 & 01 \\
01 & 01 & 11 \\
11 & 01 & 10
\end{array}
$$

and H_γ is

$$
\begin{array}{ccc}
11 & 10 & 11 \\
11 & 01 & 11
\end{array}
$$

The tautology test on H_γ is positive. The two cubes in H_γ are related to β and to δ, which are both partially redundant. The corresponding row vector is [111].

Last we consider implicant δ, and by a similar procedure we determine that the removal of γ uncovers δ. Hence the corresponding row vector is [011]. The covering matrix is:

$$
\mathbf{A} = \begin{bmatrix} 1 & 1 & 0 \\ 1 & 1 & 1 \\ 0 & 1 & 1 \end{bmatrix}
$$

A minimum row cover is given by the second column, corresponding to γ. Hence a minimum cardinality irredundant cover is $\{\alpha, \gamma, \delta\}$.

Note that in this simple example we have identified one leaf of the decomposition tree of the tautology algorithm in correspondence with each tautology test. In general, there may be more than one leaf per test (and therefore matrix **A** may not be square).

The reduced prime implicant table used in exact minimization by ESPRESSO-EXACT [23], as described in Section 7.2.2, can be derived in a similar way. The starting point is the set of all prime implicants, partitioned into *essentials*, *partially redundant* and *totally redundant*. The binary matrix **A** has a column associated with each element of R^p and each row with a leaf of the decomposition tree of the tautology algorithm, which represents the covering of (a portion of) an element in R^p by other elements of R^p. Exact minimization corresponds to selecting a subset of columns. If a row is not covered, i.e., if no column covers a 1 in that row, then the selected primes will fail covering some minterms of the function. With respect to the table used by Quine and McCluskey, the number of rows may be significantly smaller, because a row may correspond to a set of minterms covered by the same subset of primes.

Example 7.4.8. Consider again the function of Example 7.2.8, whose primes are shown in Figure 7.5 (a). Primes γ and δ are essential. No prime is totally redundant. Hence the partially redundant primes are $\{\alpha, \beta, \epsilon, \zeta\}$.

Let us apply the irredundant procedure to the partially redundant primes to construct the covering matrix **A**. Prime α is not covered when prime β is removed. (Minterms $\{0000, 0001\}$ are uncovered.) Similarly, prime β is not covered when prime α is removed, the same minterms being involved. Prime ϵ is not covered when prime ζ is removed and *vice versa*. (Minterm 1101 is uncovered.)

Thus the reduced covering matrix is:

	α	β	ϵ	ζ
0000,0010	1	1	0	0
1101	0	0	1	1

Note that there is no need of listing the minterm groups as long as all of them (i.e., all rows) are considered and covered.

7.4.4   Essentials*

The detection of essential primes is important for both exact and heuristic minimization. Indeed, essentials must be part of any cover. Hence, they can be extracted and stored while the minimizer manipulates the inessential primes. Essentials are added back to the cover in the end.

Before describing the detection of the essential primes in a formal way, we show an example to gain intuition into the problem.

Example 7.4.9. Consider a function whose cover was shown in Figure 7.4 (a) and is repeated in Figure 7.11 (a) for convenience. Let the cover be represented by the set of implicants $\{\alpha, \beta, \gamma, \delta\}$. We question whether α is essential.

We use then the following procedure. Let us remove the minterms covered by α from the cover, which reduces to $G = \{\gamma, \delta\}$, as shown in Figure 7.11 (b). Then α is not an essential if there is a prime of that function that can cover its minterms (i.e., 000 and 001). For this to be true, some cubes in G should be expandable to cover α, i.e., the cover must have minterms in one-to-one correspondence with those of α and at distance 1. This is true when $\mathcal{CONSENSUS}(G, \alpha)$ covers α. Since this is not true, α is essential. The cubes in $\mathcal{CONSENSUS}(G, \alpha)$ are shown in Figure 7.11 (c).

The detection of the essentials can be done using a corollary of Sasao's theorem [23, 24].

Theorem 7.4.2. Let $F = G \cup \alpha$, where α is a prime disjoint from G. Then α is an essential prime if and only if $\mathcal{CONSENSUS}(G, \alpha)$ does not cover α.

Proof. Consider a cube $\gamma \in G$. Its distance from α has to be larger than zero, because α and G are disjoint. Assume that the distance is larger than 1. Then, $\mathcal{CONSENSUS}(\gamma, \alpha) = \emptyset$. If all elements of G have distance larger than 1 from α, then α is essential because no other prime can cover its minterms. We can then restrict our attention to those elements of G at distance 1 from α. Now $\mathcal{CONSENSUS}(\gamma, \alpha)$ covering α implies that there exists primes, different from α, that cover all the minterms of α. Hence α is not essential. On the other hand, if $\mathcal{CONSENSUS}(\gamma, \alpha)$ does not cover α for any $\gamma \in G$, then α is an essential prime.

Corollary 7.4.1. Let F^{ON} be a cover of the *on set*, F^{DC} a cover of the *dc set* and α a prime implicant. Then α is an essential prime implicant if and only if $H \cup F^{DC}$ does not cover α, where:

$$H = \mathcal{CONSENSUS}(((F^{ON} \cup F^{DC})\#\alpha), \alpha) \qquad (7.15)$$

Proof. Let α be tested to be an essential of $F^{ON} \cup F^{DC}$. Let $G = (F^{ON} \cup F^{DC})\#\alpha$. Then $F^{ON} \cup F^{DC} = G \cup \alpha$, with α disjoint from G. Then, by Theorem 7.4.2, α is an essential if and only if $\alpha\#F^{DC}$ is not covered by H, or equivalently $H \cup F^{DC}$ does not cover α.

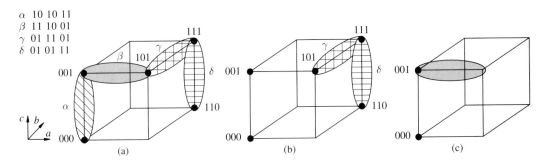

α 10 10 11
β 11 10 01
γ 01 11 01
δ 01 01 11

FIGURE 7.11
(a) Cover of F and minterms of F. (b) Cover $G = F\#\alpha$ and minterms of F. (c) $\mathcal{CONSENSUS}(G, \alpha)$ and minterms of α.

The corollary allows us to reduce the test for essentiality to a containment test, which in turn can be solved by a tautology test.

Example 7.4.10. Consider again the function of Example 7.4.9, shown in Figure 7.11 (a), whose cover is:

$$
\begin{array}{c|ccc}
\alpha & 10 & 10 & 11 \\
\beta & 11 & 10 & 01 \\
\gamma & 01 & 11 & 01 \\
\delta & 01 & 01 & 11
\end{array}
$$

Let us assume that the *dc set* is void and let us test α. Now, $F\#\alpha$ is:

$$
\begin{array}{ccc}
01 & 11 & 01 \\
01 & 01 & 11
\end{array}
$$

and $H = 11\ 10\ 01$. Then $H_\alpha = 11\ 11\ 01$, which is not a tautology. Hence α is not contained in H and is essential. Cubes of $F\#\alpha$ and of H are shown in Figures 7.11 (b) and (c), respectively.

Consider now prime implicant β. Then $F\#\beta$ is:

$$
\begin{array}{ccc}
10 & 10 & 10 \\
01 & 01 & 11
\end{array}
$$

and H is:

$$
\begin{array}{ccc}
10 & 10 & 11 \\
01 & 11 & 01
\end{array}
$$

Then H_β is:

$$
\begin{array}{ccc}
10 & 11 & 11 \\
01 & 11 & 11
\end{array}
$$

which is a tautology. Hence β is contained in H and is not essential.

7.4.5 The ESPRESSO Minimizer

ESPRESSO is a very fast and efficient two-level minimizer. It exploits the operators described above while performing an iterative refinement of a cover. The output of ESPRESSO is an irredundant cover, often minimum in cardinality.

ESPRESSO computes first the complement of a cover to be used by the *expand* operator. Then it applies the *expand* and *irredundant* operators, which make the cover prime and irredundant. After the *essentials* are extracted, ESPRESSO iterates the *reduce*, *expand* and *irredundant* operators in the search for irredundant covers of decreasing cardinalities. When no improvement can be achieved, ESPRESSO attempts to *reduce* and *expand* the cover using a different heuristic. This step is called *last_gasp* and it is fully detailed in references [4] and [23].

$ESPRESSO(F^{ON}, F^{DC})\{$

    $F^{OFF} = complement(F^{ON} \cup F^{DC});$

    $F = expand(F^{ON}, F^{OFF});$

    $F = irredundant(F, F^{DC});$

    $E = essentials(F, F^{DC});$

    $F = F - E;$ /* remove essentials */

    $F^{DC} = F^{DC} \cup E;$ /* place essentials temporarily in *don't care* set*/

    repeat {

        $\phi_2 = cost(F);$

        repeat {

            $\phi_1 = |F|;$

            $F = reduce(F, F^{DC});$

            $F = expand(F, F^{OFF});$

            $F = irredundant(F, F^{DC});$

        } **until** ($|F| \geq \phi_1$);

        $F = last\_gasp(F, F^{DC}, F^{OFF});$ /* apply different heuristics as last gasp */

    } **until** $(cost(F) \geq \phi_2);$

    $F = F \cup E;$ /* restore essentials */

    $F^{DC} = F^{DC} - E;$

    $F = make\_sparse(F, F^{DC}, F^{OFF});$

$\}$

ALGORITHM 7.4.1

Eventually ESPRESSO applies the routine *make_sparse*, which modifies the numbers of 1/0s in the array without affecting the cardinality of the cover. The rationale of this procedure is that some logic circuit implementations, like PLAs, do not benefit from primality. Indeed, a secondary goal in PLA optimization is reducing the number of transistors, which corresponds to making the implicant input part large and their output part small. This concept has been generalized to mvi-functions by defining *dense* and *sparse* variables according to the implementation style. Then *make_sparse* attempts to increase the size of the implicant fields corresponding to dense variables and to do the opposite for sparse variables.

Algorithm 7.4.1 is a simplified description of ESPRESSO. We refer the interested reader to references [4] and [23] for details.

In this description, we assume that ESPRESSO accepts as inputs the *on set* and the *dc set*, and generates the *off set* by complementation. In practice the program can accept any two sets among the *on set*, *off set* and *dc set*. The cost function *cost* is a weighted sum of the cover and literal cardinality.

Example 7.4.11. Consider again the complete differential equation integrator, as described in Example 1.5.3, and its control unit, reported in Example 4.7.6. The following table represents a two-level representation of the control unit in the format used by ESPRESSO, where loops have been opened and registers removed. Symbol "-" means *don't care*.

```
.i 9
.o 23
.ilb CLK reset c [23] [24] [25] [26] [27] [28]
.ob [39] [40] [41] [42] [43] [44] [45] s1 s2 s3 s4 s5 s6 s1s3 s2s3
 s2s4 s1s2s3 s1s3s4 s2s3s4 s1s5s6 s2s4s5s6 s2s3s4s6 s3s4s5s6
.p 45
-1------- 10000000000000000000000
-------1 10000000000000000000000
1-------- 01000000000000000000000
-0-1----- 00100000000000000000000
-01----1- 00100000000000000000000
-0--1---- 00010000000000000000000
-0---1--- 00001000000000000000000
-0----1-- 00000100000000000000000
-00----1- 00000010000000000000000
---1----- 00000001000000000000000
----1---- 00000000100000000000000
-----1--- 00000000010000000000000
------1-- 00000000001000000000000
-------1- 00000000000100000000000
--------1 00000000000010000000000
---1----- 00000000000001000000000
-----1--- 00000000000001000000000
----1---- 00000000000000100000000
-----1--- 00000000000000100000000
----1---- 00000000000000010000000
------1-- 00000000000000010000000
---1----- 00000000000000001000000
----1---- 00000000000000001000000
-----1--- 00000000000000001000000
---1----- 00000000000000000100000
-----1--- 00000000000000000100000
------1-- 00000000000000000100000
----1---- 00000000000000000010000
-----1--- 00000000000000000010000
------1-- 00000000000000000010000
---1----- 00000000000000000001000
-------1- 00000000000000000001000
--------1 00000000000000000001000
----1---- 00000000000000000000100
------1-- 00000000000000000000100
-------1- 00000000000000000000100
--------1 00000000000000000000100
----1---- 00000000000000000000010
-----1--- 00000000000000000000010
------1-- 00000000000000000000010
--------1 00000000000000000000010
-----1--- 00000000000000000000001
------1-- 00000000000000000000001
-------1- 00000000000000000000001
--------1 00000000000000000000001
.e
```

The following table shows a minimum cover computed by ESPRESSO. The cover can be shown to have minimum cardinality, by comparing it with the cover obtained by ESPRESSO-EXACT, which has the same size.

```
.i 9
.o 23
.ilb CLK reset c [23] [24] [25] [26] [27] [28]
.ob [39] [40] [41] [42] [43] [44] [45] s1 s2 s3 s4 s5 s6 s1s3 s2s3
   s2s4 s1s2s3 s1s3s4 s2s3s4 s1s5s6 s2s4s5s6 s2s3s4s6 s3s4s5s6
.p 14
-00----1- 00000010000000000000000
-01----1- 00100000000000000000000
-0--1---- 00010000000000000000000
-0---1--- 00001000000000000000000
-0----1-- 00000100000000000000000
-0-1----- 00100000000000000000000
1-------- 01000000000000000000000
-1------- 10000000000000000000000
---1----- 00000001000001001101000
-------1- 00000000000100000001101
--------1 10000000000010000001111
----1---- 00000000100000111010110
------1-- 00000000001000010110111
-----1--- 00000000010001101110011
.e
```

7.5 SYMBOLIC MINIMIZATION AND ENCODING PROBLEMS

We consider in this section extensions to two-level logic optimization to cope with problems where the choice of the representation of the inputs and/or outputs is free. Consider, for example, an instruction decoder, specified by a combinational function. If we can alter the representation of the operation codes, we can modify the size of its minimum two-level representation. Hence, an extended logic optimization problem is to find an encoding of the inputs of a function and a corresponding two-level cover of minimum cardinality. A similar search can be applied, *mutatis mutandis*, for the encoding of the outputs of a function. As a simplified case, we may just want to search for the polarity of the outputs of a function that minimizes the cardinality of its cover.

The minimization of Boolean functions where the codes of the inputs and/or outputs are not specified is referred to as *symbolic minimization*. It is tightly related to multiple-valued logic optimization, and indeed it coincides with minimization of mvi-functions in some particular cases.

Example 7.5.1. Consider an instruction decoder of Figure 7.12. It has two input and one output fields, related to the addressing mode (ad-mode), operation code (op-code)

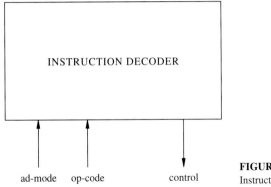

INSTRUCTION DECODER

ad-mode   op-code   control

FIGURE 7.12
Instruction decoder.

and control, respectively. It can be specified by the following symbolic truth table:

INDEX	AND	CNTA
INDEX	OR	CNTA
INDEX	JMP	CNTA
INDEX	ADD	CNTA
DIR	AND	CNTB
DIR	OR	CNTB
DIR	JMP	CNTC
DIR	ADD	CNTC
IND	AND	CNTB
IND	OR	CNTD
IND	JMP	CNTD
IND	ADD	CNTC

The ad-mode can take one of the three symbolic inputs $\{INDEX, DIR, IND\}$ (that are mnemonics for indexed, direct and indirect addressing); the op-code can indicate one of the four instructions $\{AND, OR, JMP, ADD\}$; the output corresponds to one of four control signals $\{CNTA, CNTB, CNTC, CNTD\}$. The general problem is then to replace the symbols by binary strings, so that the corresponding minimum cover is minimum over all codes.

In the following subsections we consider symbolic minimization in conjunction with the encoding of the inputs and/or outputs of a combinational function. Further extensions of these techniques are relevant for sequential logic design, as described in Section 9.2.2.

7.5.1 Input Encoding

We concentrate our attention first on the encoding of the input fields of a symbolic table. The restriction is motivated by didactic purposes, because input and output encoding exploit different properties and mechanisms and are best described separately. Thus we assume that the output codes are given.

Since we do not know the input codes, we cannot specify a bv-function that can be minimized. Trying all possible input codes would be excessively costly, because it would require an exponential number of logic minimizations. Fortunately, it is possible to optimize the function independently of the encoding and determine the codes at a later time. This requires performing the minimization at the symbolic level, before the encoding.

For input encoding, a straightforward model for symbolic functions and their optimization is to consider them as mvi-functions and apply mvi-minimization. This is possible because the input symbols can be put in one-to-one correspondence with the values of multiple-valued input variables.

Example 7.5.2. Consider again Example 7.5.1. Let us assume that the codes of the outputs are $\{1000, 0100, 0010, 0001\}$, respectively. Let us associate a three-valued variable to the ad-mode and a four-valued variable to the op-code. A positional-cube representation of the truth table is the following:

$$
\begin{array}{ccc}
100 & 1000 & 1000 \\
100 & 0100 & 1000 \\
100 & 0010 & 1000 \\
100 & 0001 & 1000 \\
010 & 1000 & 0100 \\
010 & 0100 & 0100 \\
010 & 0010 & 0010 \\
010 & 0001 & 0010 \\
001 & 1000 & 0100 \\
001 & 0100 & 0001 \\
001 & 0010 & 0001 \\
001 & 0001 & 0010 \\
\end{array}
$$

Let us consider a minimum cover of this function:

$$
\begin{array}{ccc}
100 & 1111 & 1000 \\
010 & 1100 & 0100 \\
001 & 1000 & 0100 \\
010 & 0011 & 0010 \\
001 & 0001 & 0010 \\
001 & 0110 & 0001 \\
\end{array}
$$

The first implicant means that under ad-mode $INDEX$, any op-code (full literal) implies control $CNTA$. The second implicant means that under ad-mode DIR, either op-code AND or OR imply control $CNTB$, etc.

It is very important to stress that, for the input encoding problem, the mvi-function representation is an appropriate model for symbolic minimization, because the output codes are fixed. Indeed, each mvi-implicant expresses the conjunction of literals that represent just whether the inputs match some patterns. Hence the value associated with a symbol has no other meaning than being an identifier.

Since mvi-function optimizers such as ESPRESSO can be used for this task, the only problem to be solved is the translation of the optimized mvi-representation into a binary-valued one. An important issue is the preservation of the cardinality of such a cover. This can be achieved by searching for an encoding where each mvi-implicant is mapped into one bv-implicant.

Hence the encoding of the input symbols must be such that each mv-literal corresponds to only one cube in the binary-valued representation. If an mv-literal has only one 1, then it can be mapped trivially into the cube representing the code of the corresponding symbol. Difficulties arise when an mv-literal has more than one 1, because the smallest single cube that is always TRUE when the input matches any of the corresponding symbols is the supercube of their codes. Indeed, such a cube may cover other points of the Boolean space. It is important that the supercube does not intersect the code of other symbols not in the literal; otherwise a spurious value may erroneously assert the function. It is obvious that full literals do not create problems, because any input would make them TRUE. Hence they can be translated into the universal cube and pose no constraint on the encoding.

In the sequel, we represent covers of mvi-functions in the positional-cube notation and the binary encoding of the symbols with binary digits $\{0, 1\}$. Similarly, binary-encoded covers will be shown in terms of binary digits (and the *don't care* condition symbol *), instead of using the positional-cube notation, to stress the effects of binary encoding.

Example 7.5.3. Let us consider the op-code field of the minimum cover in Example 7.5.2:

$$
\begin{array}{c}
1111 \\
1100 \\
1000 \\
0011 \\
0001 \\
0110
\end{array}
$$

The first literal is a multiple-valued *don't care*. The second, fourth and sixth literals have more than one 1, representing op-codes $\{AND, OR\}$, $\{JMP, ADD\}$ and $\{OR, JMP\}$, respectively. Hence we look for binary codes of these symbols whose supercubes do not intersect other codes. For example, the following codes are valid:

$$
\begin{array}{ll}
AND & 00 \\
OR & 01 \\
JMP & 11 \\
ADD & 10
\end{array}
$$

In this case, the supercubes are $0*$, $1*$ and $*1$, respectively. The encoding of the op-codes and the corresponding supercubes are shown in Figure 7.13 (a).

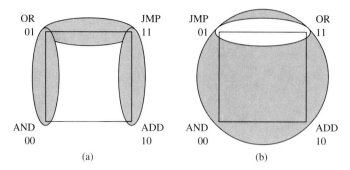

FIGURE 7.13
(a) Representation of valid op-codes and supercubes. (b) Representation of invalid op-codes and supercubes.

Instead, the following codes are not valid:

$$
\begin{array}{ll}
\text{AND} & 00 \\
\text{OR} & 11 \\
\text{JMP} & 01 \\
\text{ADD} & 10
\end{array}
$$

Now, the supercubes corresponding to $\{AND, OR\}$ and to $\{JMP, ADD\}$ are both **
and intersect the other codes. This would cause any op-code to make that mv-literal
TRUE, erroneously. This is shown in Figure 7.13 (b).

With the former valid codes, the mv-field can be replaced by the following binary-
valued field:

$$
\begin{array}{ccc}
1111 & \rightarrow & ** \\
1100 & \rightarrow & 0* \\
1000 & \rightarrow & 00 \\
0011 & \rightarrow & 1* \\
0001 & \rightarrow & 10 \\
0110 & \rightarrow & *1
\end{array}
$$

With a similar procedure it is possible to determine the codes of the ad-mode
symbols. Assume the following codes are chosen:

$$
\begin{array}{ll}
\text{INDEX} & 00 \\
\text{DIR} & 01 \\
\text{IND} & 11
\end{array}
$$

Then, the cover with encoded input symbols is:

$$
\begin{array}{lll}
00 & ** & 1000 \\
01 & 0* & 0100 \\
11 & 00 & 0100 \\
01 & 1* & 0010 \\
11 & 10 & 0010 \\
11 & *1 & 0001
\end{array}
$$

It was shown [9] that the input encoding problem has always a solution satisfying the constraints. However, the encoding of n_s symbols in a field may require more than $\lceil \log_2 n_s \rceil$ bits. Hence, the search for a minimum bit encoding is relevant. Villa and Sangiovanni showed that this problem is intractable [27]. Exact and heuristic algorithms have been developed for the input encoding problem and are described next.

ENCODING ALGORITHMS. We consider the encoding of a set of symbols S, with $n_s = |S|$. The encoding problem can be modeled by a binary[4] constraint matrix $\mathbf{A} \in \mathbf{B}^{n_r \times n_s}$ whose n_r rows are the fields of the optimized cover corresponding to the symbol set S under consideration. Note that full literals and literals with cardinality 1 can be dropped, because they do not represent encoding constraints. The matrix has as many columns as the n_s symbols. The codes of the symbols are represented by a binary matrix $\mathbf{E} \in \mathbf{B}^{n_s \times n_b}$, which is the unknown of the problem. Each row of $\mathbf{E}$ is the encoding of a symbol. The code length n_b is also an unknown.

Example 7.5.4. The constraint matrix for Example 7.5.3 is:

$$\mathbf{A} = \begin{bmatrix} 1 & 1 & 0 & 0 \\ 0 & 0 & 1 & 1 \\ 0 & 1 & 1 & 0 \end{bmatrix}$$

where the column set is in one-to-one correspondence with {*AND, OR, JMP, ADD*}. Hence the first row of $\mathbf{A}$ means {*AND, OR*}, etc.

The encoding matrix which solves the encoding problem is:

$$\mathbf{E} = \begin{bmatrix} 0 & 0 \\ 0 & 1 \\ 1 & 1 \\ 1 & 0 \end{bmatrix}$$

An encoding matrix $\mathbf{E}$ satisfies the encoding constraints represented by $\mathbf{A}$ if, for each row $\mathbf{a}^T$ of $\mathbf{A}$, the supercube of the rows of $\mathbf{E}$ corresponding to the 1s in $\mathbf{a}^T$ does not intersect any of the rows of $\mathbf{E}$ corresponding to the 0s in $\mathbf{a}^T$. An encoding matrix is *valid* if it satisfies the encoding constraints and all rows (i.e., codes) are distinct from each other. It is easy to see that a constraint matrix can incorporate the requirement for achieving disjoint codes, by appending to $\mathbf{A}$ the rows of an identity matrix of size n_s. Therefore we shall focus on the satisfaction of the constraints in the sequel.

The identity matrix is always a valid encoding, because its rows are disjoint codes and the supercube of any row subset is not intersecting the remaining ones [9, 10]. Such an encoding is often referred to as *1-hot* encoding. Unfortunately, the number of bits of 1-hot codes may be much larger than the minimum.

---

[4]The matrix can be made ternary by considering *don't care* conditions, as shown in reference [10].

Exact encoding algorithms have been proposed that transform the encoding problem into a covering or a coloring problem. They rely on the notion of *dichotomy*, which is a bipartition of a subset of the symbol set S. The rationale is that any column of matrix $\mathbf{E}$ corresponds to a bipartition induced by the assignment of bit 1 or bit 0 to each code. An encoding matrix is then equivalent to a set of bipartitions. To determine such a set, we need to consider the constraints and how they induce appropriate bipartitions. We define dichotomies in accordance with Yang and Ciesielski [29].

> **Definition 7.5.1.** The **dichotomy** associated with row $\mathbf{a}^T$ of $\mathbf{A}$ is a bipartition of a subset of S, whose first block has the symbols with the 1s in $\mathbf{a}^T$ and whose second block has the symbols with the 0s in $\mathbf{a}^T$.
>
> A **seed dichotomy** associated with row $\mathbf{a}^T$ of $\mathbf{A}$ is a bipartition of a subset of S, whose first block has the symbols with the 1s in $\mathbf{a}^T$ and whose second block has exactly one symbol with a 0 in $\mathbf{a}^T$.

The blocks of a dichotomy are called left and right blocks and are represented by a set pair $(L; R)$.

> **Example 7.5.5.** The dichotomy associated with constraint $\mathbf{a}^T = 1100$ is $(\{AND, OR\}; \{JMP, ADD\})$. The corresponding seed dichotomies are $(\{AND, OR\}; \{JMP\})$ and $(\{AND, OR\}; \{ADD\})$.

A dichotomy can be associated with an encoding column (of a matrix $\mathbf{E}$), where the entries corresponding to the elements in the left block are 1 and the remaining ones are 0s. Note that the encoding column of the dichotomy corresponding to $\mathbf{a}^T$ is $\mathbf{a}$. Any encoding matrix that contains that column satisfies the constraint $\mathbf{a}^T$, because the supercube of the codes of the symbols represented by $\mathbf{a}^T$ has a 1 in that column, while the other codes have a 0 in that column and therefore the intersection is void. This fact leads to the following theorem when taking all rows of $\mathbf{A}$ into account [9].

> **Theorem 7.5.1.** Given a constraint matrix $\mathbf{A}$, the encoding matrix $\mathbf{E} = \mathbf{A}^T$ satisfies the constraints specified by $\mathbf{A}$.

> **Example 7.5.6.** Given constraint matrix:

$$\mathbf{A} = \begin{bmatrix} 1 & 1 & 0 & 0 \\ 0 & 0 & 1 & 1 \\ 0 & 1 & 1 & 0 \end{bmatrix}$$

Then:

$$\mathbf{E} = \mathbf{A}^T = \begin{bmatrix} 1 & 0 & 0 \\ 1 & 0 & 1 \\ 0 & 1 & 1 \\ 0 & 1 & 0 \end{bmatrix}$$

satisfies the constraints expressed by $\mathbf{A}$. Consider, for example, the first row of $\mathbf{A}$. The supercube of the codes of $\{AND, OR\}$ has a 1 as the first digit, while the codes of the other symbols have a 0 as the first bit. A similar argument applies to the other rows of $\mathbf{A}$

(columns of **E**), which therefore satisfy the encoding constraints. Unfortunately matrix **A** has three columns, while two columns are sufficient, as shown in Example 7.5.4.

While Theorem 7.5.1 yields a solution to the encoding problem, it is obvious that the number of rows of **A** may be larger than the minimum number of bits required to satisfy all constraints. For this reason, we focus on the seed dichotomies and on the way of combining them into dichotomies that satisfy the constraints.

Definition 7.5.2. Two dichotomies $(L_1; R_1)$ and $(L_2; R_2)$ are **compatible** if either $L_1 \cap R_2 = \emptyset$ and $R_1 \cap L_2 = \emptyset$ or $L_1 \cap L_2 = \emptyset$ and $R_1 \cap R_2 = \emptyset$. Otherwise they are incompatible.

Note that a dichotomy is compatible with the one obtained by exchanging the left and right blocks. This symmetry is typical of the input encoding problem and is related to the fact that valid encodings are preserved under column complementation.

The constraints corresponding to any set of compatible dichotomies can be satisfied by an encoding requiring 1 bit. Hence, given the seed dichotomies, compatibility and conflict graphs can be constructed and the encoding derived by a clique covering or coloring, respectively [29].

Example 7.5.7. Consider again the constraint matrix for Example 7.5.3. The seed dichotomies (labeled $s_1 - s_6$) are:

$$
\begin{array}{c|ccc}
s_1 & (\{AND,OR\} & ; & \{JMP\}) \\
s_2 & (\{AND,OR\} & ; & \{ADD\}) \\
s_3 & (\{JMP,ADD\} & ; & \{AND\}) \\
s_4 & (\{JMP,ADD\} & ; & \{OR\}) \\
s_5 & (\{OR,JMP\} & ; & \{AND\}) \\
s_6 & (\{OR,JMP\} & ; & \{ADD\})
\end{array}
$$

A compatibility graph is shown in Figure 7.14 (a). Two cliques, denoted by the emboldened edges in the figure, cover the graph. Hence a 2-bit encoding suffices. It is determined by a dichotomy compatible with seeds $\{s_1, s_2, s_3, s_4\}$, e.g., $(\{AND, OR\}; \{JMP, ADD\})$, and by a dichotomy compatible with seeds $\{s_5, s_6\}$, e.g., $(\{OR, JMP\}; \{AND, ADD\})$. Thus:

$$
\mathbf{E} = \begin{bmatrix} 1 & 0 \\ 1 & 1 \\ 0 & 1 \\ 0 & 0 \end{bmatrix}
$$

A conflict graph is shown in Figure 7.14(b) along with a minimum vertex color leading to the same encoding.

A more efficient approach is to model the encoding problem as an exact logic minimization problem. This requires extending the notion of covering and primes to dichotomies.

Definition 7.5.3. A dichotomy (L_1, R_1) **covers** another dichotomy (L_2, R_2) if $L_1 \supseteq L_2$ and $R_1 \supseteq R_2$ or $L_1 \supseteq R_2$ and $R_1 \supseteq L_2$.

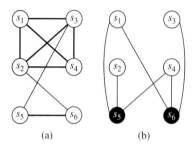

(a)   (b)

FIGURE 7.14
(a) Compatibility graph of the seed dichotomies and clique cover. (b) Conflict graph of the seed dichotomies and vertex coloring.

Definition 7.5.4. The **union** of two compatible dichotomies is a dichotomy covering both with the smallest left and right blocks.

Definition 7.5.5. A **prime dichotomy** is a dichotomy that is not covered by any compatible dichotomy of a given set.

The set of prime dichotomies represents possible bipartitions (with maximal blocks) that are compatible with the constraints. Only a subset of the prime dichotomies is necessary to satisfy the constraints, namely, a subset that covers all seed dichotomies. Hence the encoding problem can be cast as the search for all prime dichotomies that originate from a constraint matrix and as a minimum covering problem. The covering problem is unate and it can be solved by the algorithm shown is Section 2.5.3.

A procedure for computing the prime dichotomies is the following [29]. A set P is initialized with the set of seed dichotomies. All elements of P are initially unmarked. The set is updated by adding unmarked dichotomies representing the union of compatible dichotomy pairs in the set P. Those dichotomies that are covered are marked. The process is iterated until the set P stabilizes. The subset of P corresponding to unmarked dichotomies is the set of prime dichotomies.

Example 7.5.8. Consider again the constraint matrix for Example 7.5.3 and the seed dichotomies of Example 7.5.7. The union of seeds $\{s_1, s_2, s_3, s_4\}$ yields prime ($\{AND, OR\}$; $\{JMP, ADD\}$). The union of seeds $\{s_5, s_6\}$ yields prime ($\{OR, JMP\}$; $\{AND, ADD\}$). The union of seeds $\{s_3, s_5\}$ yields prime ($\{OR, JMP, ADD\}$; $\{AND\}$). The union of seeds $\{s_2, s_6\}$ yields prime ($\{AND, OR, JMP\}$; $\{ADD\}$).

Therefore the prime dichotomies (labeled $p_1 - p_4$) are:

p_1	($\{AND,OR\}$	;	$\{JMP,ADD\}$)
p_2	($\{OR,JMP\}$	;	$\{AND,ADD\}$)
p_3	($\{OR,JMP,ADD\}$	;	$\{AND\}$)
p_4	($\{AND,OR,JMP\}$	;	$\{ADD\}$)

Hence the covering matrix is:

	s_1	s_2	s_3	s_4	s_5	s_6
p_1	1	1	1	1	0	0
p_2	0	0	0	0	1	1
p_3	0	0	1	0	1	0
p_4	0	1	0	0	0	1

A minimum cover is provided by p_1 and p_2, corresponding to the encoding:

$$\mathbf{E} = \begin{bmatrix} 1 & 0 \\ 1 & 1 \\ 0 & 0 \\ 0 & 1 \end{bmatrix}$$

A few heuristic methods have been proposed. We describe briefly a column-based encoding method [10]. The algorithm does not compute the seed or the prime dichotomies. It just uses the dichotomies associated with the rows of $\mathbf{A}$. At each step, maximal sets of compatible dichotomies are computed, and the one of largest cardinality is selected. An encoding column satisfying this set of dichotomies is derived and appended to $\mathbf{E}$. The rationale of the algorithm is to satisfy as many constraints as possible in a greedy fashion. Hence it does not guarantee the global minimality of the number of columns used. The algorithm uses also a reduction strategy on the set of constraints. Indeed, as the algorithm progresses and the codes are partially assigned, some constraints become partially solved. Hence some entries of $\mathbf{A}$ are replaced by *don't care* entries and the corresponding dichotomies are partitions of smaller subsets of S. We refer the interested reader to reference [10] for further details.

Example 7.5.9. Consider again the constraint matrix for Example 7.5.3. The dichotomies (labeled $d_1 - d_3$) associated with $\mathbf{A}$ are:

d_1	({AND,OR}	;	{JMP,ADD})
d_2	({JMP,ADD}	;	{AND,OR})
d_3	({OR,JMP}	;	{AND,OR})

The first two are compatible, and the corresponding constraints can be satisfied by the encoding column $[1100]^T$. The second column is then chosen to be $[0110]^T$ to satisfy the remaining constraint.

7.5.2 Output Encoding*

The output encoding problem consists of choosing codes for the outputs of a Boolean function that minimize the cardinality of its minimum cover. Symbolic minimization is no more a straightforward application of mvi-minimization. Indeed, the output codes play a significant role in reducing the size of a function.

Consider the decoder of Example 7.5.1. When optimizing the symbolic table with mvi-minimizers, no implicant merging can take place while exploiting the implicant output parts, because output symbols are viewed as independent of each other. (Each output symbol is considered as an independent input in the positional-cube notation.) Instead, when covering relations among outputs can be used, the cover cardinality can be further reduced.

Example 7.5.10. Consider the instruction decoder of Example 7.5.1. The minimum cover could be reduced from six to four implicants, under the assumption that the code of $CNTD$ covers the codes of $CNTB$ and $CNTC$, namely:

100	1111	CNTA
011	1100	CNTB
011	0011	CNTC
001	0110	CNTD

Note that the apparent conflict between the fourth and both the second and third implicants is resolved by the covering relation. For example, the following codes could be used: $CNTA = 00$, $CNTB = 01$, $CNTC = 10$ and $CNTD = 11$. If the input codes for the ad-mode and op-codes are the same as in Example 7.5.3, the corresponding encoded cover is:

00	**	00
*1	0*	01
*1	1*	10
11	*1	11

Recall that each row of the encoded table is an implicant whose output part implies a TRUE value of each scalar output when the corresponding entry is 1. Thus, for any given input pattern, the value of the outputs is the disjunction of the output parts of the asserted implicants. Moreover, the first implicant is void and can be dropped from the table.

Some multiple-valued logic minimization techniques have been proposed [22] that optimize functions whose inputs and outputs are multiple valued. Thus they are more general in nature than the mvi-minimization described before in this chapter. In the common multiple-valued logic model, logic values are represented by integers and covering relations correspond to comparisons of integers. Therefore, multiple-valued logic minimizers can be applied to the symbolic minimization problem by assigning an integer value to each symbol. Since multiple-valued logic minimizers exploit covering relations among the implicant output parts, smaller covers may be obtained as compared to those achieved by mvi-minimizers.

Example 7.5.11. Consider again the instruction decoder of Example 7.5.1. Assign values 0,1,2,3 to $CNTA, CNTB, CNTC$ and $CNTD$, respectively. Then a multiple-valued minimizer could be invoked and would yield the result of Example 7.5.10. Note that a different integer assignment would yield a different result.

The multiple-valued minimization approach has two pitfalls when applied to minimization of symbolic functions. First, the assignment of symbols to integers affects their covering relation and the final result. The covering relation among symbols should be derived as a result of the minimization step and not prescribed before (as done implicitly by coding symbols with integers). Second, the disjunction of two binary codes may be used to represent a third one. Hence, the simultaneous assertion of two implicants may be made to mean the assertion of an output different from those of the corresponding output parts. This has no counterpart in multiple-valued minimization.

Example 7.5.12. Consider the alternative cover of the previous example:

100	1111	CNTA
010	1100	CNTB
010	0011	CNTC
001	1110	CNTB
001	0111	CNTC

Note that no implicant implies output $CNTD$. However, this output could be specified as the disjunction of the codes of $CNTB$ and $CNTC$. Hence, the simultaneous assertion of the last two implicants would obviate the problem. The following codes could be used again: $CNTA = 00$, $CNTB = 01$, $CNTC = 10$ and $CNTD = 11$.

The output encoding problem requires solving first symbolic minimization and then an encoding problem. In this case, as well as for solving the combined input and output encoding problems, symbolic minimization differs from multiple-valued logic minimization and is indeed a novel problem. Symbolic minimization optimizes a cover while exploiting *covering* and *disjunctive* relations among codes, as described above. Both exact and heuristic methods for symbolic minimization have been proposed, whose detailed description goes beyond the scope of this book. We summarize the approaches next and we refer the reader to [1] and [10] for details.

Exact symbolic minimization, as originally proposed by Ashar *et al.* [1], consists first in computing a set of *generalized prime implicants* and then in solving a constrained covering problem. The generalized prime implicants have to take into account the disjunctive relations. Hence two implicants, whose input parts have distance 1 but assert different symbolic outputs, can be merged and assert the union of the outputs. Similarly, an implicant is covered by another one only if the input part of the former is covered by the latter and the output parts are the same. Generalized prime implicants can be computed by iterating the merging of pairs of implicants with distance 1. A more efficient approach can be used, by realizing that the set of generalized primes is in one-to-one correspondence with the primes of an auxiliary bv-function, obtained by replacing the symbols by 0-hot codes (i.e., symbol i is encoded by a string of 1s, with a 0 in the ith position) [1]. Therefore efficient prime generation algorithms for bv-functions can be used to achieve a reduced prime implicant table, such as that used by Espresso-exact.

In exact symbolic minimization, the covering of the minterms must take into account the encodability constraints, i.e., the fact that a minimum subset of the generalized primes must cover the function while exploiting the covering and disjunctive relations. The unate covering algorithm (of Section 2.5.3) can be applied with the following modifications. First, covering between primes is restricted to the case in which they have the same output part. This affects the dominance relations. Second, the declaration of the "best solution seen so far" is subject to the satisfaction of a set of constraints. Details are reported in reference [1].

A heuristic method for symbolic minimization was implemented by program Cappuccino [10]. This approach exploits only the covering relations and involves invocations of an mvi-minimizer, namely Espresso, as many as the number of output symbols. Each mvi-minimization minimizes the *on set* of an output symbol, with a special *dc set* that includes the *on set* of other symbolic outputs. This allows the cover to be reduced in size while exploiting points of the *on set* of other symbols. When this occurs, these symbols are required to cover the current symbol to preserve the original meaning of the representation.

Symbolic minimization yields a minimum (or minimal) cover and a partial order, representing a covering relation. In addition, some symbolic minimizers yield a set of disjunctive relations (with possibly nested conjunctive relations [1]). The encoding of symbolic outputs must then exploit this information.

ENCODING ALGORITHMS. We consider the encoding of a set of output symbols S, with $n_s = |S|$. The *covering relations* can be modeled by a binary constraint matrix $\mathbf{B} \in \mathbf{B}^{n_s \times n_s}$ that is the adjacency matrix of the directed graph modeling the partial order of the covering relations. The *disjunctive relations* can be modeled analogously.

> **Example 7.5.13.** The constraint matrix for the problem instance of Example 7.5.10 is a set of covering relations. Namely, the fourth symbol ($CNTD$) covers the second ($CNTB$) and the third ($CNTC$) ones. Thus:
>
> $$\mathbf{B} = \begin{bmatrix} 0 & 0 & 0 & 0 \\ 0 & 0 & 0 & 0 \\ 0 & 0 & 0 & 0 \\ 0 & 1 & 1 & 0 \end{bmatrix}$$

An encoding matrix $\mathbf{E}$ satisfies the output encoding constraints if, for each column $\mathbf{e}$ of $\mathbf{E}$, the covering and disjunctive relations are satisfied bitwise. Thus satisfaction of output constraints is a property of each column of $\mathbf{E}$. A valid encoding matrix $\mathbf{E}$ is a set of columns, each satisfying the output constraints and providing a set of distinct codes (i.e., rows) for the symbols.

In practice, the output encoding problem is often considered in conjunction with an input encoding problem. Two examples are noteworthy. First, consider the design of two cascaded logic blocks, each modeled by a two-level cover and whose interconnection can be described in terms of symbolic data, as shown in Figure 7.15. Second, think of the encoding of the feedback signals of a finite-state machine, i.e., of its states. In both cases, a set of symbols S must be encoded while satisfying input and output constraints simultaneously. Note that input and output constraints may not be satisfiable simultaneously for some problem instance. Conditions for their combined satisfaction are reported in references [10], [1] and [25]. Some symbolic minimizers take these conditions into account and yield simultaneously satisfiable input and output constraints.

Dichotomy-based exact and heuristic encoding algorithms have been studied. The properties of the dichotomies must be redefined in the framework of input and output encoding. Output encoding exploits covering relations that are antisymmetric. Thus, the validity of an encoding is no more preserved by column complementation.

As a result, all definitions related to dichotomies (introduced for the input encoding problem) must be modified to be used for the input and output encoding problem. For example, two dichotomies $(L_1; R_1)$ and $(L_2; R_2)$ are compatible if and only if $L_1 \cap R_2 = \emptyset$ and $R_1 \cap L_2 = \emptyset$. Thus a dichotomy is no longer compatible with the dichotomy obtained by interchanging its blocks. Similarly a dichotomy (L_1, R_1) covers another dichotomy (L_2, R_2) if and only if $L_1 \supseteq L_2$ and $R_1 \supseteq R_2$. Dichotomies are said to satisfy the output constraints if the corresponding encoding satisfies the specified covering and disjunctive constraints. Prime dichotomies are restricted to those satisfying the output constraints and their set is constructed while considering dichotomies satisfying the constraints as well.

The following exact input and output encoding algorithm is an extension of the exact input encoding algorithm described previously. We outline briefly the major differences, and we refer the interested reader to reference [25] for details. All prime dichotomies are computed first. Whereas satisfaction of the output constraints is ensured by using the primes, satisfaction of input constraints is guaranteed by solving the covering problem. In the covering problem, we consider seed dichotomy pairs, by adding to each seed dichotomy (as defined in Section 7.5.1) the seed dichotomy with the blocks interchanged. At least one element of each seed pair must be covered. If neither dichotomy in any pair can be covered by any prime, then the input and output constraints are not simultaneously satisfiable. When contrasting the covering formulation of this problem to that of the input encoding problem, note that seed pairs are required by the redefinition of dichotomy covering. In both cases the covering problem is unate.

A heuristic algorithm for input and output encoding was implemented in program CAPPUCCINO [10]. It is an extension of the algorithm described in Section 7.5.1. The encoding matrix **E** is constructed column by column, where each column is the encoding corresponding to the largest set of maximally compatible dichotomies that satisfy the output constraints. The dichotomies that the algorithm considers are those associated with each row of the constraint matrix **A** and those obtained by exchanging the blocks.

Example 7.5.14. Consider two cascaded circuits, as shown in Figure 7.15. The former is the instruction decoder of Example 7.5.10 and the latter is another circuit, called "second stage."

We assume that the signals that link the two circuits, called "control" in Figure 7.15, can be encoded arbitrarily. Hence we want to choose an encoding of the "control" signals, corresponding to a minimal two-level cover of both the instruction decoder and the second stage.

We apply then symbolic minimization to the covers representing both circuits. In particular, we need to solve an output encoding problem for the first circuit and an input encoding problem for the second. For the sake of the example, assume that symbolic minimization of the "second stage" yields a constraint matrix:

$$\mathbf{A} = \begin{bmatrix} 1 & 1 & 0 & 0 \\ 0 & 1 & 0 & 1 \end{bmatrix}$$

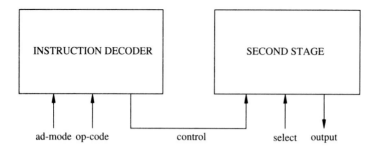

FIGURE 7.15
Two cascaded circuits.

The output covering constraints of the instruction decoder, derived in Example 7.5.10, are represented by:

$$\mathbf{B} = \begin{bmatrix} 0 & 0 & 0 & 0 \\ 0 & 0 & 0 & 0 \\ 0 & 0 & 0 & 0 \\ 0 & 1 & 1 & 0 \end{bmatrix}$$

In words, the output encoding constraints mean that the code of $CNTD$ must cover those of $CNTB$ and $CNTC$. The seed dichotomy pairs associated with **A** are:

s_{1A}	({ CNTA,CNTB } ; { CNTC })
s_{1B}	({ CNTC } ; { CNTA,CNTB })
s_{2A}	({ CNTA,CNTB } ; { CNTD })
s_{2B}	({ CNTD } ; { CNTA,CNTB })
s_{3A}	({ CNTB,CNTD } ; { CNTA })
s_{3B}	({ CNTA } ; { CNTB,CNTD })
s_{4A}	({ CNTB,CNTD } ; { CNTC })
s_{4B}	({ CNTC } ; { CNTB,CNTD })

The seed dichotomies s_{2A}, s_{3B} and s_{4B} are not compatible with **B**, because the corresponding encoding bit for $CNTD$ is 0.

Prime dichotomies that are compatible with **B** are:

p_1	({ CNTC,CNTD } ; { CNTA,CNTB })
p_2	({ CNTB,CNTD } ; { CNTA,CNTC })
p_3	({ CNTA,CNTB,CNTD } ; { CNTC })

where p_1 covers s_{1B} and s_{2B}, p_2 covers s_{3A} and s_{4A} and p_3 covers s_{1A} and s_{4A}. Hence p_1 and p_2 cover at least one element of all seed pairs, corresponding to the encoding:

$$\mathbf{E} = \begin{bmatrix} 0 & 0 \\ 0 & 1 \\ 1 & 0 \\ 1 & 1 \end{bmatrix}$$

The heuristic algorithm would consider instead the dichotomy pairs associated with each row of **A**, namely:

$$d_{1A} \quad (\{ \text{CNTC,CNTD} \} \quad ; \quad \{ \text{CNTA,CNTB} \})$$
$$d_{1B} \quad (\{ \text{CNTA,CNTB} \} \quad ; \quad \{ \text{CNTC,CNTD} \})$$
$$d_{2A} \quad (\{ \text{CNTB,CNTD} \} \quad ; \quad \{ \text{CNTA,CNTC} \})$$
$$d_{2B} \quad (\{ \text{CNTA,CNTC} \} \quad ; \quad \{ \text{CNTB,CNTD} \})$$

After having deleted the dichotomies that do not satisfy the output constraints (i.e., $\{d_{1B}, d_{2B}\}$), a solution is formed with the remaining ones (i.e., $\{d_{1A}, d_{2A}\}$). Since these two dichotomies are incompatible, they are considered one at a time, yielding two columns of the encoding matrix. The result is the same as the one before.

7.5.3 Output Polarity Assignment

The output polarity assignment problem is a form of output encoding for functions with binary-valued outputs. The goal is to decide whether it is convenient or not to implement each scalar output with the positive or negative polarity, i.e., in the uncomplemented or complemented form.

Consider first the case of a single-output function. Then, if we are free to choose the output polarity, we can minimize the *on set* and the *off set* of the function and choose the implementation with the smallest size. This approach becomes impractical for m-output functions, because there are 2^m polarity assignments and we would need an exponential number of logic minimizations.

A heuristic approach was proposed by Sasao [24]. Given a function **f** with m outputs, an auxiliary function **f** with $2m$ outputs is constructed that is called a *double-phase characteristic function*. The first m scalar components of **ff** are the same as those of **f**; the second m are their complements. The double-phase characteristic function is then minimized using a heuristic or exact minimizer. Then the minimized cover $\mathbf{ff}_{min}$ is examined, in the search for the minimum set of product terms that cover each scalar component, either in the complemented or uncomplemented form.

The computation of the minimum set of product terms can be done by using a *covering expression* that is reminiscent of Petrick's method. Each product term of $\mathbf{ff}_{min}$ is labeled by a variable. A scalar component of the original function **f** is correctly implemented if either all product terms that cover it are selected or all product terms that cover its complement are selected. This can be expressed by a sum of two products of the variables representing the product terms of $\mathbf{ff}_{min}$. Since this argument has to hold for all scalar components, the overall covering expression can be written as a *product of sums of products* form. By multiplying out this expression, we can obtain a *sum of products* expression, where a cube with fewest literals indicates the product terms of $\mathbf{ff}_{min}$ that solve our problem. This method is viable for small values of m. Otherwise, the corresponding covering problem must be solved by other (heuristic) algorithms.

Example 7.5.15. Let us consider a 2-bit adder function [24]. It has four bv-inputs and three outputs f_0, f_1, f_2. Hence the double-phase characteristic function has six outputs

f_0, f_1, f_2, f_0', f_1', f_2'. We report here the output part of a cover of $\mathbf{ff}_{min}$; its complete derivation is reported by Sasao [24].

	f_0	f_1	f_2	f_0'	f_1'	f_2'
α	1	0	0	0	0	0
β	0	1	0	1	0	0
γ	0	1	0	0	0	0
δ	1	0	0	0	1	0
ϵ	0	0	1	0	0	0
ζ	0	0	0	1	0	0
η	0	0	0	0	1	0
θ	0	0	0	0	0	1

To implement the first output (in either polarity), terms α and δ or β and ζ must be selected. This can be written as $\alpha\delta + \beta\zeta$ and similarly for the other outputs. Hence, the covering expression is:

$$(\alpha\delta + \beta\zeta) \cdot (\beta\gamma + \delta\eta) \cdot (\epsilon + \theta)$$

which, when multiplied out, yields:

$$\alpha\beta\gamma\delta\epsilon + \alpha\beta\gamma\delta\theta + \alpha\delta\epsilon\eta + \alpha\delta\eta\theta + \beta\gamma\epsilon\zeta + \beta\gamma\zeta\theta + \beta\delta\epsilon\zeta\eta + \beta\delta\zeta\eta\theta \; = \; 1$$

Hence, a solution in terms of a minimal number of product terms can be achieved by choosing, for example, $\alpha\delta\epsilon\eta$. This choice corresponds to implementing f_0, f_1', f_2, i.e., the second output has the negative polarity. This would save one product term with respect to the solution where no polarity assignment is used.

7.6   MINIMIZATION OF BOOLEAN RELATIONS*

Boolean relations are generalizations of Boolean functions, where each input pattern may correspond to more than one output pattern. As a result, a relation is specified by a subset of the Cartesian product of the input and output Boolean spaces. In general, Boolean relations are useful in specifying combinational circuits that can be implemented by different logic functions. Thus Boolean relations provide a more general interpretation of the *degrees of freedom* in logic optimization than *don't care* conditions.

Example 7.6.1.   This example is borrowed from the original paper on Boolean relations by Brayton and Somenzi [3]. Consider two interconnected circuits. The former is a 2-bit adder that takes four inputs $\{a_0, a_1, b_0, b_1\}$ and yields three outputs $\{x_0, x_1, x_2\}$. The latter circuit is a comparator, whose inputs are the outputs of the adder and whose outputs $\{z_0, z_1\}$ encode the result of a comparison of the sum with the constants 3 and 4, as shown in Figure 7.16. Namely the comparator's output denotes if the sum, represented in vector form as $\mathbf{x}$, encodes an integer smaller than 3, equal to either 3 or 4 or larger than 4.

Consider now the adder circuit, and assume that its output $\mathbf{x}$ is observed by the comparator only. Thus all patterns of $\mathbf{x}$ that encode integers smaller than 3 are equivalent

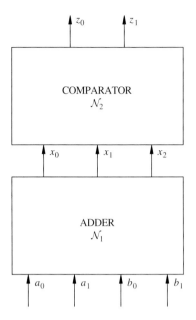

FIGURE 7.16
Example of cascaded circuits. The adder can be specified by
a Boolean relation.

as far as producing the comparator's output. Similar considerations apply to those patterns
encoding integers larger than 4 and to those encoding 3 and 4. The Boolean space spanned
by $\mathbf{x}$ can then be partitioned into three equivalent classes. The adder can be expressed
by the following relation table:

a_1	a_0	b_1	b_0	$\mathbf{x}$
0	0	0	0	{ 000, 001, 010 }
0	0	0	1	{ 000, 001, 010 }
0	0	1	0	{ 000, 001, 010 }
0	1	0	0	{ 000, 001, 010 }
1	0	0	0	{ 000, 001, 010 }
0	1	0	1	{ 000, 001, 010 }
0	0	1	1	{ 011, 100 }
0	1	1	0	{ 011, 100 }
1	0	0	1	{ 011, 100 }
1	0	1	0	{ 011, 100 }
1	1	0	0	{ 011, 100 }
0	1	1	1	{ 011, 100 }
1	1	0	1	{ 011, 100 }
1	0	1	1	{ 101, 110, 111 }
1	1	1	0	{ 101, 110, 111 }
1	1	1	1	{ 101, 110, 111 }

The more general specification of the adder by means of a Boolean relation allows
us to optimize it while taking advantage of the specific interconnection between the
adder and the comparator, which reduces the observability of the internal signals $\mathbf{x}$. The

corresponding wider degrees of freedom in optimizing the adder cannot be expressed by *don't care* conditions on variables.

The optimization problem consists of finding a minimum (or minimal) cover of a Boolean function to which this relation table can be reduced. As an example, we show next a minimum cover derived by a minimizer. (The reduction steps are not obvious at a first glance.)

a_1	a_0	b_1	b_0	x
0	*	1	*	010
1	*	0	*	010
1	*	1	*	100
*	*	*	1	001
*	1	*	*	001

It is interesting to note that an integer output value of the adder can be encoded differently according to the summands. For example, integer 4 may be computed as $2 + 2$, whose corresponding binary input patterns assert the third implicant and yield output 100. On the other hand, integer 4 may be computed as $1 + 3$ $(3 + 1)$, whose corresponding binary input patterns assert the first and fourth (second and fifth) implicants. The corresponding encodings of 4 is thus 011. Note that both encodings of 4 are equivalent, because we observe only the comparator's output.

A Boolean relation specifies implicitly many Boolean functions, i.e., all those functions whose output patterns are among those specified by the relation. Such functions are said to be *compatible* with the relation. Since there are many compatible functions, deriving and minimizing all of them is not an efficient method for determining the best implementation. Therefore, exact (and heuristic) methods for finding directly the minimum (or minimal) Boolean function compatible with a Boolean relation have been proposed.

We consider first exact optimization, which follows the strategy of exact minimization of Boolean functions.

Definition 7.6.1. A **candidate prime** (c-prime) of a Boolean relation is a prime of a function compatible with the relation.

Quine's theorem can be extended to Boolean relations and c-primes. Indeed, under the usual assumption that the cost of an implementation of an implicant increases with its literal count, a minimum cover can be found in terms of c-primes only.

Methods for generating all c-primes and weeding out redundant ones are reported by Brayton and Somenzi [3]. They do not require deriving all primes of the compatible functions. Once c-primes have been computed, we need to search for a c-prime subset of minimum cardinality that covers one function compatible with the relation.

To analyze the covering problem related to the optimization of Boolean relations, we compare the covering expression of Petrick's method in the cases of Boolean function and Boolean relation minimization. Recall how a covering expression for a Boolean function is derived: for each minterm, a clause is formed that is the disjunction of variables associated with all primes that cover it. The covering expression is the product of all clauses and is unate in all variables.

When considering Boolean relations, for each specified input pattern (i.e., minterm) we create a clause that states which combination of primes would yield a correct output. Different from the case of Boolean functions, the selection of a c-prime may imply the selection of another c-prime, as shown by the following example. This implication may be represented by a clause with a complemented variable. The covering expression is again the product of all clauses, and it may be binate in some variables. In other words, the exact Boolean relation minimization problem differs from the Boolean function optimization problem in that it requires the solution of a *binate covering* problem, rather than a unate covering problem. Binate covering can be solved exactly by the branch-and-bound algorithm described in Section 2.5.3, but it is in general harder to solve than unate covering.

Example 7.6.2. We consider a Boolean relation simpler than the one of Example 7.6.1, for the sake of describing the covering process. Note that in general the outputs of a Boolean relation need not represent a partition of the output Boolean space into equivalence classes. The Boolean relation is described by the following truth table, where a 1 in the output part of a minterm implies that the corresponding output is TRUE.

0	0	0	{ 00 }
0	0	1	{ 00 }
0	1	0	{ 00 }
0	1	1	{ 10 }
1	0	0	{ 00 }
1	0	1	{ 01 }
1	1	0	{ 00,11 }
1	1	1	{ 00,11 }

There are four Boolean functions that can be derived by taking either choice of output pattern in the last two rows of the table, in addition to the other implicants, namely:

			f^A	f^B	f^C	f^D
1	1	0	00	11	00	11
1	1	1	00	00	11	11

The c-primes are the following:

α	0	1	1	10
β	1	0	1	01
γ	1	1	0	11
δ	1	1	1	11
ϵ	*	1	1	10
ζ	1	*	1	01
η	1	1	*	11

The c-prime computation can be performed directly from the relation table, without considering the compatible functions separately [3]. In practice, the compatible functions need not even be derived explicitly. We do not describe the computation of the c-primes, but we justify their presence by considering the compatible functions.

C-primes α and β are primes of f^A and f^B. C-primes ϵ and ζ are primes of f^C, while c-primes ϵ, ζ and η are primes of f^D. C-primes γ and δ are primes of f^B and f^C, respectively; they are subsumed by η and can be discarded.

Consider now the covering problem. Take input pattern 011. The corresponding output pattern in the relation table is 10, which can be covered by either α or ϵ. The corresponding clause is then $\alpha + \epsilon$. A similar consideration applies to input pattern 101, whose corresponding clause is then $\beta + \zeta$.

When considering pattern 110, the corresponding output is 00 or 11. Thus implicant η may or may not be in the cover, yielding the neutral clause $\eta + \eta'$, which can be disregarded.

Consider now input pattern 111. The corresponding output pattern can be 00 or 11. If no implicant is selected, the output is 00 and it is correct. If either implicant η or both ϵ and ζ are selected, then the output is 11 and it is correct. However, if only one of either ϵ or ζ is selected, the output is 10 or 01, which is not correct. Hence the clause is $\eta + \epsilon\zeta + \epsilon'\zeta'$, which can be rewritten as $(\epsilon + \zeta' + \eta) \cdot (\epsilon' + \zeta + \eta)$.

The overall covering clause is:

$$(\alpha + \epsilon) \cdot (\beta + \zeta) \cdot (\epsilon + \zeta' + \eta) \cdot (\epsilon' + \zeta + \eta)$$

The corresponding covering problem is binate. A solution is given by selecting c-primes $\{\alpha, \beta\}$, yielding the following cover (corresponding to a minimum cover of f^A):

$$
\begin{array}{ccc|c}
0 & 1 & 1 & 10 \\
1 & 0 & 1 & 01
\end{array}
$$

Heuristic minimization of Boolean relations parallels heuristic minimization of Boolean functions [28]. Three operators are applied to a compatible function representation in an iterative fashion until the corresponding cost decreases. The operators *expand*, *reduce* and *irredundant* are used in this perspective with some modifications. The operators manipulate a representation of a compatible function, while keeping it consistent with the Boolean relation and exploiting the corresponding degrees of freedom. This iterative technique can cope also with multiple-valued relations that provide a powerful modeling paradigm for several problems in logic design.

7.7 PERSPECTIVES

Minimization of logic functions in two-level form is a classic problem of switching theory, probably one of the oldest [15]. Nevertheless, it has been an exciting playground of research in recent years. Heuristic logic minimization reached a plateau in the early 1980s with program ESPRESSO, which provides near-minimum solutions to most problems of practical size. Thus ESPRESSO and its derivatives are used routinely to solve different types of problems that arise in logic design and not restricted to two-level implementations.

Exact logic minimization was thought of as impractical until a decade ago. The increased speed and memory of engineering workstations, as well as the development of efficient exact covering algorithms and data structures, brought renewed interest to this problem. Despite the success of ESPRESSO-EXACT in minimizing exactly many large examples, few cases remained unsolved. A couple of recent breakthroughs have made exact minimization practical for large-scale benchmark examples, where scale is measured in terms of primes and minterms: first, the implicit enumeration of the primes [6] and the computation of reduced implicant tables with problem transformations that defeat the complexity of evaluating dominance on large-scale problems [7]; second, the use of signature cubes that allow us to represent groups of primes implicitly and thus limit the prime generation to those necessary to solve the covering problem [17]. Both these new, exciting results have opened the possibility of solving exactly many problems that can be modeled in the framework of two-level minimization.

Extensions of two-level minimization to other more general models have been shown to be useful and practical. Namely, symbolic optimization techniques are commonly used in connection with encoding problems, e.g., finite-state machine state encoding. Boolean relations, and their minimization, provide a powerful paradigm for optimizing embedded circuits and find many applications in multi-level logic optimization, as shown in the following chapter.

7.8 REFERENCES

Two-level logic optimization finds its roots in the early work on classic switching theory, well documented in several textbooks [14, 16]. Exact optimization algorithms were proposed by Quine [21] and by McCluskey [15]. Several variations on these methods have been proposed recently. Rudell and Sangiovanni [23] perfected exact optimization techniques and implemented them in program ESPRESSO-EXACT. Dagenais *et al.* [8] developed program MCBOOLE. Coudert and Madre [6] proposed a method to represent all prime implicants implicitly, and Coudert *et al.* [7] presented a minimization algorithm with an implicit representation of the prime implicant table that could minimize functions with very large sets of primes. Nguyen *et al.* [19] suggested a method for representing all primes covering a minterm that was used later by McGeer *et al.* [17] in a revolutionary minimizer that avoids the computation of all primes.

Heuristic logic minimization is also a rich field. The seminal work by Hong *et al.* [13] at IBM Research led to program MINI, the archetype of modern heuristic logic minimizers. The algorithms of program ESPRESSO, which has greatly improved the capabilities of MINI and is the most used minimizer, were jointly developed by Brayton, Hachtel, McMullen, Rudell and Sangiovanni [4, 23]. The recursive paradigm exploited previous ideas developed independently by Morreale [18], Reusch [20], Sasao [24] and others, but the unate heuristics was used first in ESPRESSO [4].

Symbolic minimization and heuristic encoding algorithms were developed first in conjunction with the solution of the state assignment problem for finite-state machines [9, 10]. Later, exact minimization and encoding techniques were proposed by Ashar *et al.* [1] and Villa and Sangiovanni [27]. The theory of dichotomies was applied in this context by Ciesielski *et al.* [5], Yang and Ciesielski [29] and Saldanha *et al.* [25] with slightly different notation.

Boolean relations and their optimization were introduced first by Brayton and Somenzi [3], who proposed an exact optimization technique. Watanabe and Brayton [28] explored heuristic methods for their optimization.

1. P. Ashar, S. Devadas and A. Newton, *Sequential Logic Synthesis*, Kluwer Academic Publishers, Boston, MA, 1992.

2. K. A. Bartlett, R. K. Brayton, G. D. Hachtel, R. M. Jacoby, C. R. Morrison, R. L. Rudell, A. Sangiovanni-Vincentelli and A. R. Wang, "Multilevel Logic Minimization Using Implicit Don't Cares," *IEEE Transactions on CAD/ICAS*, Vol. CAD-7, No. 6, pp. 723–740, June 1988.

3. R. Brayton and F. Somenzi, "An Exact Minimizer for Boolean Relations," *ICCAD, Proceedings of the International Conference on Computer Aided Design*, pp. 316–319, 1989.

4. R. Brayton, G. Hachtel, C. McMullen and A. Sangiovanni-Vincentelli, *Logic Minimization Algorithms for VLSI Synthesis*, Kluwer Academic Publishers, Boston, MA, 1984.

5. M. Ciesielski, J. Shen and M. Davio, "A Unified Approach to Input-Output Encoding for FSM State Assignment," *DAC, Proceedings of the Design Automation Conference*, pp. 176–181, 1991.

6. O. Coudert and J. Madre, "Implicit and Incremental Computation of Primes and Essential Primes of Boolean Functions," *DAC, Proceedings of the Design Automation Conference*, pp. 36–39, 1992.

7. O. Coudert, J. Madre and H. Fraisse, "A New Viewpoint on Two-Level Logic Minimization," *DAC, Proceedings of the Design Automation Conference*, pp. 625–630, 1993.

8. M. Dagenais, V. Agarwal and N. Rumin, "McBOOLE: A New Procedure for Exact Logic Minimization," *IEEE Transactions on CAD/ICAS*, Vol. CAD-5, No. 1, pp. 229–223, January 1986.

9. G. De Micheli, R. Brayton and A. Sangiovanni-Vincentelli, "Optimal State Assignment for Finite State Machines," *IEEE Transactions on CAD/ICAS*, Vol. CAD-4, No. 3, pp. 269–284, July 1985.

10. G. De Micheli, "Symbolic Design of Combinational and Sequential Logic Circuits Implemented by Two-level Logic Macros," *IEEE Transactions on CAD/ICAS*, Vol. CAD-5, No. 4, pp. 597–616, October 1986.

11. D. Dietmeyer, *Logic Design of Digital Systems*, Allyn and Bacon, Needham High, MA, 1971.

12. E. Fabricius, *Introduction to VLSI Design*, McGraw-Hill, New York, 1990.

13. S. Hong, R. Cain and D. Ostapko, "MINI: A Heuristic Approach for Logic Minimization," *IBM Journal of Research and Development*, Vol. 18, pp. 443–458, September 1974.

14. F. Hill and G. Peterson, *Switching Theory and Logical Design*, Wiley, New York, 1981.

15. E. McCluskey, "Minimization of Boolean Functions," *The Bell System Technical Journal*, Vol. 35, pp. 1417–1444, November 1956.

16. E. McCluskey, *Logic Design Principles*, Prentice-Hall, Englewood Cliffs, NJ, 1986.

17. P. McGeer, J. Sanghavi, R. Brayton and A. Sangiovanni-Vincentelli, "ESPRESSO-SIGNATURES: A New Exact Minimizer for Logic Functions," *DAC, Proceedings of the Design Automation Conference*, pp. 618–621, 1993.

18. E. Morreale, "Recursive Operators for Prime Implicant and Irredundant Normal Form Determination," *IEEE Transactions on Computers*, Vol. C-19, No. 6, pp. 504–509, June 1970.

19. L. Nguyen, M. Perkowski and N. Golstein, "Palmini-fast Boolean Minimizer for Personal Computers," *DAC, Proceedings of the Design Automation Conference*, pp. 615–621, 1987.

20. B. Reusch, "Generations of Prime Implicants from Subfunctions and a Unifying Approach to the Covering Problem," *IEEE Transactions on Computers*, Vol. C-24, No. 9, pp. 924–930, September 1975.

21. W. Quine, "The Problem of Simplifying Truth Functions," *American Mathematical Monthly*, Vol. 59, pp. 521–531, 1952.

22. D. Rine, *Computer Science and Multiple-Valued Logic*, North-Holland, Amsterdam, 1977.

23. R. Rudell and A. Sangiovanni-Vincentelli, "Multiple-valued Minimization for PLA Optimization," *IEEE Transactions on CAD/ICAS*, Vol. CAD-6, No. 5, pp. 727–750, September 1987.

24. T. Sasao, "Input-Variable Assignment and Output Phase Optimization of Programmable Logic Arrays," *IEEE Transactions on Computers*, Vol. C-33, pp. 879–894, October 1984.

25. A. Saldanha, T. Villa, R. Brayton and A. Sangiovanni-Vincentelli, "A Framework for Satisfying Input and Output Encoding Constraints," *DAC, Proceedings of the Design Automation Conference*, pp. 170–175, 1991.

26. D. Scherz and G. Metze, "A New Representation for Faults in Combinational Digital Circuits," *IEEE Transactions on Computers*, Vol. C-21, pp. 858–866, August 1972.

27. T. Villa and A. Sangiovanni-Vincentelli, "NOVA: State Assignment for Finite State Machines for Optimal Two-level Logic Implementation," *IEEE Transactions on CAD/ICAS*, Vol. C-9, No. 9, pp. 905–924, September 1990.

28. Y. Watanabe and R. Brayton, "Heuristic Minimization of Multiple-valued Relations," *IEEE Transactions on CAD/ICAS*, Vol. CAD-12, No. 10, pp. 1458–1472, October 1993.

29. S. Yang and M. Ciesielski, "Optimum and Suboptimum Algorithms for Input Encoding and Its Relationship to Logic Minimization," *IEEE Transactions on CAD/ICAS*, Vol. CAD-10, No. 1, pp. 4–12, January 1991.

7.9 PROBLEMS

1. Consider a single-output bv-function f and its cover F. Show that if F is unate in a variable, so is f.

2. Consider an mvi-function f and its cover F. Show that if F is weakly unate in a variable, so is f.

3. Consider an mvi-function f that is strongly unate. Show that its complement is also a strongly unate function. Specialize this result to bv-functions. Show, then, by counterexample, that the complement of a weakly unate function may not be unate.

4. Consider the function whose *on set* is $F^{ON} = ab'c' + a'bc' + a'bc$ and whose *dc set* is $F^{DC} = abc'$. Represent the *on set* and *dc set* in the positional-cube notation and compute the *off set* by using the sharp operator. Repeat the *off set* computation by using the disjoint sharp operator.

5. Consider the function $f = ab'c' + a'bc' + a'bc$. Determine whether the f contains cube bc by checking the tautology of the cofactor. Use covers in the positional-cube notation and show all steps. Repeat the containment check for cube $a'b$.

6. Show that $f' = x \cdot f'_x + x' \cdot f'_{x'}$.

7. Show that all primes of a unate function are essential.

8. Design a recursive algorithm to check if an expanded implicant is covered by $F^{ON} \cup F^{DC}$.

9. Consider an mvi-function f and its cover F. Show that for any p-valued variable x:

$$supercube(F') = supercube(\cup_{k=0}^{p-1} C(x^{\{k\}}) \cap supercube(F'_{x^{\{k\}}})).$$

10. Consider function $f = a'd' + a'b + ab' + ac'd$. Form a cover in the positional-cube notation and compute all primes and all essential primes using the methods outlined in Sections 7.3.4 and 7.4.4, respectively. Show all steps. Compare your results with Figure 7.5 (a).

11. Show how mvi-minimization can be done using a bv-minimizer by specifying an appropriate *don't care* set.

12. The Natchez Indian culture has four classes: *Suns, Nobles, Honoreds* and *Stinkards*. Within this culture, the allowed marriages and the resulting offspring are given by the following table:

Mother	Father	Offspring
Sun	Stinkard	Sun
Noble	Stinkard	Noble
Honored	Stinkard	Honored
Stinkard	Sun	Noble
Stinkard	Noble	Honored
Stinkard	Honored	Stinkard
Stinkard	Stinkard	Stinkard

The remaining marriages are not allowed. Represent the condition that yields a Stinkard offspring by a single symbolic implicant. Then form a minimal symbolic representation of the disallowed marriages. Label the offspring field as "none" in this case.

13. Consider an input encoding problem characterized by constraint matrix **A**. Show that a valid encoding matrix **E** remains valid for any permutation or complementation of its columns.

14. Consider the input encoding problem specified by the matrix:

$$
\mathbf{A} = \begin{bmatrix} 1 & 1 & 0 & 0 \\ 0 & 1 & 1 & 0 \\ 0 & 0 & 1 & 1 \\ 1 & 0 & 0 & 1 \end{bmatrix}
$$

where each column from left to right is associated with an element in the set (a, b, c, d). Find a minimum-length encoding $\mathbf{E}$ satisfying the constraint. Find a minimum-length encoding $\mathbf{E}$ satisfying the constraint and such that the code of b covers the code of c. Repeat the exercise for the input encoding problem specified by the matrix:

$$
\mathbf{A} = \begin{bmatrix} 1 & 1 & 0 & 1 & 0 & 1 & 0 & 0 \\ 0 & 1 & 0 & 1 & 0 & 0 & 0 & 0 \\ 1 & 0 & 0 & 0 & 0 & 1 & 0 & 0 \\ 0 & 0 & 0 & 1 & 0 & 1 & 1 & 1 \\ 0 & 0 & 0 & 0 & 0 & 1 & 0 & 1 \\ 0 & 1 & 1 & 0 & 0 & 0 & 0 & 0 \\ 1 & 0 & 0 & 0 & 1 & 0 & 0 & 0 \end{bmatrix}
$$

where each column from left to right is associated with an element in the set (a, b, c, d, e, f, g, h). Use your favorite method.

15. Consider the input encoding problem specified by the matrix:

$$
\mathbf{A} = \begin{bmatrix} 1 & 1 & 0 & 1 & 0 & 1 & 0 & 0 \\ 0 & 1 & 0 & 1 & 0 & 0 & 0 & 0 \\ 1 & 0 & 0 & 0 & 0 & 1 & 0 & 0 \\ 0 & 0 & 0 & 1 & 0 & 1 & 1 & 1 \\ 0 & 0 & 0 & 0 & 0 & 1 & 0 & 1 \\ 0 & 1 & 1 & 0 & 0 & 0 & 0 & 0 \\ 1 & 0 & 0 & 0 & 1 & 0 & 0 & 0 \end{bmatrix}
$$

Compute all seed and prime dichotomies. Formulate a covering table and solve it exactly. (You may want to write a computer program for this task.)

16. Consider the optimization of a logic unit performing $f_1 = a + b$; $f_2 = b + c$ under the assumption that the outputs feed only an OR gate. Write the Boolean relation that specifies the equivalent output patterns of the logic unit and determine a minimum (product-term and literal) compatible function. Is the minimum function unique?

CHAPTER
8

MULTIPLE-LEVEL
COMBINATIONAL
LOGIC
OPTIMIZATION

En la tardanza suele estar el peligro.
The danger is generally in the delay.

M. de Cervantes. Don Quixote.

8.1 INTRODUCTION

Combinational logic circuits are very often implemented as multiple-level networks of logic gates. The fine granularity of multiple-level networks provides us with several degrees of freedom in logic design that may be exploited in optimizing area and delay as well as in satisfying specific constraints, such as different timing requirements on different input/output paths. Thus multiple-level networks are very often preferred to two-level logic implementations such as PLAs.

The unfortunate drawback of the flexibility in implementing combinational functions as multiple-level networks is the difficulty of modeling and optimizing the networks themselves. Few exact optimization methods have been formulated [20, 35], but they suffer from high computational complexity. Exact optimization methods are not considered to be practical today, and they are not surveyed in this book. Instead, we describe some heuristic methods that have been shown effective in reducing the area and delay of large-scale networks. The overall level of understanding of multiple-level logic design is at a less mature stage than two-level logic optimization. However, the need of practical synthesis and optimization algorithms for multiple-level circuits has made this topic one of utmost importance in CAD.

We consider in this chapter multiple-level circuits that are interconnections of single-output combinational logic gates. We assume that the interconnection provides a unidirectional flow of signals from a set of primary inputs to a set of primary outputs. A consequence of the unidirectionality is the lack of feedbacks and thus the overall circuit implements strictly combinational functions. The abstract model for these multiple-level logic circuits is called a *logic network*, and it is described in detail in the next section.

Logic networks can be implemented according to different design styles that relate to the type of combinational logic gates being interconnected. Network implementations can be restricted to a single type of gate, e.g., NORs and/or NANDs with fixed (or bounded) fan-in. Alternatively, multiple-level circuits can be interconnections of logic gates that are instances of elements of a given *cell library*. This style is typical of standard-cell and array-based designs. Macro-cell implementations of networks can be in terms of gates that implement logic functions subject to some functional constraint (e.g., limitations on the fan-in or on the form of the logic expression representing the gate), because cell generators may be used to synthesize the physical view of the logic gates.

The desired design style affects the synthesis and optimization methods. Indeed the search for an interconnection of logic gates optimizing area and/or performance depends on the constraints on the choice of the gates themselves. Multiple-level logic optimization is usually partitioned into two tasks. First, a logic network is optimized while neglecting the implementation constraints on the logic gates and assuming loose models for their area and performance. Second, the constraints on the usable gates (e.g., those represented by the cell library) are taken into account as well as the corresponding detailed models. We consider in this chapter the former task only. Library binding will be presented in Chapter 10. It is important to remember that this two-stage optimization strategy is a heuristic way to cope with the overall problem complexity by breaking it into simpler subproblems at the expense of the quality of the solution. Nevertheless, the aforementioned strategy is commonly used and it is justified also by the fact that most subproblems in multiple-level optimization are solved approximatively by heuristic algorithms. An alternative approach that performs optimization and binding concurrently will be presented in Section 10.3.4.

The optimization of multiple-level logic networks has received a lot of attention. It would be impossible to report in detail on all different approaches that have been proposed. We restrict our attention to those optimization methods that are either very commonly used or representative of a class of techniques. The most general taxonomy of optimization methods is in terms of *algorithms* and *rule-based* methods. We defer the description of the latter ones to Section 8.7 and we devote the first part of this chapter to algorithms for area and delay optimization. In particular, we consider next (in Section 8.2) the modeling issues and we formulate the optimization problems. We present then in Sections 8.3 and 8.4 optimization algorithms based on the algebraic and Boolean properties of the network representation. The relations between circuit optimization and testability are surveyed in Section 8.5. Section 8.6 is devoted to algorithms for performance evaluation and optimization.

8.2 MODELS AND TRANSFORMATIONS FOR COMBINATIONAL NETWORKS

The *behavior* of an n-input, m-output combinational circuit can be expressed by an array of Boolean functions $f_i : \mathsf{B}^n \to \{0, 1, *\}$, $i = 1, 2, \ldots, m$, alternatively denoted by $\mathbf{f} : \mathsf{B}^n \to \{0, 1, *\}^m$. Such functions may be incompletely specified and represent an explicit mapping from the primary input space to the primary output space.

The *structure* of a multiple-level combinational circuit, in terms of an interconnection of logic gates, can be described by a *logic network*, as introduced in Section 3.3.2. The logic network is an incidence structure relating its *modules*, representing input/output ports and logic gates, to their interconnection *nets*. The logic network can be represented by a dag, with vertices corresponding to network modules and edges representing two-terminal nets to which the original multi-terminal nets are reduced. A logic network whose internal modules are bound logic gates is called in jargon a *bound* or *mapped network*.

A circuit behavior may be mapped into many equivalent structures. Conversely, a unique behavior of a combinational circuit can be derived from its structure. In some cases, the size of $\mathbf{f}$ may be so large as to make the representation impractical for use, when *sum of products* (*product of sums*) forms or BDDs are used. Consider, for example, a logic network of a five-level deep parity tree, with two-input EXOR gates. The corresponding behavior, expressed as a *sum of products* form, requires 2^{31} implicants. As a second example, consider a multiplier that can be represented by a Wallace-tree interconnection. The corresponding BDD representation is of exponential size.

> **Example 8.2.1.** An example of a simple bound network is shown in Figure 8.1. Its corresponding graph is shown in Figure 8.2. The input/output behavior of the network is:
>
> $$x = ab$$
> $$y = c + ab$$

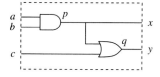

FIGURE 8.1
Example of a bound logic network.

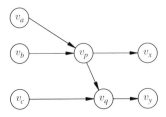

FIGURE 8.2
Example of a logic network graph.

The concept of a logic network is more general than that of an interconnection of logic gates represented by bound networks. In general, logic networks can be considered as hybrid (structural/behavioral) logic models by allowing the internal modules to be associated with arbitrary scalar (i.e., single-output) Boolean functions. Hence, they can be seen as networks of *local functions,* where the network implies a structure and the local functions model the local behavior. Logic networks are very convenient means to represent multiple-level combinational circuits and to support their optimization, because the combination of local functions with relatively small-sized representations allows us to model complex behaviors.

In this chapter, we restrict our attention to non-hierarchical logic networks, because logic optimization methods are described as applied to flat networks. Flattening the hierarchy of logic networks is trivial. Conversely, performing logic optimization on hierarchical models leads to additional complications.

Moreover, we assume that the interconnection nets are split into sets of two-terminal nets, so that the network itself is isomorphic to its graph representation. This allows us to use the notion of logic network and logic network graph interchangeably, therefore simplifying the notation and the description of the optimization methods. In other words, a logic network can be seen as a vertex-annotated directed acyclic graph, where the annotation is just the specification of the local functions.

We summarize these considerations by refining the earlier definition of logic network of Section 3.3.2.

Definition 8.2.1. A non-hierarchical combinational **logic network** is:

- A set of vertices V partitioned into three subsets called primary inputs V^I, primary outputs V^O and internal vertices V^G. Each vertex is labeled by a variable.
- A set of scalar combinational Boolean functions associated with the internal vertices. The support variables of each local function are variables associated with primary inputs or other internal vertices. The dependency relation of the support variables induces a partial order on the set of vertices and corresponds to the edge set E.
- A set of assignments of the primary outputs to internal vertices that denotes which variables are directly observable from outside the network.

Each vertex and the corresponding local function is associated with an uncomplemented variable. Since the variables in the support set of the local functions may be complemented or uncomplemented, local complementation may be needed. In the usual logic network model as defined above, where vertices are associated with variables, inverters are implicit in the model and are not represented. Equivalently, each vertex can provide signals in both polarity, and the network is called a *double-polarity logic network*. When an explicit notion of the inverters is desired, the previous definition of logic network can be modified by associating vertices with literals and edges with literal dependencies, yielding a *single-polarity* logic network. Whereas area and delay models are simpler for single-polarity networks, this model reduces some degrees of freedom for optimization. Hence the double-polarity model is often used [6]. We refer to double-polarity logic networks as logic networks for brevity. We shall consider the polarity issue in detail in Section 8.4.3.

We use the following notational conventions in the sequel. We represent a logic network by $G_n(V, E)$. We denote by $n_i = |V^I|$, $n_o = |V^O|$ and $n_g = |V^G|$ the number of input, output and internal vertices, respectively. Variables are represented by characters, possibly with subscripts, e.g., a, x_1, x_{out}. (We refrain from using character f, often used to denote a generic function.) Vertices associated with variables are denoted by a subscript corresponding to the variable, e.g., v_a, v_{x_1}, $v_{x_{out}}$. Boolean functions associated with vertices are also denoted by a subscript corresponding to the variable, e.g., f_a, f_{x_1}, $f_{x_{out}}$. In this chapter, to avoid confusion, we shall denote the cofactor operation with an extended notation, e.g., the cofactor of f w.r.t. x is denoted by $f|_{x=1}$. Since every internal vertex of the network associates a variable with a function, the logic network can also be expressed by a set of logic equations.

Example 8.2.2. Consider the logic network, with primary input variables $\{a, b, c, d, e\}$ and primary output variables $\{w, x, y, z\}$, described by the following equations:

$$p = ce + de$$
$$q = a + b$$
$$r = p + a'$$
$$s = r + b'$$
$$t = ac + ad + bc + bd + e$$
$$u = q'c + qc' + qc$$
$$v = a'd + bd + c'd + ae'$$
$$w = v$$
$$x = s$$
$$y = t$$
$$z = u$$

The network is shown in Figure 8.3 (a) and the corresponding graph in Figure 8.3 (b). The longest path involves three stages of logic. Note that the last four assignments denote explicitly the primary outputs.

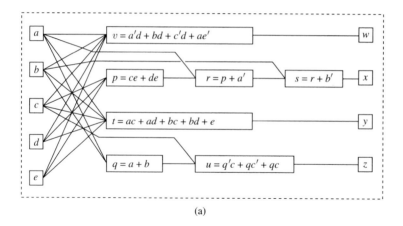

(a)

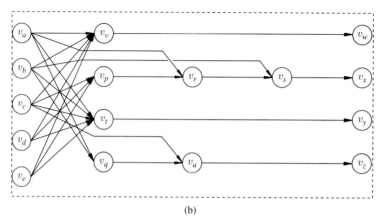

(b)

FIGURE 8.3
(a) Example of a logic network. (b) Example of an isomorphic logic network graph.

The input/output terminal behavior of the network is:

$$\mathbf{f} = \begin{bmatrix} a'd + bd + c'd + ae' \\ a' + b' + ce + de \\ ac + ad + bc + bd + e \\ a + b + c \end{bmatrix} \begin{matrix} \rightarrow w \\ \rightarrow x \\ \rightarrow y \\ \rightarrow z \end{matrix}$$

8.2.1 Optimization of Logic Networks

Major optimization objectives in combinational logic design are *area* and *delay* reduction. Maximization of the circuit *testability* is also an important goal. Let us contrast two-level logic to multiple-level logic implementations and optimization goals. When considering two-level logic implementations of *sum of products* representations, the area and delay are proportional to the size of the cover, as shown in Chapter 7 [Figure 8.4(a)]. Thus, achieving minimum (or irredundant) covers corresponds to optimizing both area and delay. Achieving irredundant covers corresponds to maximizing the circuit testability. In the case of multiple-level circuits, minimal-area implementations do not correspond in general to minimal-delay ones and *vice versa* [Figure 8.4(b)]. A simple example is provided by combinational adders. Thus, trading off area for delay is extremely important in multiple-level logic design. In addition, the relations between area, delay and testability are complex, as described in Section 8.5.

The design of multiple-level logic circuits involves a truly multiple-criteria optimization problem. It is often desirable to use optimization methods to determine area-delay trade-off points, which are solutions to the following problems:

- Area minimization (under delay constraints).
- Delay minimization (under area constraints).

Both problems address the optimization of area and delay estimates extracted from the logic network model. Delay constraints may be expressed by *input arrival times* and *output required times*, which constrain input/output path delays. A simple and common case is the one where all inputs arrive simultaneously and all outputs must satisfy the same timing requirement.

We address first the important problem of modeling area and delay in logic networks. The area occupied by a multiple-level network is devoted to the logic gates and to wires. The area of each logic gate is known in case of networks bound to library cells. Otherwise it must be estimated for a *virtual gate* implementing the corresponding logic function. A common way to estimate the area is to relate it to the number of literals of a factored form representation of the functions. (See Section 2.5.2.) The rationale is based on the fact that a cell generator for the virtual gate in MOS technology would require as many polysilicon stripes as the number of literals (plus possibly one for the gate output). If all cells have the same height and their width is roughly proportional to the number of polysilicon stripes, then the gate area is proportional to the number of literals. The total number of liter-

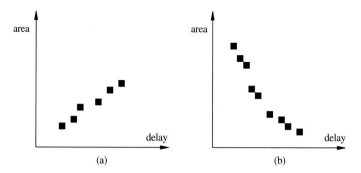

(a) (b)

FIGURE 8.4
(a) An example of the design evaluation space for a two-level logic circuit. (b) An example of the design evaluation space for a multiple-level logic circuit.

als in a network relates then to the overall gate area. Since the wiring area is often proportional to the gate area, the total number of literals is an estimate of the overall area. Even though this model should apply only to the case of a macro-cell design style with cell generators, experimental results have shown a proportionality relation between total number of literals and overall area also for other design styles.

In practice some difficulties may arise in computing the number of literals in a factored form, because this requires an optimum factorization of the logic expressions, which is a difficult problem to solve exactly. Often the number of literals in a (minimal) *sum of products* form is used to estimate area or a weighted sum of the number of literals and the number of internal vertices. In this chapter, we refer to the size (area estimate) of a local logic expression as the number of variables appearing in that expression in a minimal *sum of products* form. We refer to the size (area estimate) of a logic network as the sum of the sizes of its local expressions.

Timing optimization is the minimization of the delay of the slowest path, called the *critical path*. Delay modeling is important in computing accurately the path delays and it involves two issues: the computation of the propagation delay associated with each vertex and the path delay computation. The propagation delay associated with a vertex can be easily estimated for bound networks, because the gate delay as a function of the fanout is provided with the library specification. Otherwise, the propagation delay must be estimated for a virtual gate implementing the local function. A crude model is to assume unit delay per stage. More refined models involve relating the propagation delay to the complexity of the logic function and to its fanout [8, 21].

The path delay computation involves the derivation of a worst-case bound on the propagation time of an event along that path. A simple approximation is the sum of the propagation delay of the vertices along that path. Some paths may turn out to be *false*, i.e., unable to propagate events. Detection of false paths, as well as estimation of the critical true path, is complex and is described in Section 8.6.2. Wiring delays play also a role in the overall delay computation. Since wire lengths are not known at this stage, statistical models can be used to estimate them.

8.2.2   Transformations for Logic Networks

The optimization problems outlined above are complex and are believed to be intractable for general networks. Few exact methods have been proposed that to date are not considered practical because of their computational complexity. For example, Lawler [35] proposed a method for minimizing the literals in single-output multiple-level networks based on the computation of *multi-level prime implicants* and on the solution of a covering problem. Lawler's algorithm extends the Quine-McCluskey method for the two-level optimization of Section 7.2.2.

Heuristic methods for multiple-level logic optimization perform a stepwise improvement of the network by means of logic *transformations* that preserve the input/output network behavior. Fortunately, most logic transformations are defined so that network equivalence is guaranteed and does not need to be checked. Application of logic transformations in any sequence preserves network equivalence.

Logic transformations can have a *local* or *global* effect. In the first case, a transformation modifies a local Boolean function without affecting the structure of the network. An example would be a Boolean optimization of a local function that does not change its support set. Instead, global transformations affect also the structure of the network and can be seen as the creation/deletion of one or more vertices and/or edges. An example of a global transformation is the merging of two vertices and of their corresponding expressions. This transformation is called *elimination*.

A model for the local functions is required for describing the transformations and the related algorithms. We assume that local functions are represented by logic expressions, possibly factored. Thus local functions are sometimes referred to as expressions, and *vice versa*, by not distinguishing between functions and their representations for the sake of simplicity. Note that program implementations of logic optimization algorithms may use other models for storing the local functions, e.g., OBDDs.

We introduce now briefly some logic transformations that will be described in more detail in the following sections.

ELIMINATION. The elimination of an internal vertex is its removal from the network. The variable corresponding to the vertex is replaced by the corresponding expression in all its occurrences in the logic network.

> **Example 8.2.3.** Consider the logic network of Example 8.2.2 shown in Figure 8.3 (a). Consider vertex v_r of the network that feeds a signal to vertex v_s. An elimination of v_r corresponds to the substitution of r with $p + a'$, yielding:
>
> $$s = p + a' + b'$$
>
> The logic network after the elimination is shown in Figure 8.5.

A motivation for an elimination may be the desire to remove a vertex associated with a simple local function that can be aggregated to other local functions.

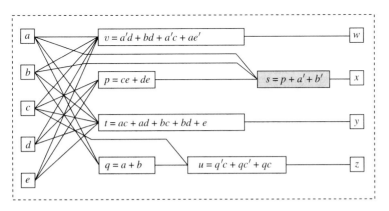

FIGURE 8.5
Example of an elimination.

DECOMPOSITION. The decomposition of an internal vertex is its replacement by two (or more) vertices that form a subnetwork equivalent to the original vertex.

> **Example 8.2.4.** Consider the logic network of Example 8.2.2 shown in Figure 8.3 (a). Consider vertex v_v of the network, where $f_v = a'd + bd + c'd + ae'$. The expression for f_v can be reduced in size by factoring it as $v = (a' + b + c')d + ae'$ and by precomputing subexpression $a' + b + c'$ at a new vertex to be introduced in the network. This corresponds to replacing the equation for v by:
>
> $$j = a' + b + c'$$
> $$v = jd + ae'$$
>
> The logic network after the decomposition is shown in Figure 8.6.

A reason to decompose a vertex is the splitting of a complex function over two (or more) internal vertices.

EXTRACTION. A common subexpression of two functions associated with two vertices can be extracted by creating a new vertex associated with the subexpression. The variable associated with the new vertex allows us to simplify the representations of the two functions by replacing the common subexpressions.

> **Example 8.2.5.** Consider the logic network of Example 8.2.2 shown in Figure 8.3 (a). Consider vertices v_p and v_t of the network. A careful examination of the corresponding equations shows that they can be factored as:
>
> $$p = (c + d)e$$
> $$t = (c + d)(a + b) + e$$

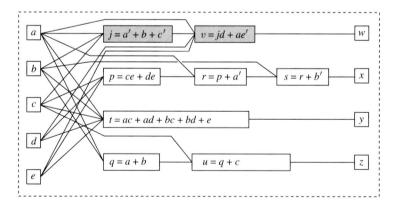

FIGURE 8.6
Example of a decomposition.

Hence $(c+d)$ is a common subexpression that can be extracted by creating an additional vertex v_k, and an additional variable k and rewriting the equations as:

$$k = c + d$$

$$p = ke$$

$$t = ka + kb + e$$

The logic network after the extraction is shown in Figure 8.7.

A reason for extracting common subexpressions is to simplify the overall network by exploiting commonalities.

SIMPLIFICATION.   A function is reduced in complexity by exploiting the properties of its representation. If the function is represented in a two-level form, then two-level optimization techniques can be applied. If the support set does not change, the transformation is local. Otherwise, simplification is a global transformation and it corresponds to deleting one or more dependencies.

Example 8.2.6. Consider the logic network of Example 8.2.2 shown in Figure 8.3 (a). Consider vertex v_u, whose function $f_u = q'c + qc' + qc$ can be simplified to $f_u = q + c$. Since the simplification does not affect the support of the function, i.e., the set of local inputs is preserved, the transformation is local. The logic network after the simplification is shown in Figure 8.8.

SUBSTITUTION.   A function is reduced in complexity by using an additional input that was not previously in its support set. The transformation requires the creation of a dependency, but it may also lead to dropping others.

Example 8.2.7. Consider the logic network of Example 8.2.2 after the extraction, as shown in Figure 8.7. Consider vertex v_t associated with function $f_t = ka + kb + e$. This

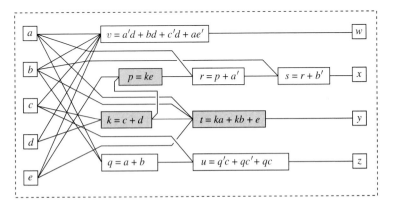

FIGURE 8.7
Example of an extraction.

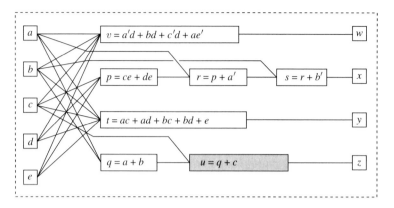

FIGURE 8.8
Example of a simplification.

function can be simplified by realizing that subexpression $a + b$ is already precomputed at vertex v_q and that the local function f_t can be rewritten as $f_t = kq + e$. This requires adding a dependency from v_q to v_t. The logic network after the substitution is shown in Figure 8.9.

Other transformations exist. The aforementioned transformations are among the most common in multiple-level logic optimization and will be described in detail in the sequel. Logic transformations affect both area and performance of a network, because they modify the number of literals, the local functions and their dependencies. Estimates are used to determine whether a transformation is locally *favorable*, and hence it should be applied to achieve the desired goal.

Multiple-level optimization follows a strategy similar to that used in heuristic two-level optimization (Section 7.2.3), where operators like *expand*, *reduce*, *irredun-*

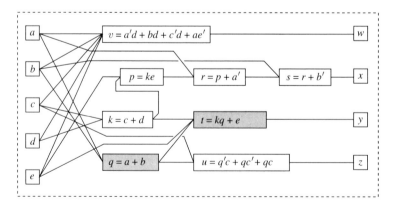

FIGURE 8.9
Example of a substitution.

dant, etc., are applied to a logic cover to reduce the number of implicants. A logic cover is declared optimal when the available operators cannot improve further the current solution. Similarly, in multiple-level optimization, logic transformations are the operators applied to the logic network in an iterative way. A network is declared to be area or performance optimal with respect to a set of transformations when none of these can improve the corresponding measure.

Example 8.2.8. Consider again the logic network of Example 8.2.2. The application of the transformations described above yields the following network:

$$j = a' + b + c'$$
$$k = c + d$$
$$q = a + b$$
$$s = ke + a' + b'$$
$$t = kq + e$$
$$u = q + c$$
$$v = jd + ae'$$
$$w = v$$
$$x = s$$
$$y = t$$
$$z = u$$

The network is shown in Figure 8.10. A comparison of the transformed network with the original one shows that the total number of literals has decreased.

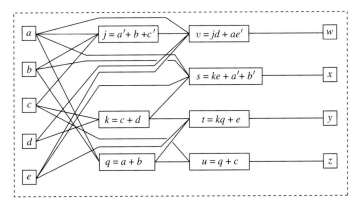

FIGURE 8.10
Example of a logic network whose area estimate has been minimized.

8.2.3 The Algorithmic Approach
to Multiple-Level Logic Optimization

There are two major limitations in optimizing networks using logic transformations. First, given a set of transformations, it is difficult, if not impossible, to claim that all equivalent network configurations can be reached by some sequence of transformations. Hence the optimum solution, or even the feasible ones in constrained optimization, may not even be reachable. Second, different sequences of transformations lead to different results that may correspond to different *local* points of optimality.

Whereas all heuristic approaches to multiple-level optimization rely on stepwise improvement by means of transformations, different flavors exist in the way in which transformations are chosen and applied. The major classification is in terms of algorithmic-based and rule-based optimization methods.

The *algorithmic approach* to multiple-level optimization consists of defining an algorithm for each transformation type. The algorithm detects when and where the transformation can be applied and it terminates when no favorable transformation of that type can be applied to the network. Hence each transformational algorithm can be seen as an operator performing a set of transformations. Notable examples of programs using this approach are MIS [8] and BOLD [3].

By contrast, in the *rule-based* approach to optimization, transformations of different types can be alternated according to a set of rules that mimic the optimization steps performed by a human designer. The rule data-base stores a set of pattern pairs representative of common network fragments. In each pair, the second pattern represents an equivalent but better implementation of the first pattern. Rule-based systems rely on the detection of situations were a subnetwork corresponding to the first pattern of a pair can be replaced by the corresponding optimal pattern. The repeated application of replacement rules improves the value of the desired objective function. Rule-based systems will be described in detail in Section 8.7 and compared to the algorithmic approach. An example of a rule-based system is IBM's LOGIC SYNTHESIS SYSTEM (LSS) [18, 19].

A major advantage of the algorithmic approach is that the transformations of a given type are systematically applied, when favorable, to the network. As a result, some network properties can be claimed after the application of one or more transformational algorithms. An example is primality and irredundancy of the local Boolean functions of the network, which play an important role in the overall network testability, as described in Section 8.5.

The application of the different algorithms related to various transformation types is usually sequenced by *scripts*, i.e., by recipes that suggest the order of application and the termination criterion. Whereas determining the properties of each algorithm and of the corresponding transformed network can be done in a rigorous way, the effects of using different sequences of transformational algorithms is often hard to predict and still the subject of investigation. Most logic optimization systems provide empirical *scripts* to the user.

Example 8.2.9. The following script is used in the MIS/SIS program [8] and is called a *rugged script*:

```
sweep; eliminate -1
simplify -m nocomp
eliminate -1

sweep; eliminate 5
simplify -m nocomp
resub -a

fx
resub -a; sweep

eliminate -1; sweep
full_simplify -m nocomp
```

The `sweep` command eliminates all single-input vertices and those with a constant local function. `Eliminate k` eliminates all vertices that would not increase the area estimate by more than k. Commands `simplify -m nocomp` and `full_simplify -m nocomp` perform Boolean simplification by invoking program ESPRESSO without computing the complete offset. (The latter command uses a larger set of *don't care* conditions than the former.) `Resub -a` performs algebraic substitution of all vertex pairs. Command `fx` extracts double-cube and single-cube expressions.

Example 8.2.10. Consider again the complete differential equation integrator, as described in Example 1.5.3, and its control unit, reported in Example 4.7.6. The following assignments denote the combinational component of the control unit after some sequential logic transformations (that will be described in Example 9.4.5 of Chapter 9). In this example, the names of the variables have been shortened for the sake of simplicity, and curly braces denote signals observable at the outputs. (Variables $y3$, $b4$ and $e4$ are associated with register inputs.)

```
{i3} = c4' f4'
{j3} = c4 f4
{k3} = z3' f4
{l3} = z3 f4'
{m3} = z3' c4
{n3} = z3 c4'
{o3} = z3' f4 + c4' f4'
{p3} = z3' f4 + c4 f4
{q3} = z3 f4' + c4 f4
{r3} = z3' f4 + c4 f4 + c4' f4'
{s3} = z3 f4' + z3' f4 + c4' f4'
{t3} = z3 f4' + z3' f4 + c4 f4
{u3} = z3 c4' + z3' c4 + c4' f4'
{v3} = z3 c4' + z3 f4' + z3' c4 + c4 f4
{w3} = z3 c4' + z3 f4' + z3' f4 + c4 f4
{x3} = z3 c4' + z3 f4' + z3' c4 + z3' f4
  y3 = z3' c4' g3' + z3' f4' g3'
  b4 = z3 f4' g3' + z3' c4' g3' + z3' h3 g3'
  e4 = z3' f4' g3' + c4 f4 g3'
```

The following expressions show the control unit as optimized by SIS with the rugged script. The number of literals has been reduced from 93 to 48:

```
{i3} = c4' f4'
{j3} = c4 f4
{k3} = z3' f4
{13} = z3 f4'
{m3} = z3' c4
{n3} = z3 c4'
{o3} = {i3} + {k3}
{p3} = f4 {n3}'
{q3} = {j3} + {13}
{r3} = {o3} + {p3}
{s3} = f4' {m3}' + {o3}
{t3} = {p3} + {q3}
{u3} = c4' {p3}' + {m3}
{v3} = z3 + c4
{w3} = z3 + f4
{x3} = {k3} + {p3}' {v3}
  y3 = z3' g3'
  b4 = z3' h3 g3' + {13} g3' + {v3}' g3'
  e4 = {j3} g3' + {w3}' y3
```

We present now a simple transformational algorithm as an example. The *elimination* algorithm iterates the elimination of variables (and vertices), with the goal of reducing the number of stages of a network. The number of *stages* of a logic network is the maximum path weight over all input/output paths, where the weight is measured in terms of internal vertices. A generic two-level circuit can then be modeled by a single-stage logic network.[1] The reduction of the number of stages is often important. Two cases are noteworthy: first, finding the maximal stage reduction such that the area (or delay) does not increase and, second, performing the unconstrained reduction to a single-stage network.

The *elimination* algorithm is used for this purpose. (See Algorithm 8.2.1.) The algorithm performs a sequence of variable eliminations by substituting them by the corresponding expressions, as in the case of Gaussian elimination for sets of linear equations. The elimination algorithm takes also a *threshold* as argument, which represents the maximum local increase in area (or delay) that an elimination can yield. The *value* associated with a vertex is the increase in area (or delay) caused by its elimination. For example, when the literal count is used as a cost function, the value of the elimination of a vertex with an l-literal expression and whose corresponding variable appears n times in the network is $nl - n - l$.

A vertex elimination is performed only if its value is lower than or equal to the threshold. To ensure no increase in area (or in delay), the threshold value k is set to zero. Often, a small positive value of the threshold is used to allow a reduction of

---

[1]The apparent mismatch between the number of levels and stages is due to the network model where each vertex is associated with a function with no constraints on the levels of its representation, i.e., a function could be represented by a cube, a two-level or a multiple-level form (e.g., a factored form).

$ELIMINATE(G_n(V, E), k)\{$

    repeat {

        v_x = selected vertex with value not larger than k;

        if $(v_x = \emptyset)$ **return**;

        Replace x by f_x in the network;

    }

}

ALGORITHM 8.2.1

stages at the expense of a limited increase in area or delay. On the other hand, a large value of the threshold allows for unconstrained stage reduction. The algorithm iterates variable eliminations in a greedy fashion by selecting the vertex with lowest value (and bounded from above by the threshold) until candidate vertices exist whose values are bounded by the threshold. Note that the candidate vertices must be internal vertices and not directly observable from outside the network, i.e., not direct predecessors of output vertices.

Example 8.2.11. Consider the network of Example 8.2.2 shown in Figure 8.3 (a). Let us consider the reduction of area, measured by the total number of literals. Hence the value of a vertex is the increase of the number of literals caused by the elimination. The application of the elimination algorithm with threshold $k = 0$ leads to removing first vertex v_r and then v_p. Thus the network is:

$$q = a + b$$

$$s = ce + de + a' + b'$$

$$t = ac + ad + bc + bd + e$$

$$u = q'c + qc' + qc$$

$$v = a'd + bd + c'd + ae'$$

$$w = v$$

$$x = s$$

$$y = t$$

$$z = u$$

Note that eliminating v_q would increase the number of literals and hence would not be performed. On the other hand, if the simplification of the function f_u is done before applying the elimination algorithm, then also v_q would be removed.

The elimination algorithm is the simplest example of a transformational algorithm. As it reduces the number of stages, it destroys also the structural information of the logic network. The opposite problem, i.e., increasing the structural information by adding stages or dependencies, is much more difficult to solve. For example, the extraction and substitution transformations require the search for common subexpressions and consequently a creative process.

The difficulty of the search for creative transformations stems from the degrees of freedom available in manipulating Boolean functions. Neglecting some Boolean properties can simplify the search for useful transformations at the expense of some quality of the result. Different heuristic optimization methods have been proposed related to different levels of complexity of the underlying models.

The complete *Boolean* model requires considering all Boolean properties of the local functions while manipulating a logic network. The *don't care* sets induced by the interconnection must be taken into account, as will be shown in Section 8.4. Logic transformations that use the full Boolean model are called in jargon *Boolean transformations*.

Brayton and McMullen [7] first suggested simplifying the Boolean model by expressing the local functions as polynomials and dropping from consideration some specific assumptions of Boolean algebra. With the resulting simplified model, the logic network can be optimized by using general properties of polynomial algebra. This leads to a simpler way of manipulating a network, especially as far as searching for common subexpressions. Transformations based on the polynomial model are referred to in jargon as *algebraic transformations*. They represent a subset of the Boolean transformations that is simpler and faster to compute.

Example 8.2.12. Consider again the substitution performed in Example 8.2.7 and shown in Figure 8.9, where function $f_t = ka + kb + e$ is simplified by substituting $q = a + b$ to yield $f_t = kq + e$. Note that $kq = k(a + b) = ka + kb$ holds regardless of any assumption specific to Boolean algebra. This is an example of an algebraic substitution.

Consider instead functions $f_h = a + bcd + e$ and $q = a + cd$. It is possible to write $f_h = a + bq + e$ by using the properties of Boolean algebra, because $a + bq + e = a + b(a + cd) + e = a(b + 1) + bcd + e = a + bcd + e$. Unfortunately such a transformation cannot be determined by using only polynomial algebra, because it relies on the Boolean property $b + 1 = 1$. This is an example of a Boolean transformation.

8.3 THE ALGEBRAIC MODEL

The algebraic model consists of representing the local Boolean functions by *algebraic expressions*. An algebraic expression is a multilinear polynomial over the set of network variables with unit coefficients. *Algebraic transformations* consist of manipulating the expressions according to the rules of polynomial algebra and by neglecting specific features of Boolean algebra. Namely, in the (non-Boolean) polynomial algebra we are considering here, only one distributive law applies [e.g., $a \cdot (b + c) = ab + ac$ but $a + (b \cdot c) \neq (a + b) \cdot (a + c)$] and complements are not defined. As a consequence, some properties like absorption, idempotence, involution and De Morgan's (see Section 2.5) as well as the identities $a + a' = 1$, $aa' = 0$ cannot be exploited. Moreover, *don't care* sets are not used.

Algebraic expressions modeling Boolean functions are obtained by representing the functions in *sum of products* form and making them minimal with respect to single-cube containment. An expression can then be viewed as a sum (or a set) of cubes, in the same way that a polynomial is a sum (or a set) of monomials. Note that the deletion of redundant cubes (even w.r.t. single-cube containment) cannot be done by algebraic transformations, due to the lack of the absorption property.

We shall use the algebraic model throughout Section 8.3 and we shall use the terms "(algebraic) expressions" and "functions" interchangeably as well as "cube" and "monomial." We denote a generic function (expression) by f and its support set by $sup(f)$. We represent a generic single-cube function, i.e., a cube, by f^C. We use sometimes the set notation for cubes (monomials) by using the capital letter C to denote the support set of f^C. With the set notation, membership and inclusion of cubes relates to their support sets. (If cube $f^C = ab$ is represented as $C = \{a, b\}$, then $a \in C$. Note that this differs from the usual Boolean meaning of inclusion, as described in Section 2.5.)

In the algebraic model, the complementation operation is not defined and complemented variables are treated as unrelated to the corresponding uncomplemented variables. Thus, we shall not distinguish between literals and variables. For the sake of simplicity, we shall not use complemented variables throughout Section 8.3.

Algebraic operations are restricted to expressions with disjoint support to preserve the correspondence of their results with *sum of products* forms that are minimal w.r.t. single-cube containment. Indeed, this prevents the generation of cubes that are covered by other cubes as well as universal (void) cubes due to the sum (product) of variables with their complements, which cannot be detected in the algebraic model. Whereas algebraic expressions can be represented in different forms (e.g., factored forms), we consider only *sum of products* algebraic expressions as inputs to the algorithms described in Section 8.3.

> **Example 8.3.1.** Consider the product of expression $a + b$ with expression $c + d$, where the supports are disjoint sets. The result, i.e., $ac + ad + bc + bd$, is an expression which denotes a minimal (w.r.t. single-cube containment) *sum of products* form of a corresponding Boolean function.
>
> Consider now the product of expression $a + b$ with expression $a + c$, where the support sets are not disjoint. Expression $aa + ac + ba + bc$ represents a non-minimal *sum of products* form. Thus this product is not considered to be a proper algebraic product.
>
> Consider then the product of expression $a + b$ with $a' + c$, where the support sets are not disjoint. The first term of $aa' + ac + ba' + bc$ is a void cube, i.e., it is not an implicant of the corresponding Boolean function. Note that aa' cannot be eliminated in the algebraic model, because a' is unrelated to a. Thus, also this product is not considered to be a proper algebraic product.

For our purposes, an important role is played by the *division* of the algebraic expressions. Indeed, divisors are related to the subexpressions that two expressions may have in common. Let $\{f_{dividend}, f_{divisor}, f_{quotient}, f_{remainder}\}$ be algebraic expressions. We define the *algebraic division* as follows.

> **Definition 8.3.1.** We say that $f_{divisor}$ is an **algebraic divisor** of $f_{dividend}$ when $f_{dividend} = f_{divisor} \cdot f_{quotient} + f_{remainder}$, $f_{divisor} \cdot f_{quotient} \neq 0$ and the support of $f_{divisor}$ and $f_{quotient}$ is disjoint. We write $f_{quotient} = f_{dividend}/f_{divisor}$.

An algebraic divisor is called a *factor* when the remainder is void. An expression is said to be *cube free* when it cannot be factored by a cube.

> **Example 8.3.2.** Let $f_{dividend} = ac + ad + bc + bd + e$ and $f_{divisor} = a + b$. Then, division yields quotient $f_{quotient} = c + d$ and remainder $f_{remainder} = e$, because $(a + b) \cdot (c + d) + e = ac + ad + bc + bd + e = f_{dividend}$ and the supports $sup(f_{divisor}) = \{a, b\}$ and $sup(f_{quotient}) = \{c, d\}$ are disjoint sets.
>
> Let now $f_i = a + bc$ and $f_j = a + b$. Expression f_j is not considered an algebraic divisor of f_i, even though $f_i = f_j \cdot f_k$ with $f_k = a + c$, because $sup(f_j) \cap sup(f_k) = \{a, b\} \cap \{a, c\} \neq \emptyset$.
>
> Expression $a + b$ is a factor of $ac + ad + bc + bd$, because their division yields no remainder. Both expressions are cube free. Examples of non-cube-free expressions are abc and $ac + ad$.

We present now an algorithm for the algebraic division of algebraic expressions. The expressions are considered as sets of elements. Let $A = \{C_j^A, \ j = 1, 2, \ldots, l\}$ be the set of cubes (monomials) of the dividend. Let $B = \{C_i^B, \ i = 1, 2, \ldots, n\}$ be the set of cubes (monomials) of the divisor. The quotient and the remainder are the sum of the monomials of sets, denoted by Q and R, respectively as shown in Algorithm 8.3.1.

Note that in the algorithm the elements of D_i and Q are sets, but they are considered as single entities as far as the intersection operation is concerned. The algorithms can be programmed to have linear or linear-logarithmic time complexity by using sorting techniques on the monomials [34].

> **Example 8.3.3.** Let $f_{dividend} = ac + ad + bc + bd + e$ and $f_{divisor} = a + b$. Hence $A = \{ac, ad, bc, bd, e\}$ and $B = \{a, b\}$.
>
> Let $i = 1$. Then $C_1^B = a$, $D = \{ac, ad\}$ and $D_1 = \{c, d\}$. Then $Q = \{c, d\}$. Let $i = 2 = n$. Then $C_2^B = b$, $D = \{bc, bd\}$ and $D_2 = \{c, d\}$. Now $Q = \{c, d\} \cap \{c, d\} = \{c, d\}$. Hence $Q = \{c, d\}$ and $R = \{e\}$. Therefore $f_{quotient} = c + d$ and $f_{remainder} = e$.

```
ALGEBRAIC_DIVISION(A, B) {
    for (i = 1 to n) {                          /* take one monomial C_i^B of the divisor at a time */
        D = {C_j^A such that  C_j^A ⊇ C_i^B};   /* monomials of the dividend that include C_i^B */
        if ( D == ∅ ) return(∅, A);
        D_i = D where the variables in sup(C_i^B) are dropped ;
        if (i = 1)
            Q = D_i;                             /* quotient is initialized to D_i */
        else
            Q = Q ∩ D_i;                         /* quotient is intersection of D_j, j = 1, 2, ..., i */
    }
    R = A − Q × B;                               /* compute remainder */
    return(Q, R);
}
```

ALGORITHM 8.3.1

Let us consider another case. Let now $f_{dividend} = axc + axd + bc + bxd + e$ and $f_{divisor} = ax + b$. Let $i = 1$. Then $C_1^B = ax$, $D = \{axc, axd\}$ and $D_1 = \{c, d\}$. Then $Q = \{c, d\}$. Let $i = 2 = n$. Then $C_2^B = b$, $D = \{bc, bxd\}$ and $D_2 = \{c, xd\}$. Now $Q = \{c, d\} \cap \{c, xd\} = \{c\}$. Therefore $f_{quotient} = c$ and $f_{remainder} = axd + bxd + e$. Note that the intersection to compute Q is a set intersection, where the monomials are atomic elements.

The following theorem provides sufficient conditions for a quotient to be empty. The first two conditions are used in the algorithm for an early exit. The others can be used as filters to detect cases where the algorithm does not need to be applied.

Theorem 8.3.1. Given two algebraic expressions f_i and f_j, f_i / f_j is empty if any of the following conditions apply:

- f_j contains a variable not in f_i.
- f_j contains a cube whose support is not contained in that of any cube of f_i.
- f_j contains more terms than f_i.
- The count of any variable in f_j is larger than in f_i.

The proof of this theorem is left as an exercise to the reader. (See Problem 1.)

8.3.1 Substitution

Substitution means replacing a subexpression by a variable associated with an existing vertex of the logic network. Namely, it targets the reduction in size of an expression f_i by using a variable j, not in the support of f_i, that is defined by the equation $j = f_j$. Algebraic substitution is a straightforward application of algebraic division. Indeed, if the algebraic division of f_i by f_j yields a non-void quotient $f_{quotient}$, then the expression f_i can be rewritten as $f_j \cdot f_{quotient} + f_{remainder}$ that is equivalent to $j \cdot f_{quotient} + f_{remainder}$.

The gain of a substitution can be measured (in terms of area and/or delay) by comparing the expressions before and after the substitution. In particular, the variation in terms of literals in a factored form model is the number of literals of f_j minus 1.

Example 8.3.4. Consider again the network of Example 8.2.7, where vertex v_t is associated with expression $f_t = ka + kb + e$ and vertex v_q to $f_q = a + b$. The algebraic division of f_t by f_q yields $f_{quotient} = k$ and $f_{remainder} = e$. Hence the expression f_t can be rewritten as $f_t = kq + e$. The corresponding network after the substitution is shown in Figure 8.9.

The search for algebraic substitutions is done by considering all unordered expression pairs, attempting algebraic division and, when a substitution is favorable, replacing the subexpression corresponding to the divisor by the related variable. The overall search requires a number of divisions that are quadratic in the size of the logic network. In practice, filters reduce heavily the number of division attempts to be made by detecting void quotients early. In addition to the conditions specified by Theorem 8.3.1, the following theorem is useful to detect empty quotients.

Theorem 8.3.2. Given two algebraic expressions f_i and f_j, $f_i/f_j = \emptyset$ if there is a path from v_i to v_j in the logic network.

Proof. Let us consider first the simple case where the path reduces to an edge. Then, variable i is in the support of f_j but not in the support of f_i. Hence there is a variable in the divisor that is not present in the dividend, and by Theorem 8.3.1 the quotient is empty.

Let us consider now the general case. The support of f_j must contain some variable, say k, corresponding to a vertex v_k on the path from v_i to v_j. Variable k cannot be in the support of f_i, because otherwise there would be an edge (v_k, v_i) and hence a cycle in the network graph. Therefore the support of the dividend lacks a variable (k) that is in the support of the divisor, and by Theorem 8.3.1 the quotient is void.

The algebraic substitution procedure is described in Algorithm 8.3.2.

The algorithm considers all possible vertex pairs in either order. The *filter test* is the verification that the assumptions of Theorems 8.3.1 and 8.3.2 are not met. The substitution is applied when favorable, i.e., if it reduces the area and/or delay. Local estimates of the variation of the objective function are required.

The overall limitation of the algebraic substitution algorithm is its inability to detect non-algebraic substitutions, such as that shown in Example 8.2.12. That substitution can be found by using Boolean methods, which will be described in Section 8.4. On the other hand, the implementation of algebraic substitution can be very fast and detect most of the possible substitutions, providing thus a very efficient transformational algorithm for logic optimization.

```
SUBSTITUTE( Gₙ(V, E) ){
    for (i = 1, 2, ..., |V|) {
        for (j = 1, 2, ..., |V|; j ≠ i) {              /* for all expression pairs {fᵢ, fⱼ}, j ≠ i */
            A = set of cubes of fᵢ;
            B = set of cubes of fⱼ;
            if (A, B pass the filter test) {            /* perform algebraic division */
                (Q, R) = ALGEBRAIC_DIVISION(A, B)
                if (Q ≠ ∅) {
                    f_quotient = sum of cubes of Q;
                    f_remainder = sum of cubes of R;
                    if (substitution is favorable)      /* perform substitution */
                        fᵢ = j · f_quotient + f_remainder;
                }
            }
        }
    }
}
```

ALGORITHM 8.3.2

8.3.2 Extraction and Algebraic Kernels

The extraction of common subexpressions relies on the search of common divisors of two (or more) expressions. If one is found, then the divisor can be extracted to represent a new local function of the network, and its corresponding variable can be used to simplify the original expressions.

Example 8.3.5. Consider again Example 8.2.5. Expressions $f_p = ce + de$ and $f_t = ac + ad + bc + bd + e$ are simplified once the common divisor $f_k = c + d$ is found and associated with the new variable k. Then, both expressions can be transformed into the product of k times the corresponding quotient plus the remainder, namely, $f_p = ke$ and $f_t = k(a + b) + e$.

The search for common algebraic divisors is done by considering appropriate subsets of the divisors of each expression in the logic network. We distinguish here two subproblems in extraction:

- the extraction of a single-cube expression, i.e., of a monomial;
- the extraction of multiple-cube expressions, i.e., of a polynomial.

The notion of the *kernel* of an expression, introduced by Brayton and McMullen [7], plays an important role for multiple-level optimization, especially in the case of the extraction of multiple-cube expressions.

Definition 8.3.2. A **kernel** of an expression is a cube-free quotient of the expression divided by a cube, which is called the **co-kernel** of the expression.

Note that single-cube quotients are not kernels, because they can be factored as products of the cube itself times 1, and hence they are not cube free. Consequently, single-cube expressions have no kernels.

The set of kernels of an expression f is denoted by $K(f)$. When an expression is cube free, it is considered to be a kernel of itself, its co-kernel being cube 1. This kernel and this co-kernel are called trivial.

Example 8.3.6. Consider expression $f_x = ace + bce + de + g$. Let us divide first f_x by variable a, yielding $f_{quotient} = ce$. Since ce is a single cube and not cube free, it is not a kernel. Similar considerations apply when dividing f_x by b, d or g (single-cube quotient) and when dividing f_x by c, yielding $f_{quotient} = ae + be$, which is not cube free.

On the other hand, dividing f_x by e yields $f_{quotient} = ac + bc + d$, which is cube free. Hence $ac + bc + d$ is a kernel of f_x and e is the corresponding co-kernel. Similarly, dividing f_x by ce yields $f_{quotient} = a + b$, which is cube free. Thus $a + b$ is a kernel of f_x and ce is the corresponding co-kernel. It is straightforward to verify that dividing f_x by other cubes does not yield other kernels. Since f_x is cube free, it is considered a kernel of itself, the corresponding co-kernel being 1. Thus the kernel set of f_x is $K(f_x) = \{(ac + bc + d), (a + b), (ace + bce + de + g)\}$.

The set of kernels of an expression provides essential information for detecting common subexpressions with multiple cubes. The kernels are sums of monomials (cubes) that form the building blocks for constructing the common subexpression to be extracted, when this is possible. We view the monomials (cubes) of each kernel as atomic. Thus, the intersection between two kernels is the largest subset of common monomials (cubes).

The relations between kernel sets and common multiple-cube expressions is stated precisely by Brayton and McMullen's theorem [7].

Theorem 8.3.3. Two expressions f_a and f_b have a common multiple-cube divisor f_d if and only if there exist kernels $k_a \in K(f_a)$ and $k_b \in K(f_b)$ such that $|k_a \cap k_b| \geq 2$.

Proof. Assume $|k_a \cap k_b| \geq 2$. Then the sum of two (or more) cubes in $k_a \cap k_b$ is necessarily a multiple-cube divisor of both f_a and f_b, because any expression obtained by dropping some cubes from a divisor is also a divisor.

Conversely, let f_d be a multiple-cube expression dividing both f_a and f_b. Since f_d has more than one cube, there exists at least a cube-free expression f_e that divides f_d. Note that f_d and f_e may coincide, when f_d is cube free, since an expression divides itself.

Therefore f_e is a divisor of f_a. We claim that f_e is a subset (possibly improper) of the cubes of a kernel $k_a \in K(f_a)$. Indeed any divisor of f_a is a subset of those cubes obtained by dividing f_a by any cube of the corresponding quotient. In addition f_e is cube free, and any expression obtained by adding cubes to f_e has the same property. Therefore there exists a kernel $k_a \in K(f_a)$ containing f_e. Note that f_e may coincide with k_a.

By the same argument, f_e is a subset (possibly improper) of the cubes of a kernel $k_b \in K(f_b)$. Hence f_e is a subset of the intersection of two kernels $k_a \cap k_b$ and it has two or more cubes, because it is cube free.

There are two important consequences of this theorem. First, when two expressions have either no kernel intersections or only single-cube kernel intersections, they cannot have any multiple-cube common subexpression. Thus, when an expression has an empty kernel set, no multiple-cube common subexpression can be extracted, and the corresponding vertex can be dropped from consideration. Second, multiple-cube kernel intersections are common multiple-cube divisors of the expressions corresponding to the kernels.

The computation of the kernel set of the expressions in the logic network is then the first step toward the extraction of multiple-cube expressions. The candidate common subexpressions to be extracted are then chosen among the kernel intersections.

Example 8.3.7. Consider a network with the following expressions:

$$f_x = ace + bce + de + g$$

$$f_y = ad + bd + cde + ge$$

$$f_z = abc$$

The kernels of f_x were computed in Example 8.3.6, namely $K(f_x) = \{(a + b), (ac + bc + d), (ace + bce + de + g)\}$. The kernel set of f_y is $K(f_y) = \{(a + b + ce), (cd + g), (ad + bd + cde + ge)\}$. The kernel set of f_z is empty.

Hence, multiple-cube common subexpressions can be extracted from f_x and f_y only. There is only one kernel intersection, namely between $(a + b) \in K(f_x)$ and $(a + b + ce) \in K(f_y)$. The intersection is $a + b$, which can be extracted to yield:

$$f_w = a + b$$
$$f_x = wce + de + g$$
$$f_y = wd + cde + ge$$
$$f_z = abc$$

KERNEL SET COMPUTATION. We consider in this section the following problem: we are given an expression f and we want to compute its kernel set $K(f)$. Note first that a multiple-term expression is a kernel of itself when it is cube free. Its co-kernel is 1. A kernel of a kernel of an expression is also a kernel of the original expression, its overall co-kernel being the product of the corresponding co-kernels. This leads to defining the following hierarchy of *levels of the kernels*. A kernel has level 0 if it has no kernel except itself. A kernel is of level n if it has at least one kernel of level $n - 1$ but no kernels of level n or greater except itself.

Example 8.3.8. Let $f_x = ace + bce + de + g$. Since it is cube free, $f_x = k_1 \in K(f_x)$. Its co-kernel is 1. Expression $k_2 = ac + bc + d$ is a kernel of f_x, its co-kernel being e. Expression $k_3 = a + b$ is a kernel of k_2, whose relative co-kernel is c. Hence k_3 is a kernel of f_x, its co-kernel being ce.

Kernel k_3 has no kernel but itself. Hence it has level 0. Kernel k_2 has a level-0 kernel and thus has level 1, while kernel k_1 has a level-1 kernel and has level 2.

A naive way to compute the kernels of an expression is to divide it by the cubes corresponding to the power set of its support set. The quotients that are not cube free are weeded out, and the others are saved in the kernel set.

This simple method can be improved upon in two ways: first, by introducing a recursive procedure that exploits the property that some kernels are kernels of other kernels and, second, by reducing the search by exploiting the commutativity of the product operation, e.g., by realizing that the kernels with co-kernels ab and ba are the same.

We consider now an intermediate recursive algorithm that exploits only the first property, for the sake of explanation. We shall describe the complete algorithm later on.

Assume that f is a cube-free expression. Recall that f^C denotes a single-cube expression and $C = sup(f^C)$. Let $CUBES(f, x)$ be a function that returns the set of cubes of f whose support includes variable x and let $CUBES(f, C)$ return the set of cubes of f whose support includes C. This is shown in Algorithm 8.3.3.

The algorithm considers all variables of $sup(f)$ one at a time. If there is only a cube whose support includes variable x, then the division by that variable will not lead to a kernel, because the quotient is a single cube as well. In this case, the variable is dropped from consideration.

```
R_KERNELS(f){
    K = ∅;
    foreach variable x ∈ sup(f) {
            if ( |CUBES(f, x)| ≥ 2 ){            /* skip cases where f/x yields one cube */
                C = largest cube containing x s.t. CUBES(f, C) = CUBES(f, x);
                K = K ∪ R_KERNELS(f/f^C);                    /* recur on f/f^C */
            }
    }
    K = K ∪ f;                                        /* add f to kernel set */
    return(K);
}
```

ALGORITHM 8.3.3

The division of f by variable x may not lead to a kernel if the quotient is not cube free. This happens when other variables are in the support of all cubes where x is present. Hence we determine the maximal set C of these variables and divide f by the corresponding single-cube expression f^C. Thus f^C is the co-kernel of the kernel corresponding to the quotient f/f^C, which is added to the kernel set at the last step of the algorithm. The recursive call allows the algorithm to compute then the kernels of lower levels.

> **Example 8.3.9.** Let us consider again expression $f = ace + bce + de + g$, which is cube free. Initially the kernel set is void. Let the variables be arranged in lexicographic order. The first variable being considered is a. Since a is in the support of only one cube of f, no action is taken. The same applies to the second variable b. The third variable, c, is in the support of two cubes, namely ace and bce. The largest subset of support variables C containing c and contained in the support set of both ace and bce is $\{c, e\}$. Hence ce is the first co-kernel being determined. The recursive call of $R\_KERNELS$ has as argument $(ace + bce + de + g)/ce = a + b$. It does not find any additional kernel, because each support variable $\{a, b\}$ appears only in one cube of $a + b$. Nevertheless, the recursive call adds $a + b$ to the kernel set at the last step.
>
> The fourth variable is then considered, namely d. Since d is in the support of only one cube of f, no action is taken. The fifth variable, e, is in the support of three cubes of f. No other variable in $sup(f)$ is in the support of the same cubes. Hence e is a co-kernel, and a recursive call is made with argument $ac + bc + d$. In the recursion, the kernel $a + b$ is rediscovered and added to the kernel set. Before returning from the recursion, $ac + bc + d$ is added to the kernel set.
>
> Eventually the sixth variable, g, is considered. Since it is in the support of only one cube of f, no action is taken. The last step adds $f = ace + bce + de + g$ to the kernel set. The algorithm returns $K = \{(ace + bce + de + g), (a + b), (ac + bc + d), (a + b)\}$. A kernel is duplicated.

While this intermediate algorithm computes all kernels, it has a pitfall. It may compute the same kernel more than once. Algorithm 8.3.4, due to Brayton and Mc-Mullen [7], uses a pointer to the variables already considered. The pointer is denoted

$KERNELS(f, j)\{$

    $K = \emptyset;$

    for $i = j$ to n {

          if$(|CUBES(f, x_i)| \geq 2)$ {   /* skip cases where f/x yields one cube */

               C = largest cube containing x s.t. $CUBES(f, C) = CUBES(f, x);$

                  if $(x_k \notin C \;\forall k < i)$ /* skip if C contains already considered variables */

                      $K = K \cup KERNELS(f/f^C, i + 1);$ /* recur on f/f^C with pointer $i + 1$ */

          }

    }

    $K = K \cup f;$ /* add f to kernel set */

    return$(K);$

}

ALGORITHM 8.3.4

by j. In the algorithm, the support of f is assumed to be $sup(f) = \{x_1, x_2, \ldots, x_n\}$ at any level of the recursion.

The algorithm is applicable to cube-free expressions. Thus, either the function f is cube free or it is made so by dividing it by its largest cube factor, determined by the intersection of the support sets of all its cubes. The pointer j is set initially to 1.

At any level of the recursion, the variables under consideration are those with index larger than or equal to j. The recursive call is made only if the cube C under consideration does not contain any variable with index lower than j, i.e., already been considered. In the recursive call, the pointer is incremented.

> **Example 8.3.10.** Let us consider again expression $f = ace + bce + de + g$. It is cube free and so no pre-processing is required. Let us consider again the variables in lexicographic order. As in the previous case, no action is taken in conjunction with the first two variables, because both are contained in one cube only.
>
> Let $i = 3$. The third variable, c, is in the support of two cubes. The largest subset C of support variables, containing c and contained in the support set of both ace and bce, is again $\{c, e\}$. The recursive call of $KERNELS$ has as arguments expression $(ace + bce + de + g)/ce = a + b$ and pointer $j = 3 + 1$. The call considers then variables $\{d, e, g\}$ but it does not find any additional kernel. Nevertheless, it adds $a + b$ to the kernel set at the last step.
>
> The fourth variable is then considered, namely d. Since d is in the support of only one cube of f, no action is taken.
>
> Then let $i = 5$. The fifth variable, e, is in the support of three cubes of f. No other variable in $sup(f)$ is in the support of the same cubes. A recursive call is done with arguments $ac + bc + d$ and pointer $j = 5 + 1$. In the recursion, only the sixth variable, g, is considered and no kernel is discovered. Before returning from the recursion, $ac + bc + d$ is added to the kernel set.
>
> Eventually the sixth variable, g, is considered. Since it is in the support of only one cube of f, no action is taken. The last step adds $f = ace + bce + de + g$ to the kernel set. The algorithm returns $K = \{(ace + bce + de + g), (ac + bc + d), (a + b)\}$.

Now no kernel is duplicated. Note also that the pruning of the search space has sped up the algorithm.

EXTRACTION OF SINGLE-CUBE AND MULTIPLE-CUBE EXPRESSIONS. Different strategies are used for the extraction of single-cube and multiple-cube subexpressions. The first problem is referred to as *cube extraction* for brevity, while the second is often called *kernel extraction* in jargon, because multiple-cube common subexpressions are constructed on the basis of kernel intersections. In this section, we present heuristic greedy algorithms that are based on the computation of common single-cube and multiple-cube subexpressions based on kernel theory. In the next section, we show other approaches based on a matrix representation. Whereas the matrix-based approach is more efficient and commonly used, the kernel-based approach is presented first because it helps in understanding the power of kernel theory.

Let us consider cube extraction first. We are interested in extracting cubes (one at a time) that are common divisors of two (or more) expressions, i.e., that are intersections of two (or more) cubes in different expressions. Obviously, such cubes should have two or more variables to be relevant for extraction. The systematic search of cube intersections is based on the following observation. The set of non-trivial co-kernels of an algebraic expression in a *sum of products* form (i.e., sum of cubes) corresponds to the intersections of two or more of its cubes [8]. Equivalently, the co-kernels are identified by the intersections of the support sets of the cubes.

> **Example 8.3.11.** Given expression $f_x = ace + bce + de + g$, its co-kernel set is $\{ce, e, 1\}$. The first co-kernel is identified by the intersection $sup(ace) \cap sup(bce) = \{c, e\}$, the second by $sup(ace) \cap sup(bce) \cap sup(de) = \{e\}$; the last one is the trivial co-kernel.

The previous observation is useful in devising a method for computing all cube intersections. Let us consider a set of auxiliary variables, each associated with an internal vertex of the logic network: $\{a_i, \forall i : v_i \in V^G\}$. Let us form an auxiliary expression $f_{aux} = \sum\limits_{i: \, v_i \in V^G} a_i \cdot f_i$ that is the sum of the products of the auxiliary variables times the corresponding local expressions, and let us represent the auxiliary expression in a *sum of products* form. Note that each local expression is minimal w.r.t. single-cube containment (by hypothesis, because we are dealing with an algebraic model) and that the auxiliary expression f_{aux} is also minimal w.r.t. single-cube containment (by construction, because each local expression is multiplied by a different auxiliary variable). Hence the algebraic model and its properties are applicable to the auxiliary expression.

The non-trivial co-kernels of this auxiliary expression denote all cube intersections. Those non-trivial co-kernels, which do not have any auxiliary variables, represent cube intersections originating from different expressions, i.e., single-cube common divisors. Relevant cube intersections are those denoted by co-kernels with two or more variables.

> **Example 8.3.12.** Given expressions $f_x = ace + bce + de + g$ and $f_s = cde + b$, the auxiliary expression is $f_{aux} = a_x ace + a_x bce + ba_x de + a_x g + a_s cde + a_s b$. The co-kernels are $\{ce, de, b, a_x ce, a_x e, a_x, a_s, 1\}$. Relevant cube intersections are those denoted

by co-kernels ce or de. By extracting ce we get:

$$t = ce$$

$$x = t(a + b) + de + g$$

$$s = td + b$$

This extraction is shown in Figure 8.11. By extracting de we obtain instead:

$$t = de$$

$$x = ace + bce + t + g$$

$$s = tc + b$$

Note that the first extraction reduces the overall number of literals in the network fragment being considered, while the second extraction does not.

Therefore the *KERNELS* algorithm can be used for determining the cube intersections for extraction. The extraction takes place if favorable, i.e., if it improves the area and/or delay measure. When the literal count is used as the objective function, an extraction of an l-variable cube with multiplicity n in the network leads to a saving of $nl - n - l$ variables. The greedy algorithm shown in Algorithm 8.3.5, performs iteratively cube extraction until favorable candidates are available. At each step, the cube leading to the biggest gain is extracted.

The selection of the candidate cube may follow different criteria [8]. In some cases the search is limited to co-kernels of level 0, for the sake of computing efficiency.

Let us consider now the kernel extraction problem. The naive way to derive the candidate common subexpressions is to compute all kernel intersections by inspecting kernel pairs, triples, etc. Unfortunately the intersections may be many and lengthy to compute.

A better way for computing the kernel intersections is based on computing the co-kernels of an appropriate auxiliary function, as in the case of cube extraction. The difference, with respect to cube extraction, is the following. In cube extraction, we were looking for cube pairs with intersecting support in a *sum of products* form. For kernel intersections we are looking for intersections of the elements of the kernels.

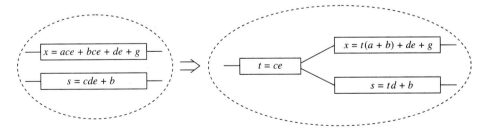

FIGURE 8.11
Example of the extraction of a single-cube subexpression.

$CUBE\_EXTRACT(\ G_n(V, E)\ )\{$

    while (some favorable common cube exist) **do** {

        C = select candidate cube to extract;

        Generate new label l;

        Add v_l to the network with expression $f_l = f^C$;

        Replace all expressions f where f_l is a divisor, by $l \cdot f_{quotient} + f_{remainder}$;

    }

}

ALGORITHM 8.3.5

These elements are cubes that are considered as atomic elements as far as their intersection is concerned, even though they are sets themselves. For this reason, we can rename each cube by a unique identifier, i.e., a new variable. A kernel is then represented by a set of these new variables.

> **Example 8.3.13.** Consider again Example 8.3.7. The kernel set of expression $f_x = ace + bce + de + g$ is $K(f_x) = \{(a+b), (ac+bc+d), (ace+bce+de+g)\}$. The kernel set of expression $f_y = ad + bd + cde + ge$ is $K(f_y) = \{(a+b+ce), (cd+g), (ad+bd+cde+ge)\}$.
>
>   Let $x_a = a$; $x_b = b$; $x_{ac} = ac$; $x_{bc} = bc$; $x_d = d$; $x_{ace} = ace$; $x_{bce} = bce$; $x_{de} = de$; $x_g = g$; $x_{ce} = ce$; $x_{cd} = cd$; $x_{ad} = ad$; $x_{bd} = bd$; $x_{cde} = cde$; $x_{ge} = ge\}$ be the new variables. Hence $K(f_x) = \{\{x_a, x_b\}, \{x_{ac}, x_{bc}, x_d\}, \{x_{ace}, x_{bce}, x_{de}, x_g\}\}$ and $K(f_y) = \{\{x_a, x_b, x_{ce}\}, \{x_{cd}, x_g\}, \{x_{ad}, x_{bd}, x_{cde}, x_{ge}\}\}$.

Let us now consider an auxiliary expression f_{aux} that is a sum of cubes, each cube being the product of the new variables corresponding to a kernel, an auxiliary variable denoting the local expression (to which the kernel belongs) and an additional auxiliary variable which is a kernel identifier. The set of non-trivial co-kernels of f_{aux} identifies the intersections among the support sets of two (or more) cubes. In addition, the co-kernels which do not have auxiliary variables represent intersections of kernels of different expressions, i.e., multiple-cube common divisors. Relevant kernel intersections are those denoted by co-kernels with two or more variables, because we are interested in multiple-cube extraction. Note that the kernel identifiers serve only the purpose to ensure minimality (w.r.t. single-cube containment) of the sum of two (or more) kernels in the same set and thus to make the co-kernel computation well defined.

> **Example 8.3.14.** Consider again Example 8.3.7. Then $f_{aux} = a_x k_{x1} x_a x_b + a_x k_{x2} x_{ac} x_{bc} x_d + a_x k_{x3} x_{ace} x_{bce} x_{de} x_g + a_y k_{y1} x_a x_b x_{ce} + a_y k_{y2} x_{cd} x_g + a_y k_{y3} x_{ad} x_{bd} x_{cde} x_{ge}$. Its co-kernels are $\{x_a x_b, x_g, a_x, a_y, 1\}$. The first co-kernel denotes the set $\{x_a, x_b\}$ corresponding to the kernel intersection $a + b$. The other co-kernels do not represent multiple-cube common subexpressions, and thus they are not useful for extraction. The extraction of $a + b$ is shown in Figure 8.12.

Therefore the search for kernel intersections can be done by invoking the $KERNELS$ algorithm after a simple variable renaming and manipulation. Whereas

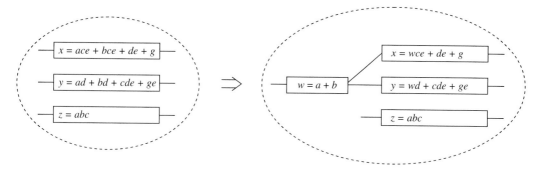

FIGURE 8.12
Example of the extraction of a multiple-cube subexpression.

appropriate co-kernels of f_{aux} identify common subexpressions, the selection of the
kernel intersections to be extracted is based on verifying that the transformation is
favorable, i.e., that the area and/or delay decreases. When the literal count is used as
an objective function, an extraction of an l-variable subexpression with n occurrences
in the network leads to a saving of $nl - n - l$ variables.

A greedy algorithm for kernel extraction is shown in Algorithm 8.3.6, as reported
in reference [8]. Two issues are noteworthy. First, the extraction of a subexpression,
and the consequent addition of a vertex to the network, may change the overall kernel
set. Hence kernels should be recomputed after an extraction, but this is computationally
expensive. A heuristic parameter n indicates the rate at which kernels are recomputed.
Namely, n means that kernels are recomputed after n extractions. Second, we may
restrict our attention to kernels of given levels. Parameter k indicates the maximum
level of the kernels being computed.

There is a trade-off in choosing the parameters. For example, a high value of
k corresponds to generating possibly large common subexpressions at the expense
of longer computing time. Similarly, a low value of n corresponds to updating the

$KERNEL\_EXTRACT(G_n(V, E), n, k)\{$
    while (some favorable common kernel intersections exist) **do** {
        Compute set of kernels of level $\leq k$;
        Compute kernel intersections;
        for ($i = 1$ to n) {
            f = select kernel intersection to extract;
            Generate new label l;
            Add v_l to the network with expression $f_l = f$;
            Replace all expressions f where f_l is a divisor by $l \cdot f_{quotient} + f_{remainder}$;
        }
    }
}

ALGORITHM 8.3.6

kernels frequently, at the expense of more computing time. Reference [8] reports a trade-off table.

Recent experimental results have shown that it is often convenient to restrict the kernel intersections to those consisting of two cubes. It was also shown that the choice of the subexpressions being extracted may affect adversely the overall circuit area if the routing area is too large. A technique based on a *lexicographic* order of the variables [1] selects subexpressions to be implemented only when the induced dependencies are compatible with this order. This correlates to a simpler wiring structure and to a smaller routing area.

MATRIX REPRESENTATION AND THE RECTANGLE COVERING PROBLEM. We describe now an alternative method for describing and computing kernels and co-kernels based on a matrix representation. Let us consider an algebraic expression in *sum of products* form f with m cubes and $n = |sup(f)|$ support variables. Let $A \in B^{m \times n}$ be the cube-variable incidence matrix. Element $a_{ij} \in A$ is 1 if and only if the jth variable belongs to the ith cube. For the sake of a simpler notation, let the rows and the columns be indexed with integers $\{1, 2, \ldots, m\}$ and $\{1, 2, \ldots, n\}$, respectively.

> **Definition 8.3.3.** A **rectangle** of a matrix A is a subset of rows and columns, where all entries are 1.

A rectangle is denoted by a pair (R, C) indicating the row and column subsets. A rectangle (R_1, C_1) *contains* another rectangle (R_2, C_2) when $R_1 \supseteq R_2$ and $C_1 \supseteq C_2$. A rectangle is *prime* if it is not strictly contained by another rectangle.

> **Definition 8.3.4.** The **co-rectangle** of a rectangle (R, C) is the pair (R, C'), where C' is the complement of the column subset C.

> **Example 8.3.15.** Consider expression $f_x = ace + bce + de + g$, with $m = 4$ cubes and $n = 6$ variables. The corresponding matrix (with the identifiers) is:
>
> | | var | a | b | c | d | e | g |
> |---|---|---|---|---|---|---|---|
> | Cube | $R \backslash C$ | 1 | 2 | 3 | 4 | 5 | 6 |
> | ace | 1 | 1 | 0 | 1 | 0 | 1 | 0 |
> | bce | 2 | 0 | 1 | 1 | 0 | 1 | 0 |
> | de | 3 | 0 | 0 | 0 | 1 | 1 | 0 |
> | g | 4 | 0 | 0 | 0 | 0 | 0 | 1 |
>
> A rectangle is $(\{1, 2\}, \{3, 5\})$. It is prime because there is no other rectangle containing it. Its co-rectangle is $(\{1, 2\}, \{1, 2, 4, 6\})$. Another prime rectangle is $(\{1, 2, 3\}, \{5\})$. Its co-rectangle is $(\{1, 2, 3\}, \{1, 2, 3, 4, 6\})$.

The cube-variable matrix associated with an algebraic expression allows us to describe non-trivial kernels and co-kernels in terms of rectangles. Indeed, a co-kernel corresponds to a prime rectangle with $|R| \geq 2$. The corresponding kernel is identified by the co-rectangle (R, C') by summing the terms corresponding to R restricted to the variables in C'. This observation can be derived directly from the definitions of kernels and rectangles.

> **Example 8.3.16.** Consider the cube-variable matrix of Example 8.3.15. Prime rectangle $(\{1, 2\}, \{3, 5\})$ identifies co-kernel ce, whose kernel is identified by $(\{1, 2\}, \{1, 2, 4, 6\})$ by summing the first two cubes (i.e., $ace + bce$) and restricting the sum to variables $\{a, b, d, g\}$. Hence the kernel is $a + b$.
>
> The other prime rectangle $(\{1, 2, 3\}, \{5\})$ identifies co-kernel e. The kernel is $ac + bc + d$, i.e., the sum of the first three cubes restricted to variables $\{a, b, c, d, g\}$, as indicated by the co-rectangle.

This matrix representation supports other methods for computing kernels and co-kernels [6, 9, 22, 37]. In addition, it provides a good framework for solving directly the cube and kernel extraction problems. The power of the matrix representation of an expression stems from the fact that it is not necessarily bound to satisfy all assumptions of the algebraic model. Thus non-minimal expressions (w.r.t. single-cube containment) can be represented and manipulated.

Let us consider cube extraction first. Let us form the cube-variable matrix corresponding to an auxiliary expression f_{aux} that is the sum of all expressions in the network. An identifier denotes the membership of the cube to the original expressions. A rectangle with $|R| \geq 2$ identifies a common cube, which is a local best choice when it is prime. The cube is common to two (or more) different expressions if the rows have two (or more) different identifiers. Note that auxiliary variables are not needed, because expression f_{aux} does not need to be minimal (w.r.t. single-cube containment). Indeed, cube intersections are determined by rectangles and a co-kernel computation is not required.

> **Example 8.3.17.** Consider again Example 8.3.12. Given expressions $f_x = ace + bce + de + g$ and $f_s = cde + b$, the auxiliary function is $f_{aux} = ace + bce + de + g + cde + b$. The cube-variable matrix is:

		var	a	b	c	d	e	g
Cube	ID	$R\backslash C$	1	2	3	4	5	6
ace	x	1	1	0	1	0	1	0
bce	x	2	0	1	1	0	1	0
de	x	3	0	0	0	1	1	0
g	x	4	0	0	0	0	0	1
cde	s	5	0	0	1	1	1	0
b	s	6	0	1	0	0	0	0

> A prime rectangle is $(\{1, 2, 5\}, \{3, 5\})$, corresponding to cube ce, which is present in both expressions f_x (row 1,2) and f_s (row 5). Cube ce can be extracted, yielding the same network shown in Figure 8.11.

Let us consider now kernel intersections. We need to construct now a kernel-cube incidence matrix that has as many rows as the kernels and as many columns as the cubes forming the kernels. If such cubes are re-labeled by new variables, as done in the previous section, then the matrix can be seen again as a cube-variable matrix, where the cubes correspond to the sets of new variables defining the kernels.

We can then form again an auxiliary function f_{aux} that is the sum of all kernels and a corresponding incidence matrix with identifiers denoting the membership of a kernel to a kernel set. A rectangle with $|R| \geq 2$ identifies a kernel intersection, which is again a local best choice when it is prime. The kernel intersection is common to two (or more) kernels in different kernel sets if the rows have two (or more) different identifiers. Note again that the auxiliary variables are not needed.

Example 8.3.18. Consider again the network of Example 8.3.7, with the following expressions:

$$f_x = ace + bce + de + g$$
$$f_y = ad + bd + cde + ge$$
$$f_z = abc$$

The kernel sets are $K(f_x) = \{(a + b), (ac + bc + d), (ace + bce + de + g)\}$, $K(f_y) = \{(a + b + ce), (cd + g), (ad + bd + cde + ge)\}$, $K(f_z) = \emptyset$. By labeling cubes as in Example 8.3.14, $f_{aux} = x_a x_b + x_{ac} x_{bc} x_d + x_{ace} x_{bce} x_{de} x_g + x_a x_b x_{ce} + x_{cd} x_g + x_{ad} x_{bd} x_{cde} x_{ge}$. The kernel-cube matrix is shown in Table 8.1.

A prime rectangle is $(\{1, 4\}, \{1, 2\})$, corresponding to kernel intersection $a + b$. Since the identifiers of rows 1 (i.e., x) and 4 (i.e., y) are different, the intersection corresponds to a subexpression that can be extracted from two different expressions.

The "greedy" approach can be applied to both cube and kernel extraction with the matrix notation. The major steps are the rectangle selection and the matrix update. The former step involves associating a value with each rectangle that measures the local gain by choosing that rectangle. The update of the matrix must reflect the presence of a new variable in the network after the extraction. Hence a column must be added in correspondence to that variable. For cube extraction, the new cube corresponds to a new row in the cube-variable matrix. For kernel extraction, the kernels of the new expression being extracted must be appended to the matrix.

The greedy approach to extraction may be myopic, because only the local gain of one extraction is considered at a time. Algorithms for the simultaneous extraction of cubes and kernels have also been proposed based on the matrix formulation [9, 37]. The idea lies in the search for a cover of the matrix by rectangles that optimizes an objective function related to the total gain of the extraction.

If the gain in area and/or delay related to prime rectangles were always superior to the gain related to rectangles covered by primes, then we could devise a procedure similar to the Quine-McCluskey algorithm for exact two-level logic minimization (Section 7.2.2). That is, we would search for the set of prime rectangles and then determine a minimum-cost cover. Unfortunately, it was shown that Quine's theorem cannot be applied to rectangles, because non-prime rectangles can contribute to lower-cost covers [9, 37]. Nevertheless, heuristic algorithms have been developed for the

TABLE 8.1
Kernel-cube matrix.

			a x_a 1	b x_b 2	d x_d 3	g x_g 4	ac x_{ac} 5	ad x_{ad} 6	bc x_{bc} 7	bd x_{bd} 8	cd x_{cd} 9	ce x_{ce} 10	de x_{de} 11	ge x_{ge} 12	ace x_{ace} 13	bce x_{bce} 14	cde x_{cde} 15
$a+b$	$x_a x_b$	x 1	1	1	0	0	0	0	0	0	0	0	0	0	0	0	0
$ac+bc+d$	$x_{ac} x_{bc} x_d$	x 2	0	0	1	0	1	0	1	0	0	0	0	0	0	0	0
$ace+bce+de+g$	$x_{ace} x_{bce} x_{de} x_g$	x 3	0	0	0	1	0	0	0	0	0	0	1	0	1	1	0
$a+b+ce$	$x_a x_b x_{ce}$	y 4	1	1	0	0	0	0	0	0	0	1	0	0	0	0	0
$cd+g$	$x_{cd} x_g$	y 5	0	0	0	1	0	0	0	0	1	0	0	0	0	0	0
$ad+bd+cde+ge$	$x_{ad} x_{bd} x_{cde} x_{ge}$	y 6	0	0	0	0	0	1	0	1	0	0	0	1	0	0	1

simultaneous cube and kernel extraction, where a prime cover is iteratively refined. We refer the interested reader to references [6] and [37] for the details.

We close the section by recalling that the expressive power of the matrix formulation goes beyond the algebraic model. Indeed cube-variable matrices can include *don't care* entries and rectangle covering methods can exploit some Boolean properties of the functions that lead to lower-cost solutions. We again refer the interested reader to reference [37] for the details.

8.3.3 Decomposition

The decomposition of logic functions has been the object of intensive investigation that traces back to the methods proposed by Ashenhurst [2] and Curtis [16]. Davidson [20] and Dietmeyer with Su [26] developed methods for transforming two-level covers into networks of NAND (or NOR) gates. We do not survey these classical methods here, because they are reported in textbooks on switching theory [16, 25] and they use Boolean operations. We present instead algebraic methods for decomposing large expressions.

Such a decomposition has two goals. First it reduces the size of the expressions to that typical of library cells used in the implementation or to that acceptable by cell layout generators. Second, small-sized expressions are more likely to be divisors of other expressions. Hence decomposition may enhance the ability of the $SUBSTITUTE$ algorithm to reduce the size of the network.

We consider here decompositions based on algebraic division. Given an expression f, let $f_{divisor}$ be one of its divisors. The decomposition associates a new variable, say t, with the divisor and reduces the original expression to $f = t \cdot f_{quotient} + f_{remainder}$; $t = f_{divisor}$. Then decomposition can be applied recursively to the divisor, quotient and remainder.

> **Example 8.3.19.** Consider expression $f_x = ace + bce + de + g$. Let us select the following divisor: $f_{divisor} = ac + bc + d$. Then, by introducing a new variable t, we have:
>
> $$f_x = te + g$$
> $$f_t = ac + bc + d$$
>
> The decomposition can be applied recursively only to f_t with success in this case. Indeed, by choosing $f_{divisor} = a + b$ as a divisor of f_t and by associating it with a new variable s, we obtain:
>
> $$f_x = te + g$$
> $$f_t = sc + d$$
> $$f_s = a + b$$
>
> The decomposition is shown in Figure 8.13.

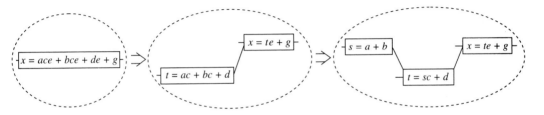

FIGURE 8.13
Example of a decomposition.

An important issue is the choice of the divisor. An obvious choice is a kernel of the expression to be decomposed. Different heuristics have been tried along this line. A fast method is to pick one level-0 kernel. A more accurate one is to evaluate all kernels and select the most promising one.

A slight variation of this method is to choose a divisor that yields either a cube-free or a single-variable quotient. Such a divisor can be computed by performing first a division by a kernel. Then the quotient is reduced to a cube-free expression (if a multi-cube expression) or to a single variable (if a cube) and is used as a divisor, as shown in the following example. (Note that two divisions are involved and that the quotient and divisor exchange roles.) This trick can compensate a poor choice of a kernel. Comparisons of different decomposition strategies, as well as extensions to non-algebraic decompositions (by replacing the algebraic division by Boolean methods), have been reported in reference [8].

Example 8.3.20. Consider again expression $f_x = ace + bce + de + g$. Let us select kernel $a + b$, which yields $f_x = tce + de + g$; $f_t = a + b$. Whereas the choice of another kernel would have given a more balanced decomposition (as shown by Example 8.3.19), we can improve the decomposition by taking advantage of extracting also variable e from the original expression.

This is exactly what the refined method does. The first divisor is $a + b$ and the quotient ce. We reduce the quotient to a single variable, say c, and we divide again the original expression by that variable. In this case, the second quotient is $e(a + b)$, which is used as the new expression, namely:

$$f_x = tc + de + g$$
$$f_t = e(a + b)$$

Decomposition can be applied to all expressions of the network whose size is above a given threshold k. In general, decomposition increases the area of the network, because the number of variables and vertices increases. The number of stages can also increase and consequently the delay increases, even though under some delay models more stages related to simpler expressions are faster than fewer stages with more complex expressions.

$DECOMPOSE(G_n(V,E),k)\{$

 repeat {

   v_x = selected vertex with expression whose size is above k;

   if $(v_x = \emptyset)$ **return**;

   Decompose expression f_x;

 }

}

ALGORITHM 8.3.7

A general framework of the decomposition algorithm is shown in Algorithm 8.3.7.

8.4 THE BOOLEAN MODEL

We consider now transformations on logic networks that use the full power of the Boolean model. Hence, each vertex of the network is associated with a local Boolean function and a local *don't care* set.

The fundamental transformation is the Boolean simplification of one or more local functions. If the functions are expressed in *sum of products* form, then exact (or heuristic) two-level logic optimization techniques can be used for this purpose, with slight modifications, such as the search for minimal-literal (instead of minimal-term) forms.

We shall revisit two-level logic optimization in Section 8.4.2, where we consider also the relations between simplification and Boolean substitution. Since all Boolean techniques rely on the use of the *don't cares* as degrees of freedom for optimization, we shall consider first the problem of their evaluation in Section 8.4.1. We suggest the novice reader to skip the sections on the algorithms for computing the *don't care* conditions in a first reading and concentrate on the meaning of *don't cares* and their use in logic optimization without delving into complex mathematical models.

Eventually we shall survey other Boolean transformations for multiple-level logic optimization, such as redundancy removal, transduction, global flow and polarity assignment.

8.4.1 *Don't care* Conditions and Their Computation

Generally speaking, *don't care* conditions are related to the embedding of a Boolean function in an environment. The knowledge of the environment can be captured by means of the relevant patterns at the primary inputs and outputs. The *don't care* conditions related to the environment, called collectively *external don't care* conditions, consist of a *controllability* and an *observability* component.

Whereas combinational *don't cares* are specified as sets, they can also be represented by Boolean functions. For the sake of simplicity, we do not distinguish in this chapter between *don't care* sets and their corresponding functions. Therefore logic operators like $(+, \cdot, \oplus, ')$ are also applied to sets with an abuse of notation. When different *don't care* conditions are applicable to the various network outputs, we use the vector notation to represent them collectively. Vectors of *don't care* conditions can be seen as vectors of representative functions, or equivalently as a linear array of *don't care* sets. When using the vector notation, we assume that the dimensions match the number of primary outputs n_o of the network, unless otherwise stated.

> **Definition 8.4.1.** The **input controllability** *don't care* set includes all input patterns that are never produced by the environment at the network's inputs.

Input *controllability don't cares* are denoted by the set CDC_{in}.

> **Definition 8.4.2.** The **output observability** *don't care* sets denote all input patterns that represent situations when an output is not observed by the environment.

The output *observability don't care* sets are represented by a vector $\mathbf{ODC}_{out}$, with as many elements as the number n_o of primary outputs. Each element denotes when a primary output is not observed.

The *external don't care* conditions are denoted by $\mathbf{DC}_{ext} = \mathbf{CDC}_{in} \cup \mathbf{ODC}_{out}$, where $\mathbf{CDC}_{in}$ is a vector with n_o entries equal to CDC_{in}. Note that the vector notation is necessary to distinguish the *don't care* components related to different outputs.

> **Example 8.4.1.** Consider network $\mathcal{N}_1$ of Figure 8.14, with inputs and outputs denoted by **x** and **y**, respectively. The remaining circuitry, shaded in the picture, is considered the environment. Network $\mathcal{N}_1$ is fed by a demultiplexer, so that one and only one of the signals $\{x_1, x_2, x_3, x_4\}$ is TRUE at any given time. For example, pattern 0000 can never be an input to $\mathcal{N}_1$. Thus, the input controllability *don't care* set is $CDC_{in} = x_1'x_2'x_3'x_4' + x_1x_2 + x_1x_3 + x_1x_4 + x_2x_3 + x_2x_4 + x_3x_4$.
>
> The output of network $\mathcal{N}_2$ is in conjunction with x_1. Thus the values of y_1 and y_2 are irrelevant when x_1 is FALSE. The observability *don't care* set of y_1 and y_2 is represented by x_1'. Similarly, the observability *don't care* set of y_4 and y_3 is represented by x_4'. When considering network $\mathcal{N}_1$ with $n_o = 4$ outputs, its *ODC* is the vector:

$$\mathbf{ODC}_{out} = \begin{bmatrix} x_1' \\ x_1' \\ x_4' \\ x_4' \end{bmatrix}$$

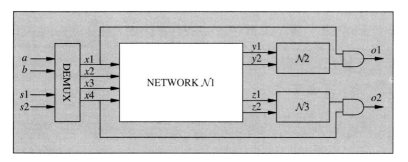

FIGURE 8.14
Example of a logic network $\mathcal{N}_1$ and its environment (shaded).

The overall external *don't care* set is:

$$\mathbf{DC}_{ext} = \mathbf{CDC}_{in} + \mathbf{ODC}_{out}$$

$$= \begin{bmatrix} x_1' x_2' x_3' x_4' + x_1 x_2 + x_1 x_3 + x_1 x_4 + x_2 x_3 + x_2 x_4 + x_3 x_4 + x_1' \\ x_1' x_2' x_3' x_4' + x_1 x_2 + x_1 x_3 + x_1 x_4 + x_2 x_3 + x_2 x_4 + x_3 x_4 + x_1' \\ x_1' x_2' x_3' x_4' + x_1 x_2 + x_1 x_3 + x_1 x_4 + x_2 x_3 + x_2 x_4 + x_3 x_4 + x_4' \\ x_1' x_2' x_3' x_4' + x_1 x_2 + x_1 x_3 + x_1 x_4 + x_2 x_3 + x_2 x_4 + x_3 x_4 + x_4' \end{bmatrix}$$

$$= \begin{bmatrix} x_1' + x_2 + x_3 + x_4 \\ x_1' + x_2 + x_3 + x_4 \\ x_4' + x_2 + x_3 + x_1 \\ x_4' + x_2 + x_3 + x_1 \end{bmatrix}$$

It is obvious that any logic network does not have *per se* a set of external *don't cares* but that these are due to the circuits feeding (and being fed by) the network itself. We can then extend this reasoning to the internals of a network. Let us assume we are considering a connected subnetwork of a logic network. The boundary of the subnetwork has its own ports, i.e., inputs and outputs. Then, we can define the external controllability and observability *don't cares* for the subnetwork that are internal *don't cares* for the overall network. This is shown in Figure 8.15.

In the limit, we may consider the subnetwork as consisting of only one internal vertex. The controllability *don't cares* for that vertex are the patterns that are never possible as inputs to the corresponding expression. Similarly, the observability *don't cares* are those determining when the variable related to the vertex is never observed at any output. The combination of the controllability and observability *don't cares* at a vertex is called the *local don't care* set. The controllability, observability and local *don't care* sets (in short CDC, ODC and DC) of a variable are represented by using that variable as a subscript, e.g., these sets for variable x are denoted as CDC_x, ODC_x and DC_x, respectively, when the network has a single output and as $\mathbf{CDC}_x$, $\mathbf{ODC}_x$ and $\mathbf{DC}_x$, respectively, when the network has multiple outputs. Boolean simplification of a local expression can exploit these *don't care* conditions.

Example 8.4.2. Consider the portion of the network shown in Figure 8.16 (a) and defined by the expressions:

$$x = a' + b$$

$$y = abx + a'cx$$

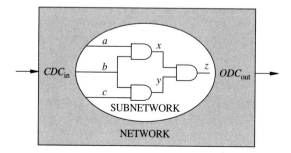

SUBNETWORK

NETWORK

FIGURE 8.15
Example of a logic network and a subnetwork.

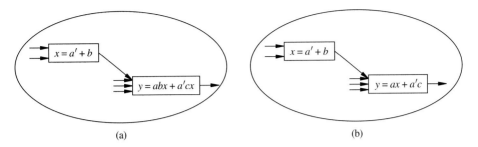

FIGURE 8.16
(a) Example of a logic network. (b) Example of a logic network after simplification.

Let us focus on vertex v_y. We claim that the pattern $ab'x$ can never be an input to f_y. This is true, because $ab' = 1$ implies $x = a' + b = 0$. Hence $ab'x$ is part of the CDC of vertex v_y. Similarly, $a'x'$ can never be an input to f_y, because $a = 0 \Rightarrow x = 1$.

We can now use the CDC set to optimize f_y. Let us minimize $f_y = abx + a'cx$ with $ab'x + a'x'$ as the *don't care* set. We obtain $\tilde{f}_y = ax + a'c$. The optimized network is shown in Figure 8.16 (b). Note that this simplification would not have been possible without considering the *don't care* conditions.

We use in the sequel the concept of *perturbed network* for both network analysis and optimization.

Definition 8.4.3. Given a logic network $G_n(V, E)$ and a vertex v_x, the **perturbed network** at v_x is the one obtained by replacing the local function f_x by $f_x \oplus \delta$, where δ is an additional input called perturbation.

We represent the input/output behavior of the perturbed network by $\mathbf{f}^x(\delta)$, where the superscript on $\mathbf{f}$ denotes the perturbation placed at v_x. When $\delta = 0$ the perturbed network has behavior equal to the original network. When $\delta = 1$, the perturbed network differs from the original network in that variable x has changed polarity.

A variable is observable if a change in its polarity is perceived at one of the network's output. Thus, the perturbed network model gives us a mechanism to define and compute ODC sets. Namely, the observability *don't care* of variable x is just $\mathbf{f}^x(0) \; \overline{\oplus} \; \mathbf{f}^x(1)$.

Example 8.4.3. Consider the network of Figure 8.17 (a). Since the network has a single output, we use the scalar notation. The input/output behavior of the network is $f = abc$.

Figure 8.17 (b) shows the network perturbed at x. The input/output behavior of the perturbed network is $f^x(\delta) = bc(\delta \oplus ab)$. Note that $f = f^x(0)$ and that $f^x(1) = a'bc$.

The observability *don't care* set for variable x corresponds to those input patterns where a change in polarity of x does not affect the output, namely, $ODC_x = (abc) \oplus (a'bc) = abc \cdot a'bc + (abc)' \cdot (a'bc)' = b' + c'$. Indeed, when either b or c is FALSE, $y = 0$ and $z = 0$ independently of x.

Let us optimize then the local function $f_x = ab$ with the *don't care* set $ODC_x = b' + c'$. A minimum solution is $\tilde{f}_x = a$, which corresponds to replacing the AND gate by a wire, as shown in Figure 8.17 (c).

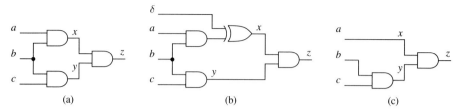

FIGURE 8.17
(a) Example of a logic network. (b) Perturbed network. (c) Optimized network.

It is interesting to compare the *don't care* sets associated with multiple-level networks to those associated with Boolean functions. External *don't care* conditions can be defined in both cases. Nevertheless internal *don't care* conditions are specific to logic networks, because they are related to the structure of the network itself and to the interplay of the local functions.

Let us consider now a generic logic network. Since the network can be seen as a set of equations, these equations state the possible relations among the network variables. Consider any vertex of the network, say v_x. The assignment $x = f_x$ relates x with the variables in $sup(f_x)$. Conditions described by $x \neq f_x$, or equivalently by $x \oplus f_x$, are not possible. Hence they can be considered as *don't cares*.

> **Example 8.4.4.** Consider the expression $x = a' + b$. Since $x \neq a' + b$ is not possible, $x \oplus (a' + b) = x'a' + x'b + xab'$ is a *don't care* condition.

The *don't care* conditions induced by the equations of a logic network are called *satisfiability don't care* conditions.

> **Definition 8.4.4.** The **satisfiability** *don't care* conditions of a network $G_n(V, E)$ is the set:
> $$SDC = \sum_{v_x \in V^G} x \oplus f_x \tag{8.1}$$

> **Example 8.4.5.** Consider again the network of Figure 8.16 (a). Then $SDC = x \oplus (a' + b) + y \oplus (abx + a'cx) = x'a' + x'b + xab' + y'abx + y'a'cx + ya'c' + yb'a + yb'c' + yx'.$

The SDC set can be seen as a Boolean function over the network variables. The computation of the satisfiability *don't cares* is straightforward, even though representations of the SDC set may be large at times. The evaluation of the controllability and observability *don't cares* is more involved and is detailed in the following sections.

ALGORITHMS FOR COMPUTING THE CDC SET.* The knowledge of the impossible patterns at the network's input CDC_{in} and of the satisfiability *don't cares* SDC allows us to compute the set of output patterns that the network cannot produce, termed *output controllability don't care* conditions and denoted by CDC_{out}. Similarly, this information is sufficient to compute the patterns that are not fed as input to any subnetwork, and in particular to any single vertex, i.e., the *internal controllability don't care* sets.

It is important to recognize that the complement of the set CDC_{out}, i.e., $(CDC_{out})'$, is the set of all possible patterns that the network can produce for any possible input $(CDC_{in})'$. Therefore $(CDC_{out})'$ is the *image* of $(CDC_{in})'$ under the function **f** representing the input/output behavior of the network.

There are two major approaches for computing internal and output CDC sets. The first relies on network traversal and the second on image computation.

The first method consists of computing the controllability *don't care* sets associated with successive *cuts* of the networks. Cuts are vertex separation sets in the network graph and correspond also to subsets of variables. The cut is moved from the primary inputs to the primary outputs. Namely, the initial cut is the set of primary input variables. To move the cut, vertices are sorted in a sequence compatible with the partial order induced by the network graph and are added to the cut one at a time. Vertices whose direct successors are all in the cut can be dropped from the cut.

For each vertex under consideration, its contribution to the SDC set is taken into account, by adding it to the CDC set of the current cut. For all vertices being dropped from the cut, the corresponding variables are removed by applying repeatedly the *consensus* operation on the local CDC set, i.e., by preserving the component of the CDC set that is independent of those variables. In Algorithm 8.4.1, the cut is denoted by set C and the vertices being dropped are denoted by D.

The complexity of the algorithm is linear in the size of the graph if we bound the complexity of performing the *consensus* operation, which unfortunately may be potentially explosive.

For any given cut CDC_{cut} is expressed in the local variables of the cut, and hence it can be directly used for logic optimization of the subnetwork being driven.

CONTROLLABILITY($G_n(V, E)$, CDC_{in}) {
    $C = V^I$;
    $CDC_{cut} = CDC_{in}$;
    foreach vertex $v_x \in V^G$ in topological order {
        $C = C \cup v_x$;   /* add v_x to the cut */
        $CDC_{cut} = CDC_{cut} + f_x \oplus x$;   /* add local SDC component */
        $D = \{v \in C \text{ s.t. all direct successors of } v \text{ are in } C\}$;
        foreach vertex $v_y \in D$
            $CDC_{cut} = C_y(CDC_{cut})$;   /* remove dependencies from variables in D */
        $C = C - D$;   /* drop D from the cut */
    }
    $CDC_{out} = CDC_{cut}$;
}

ALGORITHM 8.4.1

Example 8.4.6. Consider the network shown in Figure 8.18. Let us determine the impossible assignments for the variables $z_1 = d$ and $z_2 = e$, given the input *don't care* set $CDC_{in} = x_1' x_4'$.

The first vertex being selected is v_a. Its contribution to CDC_{cut} is $a \oplus (x_2 \oplus x_3)$. Now variables $D = \{x_2, x_3\}$ can be dropped from the cut. The consensus operation leaves $CDC_{cut} = x_1' x_4'$.

Vertex b is selected next, adding $b \oplus (x_1 + a)$ to CDC_{cut}. Now variable x_1 can be dropped from the cut. The consensus operation leaves $CDC_{cut} = b' x_4' + b' a$.

Then vertex c is chosen. For the *cut* $\{b, c\}$ we have $CDC_{cut} = b'c'$. Eventually the algorithm reaches the primary outputs. The output *controllability don't care* set is $CDC_{out} = e' = z_2'$.

This means that the network can never yield a 0 on output z_2. From an intuitive standpoint, this is guaranteed by noticing that $(x_1 = 1$ or $x_4 = 1)$ are sufficient to imply $z_2 = 1$. But both x_1 and x_4 cannot be FALSE, because $x_1' x_4'$ is part of CDC_{in}. The different cuts are shown in Figure 8.18 (c).

A more efficient method to compute *controllability don't care* sets is by means of *image* computation. This method, devised first for the implicit enumeration of the states of a finite-state machine [15], was adapted later to the computation of *don't cares* in combinational networks [42].

For a given *cut* of the network (or of a subnetwork) with n_c variables, let $\mathbf{f} : B^{n_i} \to B^{n_c}$ be the vector function that maps the primary inputs into the variables of the cut. Note that when the *cut* is the set of primary outputs, $CDC_{cut} = CDC_{out}$ and $\mathbf{f}$ is the network input/output behavior. Then CDC_{cut} is just the complement of the *image* of $(CDC_{in})'$ under $\mathbf{f}$. For the sake of simplicity, assume first that $CDC_{in} = \emptyset$. Then CDC_{cut} is just the complement of the *range* of $\mathbf{f}$.

We show now how to compute the range of a function in terms of its output variables. Let us consider first the range of a single-output function $y = f(\mathbf{x})$. In general the function can evaluate to TRUE or FALSE according to the input values $\mathbf{x}$.

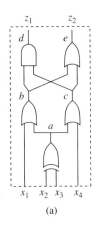

(a)

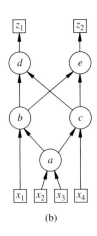

(b)

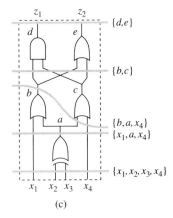

(c)

FIGURE 8.18
(a) Example of a logic network. (b) Network graph. (c) Cuts in a logic network.

Hence the range can be expressed in terms of the output variable as $y + y'$. Exceptions are the cases when f is a tautology ($f = 1$) and its range is y, or f' is a tautology ($f = 0$) and its range is y'.

Let us consider now the range of an m-output function $\mathbf{y} = \mathbf{f(x)} = [f^1, f^2, \ldots, f^m]^T$. The range of $\mathbf{f}$ can be expanded using the output variables as a basis as follows:

$$range\,(\mathbf{f(x)}) = y_1\ range\left(\begin{bmatrix} f^2 \\ f^3 \\ \ldots \\ f^m \end{bmatrix}_{|f^1=1}\right) + y_1'\ range\left(\begin{bmatrix} f^2 \\ f^3 \\ \ldots \\ f^m \end{bmatrix}_{|f^1=0}\right) \qquad (8.2)$$

In words, the identity means that the *range* is the sum of two subsets. The former corresponds to y_1 TRUE in conjunction with the *range* of the other components restricted to the subset of the domain where y_1 is TRUE. The latter corresponds to y_1 FALSE with the *range* of the other components restricted to the subset of the domain where y_1 is FALSE. Note that the restriction is the *generalized cofactor*, as described in Section 2.5.1, with respect to the orthonormal basis provided by f^1 and its complement. Expansion 8.2 can be applied recursively by considering all output variables, until the function is scalar and the range can be evaluated. The trick is to use the output variables for the expansion, while $\mathbf{f}$ is expressed in terms of the input variables. Note that the method is applicable to the range computation of any set of *cut* variables $\mathbf{y}$.

Example 8.4.7. Let us consider again the network of Example 8.4.6, shown in Figure 8.18. Assume that $CDC_{in} = \emptyset$. Without loss of generality, let us use the identity $a = x_2 \oplus x_3$. Variables d and e can be expressed in terms of $\{x_1, a, x_4\}$ by the vector function:

$$\mathbf{f} = \begin{bmatrix} f^1 \\ f^2 \end{bmatrix} = \begin{bmatrix} (x_1 + a)(x_4 + a) \\ (x_1 + a) + (x_4 + a) \end{bmatrix} = \begin{bmatrix} x_1 x_4 + a \\ x_1 + x_4 + a \end{bmatrix}$$

The range of $\mathbf{f}$ is evaluated in terms of the output variables $\{d, e\}$ as follows:

$$\begin{aligned}
range(\mathbf{f}) &= d\ range(f^2|_{(x_1 x_4 + a)=1}) + d'\ range(f^2|_{(x_1 x_4 + a)=0}) \\
&= d\ range((x_1 + x_4 + a)|_{(x_1 x_4 + a)=1}) + d'\ range((x_1 + x_4 + a)|_{(x_1 x_4 + a)=0}) \\
&= d\ range(1) + d'\ range(a'(x_1 \oplus x_4)) \\
&= de + d'(e + e') \\
&= d' + e
\end{aligned}$$

where the domain restriction is easily computed, as shown in Figure 8.19.

Thus $d' + e$ are the possible patterns, while $CDC_{out} = (d' + e)' = de' = z_1 z_2'$ represents the impossible pattern. This is justified intuitively by noticing that an OR gate cannot have a FALSE output (z_2') when an AND gate with the same inputs has a TRUE one (z_1).

The effectiveness of the method stems from the fact that $\mathbf{f}$ can be represented efficiently when the local functions in a logic network are modeled by BDDs. In

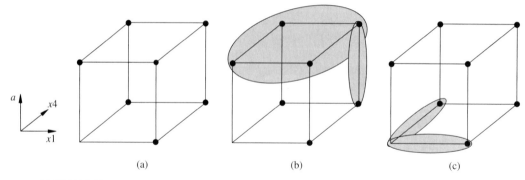

FIGURE 8.19
(a) Minterms of f^2. (b) The shaded region denotes $f^1 = 1$. Function f^2 is always true in the shaded region. (c) The shaded region denotes $f^1 = 0$. Function f^2, restricted to the shaded region, is $a'(x_1 \oplus x_4)$.

addition, the range computation can be transformed into an existential quantification of the characteristic equation $\chi(\mathbf{x}, \mathbf{y}) = 1$ representing $\mathbf{y} = \mathbf{f}(\mathbf{x})$, namely $range(\mathbf{f}) = \mathcal{S}_\mathbf{X}(\chi(\mathbf{x}, \mathbf{y}))$ [15, 42]. BDD representations of characteristic functions and applying the smoothing operator to BDDs are simple and effective, thus making this method practical for large circuits.

We would like to comment now on the case in which CDC_{in} is not empty. Then, the *range* computations must be replaced by the *image* computation of CDC'_{in} under $\mathbf{f}$.

> **Example 8.4.8.** Let us consider again the network of Example 8.4.7, with $CDC_{in} = x'_1 x'_4$. The same expansion used before applies, having changed the *range* computation into an *image* computation of $(CDC_{in})' = x_1 + x_4$.
>
> Thus, function $f^2 = x_1 + x_4 + a$ has to be considered first in the restricted domain delimited by both $x_1 x_4 + a = 1$ and $x_1 + x_4 = 1$. Function f^2 is always TRUE in this restricted domain. Then, function $f^2 = x_1 + x_4 + a$ has to be restricted to both $x_1 x_4 + a = 0$ and $x_1 + x_4 = 1$. Function f^2 is always TRUE also in this second restricted domain. The restriction is shown graphically in Figure 8.20.
>
> Expressions similar to those of Example 8.4.7 hold, after having changed *range* into *image*. By using an expansion in terms of variable d, since both restrictions of f^2 yield a tautology, we get $image(CDC'_{in}, \mathbf{f}) = de + d'e = e$. As a result, $CDC_{out} = e' = z'_2$, as computed by the CONTROLLABILITY algorithm before.

The image computation CDC'_{in} under $\mathbf{f}$ can be transformed into a range computation of the generalized cofactor of $\mathbf{f}$ w.r.t. CDC'_{in}. Thus, the image computation can be derived by an existential quantification of the characteristic equation $\chi(\mathbf{x}, \mathbf{y})$ intersected with CDC'_{in}.

> **Example 8.4.9.** Let us consider again the network of Example 8.4.7, with $CDC_{in} = x'_1 x'_4$. The characteristic equation is:
>
> $$(d \ \overline{\oplus} \ (x_1 x_4 + a)) \ (e \ \overline{\oplus} \ (x_1 + x_4 + a)) = 1$$

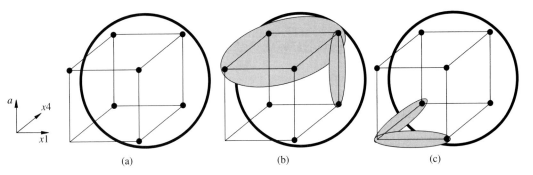

(a) (b) (c)

FIGURE 8.20
(a) Minterms of f^2 and a ring showing restriction of the domain by excluding CDC_{in}. (b) Minterms of f^2. The shaded region ($f^1 = 1$) inside the ring denotes the restricted domain of interest. Function f^2 is always true in the shaded region. (c) Minterms of f^2. The shaded region ($f^1 = 0$) inside the ring denotes the restricted domain of interest. Function f^2 is always true in the shaded region.

by expanding the expression:

$$\chi = de(x_1x_4 + a) + d'ea'(x_1'x_4 + x_1x_4') + d'e'a'x_1'x_4'$$

The range of **f** is $\mathcal{S}_{ax_1x_4}(\chi) = de + d'e + d'e' = d' + e$ as in Example 8.4.8. The image of $CDC_{in}' = x_1 + x_4$ is equal to the range of $\chi(x_1 + x_4)$, i.e., to the range of:

$$de(x_1x_4 + a)(x_1 + x_4) + d'ea'(x_1'x_4 + x_1x_4')$$

which is $\mathcal{S}_{ax_1x_4}(de(x_1x_4 + a)(x_1 + x_4) + d'ea'(x_1'x_4 + x_1x_4')) = de + d'e = e$ as in Example 8.4.8.

The size of the CDC set may be large at times, and therefore it may be convenient to use subsets of the CDC set in logic optimization. The subsets can be selected by using filters that drop some terms that are useless or unlikely in the optimization of one or more local functions of interest. Filters will be described in Section 8.4.2. As far as CDC conditions are concerned, the use of any local subset is always safe, because it can lead only to underestimating other local CDC subsets.

ALGORITHMS FOR COMPUTING THE ODC SETS.* The observability of a network variable can be tested by assuming that the variable can be changed in polarity. The conditions under which such a change is not perceived at an output is the corresponding *observability don't care* set. As an example, let us consider a specific internal variable x and an output variable y. Consider the behavior of a (single-output) network as expressed by a logic function $y = f(x)$. Then, the *observability don't care* set is just the complement of the Boolean difference $\partial f / \partial x = f|_{x=1} \oplus f|_{x=0}$, i.e., $ODC_x = f|_{x=1} \overline{\oplus} f|_{x=0}$. If the network has multiple outputs, the ODC set is a vector whose elements are the complemented Boolean differences of each scalar output function with respect to the variable of interest.

The observability of a variable can be redefined by using the concept of *perturbed* network. Namely, if we represent the input/output behavior of the perturbed network by $\mathbf{f}^x(\delta)$, the observability *don't care* set of variable x is just $\mathbf{f}^x(0) \oplus \mathbf{f}^x(1)$.

From a practical standpoint, the definition of ODC sets does not help its computation, because it is often not possible to express outputs as explicit functions of internal variables for size reasons. Similarly, the computation of the Boolean difference by means of the *chain rule* [6] can be computationally explosive. We describe here an incremental method for the computation of the internal ODC set that exploits a traversal of the logic network.

Consider first, as a simple example, a single-output network with a tree structure. The scalar notation is used, being only one output is involved. Assume we know the internal ODC conditions for a subset of the vertices. This subset contains initially the output vertex, being ODC_{out} determined by the environment and hence known. We want to compute the ODC set for a vertex v_x whose direct successor v_y has a known ODC set.

The condition under which a change in polarity of x is not perceived at the output is the sum of two components:

$$ODC_x = (\partial f_y / \partial x)' + ODC_y \qquad (8.3)$$

The first term spells the conditions under which x is not observed at v_y, while the second term denotes the conditions under which y itself is not observed. Therefore a simple network traversal, from the output to the inputs, allows us to compute the ODC sets at the internal vertices and at the primary inputs.

Example 8.4.10. Let us consider the network shown in Figure 8.21 (a) and specified by the equations:

$$e = b + c$$

$$b = x_1 + a_1$$

$$c = x_4 + a_2$$

Let us assume that the environment does not restrict the observability of the output, i.e., $ODC_{out} = ODC_e = 0$.

Then, ODC_b is just equal to $(\partial f_e / \partial b)' = (b+c)|_{b=1} \oplus (b+c)|_{b=0} = c$. Similarly, $ODC_c = (\partial f_e / \partial c)' = b$.

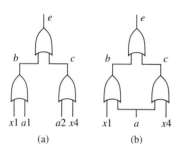

(a)   (b)

FIGURE 8.21

(a) Tree-shaped network. (b) Network with fanout reconvergence.

The observability *don't care* set for x_1 is equal to the sum of ODC_b with $(\partial f_b/\partial x_1)'$. Hence $ODC_{x_1} = c + a_1 = a_1 + a_2 + x_4$. The ODC sets at the other inputs can be computed analogously, yielding $ODC_{a_1} = c + x_1 = x_1 + a_2 + x_4$, $ODC_{a_2} = b + x_4 = x_1 + a_1 + x_4$ and $ODC_{x_4} = b + a_2 = x_1 + a_1 + a_2$.

For networks with general topology, a vertex may have multiple fanout and multiple paths to any output. This situation, called *fanout reconvergence* in jargon, is known to be hard to handle in both logic optimization and testing problems. The problem lies in combining the observability of a variable along different paths. A naive, but wrong, assumption is that the ODC set is the intersection of the ODC sets related to the different fanout stems. Indeed, if a vertex has two (or more) paths to a given output, we must take into account the interplay of the ability of the paths to propagate to the output a change of polarity of the corresponding variable.

Example 8.4.11. Let us consider the network shown in Figure 8.21 (b) and specified by the equations:

$$e = b + c$$

$$b = x_1 + a$$

$$c = x_4 + a$$

By using the arguments of Example 8.4.10, we get $ODC_b = c$ and $ODC_c = b$.

The ODC set of a related to the path (a, b, e) is $x_1 + c = x_1 + a + x_4$, while the ODC set of a related to the path (a, c, e) is $x_4 + b = x_4 + a + x_1$. Their intersection contains a, which would lead us to believe that a is not observed at the output, which is false. Indeed, one of the edges $(v_a, v_b), (v_a, v_c)$ is redundant, but not variable a.

Let us consider now multiple-output networks for the sake of generality. To compute the internal ODC sets correctly, we use an extended annotation of the logic network, where variables and ODC sets are associated with both vertices and edges. We denote the ODC set related to edge (v_x, v_y) as $\mathbf{ODC}_{x,y}$, which is computed by a formula similar to 8.3 but in vector notation, namely:

$$\mathbf{ODC}_{x,y} = (\partial f_y/\partial x)'\mathbf{1} + \mathbf{ODC}_y \tag{8.4}$$

In words, each component of vector $\mathbf{ODC}_{x,y}$ is the sum of the corresponding component of $\mathbf{ODC}_y$ and the complement of the local Boolean difference.

Whereas the computation of the edge ODCs is straightforward, the derivation of the ODCs at vertices that are the tail of multiple edges (i.e., multiple-fanout vertices) is more complex. For simplicity, let us consider first the computation of the ODC set at vertex v_x, denoted as $\mathbf{ODC}_x$, that is the tail of two edges $(v_x, v_y), (v_x, v_z)$, as shown in Figure 8.22.

A method for analyzing this situation proposed by Damiani et al. [17] is based on considering first the observability of x along each path independently. For this reason, variable x is copied into variables x_1 and x_2, which are associated with edges (v_x, v_y) and (v_x, v_z), respectively. Let us assume now perturbations of the network

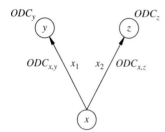

FIGURE 8.22
Example of a two-way fanout stem.

at x_1 and x_2 and let $\mathbf{f}^{x_1,x_2}(\delta_1, \delta_2)$ denote the input/output behavior of the perturbed network.

The observability along the path including (v_x, v_y) is computed assuming that δ_1 and δ_2 are independent perturbations. In particular, $\mathbf{ODC}_{x,y} = \mathbf{f}^{x_1,x_2}(1, \delta_2) \oplus \mathbf{f}^{x_1,x_2}(0, \delta_2)$. Similarly $\mathbf{ODC}_{x,z} = \mathbf{f}^{x_1,x_2}(\delta_1, 1) \oplus \mathbf{f}^{x_1,x_2}(\delta_1, 0)$. Note that $\mathbf{f}^{x_1,x_2}(1, 1) \oplus \mathbf{f}^{x_1,x_2}(0, 0)$ is the ODC set of variable x in the original network, because it captures the conditions under which a consistent change in polarity of x is propagated to the outputs along the paths through v_y and v_z.

We use now an algebraic manipulation[2] to relate the previous expressions for $\mathbf{ODC}_{x,y}$ and $\mathbf{ODC}_{x,z}$ to $\mathbf{ODC}_x$, namely:

$$\mathbf{ODC}_x = \mathbf{f}^{x_1,x_2}(1, 1) \;\overline{\oplus}\; \mathbf{f}^{x_1,x_2}(0, 0) \tag{8.5}$$

$$= \mathbf{f}^{x_1,x_2}(1, 1) \;\overline{\oplus}\; \mathbf{f}^{x_1,x_2}(0, 0) \;\overline{\oplus}\; (\mathbf{f}^{x_1,x_2}(0, 1) \;\overline{\oplus}\; \mathbf{f}^{x_1,x_2}(0, 1)) \tag{8.6}$$

$$= (\mathbf{f}^{x_1,x_2}(1, 1) \;\overline{\oplus}\; \mathbf{f}^{x_1,x_2}(0, 1)) \;\overline{\oplus}\; (\mathbf{f}^{x_1,x_2}(0, 1) \;\overline{\oplus}\; \mathbf{f}^{x_1,x_2}(0, 0)) \tag{8.7}$$

$$= \mathbf{ODC}_{x,y}|_{\delta_2=1} \;\overline{\oplus}\; \mathbf{ODC}_{x,z}|_{\delta_1=0} \tag{8.8}$$

$$= \mathbf{ODC}_{x,y}|_{x_2=x'} \;\overline{\oplus}\; \mathbf{ODC}_{x,z}|_{x_1=x} \tag{8.9}$$

This formula yields a way of constructing the ODC set of vertex v_x from those of the edges $(v_x, v_y), (v_x, v_z)$.

By symmetry, by exchanging y with z (and the corresponding ODC sets), a similar formula can be derived:

$$\mathbf{ODC}_x = \mathbf{ODC}_{x,y}|_{x_2=x} \;\overline{\oplus}\; \mathbf{ODC}_{x,z}|_{x_1=x'} \tag{8.10}$$

Note that for two-way fanout stems there is no need for explicit edge variables. Indeed, any edge ODC set can depend only on the copy of the variable associated with the other edge. Since no confusion can arise, the edge variables can be the same as the corresponding vertex variable. Hence we can rewrite Equations 8.9 and 8.10 as:

$$\mathbf{ODC}_x = \mathbf{ODC}_{x,y}|_{x=x'} \;\overline{\oplus}\; \mathbf{ODC}_{x,z} \tag{8.11}$$

$$= \mathbf{ODC}_{x,y} \;\overline{\oplus}\; \mathbf{ODC}_{x,z}|_{x=x'} \tag{8.12}$$

---

[2]Recall that $a \;\overline{\oplus}\; 1 = a$ and that $b \;\overline{\oplus}\; b = 1$ for any variable or function a, b. Hence $a \;\overline{\oplus}\; (b \;\overline{\oplus}\; b) = a$.

This is not the case for multi-way fanout stems, because an edge ODC set may depend on different copies of the related vertex variable. Hence, edge variables must be recorded and distinguished.

Example 8.4.12. Let us consider again the network of Example 8.4.11 and shown in Figure 8.21 (b). Since there is a single output, we use the scalar notation $ODC_{a,b} = x_1 + a + x_4$ and $ODC_{a,c} = x_1 + a + x_4$.

Then $ODC_a = ODC_{a,b}|_{a=a'} \overline{\oplus} ODC_{a,c}$. Hence $ODC_a = (x_1 + a' + x_4) \overline{\oplus} (x_1 + a + x_4) = x_1 + x_4$, which represents exactly the conditions under which variable a is not observed at the output.

This result can be generalized to more than two fanout stems. The following theorem shows how the ODC of a vertex can be computed from those of its outgoing edges in the general case.

Theorem 8.4.1. Let $v_x \in V$ be any internal or input vertex of a logic network. Let $\{x_i, i = 1, 2, \ldots, p\}$ be the variables associated with edges $\{(x, y_i); i = 1, 2, \ldots, p\}$ and $\mathbf{ODC}_{x,y_i}$, $i = 1, 2, \ldots, p$, the corresponding ODC sets. The observability *don't care* set at a vertex v_x is given by:

$$\mathbf{ODC}_x = \overline{\bigoplus_{i=1}^{p}} \mathbf{ODC}_{x,y_i}|_{x_{i+1}=\cdots=x_p=x'} \tag{8.13}$$

Proof. Let $\mathbf{f}^{\mathbf{X}}(x_1, x_2, \ldots, x_p)$ describe the modified network with independent perturbations $\{\delta_i, i = 1, 2, \ldots, p\}$ on variables $\{x_i, i = 1, 2, \ldots, p\}$. The ODC set for v_x is:

$$\mathbf{ODC}_x = \mathbf{f}^{\mathbf{X}}(1, 1, \cdots, 1, 1) \overline{\oplus} \mathbf{f}^{\mathbf{X}}(0, 0, \cdots, 0, 0) \tag{8.14}$$

It can be rewritten as:

$$\mathbf{ODC}_x = (\mathbf{f}^{\mathbf{X}}(1, 1, \ldots, 1) \overline{\oplus} \mathbf{f}^{\mathbf{X}}(0, 1, \ldots, 1)) \overline{\oplus} \cdots$$
$$\overline{\oplus} (\mathbf{f}^{\mathbf{X}}(0, 0, \ldots, 1) \overline{\oplus} \mathbf{f}^{\mathbf{X}}(0, 0, \ldots, 0)) \tag{8.15}$$

Equivalently:

$$\mathbf{ODC}_x = \mathbf{ODC}_{x,y_1}|_{\delta_2=\cdots=\delta_p=1} \overline{\oplus} \cdots \overline{\oplus} \mathbf{ODC}_{x,y_p}|_{\delta_1=\cdots=\delta_{p-1}=0} \tag{8.16}$$
$$= \mathbf{ODC}_{x,y_1}|_{x_2=\cdots=x_p=x'} \overline{\oplus} \cdots \overline{\oplus} \mathbf{ODC}_{x,y_{p-1}}|_{x_{p-1}=x'} \overline{\oplus} \mathbf{ODC}_{x,y_p} \tag{8.17}$$

which can be rewritten as:

$$\mathbf{ODC}_x = \overline{\bigoplus_{i=1}^{p}} \mathbf{ODC}_{x,y_i}|_{x_{i+1}=\cdots=x_p=x'} \tag{8.18}$$

This theorem allows us to devise an algorithm, shown in Algorithm 8.4.2, for the computation of the internal and input ODC sets, starting from $\mathbf{ODC}_{out}$. The algorithm traverses the network from the primary outputs to the primary inputs. For each vertex being considered, the ODC sets associated with its outgoing edges is computed first as the union of the ODC sets of the direct successors and the complement of the local Boolean difference. Then, the ODC set for the vertex is computed using Theorem 8.4.1.

OBSERVABILITY($G_n(V, E)$, $\mathbf{ODC}_{out}$) {

    foreach vertex $v_x \in V$ in reverse topological order {

        for ($i = 1$ to p)   /* consider all direct successors of v_x */

                $\mathbf{ODC}_{x,y_i} = (\partial f_{y_i}/\partial x)'\mathbf{1} + \mathbf{ODC}_{y_i};$ /* compute edge ODC set */

        $\mathbf{ODC}_x = \bigoplus_{i=1}^{p} \mathbf{ODC}_{x,y_i}|_{x_{i+1}=\cdots=x_p=x'};$ /* compute vertex ODC set */

    }

}

ALGORITHM 8.4.2

Note that this step is trivial, when the vertex has only one direct successor, because its ODC set equals that of the outgoing edge. Vertices are sorted in a reverse topological order consistent with the partial order induced by the network graph.

    The complexity of the algorithm is linear in the size of the graph if we bound the complexity of performing the *cofactor* operation, which may require re-expressing local functions with different variables. For each variable, the intersection of the elements of the corresponding ODC vector yields the conditions under which that variable is not observed at any output. This set is called the *global* ODC of a variable. The vector of dimension n_i whose entries are the global ODC conditions of the input variables is the *input observability don't care* set $\mathbf{ODC}_{in}$.

> **Example 8.4.13.** Consider again the network of Figure 8.18. The ODC sets at vertices v_d and v_e are:
>
> $$\mathbf{ODC}_d = \begin{pmatrix} 0 \\ 1 \end{pmatrix}; \quad \mathbf{ODC}_e = \begin{pmatrix} 1 \\ 0 \end{pmatrix}$$
>
> which means that the d is fully observable at the first output (i.e., empty ODC) and not observable at the second one (full ODC). Opposite considerations apply to e.
>
>   First, consider vertex v_b and edge (v_b, v_d). $\mathbf{ODC}_{b,d} = \begin{pmatrix} 0 \\ 1 \end{pmatrix} + \begin{pmatrix} c' \\ c' \end{pmatrix} = \begin{pmatrix} c' \\ 1 \end{pmatrix},$ because $c' = (\partial f_d/\partial b)'$. Similarly $\mathbf{ODC}_{b,e} = \begin{pmatrix} 1 \\ 0 \end{pmatrix} + \begin{pmatrix} c \\ c \end{pmatrix} = \begin{pmatrix} 1 \\ c \end{pmatrix}$. Therefore $\mathbf{ODC}_b = \mathbf{ODC}_{b,d}|_{b=b'} \oplus \mathbf{ODC}_{b,e} = \begin{pmatrix} c' \\ c \end{pmatrix}$.
>
>   Next, consider vertex v_c. Now $\mathbf{ODC}_{c,d} = \begin{pmatrix} 0 \\ 1 \end{pmatrix} + \begin{pmatrix} b' \\ b' \end{pmatrix} = \begin{pmatrix} b' \\ 1 \end{pmatrix}$. Similarly, $\mathbf{ODC}_{c,e} = \begin{pmatrix} 1 \\ 0 \end{pmatrix} + \begin{pmatrix} b \\ b \end{pmatrix} = \begin{pmatrix} 1 \\ b \end{pmatrix}$. Therefore $\mathbf{ODC}_c = \mathbf{ODC}_{c,d}|_{c=c'} \oplus \mathbf{ODC}_{c,e} = \begin{pmatrix} b' \\ b \end{pmatrix}$. Consider now vertex v_a:
>
> $$\mathbf{ODC}_{a,b} = \begin{pmatrix} c' \\ c \end{pmatrix} + \begin{pmatrix} x_1 \\ x_1 \end{pmatrix}$$
>
> $$= \begin{pmatrix} c' + x_1 \\ c + x_1 \end{pmatrix}$$
>
> $$= \begin{pmatrix} x_4'a' + x_1 \\ x_4 + a + x_1 \end{pmatrix}$$

$$\mathbf{ODC}_{a,c} = \begin{pmatrix} b' \\ b \end{pmatrix} + \begin{pmatrix} x_4 \\ x_4 \end{pmatrix}$$

$$= \begin{pmatrix} b' + x_4 \\ b + x_4 \end{pmatrix}$$

$$= \begin{pmatrix} x_1' a' + x_4 \\ x_1 + a + x_4 \end{pmatrix}$$

Therefore:

$$\mathbf{ODC}_a = \mathbf{ODC}_{a,b}|_{a=a'} \ \overline{\oplus} \ \mathbf{ODC}_{a,c}$$

$$= \begin{pmatrix} x_4' a' + x_1 \\ x_4 + a + x_1 \end{pmatrix}\Big|_{a=a'} \ \overline{\oplus} \ \begin{pmatrix} x_1' a' + x_4 \\ x_1 + a + x_4 \end{pmatrix}$$

$$= \begin{pmatrix} x_4' a + x_1 \\ x_4 + a' + x_1 \end{pmatrix} \ \overline{\oplus} \ \begin{pmatrix} x_1' a' + x_4 \\ x_1 + a + x_4 \end{pmatrix}$$

$$= \begin{pmatrix} x_1 x_4 \\ x_1 + x_4 \end{pmatrix}$$

The global observability of a is $(x_1 x_4)(x_1 + x_4) = x_1 x_4$. By progressing in a similar way, we can then determine the observability of the inputs, which may be used as external ODC sets for the optimization of a network feeding the one under consideration.

In general, ODC sets may be large functions. Therefore ODC subsets may be useful for logic optimization. Differently from the CDC set case, we cannot drop arbitrary elements of the ODC set of a variable and then use the subset to compute the ODC set for some other variable. Indeed, the $\overline{\oplus}$ operator used to compute the vertex ODC set involves implicitly the complement of the edge ODC sets. Hence, some ODC subset may lead to computing ODC supersets at other vertices, which is not a safe approximation.

Example 8.4.14. Consider again the network of Figure 8.18. Let the ODC sets of edges (v_a, v_b) and (v_a, v_c) be approximated by subsets as $\widetilde{\mathbf{ODC}}_{a,b} = \begin{pmatrix} x_1 \\ x_1 \end{pmatrix}$ and $\widetilde{\mathbf{ODC}}_{a,c} = \begin{pmatrix} x_4 \\ x_4 \end{pmatrix}$. Then:

$$\widetilde{\mathbf{ODC}}_a = \begin{pmatrix} x_1 \\ x_1 \end{pmatrix}\Big|_{a=a'} \ \overline{\oplus} \ \begin{pmatrix} x_4 \\ x_4 \end{pmatrix}$$

$$= \begin{pmatrix} x_1 x_4 + x_1' x_4' \\ x_1 x_4 + x_1' x_4' \end{pmatrix}$$

Note that both components of $\widetilde{\mathbf{ODC}}_a$ are not subsets of the corresponding components of $\mathbf{ODC}_a$.

Different safe approximations of the ODC sets have been proposed and they are surveyed and contrasted in reference [17]. The simplest [8] is to approximate the local ODC set by just the intersection of the complement of the Boolean difference of the immediate successors with respect to the variable under consideration. In other words, the ODC subsets of the successors are neglected.

Other approximations yield subsets of larger size. Among these, we mention two that are noteworthy and we show here the formulae for two-way fanout stems, the extension being straightforward. The first can be obtained by expanding the $\overline{\oplus}$ operator appearing in Equations 8.11 and 8.12 and summing them:

$$\mathbf{ODC}_x = (\mathbf{ODC}_{x,y}|_{x=x'}) \cdot (\mathbf{ODC}_{x,z}) + (\mathbf{ODC}_{x,y}|_{x=x'})' \cdot (\mathbf{ODC}_{x,z})' +$$

$$+ (\mathbf{ODC}_{x,y}) \cdot (\mathbf{ODC}_{x,z}|_{x=x'}) + (\mathbf{ODC}_{x,y})' \cdot (\mathbf{ODC}_{x,z}|_{x=x'})' \quad (8.19)$$

If subsets of the edge ODCs and of their complements are available, then the above equation can provide a subset of the actual $\mathbf{ODC}_x$, by substituting the subsets in place of the corresponding full sets. Note that subsets of the ODC sets and of their complements must be computed and stored.

A second approximation method is based on disregarding the complements of the ODC sets and hence on the formula:

$$\widetilde{\mathbf{ODC}}_x = (\mathbf{ODC}_{x,y}|_{x=x'}) \cdot (\mathbf{ODC}_{x,z}) + (\mathbf{ODC}_{x,y}) \cdot (\mathbf{ODC}_{x,z}|_{x=x'}) \quad (8.20)$$

If subsets of the edge ODCs are used as arguments of the above formula, a subset of the ODC is returned.

Example 8.4.15. Consider again the network of Figure 8.18. Again let the ODC sets of edges (v_a, v_b) and (v_a, v_c) be approximated by subsets as $\widetilde{\mathbf{ODC}}_{a,b} = \begin{pmatrix} x_1 \\ x_1 \end{pmatrix}$ and $\widetilde{\mathbf{ODC}}_{a,c} = \begin{pmatrix} x_4 \\ x_4 \end{pmatrix}$. Then:

$$\widetilde{\mathbf{ODC}}_a = \begin{pmatrix} x_1 \\ x_1 \end{pmatrix}_{|a=a'} \cdot \begin{pmatrix} x_4 \\ x_4 \end{pmatrix} + \begin{pmatrix} x_1 \\ x_1 \end{pmatrix} \cdot \begin{pmatrix} x_4 \\ x_4 \end{pmatrix}_{|a=a'} = \begin{pmatrix} x_1 x_4 \\ x_1 x_4 \end{pmatrix} \quad (8.21)$$

Both components of $\widetilde{\mathbf{ODC}}_a$ are now subsets of the corresponding components of $\mathbf{ODC}_a$.

8.4.2 Boolean Simplification and Substitution

We consider now the problem of simplifying a logic function with the degrees of freedom provided by the local *don't care* set. We define simplification informally as the manipulation of a representation of the local function that yields an equivalent one with smaller size. Equivalence will be defined rigorously later. We shall then extend this problem to the simultaneous optimization of more than one local function.

Two-level logic optimization algorithms are applicable to this problem under the assumption that local functions are represented by two-level *sum of products* forms. Nevertheless we must consider a few subtle differences with respect to the problems addressed in Chapter 7. First, the objective of local function simplification in the frame of multiple-level logic optimization should not be the reduction of the number of terms. A more relevant goal is reducing the number of literals. It may also be important to minimize the cardinality of the support set of the local function, which is directly related to the required local interconnections.

Exact and heuristic algorithms for two-level minimization can be modified to optimize the number of literals. In the exact case, prime implicants can be weighted by the number of literals they represent as a product, leading to a minimum weighted (unate) covering problem that can be solved with the usual methods. In the heuristic case, the *expand* operator yields a prime cover that corresponds to a locally minimum-literal solution. The minimizer may be steered by heuristic rules toward a solution that rewards a maximal reduction of literals (instead of product terms). A method for minimizing the cardinality of the support set is summarized in reference [6].

The second noteworthy issue in simplifying a local function is that the reduction in literals can be achieved by using a variable which was not originally in the support set. Roughly speaking, this corresponds to "substituting a portion of the function" by that variable. Indeed, this is exactly the same as performing the Boolean substitution transformation that was mentioned in Section 8.2.3. Note that the simplification of a local function in isolation (i.e., without the *don't care* conditions induced by the network interconnection) would never lead to adding a variable to the support set, because such a variable is apparently unrelated to the function. The *don't care* conditions help to bridge this gap, because they encapsulate global information of the network and represent the interplay among the different local functions.

Example 8.4.16. Consider again the second case of substitution mentioned in Example 8.2.12, where we try to substitute $q = a + cd$ into $f_h = a + bcd + e$ to get $f_h = a + bq + e$. Let us consider the SDC set $q \oplus (a + cd) = q'a + q'cd + qa'(cd)'$. The simplification of $f_h = a + bcd + e$ with $q'a + q'cd + qa'(cd)'$ as a *don't care* set yields $\widetilde{f_h} = a + bq + e$. (Cube bq can replace cube bcd because the minterms where the cubes differ are part of the *don't care* set.) One literal is saved by changing the support set of f_h.

Thus Boolean simplification encompasses Boolean substitution. Therefore we can concentrate on Boolean simplification in the sequel.

We can envision two major strategies for performing Boolean simplification: optimizing the local functions one at a time or optimizing simultaneously a set of local functions. Consequently, the target of each simplification is a vertex in the first case and a subnetwork in the second case. We refer to the two strategies as *single-vertex* and *multiple-vertex* optimization, respectively.

The two approaches differ substantially, because optimizing a local function affects the *don't care* sets of other local functions. Thus, when optimizing the local functions one at a time, the *don't care* sets change as the network is being optimized. Conversely, when optimizing more than one local function at a time, the corresponding degrees of freedom cannot be expressed completely by independent sets.

SINGLE-VERTEX OPTIMIZATION. Let us consider a logic network $G_n(V, E)$ and a vertex v_x that is chosen to be a target of optimization. We must consider first which functions are the feasible replacements of function f_x. This is obviously dependent on the local environment, i.e., on the interconnections of v_x in the network.

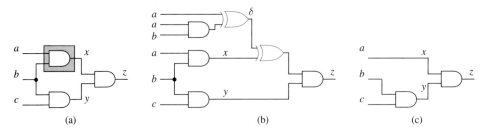

FIGURE 8.23
(a) Example of a logic network. (b) Perturbed network at x to show the replacement of $f_x = ab$ with $g_x = a$. (c) Optimized network.

We use again the concept of a perturbed network. The replacement of a local function f_x with another one, say g_x, can be modeled by means of a perturbation $\delta = f_x \oplus g_x$, as shown in Figure 8.23. This allows us to formalize the equivalence of two networks, the latter obtained from the former by replacing f_x with g_x. Recall that $\mathbf{f}, \mathbf{f}^x$ and $\mathbf{DC}_{ext}$ are functions of the primary input variables.

Definition 8.4.5. Consider a logic network and another network obtained by replacing a local function f_x with g_x. The two logic networks are **equivalent** if the vector equality:

$$\mathbf{f}^x(0) = \mathbf{f}^x(f_x \oplus g_x) \tag{8.22}$$

is satisfied for all observable components of $\mathbf{f}$ and all possible primary input assignments.

In other words, the difference in input/output behavior of the two networks must be included in the *external don't care* set, or equivalently:

$$\mathbf{f}^x(0) \oplus \mathbf{f}^x(f_x \oplus g_x) \subseteq \mathbf{DC}_{ext} \tag{8.23}$$

If we define $\mathbf{e}(\delta) = \mathbf{f}^x(0) \oplus \mathbf{f}^x(\delta)$, then $\mathbf{e}(\delta)$ represents the error introduced by replacing f_x with g_x. The *don't care* set represents the tolerance on the error. We say that a perturbation is *feasible* if 8.23 holds.

Additional insight into the problem can be gained by the following theorem.

Theorem 8.4.2. Consider a logic network obtained by replacing a local function f_x with g_x. A sufficient condition for equivalence is that the perturbation $\delta = f_x \oplus g_x$ is contained in all components of $\mathbf{DC}_{ext} + \mathbf{ODC}_x$.

Proof. By definition of equivalence, $\mathbf{f}^x(0) \oplus \mathbf{f}^x(\delta) \subseteq \mathbf{DC}_{ext}$, where $\delta = f_x \oplus g_x$. Let $\mathbf{e}(\delta) = \mathbf{f}^x(0) \oplus \mathbf{f}^x(\delta)$. By performing an expansion on variable δ, we have:

$$\mathbf{e}(\delta) = \delta(\mathbf{f}^x(0) \oplus \mathbf{f}^x(1)) + \delta'(\mathbf{f}^x(0) \oplus \mathbf{f}^x(0))$$

Hence $\mathbf{e}(\delta) = \delta \, \mathbf{ODC}'_x$, which implies that:

$$\delta \mathbf{ODC}'_x \subseteq \mathbf{DC}_{ext} \tag{8.24}$$

The above equality holds if and only if:

$$\delta 1 \subseteq \mathbf{DC}_{ext} + \mathbf{ODC}_x \tag{8.25}$$

Theorem 8.4.2 does not assume anything about the support set of g_x. If we assume that $sup(g_x)$ is a subset of the primary input variables, the condition is also necessary, because otherwise there would be some *don't care* condition not accounted for.

Assume now that f_x and g_x have the same support set $S(x)$ that includes variables associated with internal vertices. Then the impossible input patterns for f_x and g_x are additional degrees of freedom for the replacement of f_x with g_x. Therefore the local *don't care* set of variable x includes also the controllability component, i.e.:

$$\mathbf{DC}_x = \mathbf{DC}_{ext} + \mathbf{ODC}_x + \mathbf{CDC}_{S(x)} \tag{8.26}$$

Finally, assume that $sup(g_x)$ is included in the set of network variables, as in the case in which we attempt Boolean substitution. Again the impossible input patterns at the inputs of f_x and g_x also have to be accounted for. Therefore the local *don't care* set includes also the SDC set for all variables but x, i.e., $SDC_x = \sum\limits_{v_y \in V^G:\ v_y \neq v_x} y \oplus f_y$.

Therefore:

$$\mathbf{DC}_x = \mathbf{DC}_{ext} + \mathbf{ODC}_x + \mathbf{SDC}_x \tag{8.27}$$

where $\mathbf{SDC}_x = SDC_x \mathbf{1}$. Whereas this definition of SDC_x is compatible with the literature [6], the astute reader will notice that variables corresponding to successor vertices of v_x cannot be added to the support of the local function f_x, and therefore the corresponding contribution to SDC_x can be ignored.

Corollary 8.4.1. Let f_x be replaced by g_x, where $sup(g_x)$ is included in the set of all variables excluding x. A necessary and sufficient condition for equivalence is that the perturbation $\delta = f_x \oplus g_x$ is contained in all components of $\mathbf{DC}_x = \mathbf{DC}_{ext} + \mathbf{ODC}_x + \mathbf{SDC}_x$.

The corollary states that $\mathbf{DC}_x = \mathbf{DC}_{ext} + \mathbf{ODC}_x + \mathbf{SDC}_x$ represents the *don't care* conditions for any vertex $v_x \in V^G$. Note that since the perturbation must be contained in all components of $\mathbf{DC}_x$, it is convenient to check its containment in their intersection, denoted by DC_x.

Example 8.4.17. Consider the network of Figure 8.23 (a). Assume that there are no external *don't care* sets and that we want to optimize f_x by replacing the AND gate by a straight connection, i.e., $g_x = a$. Hence the perturbation is $\delta = f_x \oplus g_x = ab \oplus a = ab'$. A feasible replacement is one where the perturbation is contained in the local *don't care* set DC_x. A simple analysis yields $ODC_x = y' = b' + c'$. Then $DC_x = b' + c'$ because the inputs of g_x are primary inputs. Since $\delta = ab' \subseteq DC_x = b' + c'$, the replacement is feasible. The optimized network is shown in Figure 8.23 (c).

The local *don't care* sets encapsulate all possible degrees of freedom for replacing f_x with g_x. In practice, the external *don't care* sets are specified along with the

$SIMPLIFY\_SV(\ G_n(V, E)\ )\{$

> **repeat** {
>> v_x = selected vertex;
>> Compute the local *don't care* set DC_x;
>> Optimize the function f_x;
> } **until** (no more reduction is possible)

}

ALGORITHM 8.4.3

logic network and the SDC, CDC and ODC sets can be computed by the algorithms presented before. Note that the local *don't care* set for a variable may be large and contain components that are irrelevant to the simplification of the corresponding local function. Corollary 8.4.1 just states that the local *don't care* sets contain all relevant components.

Filters are used to reduce the size of the local *don't care* sets. Filters may drop terms of the local *don't care* sets while considering the uselessness, or unlikelihood, that they contribute to minimizing those functions. In the first case the filters are called *exact*, otherwise they are called *approximate*. Both types of filters exploit the topology of the network.

An example of an exact filter is the following. The filter considers the support sets of the local function to be simplified and of the cubes of a *sum of products* representation of the corresponding local *don't care* set. Consider a vertex v_x. Let $DC_x = D \cup E$ such that $|(sup(F) \cup sup(D)) \cap sup(E)| \leq 1$. Then, it can be shown formally that the cubes of E are useless for optimizing f_x [38, 43]. (See Problem 9.)

Approximate filters are based on heuristics. An example is discarding those cubes of the *don't care* set whose support is disjoint from that of the function to be minimized. Another example is to disregard those cubes whose support is larger than a given threshold, i.e., corresponding to Boolean subspaces too small to be relevant. We refer the interested reader to references [38] and [43] for details.

A simple framework for single-vertex optimization is given in Algorithm 8.4.3. Different heuristics can be used to select vertices and filters to limit the size of the local *don't care* set.

MULTIPLE-VERTEX OPTIMIZATION.* We consider now the problem of the simultaneous optimization of the functions associated with a set of vertices of a logic network. We use perturbation analysis as a way of describing the replacement of local functions with new ones at the vertices of interest.

Let us assume that we want to simplify n local functions, related to variables $\{x_i, i = 1, 2, \ldots, n\}$. Hence we introduce multiple perturbations $\{\delta_i, i = 1, 2, \ldots, n\}$ in the network, one for each variable, as shown in Figure 8.24. Each perturbation is the difference between the old and new functions to be assigned to each vertex, i.e., $\{\delta_i = f_{x_i} \oplus g_{x_i}, i = 1, 2, \ldots, n\}$. We use the vector notation to denote the set of perturbations δ and perturbed variables $\mathbf{x}$.

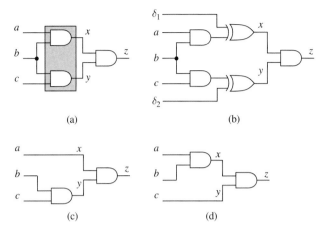

FIGURE 8.24
(a) Example of a logic network. (b) Network with multiple perturbations. (c) Optimized network. (d) Another optimized network.

By definition of equivalence, the difference in input/output behavior of the original and perturbed network must be contained in the external *don't care* set, namely:

$$\mathbf{f^X}(\mathbf{0}) \oplus \mathbf{f^X}(\delta) \subseteq \mathbf{DC}_{ext} \tag{8.28}$$

It would be useful to transform this inequality into bounds on the individual perturbations $\{\delta_i, = 1, 2, \ldots, n\}$, in a fashion similar to Theorem 8.4.2. Unfortunately the bounds are more complex than those used in single-vertex optimization. Note first that a simultaneous optimization of multiple vertices using the local *don't care* sets, computed as in the previous section, would lead to erroneous results.

> **Example 8.4.18.** Consider the circuit of Figure 8.24 (a). Assume the external *don't care* set is empty, for the sake of simplicity. Let us compute the ODC sets $ODC_x = y' = b' + c'$, $ODC_y = x' = b' + a'$. Assume we want to replace f_x by $g_x = a$ and f_y by $g_y = c$. Even though both perturbations $\delta_1 = ab'$ and $\delta_2 = b'c$ are included in the corresponding local *don't care* sets, the simultaneous optimization leads to an erroneous network, implementing $z = ac$. Only one perturbation (corresponding to replacing one AND gate by a wire) is applicable at a time.

It is interesting to look at the multiple-vertex optimization problem from a different perspective. Let **x** be the set of variables under consideration. Then, some patterns of **x** may be equivalent as far as determining the value of the primary outputs. Thus, the mapping from the primary input variables to the variable **x**, which satisfies the equivalence constraint 8.28, can be characterized by a Boolean relation. Therefore multiple-vertex optimization can be modeled as finding a minimal implementation of a multiple-output function compatible with a Boolean relation.

Example 8.4.19. Consider again the circuit of Figure 8.24 (a) with no external *don't cares*. Let x, y represent the perturbed variables. The primary output is TRUE only when $x, y = 1, 1$. The remaining patterns for x, y, i.e., { 0,0; 0,1; 1,0 }, are equivalent. Hence the subnetwork with inputs a, b, c and outputs x, y can be represented by the following Boolean relation:

a	b	c	x, y
0	0	0	{ 00, 01, 10 }
0	0	1	{ 00, 01, 10 }
0	1	0	{ 00, 01, 10 }
0	1	1	{ 00, 01, 10 }
1	0	0	{ 00, 01, 10 }
1	0	1	{ 00, 01, 10 }
1	1	0	{ 00, 01, 10 }
1	1	1	{ 11 }

A minimum-literal solution is:

a	b	c	x, y
1	*	*	10
*	1	1	01

corresponding to $x = a$ and $y = bc$ as in the implementation shown in Figure 8.24 (c). Note that this solution is not unique. For example, another minimum-literal solution is $x = ab$ and $y = c$ as shown in Figure 8.24 (d).

Thus, the feasible perturbations can be expressed as solutions to a Boolean relation. Two points are noteworthy. First, Boolean relations capture more degrees of freedom than *don't care* sets, and thus multiple-vertex optimization is a more powerful paradigm for optimization than single-vertex optimization. Second, Boolean relations reduce to Boolean functions in the limiting case of single-vertex optimization. Indeed, single-output relations are logic functions, possibly incompletely specified. Thus, in the single-output case it is possible to specify feasible perturbations as incompletely specified functions, as stated by Theorem 8.4.2.

Example 8.4.20. Consider the circuit of Figure 8.24 (b). Assume again that the external *don't care* set is empty. The constraints on the equivalence of the perturbations are:

$$f^{x,y}(a, b, c, \delta_1, \delta_2) \; \overline{\oplus} \; f^{x,y}(a, b, c, 0, 0) \; = \; 1$$

Therefore:

$$(ab \oplus \delta_1)(bc \oplus \delta_2) = abc$$

A trivial solution is obviously provided by $\delta_1 = 0, \delta_2 = 0$. Non-trivial solutions are $\delta_1 = ab', \delta_2 = 0$ and $\delta_1 = 0, \delta_2 = b'c$, whose validity can be easily checked by back substitution. The first non-trivial perturbation corresponds to choosing $g_x = ab \oplus ab' = a$, and its implementation is shown in Figure 8.23 (c).

There are then two major avenues for multiple-vertex optimization. The first is to extract the subnetwork target of optimization, model its input/output behavior as a

Boolean relation and apply corresponding exact (or heuristic) minimization techniques. The second method is to search for bounds for feasible simultaneous perturbations. These bounds can be used to compute mutually compatible *don't care* sets that reflect the degrees of freedom of each local function in the simultaneous optimization problem. Whereas this approach leads to formulating the optimization problem as a Boolean function minimization problem (and hence simpler to solve than the corresponding Boolean relation minimization problem), it is possible to show that some degrees of freedom in logic optimization are not captured by the compatible *don't care* sets. Hence, this approach is less general than Boolean relation optimization. It should be noted, though, that compatible *don't care* sets can be computed for arbitrary subsets of vertices whose local functions may have arbitrary support sets. On the other hand, the specification of a Boolean relation entails identifying a set of inputs and a set of outputs, the latter being associated with the vertices that are the target of the optimization. Therefore this approach is well-suited for optimizing cascaded subnetworks.

We describe multiple-vertex optimization by means of compatible *don't care* sets first, to draw the parallel with single-vertex optimization. We comment on relation-based optimization later.

COMPATIBLE *DON'T CARE* CONDITIONS.* Single-vertex optimization relies on the results of Theorem 8.4.2 and Corollary 8.4.1, which state that a single perturbation is valid when it is bounded from above by the local *don't care* set. This result does not generalize to multiple-vertex optimization, as shown by Example 8.4.18. Indeed it is possible to show formally [17] that multiple perturbations are feasible when they satisfy bilateral (upper and lower) bounds. The bounds for one perturbation depends in general on the other perturbations.

To avoid the complexity of computing and deriving the exact bounds, a conservative approach can be used, where only perturbation-independent upper bounds are used. Satisfaction of these bounds is a sufficient (but not necessary) condition for equivalence. The implementation of this approach bears some similarities with single-vertex optimization, where a local *don't care* set is computed and used for simplification. Nevertheless some feasible transformations may not be found, because of the restriction applied to the bounds.

Note that the external *don't care* sets are invariant on the perturbations. Conversely, the satisfiability *don't cares* related to the vertices being optimized are obviously subject to change, and therefore are excluded from the local *don't care* sets. We concentrate then on the analysis of the ODC sets and on the computation of those subsets that are invariant under simultaneous independent perturbations.

Definition 8.4.6. Consider a set of variables x_i, $i = 1, 2, \ldots, n$. Observability *don't care* sets are called **compatible** and denoted $\mathbf{CODC}_{x_i}$, $i = 1, \ldots, n$, when they do not depend on the perturbations themselves and

$$\delta_i \mathbf{1} \subseteq \mathbf{DC}_{ext} + \mathbf{CODC}_{x_i}, \quad i = 1, \ldots, n \tag{8.29}$$

represents a sufficient condition for equivalence.

A compatible ODC (CODC) set is said to be *maximal* if no other cube can be added to it while preserving compatibility.

From a practical standpoint, the derivation of CODC sets is very important, because it enables us to simplify simultaneously and correctly a set of local functions. Thus CODCs play the role that ODCs have in single-vertex optimization. Maximal CODC sets are relevant because they provide more degrees of freedom than non-maximal ones.

From an intuitive point of view, we can justify the computation of the CODC sets as follows. Let us order the vertices of interest in an arbitrary sequence. The first vertex would have its CODC equal to its ODC set. The second vertex would have a CODC smaller than its ODC by a quantity that measures how much a permissible perturbation at the first vertex would increase the observability of the second (i.e., deprive the second vertex of some degrees of freedom). And so on.

Two simple observations are obvious. First, the CODC sets depend on the order of the variables being considered. This is because we remove iteratively the dependency of the ODC set on the other perturbations. Since it is not efficient to consider all orders, we must settle for a possible underestimation of the degrees of freedom due to observability.

The second observation is that CODC sets can still be derived by network traversal algorithms, such as the ODCs. Attention has to be paid to restricting the ODC set in an appropriate way. Unfortunately, in the general case, maximal-vertex CODC sets cannot be derived directly from edge CODC sets, and full ODC sets must also be derived. Approximation methods are also possible that compute non-maximal CODC sets by avoiding the computation of full ODC sets.

The calculation of the CODC sets by network traversal entails three major steps:

1. The computation of the ODC sets related to the edges.
2. The derivation of the corresponding CODC sets by restricting the ODC sets.
3. The computation of the CODC sets related to the vertices by combining the edge CODC sets.

The first and last steps parallel those used for the full ODC sets. The intermediate step is illustrated here for the case for $n = 2$ perturbations and a single-output network, namely:

$$CODC_{x_1} = ODC_{x_1} \tag{8.30}$$

$$CODC_{x_2} = \mathcal{C}_{x_1}(ODC_{x_2}) + ODC_{x_2}ODC'_{x_1} \tag{8.31}$$

Intuitively, the first CODC set is the full ODC. The second CODC set consists of two terms: the component of ODC_{x_2} independent of x_1 and the restriction of ODC_{x_2} to when x_1 is observable (and thus no optimization can take advantage of $CODC_{x_1}$). We refer the reader to reference [17] for the general case.

Example 8.4.21. Consider computing the CODC sets for the circuit of Figure 8.24 (a). The ODC sets are $ODC_x = y' = b' + c'$, $ODC_y = x' = b' + a'$. Let us now derive the CODC sets. Assume (arbitrarily) that the first vertex is x and the second is

$SIMPLIFY\_MV(\ G_n(V, E) \ )\{$

    repeat {

        U = selected vertex subset;

        foreach vertex $v_x \in U$

            Compute **CODC**$_x$ and the corresponding local *don't care* subset $\widetilde{\text{DC}}_x$;

        Optimize simultaneously the functions at vertices U;

    } **until** (no more reduction is possible)

}

ALGORITHM 8.4.4

    y. Then $CODC_x = ODC_x = y'$ and $CODC_y = C_x(ODC_y) + ODC_y(ODC_x)' = C_x(x') + x'y = x'y = (b' + a')bc = a'bc$. Indeed $CODC_y \subseteq ODC_y$. Note that in this example the AND gates have single fanout. Hence the edge observability is the same as the vertex observability and step 3 is skipped.

    The significance of the increased observability of y is due to the fact that f_x may be subject to change, for example, being replaced by $g_x = a$. In this situation, f_y cannot be simultaneously reduced to $g_y = c$, because this would not yield equivalent networks.

    Thus, the simultaneous optimization of $f_x = ab$ and $f_y = bc$ with $\widetilde{DC}_x = b' + c'$ and $\widetilde{DC}_y = a'bc$ yields $g_x = a$ and $g_y = bc$, as shown in Figure 8.24 (c). Note that the multiple optimization specified by $g_x = ab$ and $g_y = c$ cannot be found with these CODC sets, while it is feasible in principle [Figure 8.24 (d)]. Conversely, the computation of the CODC sets with a different variable sequence would allow the simplifications $g_x = ab$ and $g_y = c$, while it would disallow $g_x = a$ and $g_y = bc$.

The general approach to multiple-vertex simplification using compatible *don't care* sets and logic minimization is based on iterating the steps shown in Algorithm 8.4.4.

Often, all internal vertices are considered at once (i.e., $U = V^G$) and one iteration is used. Thus the local *don't care* sets reduce to the combination of the CODC sets and the external DC sets. Multiple-output logic minimizers can optimize simultaneously all local functions.

BOOLEAN RELATIONS AND MULTIPLE-VERTEX OPTIMIZATION.* Multiple-vertex optimization can be modeled by associating the vertices (that are the target of optimization) with the outputs of a Boolean relation. This approach is more general than using compatible *don't cares*, because Boolean relations model implicitly the mutual degrees of freedom of several functions.

The difficulties related to using a relational model are the following. First, the equivalence classes of the Boolean relation have to be determined. Second, an optimal compatible (multiple-output) function must be derived. Boolean relation optimization (see Section 7.6) involves solving a binate covering problem and is more difficult than minimization of Boolean functions.

We show next how it is possible to derive the equivalence classes of a Boolean relation from the ODC sets of the vertices that are the target of multiple-vertex optimization. We introduce the method by elaborating on an example.

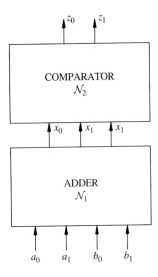

FIGURE 8.25
Example of two cascaded logic networks.

Example 8.4.22. Consider again the circuit of Figure 8.25 related to Example 7.6.1. We want to determine the equivalence classes of the relation that can model subnetwork $\mathcal{N}_1$, whose observability is limited by network $\mathcal{N}_2$.

Let $\mathbf{f}$ denote the input/output behavior of network $\mathcal{N}_2$. Without loss of generality, assume that the comparator's output $\mathbf{z}$ is 0,0 when $\mathbf{x}$ is smaller than 3, 1,0 when $\mathbf{x}$ is 3 or 4 and 1,1 when $\mathbf{x}$ is larger than 4. Then:

$$\mathbf{f} = \begin{pmatrix} x_2 x_0 + x_2 x_1 \\ x_1 x_0 + x_2 \end{pmatrix}$$

The observability *don't care* sets of variables $\mathbf{x}$, found by the OBSERVABILITY algorithm, are:

$$\mathbf{ODC}_{x_0} = \begin{pmatrix} x_2' + x_1 \\ x_1' + x_2 \end{pmatrix}; \quad \mathbf{ODC}_{x_1} = \begin{pmatrix} x_2' + x_0 \\ x_0' + x_2 \end{pmatrix}; \quad \mathbf{ODC}_{x_2} = \begin{pmatrix} x_1' x_0' \\ x_1 x_0 \end{pmatrix}$$

The equivalence class of pattern $x_1, x_2, x_3 = 0, 0, 0$ is the set of patterns x_0^r, x_1^r, x_2^r that satisfies $\mathbf{f}(x_2, x_1, x_0) = \mathbf{f}(0, 0, 0)$ and is therefore described by the function:

$$\mathbf{EQV}_{0,0,0} = \mathbf{f}(x_2, x_1, x_0) \; \overline{\oplus} \; \mathbf{f}(0, 0, 0)$$

We use now an algebraic manipulation similar to that used for computing the ODC sets. Since $\mathbf{f}(x_2, x_1, 0) \overline{\oplus} \mathbf{f}(x_2, x_1, 0) = \mathbf{1}$ and $\mathbf{f}(x_2, 0, 0) \overline{\oplus} \mathbf{f}(x_2, 0, 0) = \mathbf{1}$, we can rewrite the above equation as:

$$\mathbf{EQV}_{0,0,0} = \Big(\mathbf{f}(x_2, x_1, x_0) \; \overline{\oplus} \; \mathbf{f}(x_2, x_1, 0)\Big) \; \overline{\oplus} \; \Big(\mathbf{f}(x_2, x_1, 0) \; \overline{\oplus} \; \mathbf{f}(x_2, 0, 0)\Big) \; \overline{\oplus} \; \Big(\mathbf{f}(x_2, 0, 0)$$

$$\overline{\oplus} \; \mathbf{f}(0, 0, 0)\Big)$$

$$= \Big(x_0' + \mathbf{ODC}_{x_0}\Big) \; \overline{\oplus} \; \Big(x_1' + \mathbf{ODC}_{x_1}|_{x_0=0}\Big) \; \overline{\oplus} \; \Big(x_2' + \mathbf{ODC}_{x_2}|_{x_1=0, \, x_0=0}\Big)$$

$$= \begin{pmatrix} x_0' + x_1 + x_2' \\ x_0' + x_1' + x_2 \end{pmatrix} \; \overline{\oplus} \; \begin{pmatrix} x_1' + x_2' \\ 1 \end{pmatrix} \; \overline{\oplus} \; \begin{pmatrix} 1 \\ x_2' \end{pmatrix} = \begin{pmatrix} x_2' + x_1' x_0' \\ x_2' x_1' + x_2' x_0' \end{pmatrix}$$

where we used the identity:

$$\mathbf{f}(a, b) \overline{\oplus} \mathbf{f}(a, 0) = (b \cdot \partial \mathbf{f}/\partial b)' = b'\mathbf{1} + \mathbf{ODC}_b$$

Each component of $\mathbf{EQV}_{0,0,0}$ describes the equivalence class of the reference pattern $0, 0, 0$ with respect to an output. We are interested in the equivalence class with respect to all outputs, described by the product of all components of $\mathbf{EQV}_{0,0,0}$, which in this example is $x_2'x_0' + x_2'x_1'$. This corresponds to the patterns $\{000, 001, 010\}$ of Example 7.6.1. The other equivalence classes can be derived in a similar way.

The derivation of the equivalence classes is applicable to any subnetwork with outputs x_i, $i = 1, 2, \ldots, p$. Let $\mathbf{f}$ denote the input/output behavior of the driven network (e.g., $\mathcal{N}_2$ in Figure 8.25). Let $\mathbf{x}^r \in \mathsf{B}^p$ be a reference pattern.

Theorem 8.4.3. The equivalence class of any given configuration pattern $\mathbf{x}^r$ is:

$$\mathbf{EQV}_{\mathbf{x}^r} = \overline{\bigoplus}_{i=1}^{p} \left(x_i \overline{\oplus} x_i^r + ODC_{x_i}|_{x_1 = x_1^r \ldots x_{i-1} = x_{i-1}^r} \right)$$

Proof. The equivalence class of any given configuration $\mathbf{x}^r = [x_1^r, x_2^r \ldots, x_p^r]^T$ is the set of configurations that satisfies:

$$\mathbf{EQV}_{\mathbf{x}^r} = \mathbf{f}(\mathbf{x}) \overline{\oplus} \mathbf{f}(\mathbf{x}^r) \tag{8.32}$$

The following identity holds:

$$\begin{aligned}
\mathbf{EQV}_{\mathbf{x}^r} &= \left(\mathbf{f}(x_1, x_2, \ldots, x_p) \overline{\oplus} \mathbf{f}(x_1^r, x_2, \ldots, x_p) \right) \\
&\quad \overline{\oplus} \left(\mathbf{f}(x_1^r, x_2, \ldots, x_p) \overline{\oplus} \mathbf{f}(x_1^r, x_2^r, \ldots, x_p) \right) \tag{8.33} \\
&= \overline{\oplus} \cdots \overline{\oplus} \left(\mathbf{f}(x_1^r, x_2^r, \ldots, x_{p-1}^r, x_p) \overline{\oplus} \mathbf{f}(x_1^r, x_2^r, \ldots, x_p^r) \right) \\
&= \overline{\bigoplus}_{i=1}^{p} \left(\mathbf{f}(x_1^r, x_2^r, \ldots, x_{i-1}^r, x_i, \ldots, x_p) \overline{\oplus} \mathbf{f}(x_1^r, x_2^r, \ldots, x_i^r, x_{i+1}, \ldots, x_p) \right)
\end{aligned}$$

$$\tag{8.34}$$

The theorem follows from the fact that the ith term of the summation is:

$$x_i \overline{\oplus} x_i^r + ODC_{x_i}|_{x_1 = x_1^r \ldots x_{i-1} = x_{i-1}^r} \tag{8.35}$$

where we used the identity:

$$\mathbf{f}(a, b) \overline{\oplus} \mathbf{f}(a, 0) = (b \cdot \partial \mathbf{f}/\partial b)' = b'\mathbf{1} + \mathbf{ODC}_b$$

The significance of this theorem is that we can express each equivalence class for the outputs of the subnetwork to be optimized by means of the information of the ODC sets. Hence, Boolean relation-based optimization of multiple-level networks can be organized as shown in Algorithm 8.4.5.

Example 8.4.23. Consider again the network of Figure 8.24 (a). Assume that a subnetwork is selected, such as that corresponding to the shaded gates in the figure. The scalar notation is used, because there is a single output.

$SIMPLIFY\_MVR(\ G_n(V, E)\ )\{$

    repeat {

        U = selected vertex subset;

        foreach vertex $v_x \in U$

            Compute OCD_x;

        Determine the equivalence classes of the Boolean relation of the subnetwork induced by U;

        Find an optimal function compatible with the relation using a relation minimizer;

    } **until** (no more reduction is possible)

}

ALGORITHM 8.4.5

        Let us compute first the ODC sets at the output of the subnetwork: $ODC_x = y'$ and $ODC_y = x'$. Let us now derive the equivalence classes. Let us consider first pattern 00. Hence:

$$
\begin{aligned}
EQV_{00} &= (x \ \overline{\oplus} \ 0 + ODC_x) \ \overline{\oplus} \ (y \ \overline{\oplus} \ 0 + ODC_y|_{x=0}) \\
&= (x' + y') \ \overline{\oplus} \ (y' + x'|_{x=0}) \\
&= (x' + y') \ \overline{\oplus} \ 1 \\
&= x' + y'
\end{aligned}
$$

This means that patterns $\{00, 01, 10\}$ are equivalent at the primary output. This can be understood by considering that only the pattern 11 can yield a 1 at the primary output. Since there are only four possible patterns, this completes the computation of the equivalence class. The relation can then be expressed by the table shown by Example 8.4.19. The corresponding minimum compatible functions are $g_x = a$; $g_y = bc$ [Figure 8.24 (c)] and $g_x = ab$; $g_y = c$ [Figure 8.24 (d)].

8.4.3   Other Optimization Algorithms Using Boolean Transformations*

We review briefly here other methods for multiple-level logic optimization that use Boolean operations and transformations.

REDUNDANCY IDENTIFICATION AND REMOVAL.* Techniques based on redundancy identification and removal originated in the testing community [10, 12]. The underlying concept is that an untestable fault corresponds to a redundant part of the circuit which can be removed. These methods can be explained in terms of *don't care* sets and perturbations with the formalism of the previous section.

    We consider tests for single stuck-at faults at the inputs of a gate, modeled by a local function. We restrict our attention to single-output networks for the sake of simplicity. Let us assume that the gate corresponds to vertex v_x, with input variable y related to an input connection to be tested. The input connection y corresponds to edge (v_y, v_x) in the network, whose ODC set is $ODC_{y,x}$.

An input test pattern **t** that tests for a stuck-at-0 (or for a stuck-at-1) fault on y must set variable y to TRUE (or to FALSE) and ensure observability of y at the primary output. Namely, the set of all input vectors that detect a stuck-at-0 fault on y is $\{\mathbf{t}|y(\mathbf{t}) \cdot ODC'_{y,x}(\mathbf{t}) = 1\}$ ($\{\mathbf{t}|y'(\mathbf{t}) \cdot ODC'_{y,x}(\mathbf{t}) = 1\}$ for stuck-at-1 faults). If either set is empty, y has an untestable fault.

Assume now that a test pattern generator detects an untestable stuck-at-0 (or stuck-at-1) fault on variable y. Then, variable y can be set to a permanent FALSE (or TRUE) value in the local function f_x. This leads to removal of a connection and to simplification of the corresponding local function.

> **Example 8.4.24.** Consider the circuit of Figure 8.26 (a). A test pattern generator has detected an untestable stuck-at-0 fault (on variable y). The circuit is modified first, by asserting a 0 on that connection, as shown in Figure 8.26 (b). The circuit can then be simplified further to that of Figure 8.26 (c).

We justify redundancy removal by relating it to perturbation analysis. We consider the case of variable y stuck at 0. (The case of variable y stuck at 1 is analogous.) Let then y be set to 0, i.e., $g_x = f_x|_{y=0}$. This can be seen as a perturbation $\delta = f_x \oplus f_x|_{y=0} = y \cdot \partial f_x/\partial y$. We claim that the perturbation is feasible whenever y is untestable for stuck-at-0 faults. By definition of an untestable fault, no input vector **t** can make $y(\mathbf{t}) \cdot ODC'_{y,x}(\mathbf{t})$ TRUE, or equivalently no input vector can make

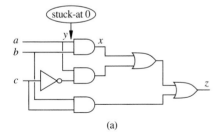

(a)

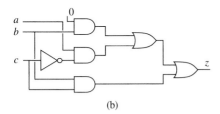

(b)

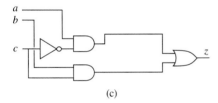

(c)

FIGURE 8.26
(a) Circuit with untestable stuck-at-0 fault. (b) Modified circuit. (c) Simplified circuit.

$y(\mathbf{t}) \cdot ODC'_x(\mathbf{t}) \cdot \partial f_x / \partial y$ TRUE, because $ODC_{y,x} = ODC_x + (\partial f_x / \partial y)'$. The quantity $y \cdot ODC'_x \cdot \partial f_x / \partial y$ is then part of the satisfiability *don't care* set. The local *don't care* set at v_x is then $DC_x \supseteq ODC_x + y \cdot ODC'_x \cdot \partial f_x / \partial y = ODC_x + y \cdot \partial f_x / \partial y$. Hence $\delta = y \cdot \partial f_x / \partial y$ is included in the local *don't care* conditions.

> **Example 8.4.25.** Consider Figure 8.26 (a). Let $f_x = ab$ and $g_x = 0$. Then $\delta = ab \oplus 0 = ab$. $ODC_x = ac' + bc$ and $DC_x \supseteq ODC_x \supseteq \delta$.

In practice, redundancy removal is applied as a byproduct of *automatic test pattern generation* (ATPG), which returns a list (possibly empty) of untestable faults. Note that simplifying a local function in connection with an untestable fault may make other untestable faults become testable. Hence, for each untestable fault (excluding the first), an additional redundancy test is applied. If the fault is still untestable, the local function is simplified. The entire process can be repeated until there are no untestable faults.

TRANSDUCTION.* The transduction method is an iterative optimization method proposed by Muroga *et al.* [36]. *Transduction* means *transformation* plus *reduction*. A different terminology was used in the description of the transduction method. We relate the original terminology to our definitions of local functions and *don't care* sets.

Consider a vertex v_x of the logic network. (Vertices are referred to as *gates*.) A *permissible function* is a function g_x such that $f_x \cap (DC_x)' \subseteq g_x \subseteq f_x \cup DC_x$. Hence it is a feasible replacement for function f_x. The *maximum set of permissible functions* (MSPF) is the largest set of functions g_x satisfying the bounds. The *compatible set of permissible functions* (CSPF) is the largest set of functions g_x satisfying the bounds when simultaneous optimizations are performed. Hence CSPF sets are related to compatible observability *don't care* sets.

The original transduction method dealt with networks of NOR gates only. Later enhancements removed this restriction. Similarly, the original methods for computing the MSPF and CSPF of each gate were based on manipulation of tabular forms, in terms of primary input variables. More recent implementations of the transduction method exploit BDDs.

Several optimization techniques have been proposed for the transduction framework [36]. The most important algorithms are based on the CSPF computation and on the iterative application of the following transformations:

- *Pruning* of redundant subcircuits, based on their observability. When using CSPFs, redundancy removal can be applied simultaneously to several gates.

- Gate *substitution*. When a local function associated with vertex v_x is included in the CSPF (or MSPF) of another gate v_y and v_x is not a successor of v_y, then gate v_x can substitute v_y. The output of v_x is connected to all direct successors of v_y, and v_y can be deleted. This is a form of Boolean substitution.

- *Connection addition and deletion*. This corresponds to doing simplification (and Boolean substitution) by increasing (and/or decreasing) the support set of a local function. The addition of a connection is justified when other connections can be dropped or gates can be merged (as described next).

- Gate *merging*. Consider two gates and their CSPFs. If the intersection of the two CSPFs is not void, and feasible (i.e., acyclic) connections can be made to a new gate implementing a function in the intersection of the CSPFs, then the two gates can be replaced by this new gate.

- *Error compensation*. Remove a gate and compute the corresponding error at the primary outputs. Then restructure the network to compensate for the error.

Performance constraints can be added to the network. Overall, the transduction method can be seen as a family of Boolean transformations that exploit a different organization of the logic information than that described in Section 8.4 but that also have intrinsic close resemblance.

GLOBAL FLOW.* The *global flow* method was proposed by Berman and Trevillyan and incorporated in IBM's LOGIC SYNTHESIS SYSTEM [4]. It consists of two steps: first, collecting information about the network with a method similar to those used for data-flow analysis in software compilers; second, optimizing the circuit iteratively using the gathered information. Overall, the method exploits global transformations and the Boolean model.

Global flow information is expressed in terms of *forcing sets* for each variable. Let S denote the set of variables. Then:

$$F_{ij} = \{s \in S : x = i \implies s = j; \quad i, j \in \{0, 1\}\} \tag{8.36}$$

Example 8.4.26. Consider the circuit of Figure 8.27 (a). If $x = 1$, then $c = 0$, $d = 0$, $e = 1$ and $z = 0$. Hence $\{c, d, z\} \subseteq F_{10}(x)$ and $\{e\} \subseteq F_{11}(x)$.

It is important to remark that forcing sets express implicitly the information of the SDC set. Indeed, the implication $x = i \implies s = j$ can be stated as $(x \oplus i) + (s \overline{\oplus} j)$ and its complement [i.e., $(x \overline{\oplus} i)(s \oplus j)$] is part of the SDC set. Note that since variables x and s may not be related by a local function, the implication may not lead to a cube that appears in the SDC as computed according to Definition 8.4.4, but it may still be derived by performing variable substitution and consensus.

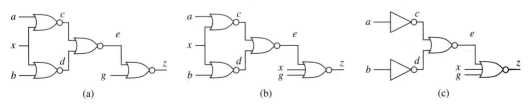

FIGURE 8.27
(a) Logic network fragment. (b) Logic network fragment after reduction. (c) Logic network fragment after reduction and expansion.

The computation of the forcing sets is complex, because it involves a solution to a satisfiability problem. Therefore, subsets are often used that are called *controlling sets*. Such sets can be efficiently computed as fixed points of monotonic recurrence relations. As an example, we consider the controlling sets of reference [4], where it is assumed that all local functions are NORs.

Consider the conditions that apply when a variable $x = 1$ forces $s = 0$. If x forces variable y to 1, then it must force to 0 those variables in the support of f_y (NORs inputs). Similarly, if $x = 1$ forces $y = 1$ and v_y is a direct predecessor of v_s, then $s = 0$. Eventually, $s = 1 \Rightarrow x = 0$ implies that $x = 1 \Rightarrow s = 0$, which is called the *contrapositive*. Let the direct predecessor relation among variables be denoted by $p(\cdot)$, i.e., $y \in p(x)$ when (v_y, v_x) is an edge of the network graph. Hence the controlling set can be defined recursively as follows:

$$C_{10}(x) = \{s \in S : \exists y \in C_{11}(x); s \in p(y)\} \cup \{s \in S : \exists y \in p(s); y \in C_{11}(x)\}$$

$$\cup \{s : x \in C_{10}(s)\} \cup C_{10}(x) \tag{8.37}$$

Similar definitions apply to C_{11}, C_{01} and C_{00}. (See Problem 12.)

The controlling sets can then be computed by iteration until the sets stabilize. Weaker conditions can also be used to allow fast incremental recomputation of the sets as the network is modified [4].

Let us consider now the use of the controlling sets $C_{11}(x)$ or $C_{10}(x)$. Similar properties and transformations can be defined for controlling sets $C_{01}(x)$ and $C_{00}(x)$. Transformations using controlling sets aim at adding/reducing connections in the circuit, with the overall goal of reducing the circuit area and/or delay.

A first transformation, called *reduction* in [6], uses the following straightforward rules:

- If $y \in C_{11}(x)$, then replace f_y by $f_y + x$.
- If $y \in C_{10}(x)$, then replace f_y by $f_y x'$.

Example 8.4.27. Consider the circuit of Figure 8.27 (a). Since $z \in C_{10}(x)$, we can modify $f_z = (e + g)'$ to $\widetilde{f_z} = (e + g)'x' = (e + g + x)'$. This corresponds to propagating the blocking effect of $x = 1$ forward in the network. The modified network is shown in Figure 8.27 (b).

The reduction step increases the number of literals, as in two-level logic minimization. A transformation in the opposite direction, called *expansion* in [6], targets the replacement of a local function f_y by its cofactor $f_y|_{x=i}$ $i \in \{0, 1\}$, thus reducing the number of literals. For this transformation to be feasible, a specific relation must exist between x and y. This relation can be captured by means of the controlling sets. Since the formal justification of this transformation is complex, we refer the reader to reference [4] for details, and we just give an example to understand the intuitive features of the method.

Example 8.4.28. Consider the circuit of Figure 8.27 (a). A reduction of f_z, as shown in Figure 8.27 (b), represents all implications $x = 1 \Rightarrow z = 0$. Thus, other vertices that are in $C_{1i}(x)$, $i = \{0, 1\}$, can be expanded (i.e., made insensitive to x) when they are predecessors of v_z on all paths to primary outputs. In this case, vertices $\{v_c, v_d\}$ satisfy this property, and their local functions can be replaced by the cofactors with respect to x. The resulting network is shown in Figure 8.27 (c).

The features of the method can be summarized as follows. Two relevant subgraphs can be isolated in the logic network graph for any variable x whose corresponding variables belong to appropriate controlling sets of x. We call them $A(x)$ and $B(x)$. The local functions associated with $B(x)$ have the property that they can be replaced by their cofactor with respect to x (or to x') when all functions in $A(x)$ are reduced as described above.

Hence the global flow transformation consists of selecting a variable x and determining its corresponding sets $A(x)$ and $B(x)$ that are reduced and expanded. The process can be iterated while using heuristics to select the variables.

POLARITY ASSIGNMENT. Polarity assignment problems arise in the context of single-rail circuit design, which is the most common case (e.g., static CMOS logic design). When dual-rail circuit styles are used, such as *emitter-coupled*, *current-mode* and *differential cascade logic*, signals are always present along with their complements. In this case, the logic network (as defined in Section 8.2) models accurately the area and the delay, because no inverters are required. Unfortunately, double-rail circuit design has overheads that limit its use to specific applications.

We consider here single-rail circuit design. When the logic network model of Section 8.2 is used, the area and delay cost of inverters is implicit in the representation, but it eventually has to be taken into account. Whereas this task is easy, optimizing a network with this additional cost component is more difficult.

We address here the reduction of the size (or delay) of the network by choosing the polarity of the local functions. Note that the complementation of a local function requires the input signals with complemented polarity. For this reason this problem is called *global polarity assignment*. An additional degree of freedom may be the possible choice in polarity of the primary inputs and/or outputs, allowed by several design styles. Indeed, combinational networks are often components of sequential circuits. When static latches (or registers) are used, they often provide double-polarity output signals. Hence it is possible to optimize a design with the freedom of selecting the polarity of the input and/or output signals.

A specific global polarity assignment problem is one focusing on the inverter cost only. When local functions are modeled by BDDs or factored forms, their complementation is straightforward and size preserving. So it is reasonable to assume that the area (and/or delay) is invariant under complementation. Then the global polarity assignment problem corresponds to reducing the number of inverters in a network.

We now briefly survey methods for inverter reduction, and we refer the interested reader to reference [43] for details. The problem of choosing the polarity of the network's local functions so that the required inverters are fewer than a given bound is intractable [43]. The inverter minimization problem can still be solved by formulating it as a ZOLP, even though this approach may be practical only for networks of limited size. For tree or forest networks, a dynamic programming approach to the inverter minimization problem can yield the optimum choice in linear time. Unfortunately, most networks have reconverging paths. For these reasons, heuristic methods have been used [8, 43].

A heuristic algorithm is based on a local search that measures the variations in inverter count due to the complementation of a local function. The variation, called *inverter saving*, can be measured by local inspection. Let v_x be the vertex under consideration, and let $PFO(x)$ and $NFO(x)$ be the total number of direct successors of v_x whose functions depend on literals x and x', respectively. Note that $PFO(x) \cap NFO(x)$ may not be void. An example is shown in Figure 8.28. The inverter count variation depends on four factors:

- An inverter is removed at the output of v_x when $PFO(x) = \emptyset$. (Before complementation, all direct successors of v_x depend on x'. After complementation, all direct successors of v_x depend on x, and no inverter is required at the output.)

- An inverter is added at the output of v_x when $NFO(x) = \emptyset$. (Before complementation, all direct successors of v_x depend on x, and no inverter is present at the output. After complementation, all direct successors of v_x depend on x'.)

- For each direct predecessor v_y of v_x, an inverter is removed when $NFO(y) = x$ and $x \notin PFO(y)$. (Variable y is in the support of f_x only. Before complementation, y' is used and an inverter is present. The inverter is not needed after complementation of f_x, which requires only literal y.)

- For each direct predecessor v_y of v_x, an inverter is added when $NFO(y) = \emptyset$. (Before complementation, no inverter is present at the output of f_y. After complementation, f_x depends on literal y'.)

With these rules, the inverter savings can be readily evaluated.

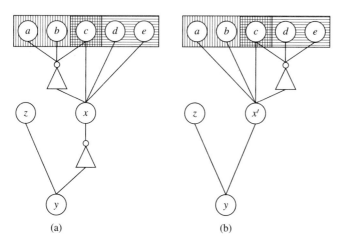

(a)   (b)

FIGURE 8.28
Example of the inverter savings due to polarity assignment. (a) Before complementation of f_x. (b) After complementation of f_x.

Example 8.4.29. Consider the fragment of a logic network shown in Figure 8.28. Let us consider the complementation of f_x. Here $NFO(x) = \{a, b, c\}$ and $PFO(x) = \{c, d, e\}$. Since both fanout subsets are not empty, the complementation of the function would not change the number of inverters at its output. On the other hand, an inverter is saved at the input of v_x, because $NFO(y) = x$ and $x \notin PFO(y)$.

A simple greedy algorithm for phase assignment would complement those functions with positive savings, starting from those with the largest ones. It would terminate when no local complementation can lead to positive savings. The algorithm can be perfected by doing a group complementation that is reminiscent of the group exchange of the Kernighan and Lin partitioning algorithm [30]. In this case, complementations with a local increase of inverters are accepted to escape low-quality local minima. The sequence of complementations that globally saves most inverters is accepted. The sequence is constructed in a greedy fashion, and functions are complemented only once. The process is repeated until an improvement is achieved. Comparisons of different heuristics for polarity assignments are reported in references [8] and [43].

It is important to remark that the polarity assignment problem is solved best in conjunction with library binding, where the real cost of the implementation of the logic functions by means of cells is commensurable with the inverter cost. This issue will be addressed in Section 10.3.3. Nevertheless, the inverter savings problem is also relevant for macro-cell-based design styles, because each local function is directly implemented as a logic gate. For macro-cell design, the polarity assignment problem may have the additional constraint of requiring that each local function is negative (or positive) unate, as in the case of static (or domino) CMOS designs.

8.5 SYNTHESIS OF TESTABLE NETWORKS

Optimization of multiple-level logic circuits is tightly connected to their testability properties, as in the case of two-level logic circuits. We summarize in this section the major findings on the relations between testable and optimal designs. We use the definition of testability introduced in Section 7.2.4. Namely, a fully testable circuit is one that has test patterns which can detect all faults. We restrict our analysis to stuck-at faults, which represent the most common fault model. We refer the reader to reference [23] for synthesis of circuits that are *delay-fault* testable.

Whereas the major goals in logic synthesis are to provide minimum-area (or minimum-delay) circuits possibly under delay (or area) constraints, the synthesis of fully (or at least highly) testable circuits for single/multiple stuck-at faults is also highly desirable. We recall that a necessary and sufficient condition for full testability for single stuck-at faults of an AND-OR implementation of a two-level cover is that the cover is prime and irredundant. We now extend this result to multiple-level networks. Therefore we extend the notion of primality and irredundancy to logic networks.

Definition 8.5.1. A logic network $G_n(V, E)$ whose local functions are expressed in *sum of products* form is **prime and irredundant** if no literal or implicant of any local function can be dropped without changing the input/output behavior of the network.

Definition 8.5.2. A logic network $G_n(V, E)$ whose local functions are expressed in *sum of products* form is **simultaneously prime and irredundant** if no subset of literals and/or implicants can be dropped without changing the input/output behavior of the network.

A logic network can be put in one-to-one correspondence with a circuit implementation by replacing each local *sum of products* expression by some AND gates and one OR gate. This is consistent with our logic network model where signals are available along with their complements and inverters are implicit in the representation. Note that a complete circuit implementation would require inverters, but these do not alter the testability analysis and thus can be neglected.

Theorem 8.5.1. A logic network is prime and irredundant if and only if its AND-OR implementation is fully testable for single stuck-at faults.

Proof. Consider a vertex of the network and the AND-OR implementation of its local function in a *sum of products* form. Faults can happen at the input of the ANDs or of the OR. Assume that there is an untestable stuck-at-x, $x \in \{0, 1\}$, fault on an input of an AND gate corresponding to variable y. Then, that input can be set to x without affecting the network input/output behavior. As a result, either the implicant can be dropped (and the network is not irredundant) or the variable can be dropped (and the network is not prime). Assume next that there is a untestable stuck-at-x, $x \in \{0, 1\}$, fault on an input of the OR gate related to vertex v_x. Then, we can set the input of the OR gate to x with no change in the circuit input/output behavior. Thus, either an implicant in the *sum of products* form is redundant or an implicant can be expanded to a tautology and hence it is not prime.

Conversely, assume an implicant is not prime. Hence, at least a variable, say y, can be dropped while preserving the network input/output behavior. Consider the connection related to variable y. If a stuck-at-1 fault is present on that connection, it is not possible to detect it at the primary outputs. Hence the circuit is not fully testable. Assume last that an implicant is redundant. Then, if the corresponding input to the OR gate has a stuck-at-0 fault, this fault is not detectable. Thus the circuit is not fully testable.

Corollary 8.5.1. A logic network is simultaneously prime and irredundant if and only if its AND-OR implementation is fully testable for multiple stuck-at faults.

The key role in relating testability to primality and irredundancy is played by the *don't care* sets. Indeed, when considering each local function without *don't care* sets, their primality and irredundancy are just necessary conditions for full testability. The *don't care* conditions capture the interplay among the local functions in the logic network. It can be shown by similar arguments [39] that when all local functions of a logic network are prime and irredundant *sum of products* forms with respect to their local (and complete) *don't care* sets, then the logic network is fully testable for single stuck-at faults.

Example 8.5.1. Consider the network of Figure 8.26 (a). Let $DC_x = ODC_x = ac' + bc$. Then $f_x = ab$ consists of a redundant term when considering its *don't care* set. Hence the circuit is not fully testable. Indeed it may be made testable by replacing f_x with $g_x = 0$.

This suggests a way to construct fully testable networks by computing *don't care* sets and minimizing the local functions. Unfortunately, complete local *don't care* sets cannot be used for the simultaneous (multiple-vertex) optimization of the local functions. The use of compatible *don't care* sets would allow us to perform simultaneous minimization, but it would not guarantee full testability because only subsets of the local *don't care* sets are taken into account.

On the other hand, it is possible to iterate (single-vertex) optimization of the local functions using the complete local *don't care* sets. Note that the *don't care* conditions change as the local functions change. This approach has been implemented in program ESPRESSO_MLD [3], which simplifies the local functions and updates the internal *don't care* sets repeatedly until all local functions are prime and irredundant (w.r.t. these *don't care* sets). Experiments have shown that the method is effective only for networks of limited size, due to the possibility that the iteration does not terminate in a reasonable amount of time.

An alternative way for constructing testable networks is based on flattening multiple-level networks to two-level representations, making these fully testable, and then retransforming the two-level covers into multiple-level networks while preserving testability. The premise of this approach is that flattening does not lead to an explosion in size, which indeed may happen. If a network can be flattened, then it can be made fully testable for multiple stuck-at faults by computing prime and irredundant covers of each individual output with a logic minimizer (see Section 7.2.4). We need then transformations that are testability preserving and that allow us to construct a (possibly area/delay minimal) multiple-level network from a two-level representation.

Hachtel *et al.* [29] showed that some algebraic transformations are just *syntactic* rewriting of the network and that they preserve multiple-fault testability. In particular, they proved formally that algebraic factorization, substitution and cube and kernel extraction are testability preserving. These transformations, and others built on these, can then be used to transform a two-level cover into a multiple-level network with the desired area/delay properties and full testability for multiple faults. It is interesting to note that single-fault testability may not be preserved by algebraic transformations, as shown by some counterexamples [29].

No general result applies to Boolean transformations as far as testability preservation is concerned. Some results, reported in reference [6], relate transformations to the potential change of observability of one or more variables. Thus instead of searching for transformations that preserve testability, it is possible to search for observability invariance as a result of a transformation.

We would like to summarize this section by stating that optimizing circuit area correlates positively to maximizing testability. Indeed, (simultaneous) prime irredundant networks that guarantee full testability for (multiple) stuck-at faults are also local minima as far as area reduction, using literals as an area measure. As far as the relation between delay and testability is concerned, Keutzer *et al.* [31] proposed algorithms that transform any logic network into a fully testable one (for single or multiple stuck-at faults) with equal or less delay. Hence a minimal-delay circuit may correspond to a fully testable one. Unfortunately, we still ignore the triangular relations among area, delay and testability. Namely, given a partially testable circuit with a given area and delay, we do not know how to make it fully testable while guaranteeing no increase in both area and delay. This is the subject of future research.

8.6 ALGORITHMS FOR DELAY EVALUATION AND OPTIMIZATION

A major problem in circuit design is achieving the maximum performance for the given technology. In the context of logic synthesis and optimization, maximizing performance means reducing the maximum propagation delay from inputs to outputs. When combinational networks are part of sequential circuits, the maximum propagation delay in the combinational component is a lower bound on the *cycle-time*, whose reduction is one of the major design goals.

Performance evaluation is an important task *per se*, even when it is not coupled to optimization. Whereas performance can be determined accurately by circuit simulation, this becomes infeasible for large circuits, due to the large number of possible input/output responses. Timing analysis consists of extracting the delay information directly from a logic network.

Multiple-level networks offer interesting trade-offs between area and delay, as mentioned in Section 8.2 and shown in Figure 8.4. Very often the fastest circuit is not the smallest, as it is in the case of two-level circuits. An example is given by combinational adders. Ripple-carry implementations tend to be compact and slow, while carry look-ahead adders are larger and faster (see Figure 1.16). Thus circuit transformations can target the decrease of area or delay, possibly while satisfying bounds on delay or area. This leads eventually to circuit implementations that are non-inferior points in the area-delay evaluation space.

We shall consider first delay modeling issues for logic networks in Section 8.6.1. The modeling framework will enable us to define topological critical paths whose delay measure provides useful, but loose, information on performance. We shall then consider the problem of determining genuine critical paths, i.e., paths that can propagate events and have the longest delays. This involves weeding out the *false paths*, by considering the logic information of the local functions while computing the path delays, as shown in Section 8.6.2. Eventually we shall consider circuit transformations and algorithms for delay (or area) reduction, possibly under (area) or delay constraints in Section 8.6.3.

8.6.1 Delay Modeling

Delay modeling of digital circuits is a complex issue. We present here simple models and the fundamental concepts that provide sufficient information to understand the algorithms. The extension of the algorithms presented here to more elaborate models is often straightforward.

Assume we are given a bound network. Then every vertex of the network is associated with a cell, whose timing characteristics are known. Often propagation delays are functions of the load on the cell, which can be estimated by considering the load for each fanout stem. Precise models include also a differentiation for rising and falling transition delays.

For unbound networks, such as those considered in this chapter, less precise measures of the propagation delays are possible. Different models have been used for estimating the propagation delay of the virtual gate implementing a local Boolean function. The simplest model is to use unit delays for each stage. A more refined model is to relate the delay to a minimal factored form. The underlying principles are that a cell generator can synthesize logic gates from any factored form and the decomposition tree of the factored form is related to the series and parallel transistor interconnection, which affects the propagation delay. More detailed models include also a load dependency factor and separate estimates for rising and falling transitions. Empirical formulae exist for different technologies that relate the parameters of the factored form to the delay [8, 21]. Good correlation has been shown between the predicted delay and the delay measured by circuit simulation.

We are concerned here with delay evaluation in a network. For our purposes, we can just assume that the *propagation delay* assigned to a vertex of the logic network is a positive number. In general, a certain margin of error exists in propagation delay estimates. Hence worst-case estimates are used that include the possible worst-case operating conditions (e.g., temperature, power supply, etc.) and fabrication parameters. The worst-case assumption is motivated by the desire of being on the safe side: overestimating delays may lead to circuits that do not exploit a technology at its best; underestimating delays may lead to circuits that do not operate correctly.

We assign also to each vertex an estimate of the time at which the signal it generates would settle, called *data-ready* time or *arrival* time. Data-ready times of the primary inputs denote when these are stable, and they represent the reference points for the delay computation in the circuit. Often the data-ready times of the primary inputs are zero. This is a convenient convention for sequential circuits, because the delay computation starts at the clock edge that makes the data available to the combinational component. Nevertheless, positive-input data-ready times may be useful to model a variety of effects in a circuit, including specific delays through the input ports or circuit blocks that are not part of the logic network abstraction.

The data-ready computation can be performed in a variety of ways. We consider in this section a model that divorces the circuit topology from the logic domain; i.e., data-ready times are computed by considering the dependencies of the logic network graph only and excluding the possibility that some paths would never propagate events due to the specific local Boolean functions. (This assumption will be removed in Section 8.6.2.) We can assume then that the data-ready time at each internal and output vertex of the network is the sum of the propagation delay plus the data-ready time of the latest local input. Note that each output vertex depends on one internal vertex. A non-zero propagation delay can be associated with the output vertices to model the delay through the output ports, in the same way as non-zero input data-ready times can model delays through the input ports.

Let the propagation delay and the data-ready time be denoted by $\{d_i | v_i \in V\}$ and $\{t_i | v_i \in V\}$, respectively. Then:

$$t_i = d_i + \max_{j:(v_j, v_i) \in E} t_j \qquad (8.38)$$

The data-ready times can be computed by a forward traversal of the logic network in $O(|V| + |E|)$ time. The maximum data-ready time occurs at an output vertex, and it is called the *topological critical delay* of the network. It corresponds to the weight of the longest path in the network, where weights are the propagation delays associated with the vertices. Such a path is called a *topological critical path*. Note that such a path may not be unique.

> **Example 8.6.1.** Consider the logic network of Figure 8.29. Assume that the data-ready times of the primary inputs are $t_a = 0$ and $t_b = 10$, respectively. Let the propagation delay of the internal vertices be $d_g = 3; d_h = 8; d_m = 1; d_k = 10; d_l = 3; d_n = 5; d_p = 2; d_q = 2; d_x = 2; d_y = 3$.
> Then, the data-ready times are as follows:
>
> $$t_g = 3 + 0 = 3$$
> $$t_h = 8 + 3 = 11$$
> $$t_k = 10 + 3 = 13$$
> $$t_n = 5 + 10 = 15$$
> $$t_p = 2 + \max\{15, 3\} = 17$$
> $$t_l = 3 + \max\{13, 17\} = 20$$
> $$t_m = 1 + \max\{3, 11, 20\} = 21$$
> $$t_x = 2 + 21 = 23$$
> $$t_q = 2 + 20 = 22$$
> $$t_y = 3 + 22 = 25$$

The maximum data-ready time is $t_y = 25$. The topological critical path is $(v_b, v_n, v_p, v_l, v_q, v_y)$.

Timing optimization problems for multiple-level networks can be formulated by restructuring the logic network to satisfy bounds on output data-ready times or to minimize them. We denote the *required data-ready* time at an output by a bar, e.g., the required data-ready time at output v_x is $\bar{t}_x$. The required data-ready times can be propagated backwards, from the outputs to the inputs, by means of a backward network traversal, namely:

$$\bar{t}_i = \min_{j:(v_i, v_j) \in E} \bar{t}_j - d_j \qquad (8.39)$$

It is usual to record at each vertex the difference between the required data-ready time and the actual data-ready time. This quantity is called the timing *slack*:

$$s_i = \bar{t}_i - t_i \quad \forall v_i \in V \qquad (8.40)$$

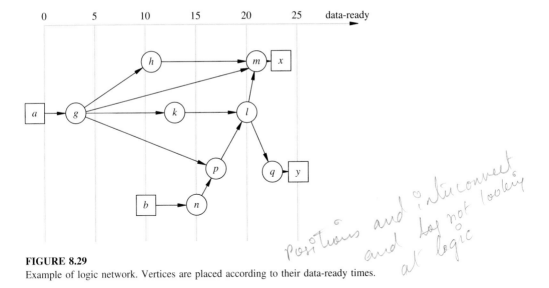

FIGURE 8.29
Example of logic network. Vertices are placed according to their data-ready times.

Example 8.6.2. Consider again the circuit of Example 8.6.1, shown in Figure 8.29. Let us assume that the required data-ready time at both outputs is 25, namely $\bar{t}_x = 25$ and $\bar{t}_y = 25$. Then the slacks are $s_x = 2$ and $s_y = 0$, which means that signal x could be delayed up to 2 units without violating the bound, while signal y cannot.

The required data-ready times and slacks at v_m and v_q are $\bar{t}_m = 25 - 2 = 23$ and $\bar{t}_q = 25 - 3 = 22$ and $s_m = 2$ and $s_q = 0$, respectively. The required data-ready time at v_l is $\min\{23 - 1; 22 - 2\} = 20$. Hence $s_l = 0$.

The remaining required data-ready times and slacks are $\bar{t}_h = 23 - 1 = 22$; $s_h = 22 - 11 = 11$; $\bar{t}_k = 20 - 3 = 17$; $s_k = 17 - 13 = 4$; $\bar{t}_p = 20 - 3 = 17$; $s_p = 17 - 17 = 0$; $\bar{t}_n = 17 - 2 = 15$; $s_n = 15 - 15 = 0$; $\bar{t}_b = 15 - 5 = 10$; $s_b = 10 - 10 = 0$; $\bar{t}_g = \min\{22 - 8; 17 - 10; 17 - 2\} = 7$; $s_g = 7 - 3 = 4$; $\bar{t}_a = 7 - 3 = 4$; $s_b = 4 - 0 = 4$.

The topological critical path $(v_b, v_n, v_p, v_l, v_q, v_y)$ corresponds to the vertices with zero slacks.

The required data-ready times and the slacks of the internal and input vertices can be computed by backward network traversal in $O(|V| + |E|)$ time. Critical paths are identified by vertices with zero slack, when the required data-ready times at the outputs are set equal to the maximum data-ready time. Note the similarities and differences with scheduling (see Section 5.3): in scheduling we use the *start* time for a computation, while in logic design the data-ready time denotes the *end* time of a logic evaluation.

8.6.2 Detection of False Paths*

Topological critical paths can be determined easily from the logic network graph only. Unfortunately, they may lead to overestimating the delay of the circuit. It may be possible that a topological critical path is a *false* path, when no *event* (i.e., signal transition) can propagate along it. This can be explained by considering the interplay of the local functions of the network with the network topology.

Example 8.6.3. Consider the circuit of Figure 8.30. Assume that the propagation delay of each gate is 1 unit of delay and that inputs are available at time 0.

The longest topological path is $(v_a, v_c, v_d, v_y, v_z)$ and has a total of 4 units of delay. It is easy to see that no event can propagate along this path, because the AND gate requires $e = 1$ and the OR gate requires $e = 0$ to propagate the event. On the other hand, the value of e settles at time 1 and hence it is either 0 or 1. The longest true path is (v_a, v_c, v_d, v_y) and has a total of 3 units of delay.

Obviously false paths do not affect the circuit performance. Therefore, it is very important to weed out false paths and detect the true critical paths. The true critical paths are called critical paths for brevity and may have smaller delay than the topological critical paths. The detection of the critical path is important for timing verification, i.e., checking if a circuit can operate at a given speed. It is also important for delay optimization, because logic transformations should be applied to true critical paths, which are responsible for the circuit speed.

Definition 8.6.1. A path of a logic network is **sensitizable** if an event can propagate from its tail to its head. A **critical path** is a sensitizable path of maximum weight.

In the sequel, we do not distinguish between a network of (virtual) gates and its representative graph. Let us consider a vertex along a sensitizable path. The direct predecessors of the vertex under consideration are called the *inputs* to the vertex. The inputs that are not along the path are called *side inputs*.

Let us consider now conditions for sensitization of a path $P = (v_{x_0}, v_{x_1}, \ldots, v_{x_m})$. An event propagates along P if $\partial f_{x_i}/\partial x_{i-1} = 1$, $\forall i = 1, 2, \ldots, m$. Since the Boolean differences are functions of the side inputs and values on the side inputs may change, the sensitization condition is that the Boolean differences must be true at the time that the event propagates. For this reason, this form of sensitization is called *dynamic sensitization*.

Example 8.6.4. Consider again Example 8.6.3 and path $(v_a, v_c, v_d, v_y, v_z)$ shown in Figure 8.30. For an event to propagate along that path, assuming that the event leaves v_a

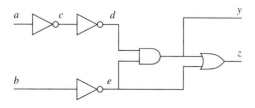

FIGURE 8.30
Example of false path.

at time 0, we need:

$\partial f_y / \partial d = e = 1$ at time 2;
$\partial f_z / \partial y = e' = 1$ at time 3.

It is impossible to meet both the first and last conditions, because e will settle to a final value at time 1.

A weaker condition is *static sensitization*, which is not sufficient to characterize the false-path problem but deserves an analysis because it is related to other sensitization criteria. A path is *statically sensitizable* if there is an assignment of primary inputs $\mathbf{t} \in \mathsf{B}^n$ such that $\partial f_{x_i}(\mathbf{t}) / \partial x_{i-1} = 1$, $\forall i = 1, 2, \ldots, m$. In other words, in static sensitization we assume that the Boolean differences must be true at time infinity, i.e., when all signals settle, instead of when the event propagates. As a consequence any approach based on static sensitization may lead to underestimate the delay, because there may be paths that are not statically sensitizable but that can still propagate events.

Example 8.6.5. Consider first the circuit of Figure 8.30. Path $(v_a, v_c, v_d, v_y, v_z)$ is not statically sensitizable because no input vector can make $e = 1$ and $e' = 1$. The path is false as we argued before.

Consider next the circuit of Figure 8.31. For the sake of simplicity, all input data-ready times are zero and all vertices (gates) have unit delay. The topological critical paths are $\{(v_a, v_d, v_g, v_o); (v_b, v_d, v_g, v_o)\}$ with delay 3. For a to propagate to d we must have $b = 1$. For g to propagate to o we must have $e = 1$, which implies $a = b = 0$. Hence there is no input vector that statically sensitizes path (v_a, v_d, v_g, v_o) and similarly for (v_b, v_d, v_g, v_o). Both paths appear to be false. On the other hand, path (v_a, v_e, v_o) is statically sensitizable, with $b = 0$ and $c = 1$, and has weight 2.

Let us assume that $c = 0$ in the interval of interest. Let the event be a simultaneous transition of a and b from 1 to 0. Then e rises to 1 and d drops to 0 after 1 unit of time. Signal g drops to 0 after 2 units of time. Eventually signal o raises to 1 and drops to 0 after 2 and 3 units of time, respectively. Hence the critical path delay is 3. Any sampling of the output after 2 units of delay would yield a wrong result.

Thus both paths $\{(v_a, v_d, v_g, v_o); (v_b, v_d, v_g, v_o)\}$ are true critical paths, even though they are not statically sensitizable.

Another important criterion for the detection of critical paths is robustness. The robustness of the method relies on interpreting correctly the underlying assumptions. Consider the propagation delay of a vertex. It may be expressed by one number, representing the worst-case condition, or by an interval, representing the possible delays of the gate implementation under all possible operating conditions. The former delay model is called a *fixed* delay model and the latter a *bounded* delay model.

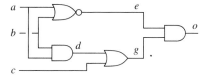

FIGURE 8.31
Example of a true path that is not statically sensitizable.

In addition, a delay computation requires assumptions on the "memory" that circuit nets can have, due to parasitic capacitances in practical circuits. In the *transition* mode of operation, variables are assumed to hold their previous values until they respond to an input stimulus vector **t**. This model abstracts accurately the operation of the circuit. Any circuit delay measure involves then two input vectors, the first devoted to setting the circuit in a particular state. Needless to say, delay computation with this mode can be lengthy. Conversely, in the *floating* mode of operation, the circuit under consideration is memoryless. Thus, the values of the variables are assumed to be unknown until they respond to an input stimulus vector **t**. Whereas the delay computation with the floating mode of operation is simpler to perform, it results in pessimistic estimates of the performance; i.e., the critical path delay is never inferior to that computed with the transition mode [14].

It is interesting to remark that the transition mode of operating a circuit may not have the *monotone speed-up* property [32]. Namely, the replacement of a gate with a faster one may lead to a slow-down of the overall circuit.

Example 8.6.6. The following example is adapted from reference [32]. Consider the circuit in Figure 8.32 (a). Assume that all propagation delays are 2 units, except for

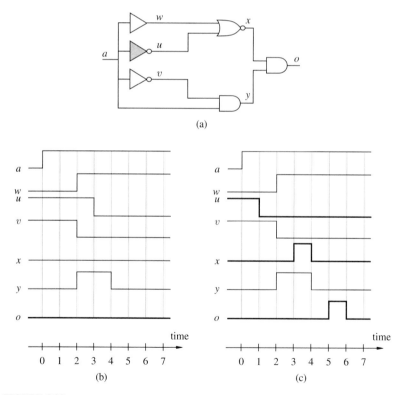

(a)

(b)

(c)

FIGURE 8.32
Example of monotone speed-up failure.

the shaded gate, whose delay is 3 units. The waveforms for a rising transition at v_a are shown in Figure 8.32 (b). The circuit has no sensitizable paths and its critical path delay is zero.

Assume now that the propagation delay of the shaded gate is 1. The waveforms for a rising transition at v_a are shown now in Figure 8.32 (c). Note that the circuit output settles after 6 delay units. Hence speeding up the shaded gate makes the overall circuit slower.

Thus, when considering the transition mode of operation, the fixed gate delay model and the bounded gate delay model may yield different delay estimates. As a consequence, any robust method for false-path detection using the transition mode must take into account the monotone speed-up effect. On the other hand, in the floating mode of operation, the critical path delay of a circuit is the same when considering either bounded or worst-case propagation delays [14].

Example 8.6.7. Consider the circuit in Figure 8.32. Regardless of the value of the propagation delay of v_u, the floating mode delay of v_o is 6, because it is undetermined until that time.

There are several theories for computing the critical path delays that use different underlying assumptions [14, 24, 27, 32, 33]. We summarize here an approach that uses the floating mode of operation [24]. Without loss of generality, assume that the circuit can be represented in terms of AND and OR gates, in addition to inverters that do not affect our analysis. We say that 0 is a *controlling value* for an AND gate and that 1 is a controlling value for an OR gate. The controlling value determines the value at the gate output regardless of the other values. We say that a gate has a *controlled value* if one of its inputs has a controlling value.

Consider now a path $P = (v_{x_0}, v_{x_1}, \ldots, v_{x_m})$. A vector *statically co-sensitizes* a path to 1 (or to 0) if $x_m = 1$ (or 0) and if $v_{x_{i-1}}$ has a controlling value whenever v_{x_i} has a controlled value, $\forall v_{x_i}, i = 1, 2, \ldots, m$. Note that this criterion differs from static sensitization, which requires $v_{x_{i-1}}$ to present the only controlling value.

Example 8.6.8. Consider the circuit of Figure 8.31. Path (v_a, v_d, v_g, v_o) is statically co-sensitizable to 0 by input vector $a = 0, b = 0, c = 0$, because the gates with controlled values (corresponding to v_o and v_d) have a controlling input along the path, i.e., $g = 0, a = 0$.

On the other hand, static sensitization requires $e = 1, c = 0$ and $b = 1$, and since the first and last conditions are incompatible, the path (v_a, v_d, v_g, v_o) is not statically sensitizable.

Static co-sensitization is a necessary condition for a path to be true, whereas static sensitization is not, as shown by Example 8.6.5. Necessary and sufficient conditions for a path to be true can be formally stated, as reported in reference [24]. We present here the intuitive features of that analysis.

The static co-sensitization criterion is based on logic properties only and does not take delays into account. When a vector statically co-sensitizes a path to 1 (or 0), there may be some other path that sets the path output to that value earlier. Thus, for a path to cause the output transition, the following must hold for all gates along the path. If a gate has a controlled value, then the path must provide the first of the controlling values. If a gate does not have a controlled value, the path must provide the last of the non-controlling values. The reason is that when a gate has a controlled value, the delay up to its output is determined by the first controlling input, while when it does not have a controlled value, the delay is determined by the last (non-controlling) input.

As a result, a test for detecting a false path can be done as follows. A path is false if, for all possible input vectors, one of the following conditions is true:

1. A gate along the path is controlled, the path provides a non-controlling value and a side input provides a controlling value.

2. A gate along the path is controlled and both the path and a side input have controlling values, but the side input presents the controlling value first.

3. A gate along the path is not controlled but a side input presents the controlling value last.

> **Example 8.6.9.** Consider the circuit of Figure 8.30. We question the falsity of path $(v_a, v_c, v_d, v_y, v_z)$.
> For input vector $a = 0, b = 0$ condition 1 occurs at the OR gate.
> For input vector $a = 0, b = 1$ condition 2 occurs at the AND gate.
> For input vector $a = 1, b = 0$ condition 2 occurs at the OR gate.
> For input vector $a = 1, b = 1$ condition 1 occurs at the AND gate.
> Therefore the path is false.

There are different interesting design problems related to false-path detection. A very important question is to determine if a circuit works at a given speed, i.e., if its critical path delay is no larger than $\bar{t}$. This problem can be rephrased as checking if all paths with delay larger than $\bar{t}$ are false. Another important issue is to determine the critical path delay of a circuit. This problem can be reduced to the previous problem, by making a sequence of tests if the critical path delay is no larger than $\bar{t}$, where $\bar{t}$ is the outcome of a binary search in the set of path delays sorted in decreasing order.

Therefore it is important to have methods that can detect groups of false paths to avoid checking paths one at a time. Indeed, the number of paths in a network grows exponentially with the number of vertices, and false-path detection is particularly important for those circuits with many paths with similar delays. Multiple false-path detection methods have been recently developed [24], and leverage concepts developed in the testing community, such as the D-calculus. The description of these techniques goes beyond the scope of this book. We refer the interested reader to reference [24] for further details.

8.6.3   Algorithms and Transformations for Delay Optimization

We consider in this section algorithms for reducing the critical delay (possibly under area constraints) and reducing the area under input/output delay constraints. We shall refer to critical paths without distinguishing among topological or sensitizable paths,

$REDUCE\_DELAY(G_n(V, E), \epsilon)\{$

    repeat {

          Compute critical paths and critical delay τ;

          Set output required data-ready times to τ;

          Compute slacks;

          U = vertex subset with slack lower than ϵ;

          W = select vertices in U;

          Apply transformations to vertices W;

    } **until** (no transformation can reduce τ)

}

ALGORITHM 8.6.1

because the techniques are independent of the criterion used. Obviously the quality of the results reflects the choice.

SYNTHESIS OF MINIMAL-DELAY CIRCUITS. Let us consider first the problem of minimizing the critical delay that arises often in synchronous circuit design, where the logic network under consideration is the combinational portion of a sequential circuit. Recall that the critical delay is a lower bound for the cycle-time.

Most critical delay optimization algorithms have the frame shown in Algorithm 8.6.1.

The parameter ϵ is a threshold that allows us to consider a wider set of paths than the critical ones. When $\epsilon = 0$, the vertices in U are critical, and they induce a path if the critical path is unique. Quasi-critical paths can be selected with $\epsilon > 0$, so that more transformations can be applied between two recomputations of the critical path. It is obvious that the smaller the difference between the critical delay and the largest delay of a non-critical path, the smaller the gain in speeding up the critical paths only.

Transformations target the reduction of the data-ready time of a vertex. There are two possibilities: reducing the propagation delay and reducing the dependency of the vertex under consideration on some critical input. The transformation must not have as a side effect the creation of another critical path with equal or larger delay. When a single transformation is applied at each iteration of the algorithm, this can be ensured by monitoring the data-ready variation at the neighboring vertices that should be bounded from above by their corresponding slacks. When area constraints are specified, the marginal area variation of the transformation should be recorded and checked against a bound. Note that recomputation of the critical paths and slacks is also an important issue for the computational efficiency of the algorithm. Several schemes for determining the cone of influence of a transformation have been devised, so that the data-ready times and slacks can be updated without redundant computation.

Most algorithms differentiate in the selection of the vertex subset and in the transformations that are applied. The propagation delay model of the local functions plays an important role because they are tightly coupled to the transformations.

We consider first, as a simple example, the case of unit propagation delays. Hence, a minimum delay network is a single-stage network that can be obtained by the *ELIMINATE* algorithm. Single-stage networks may have local functions so complex that they are not practical. To cope with the problem of literal and area explosion, vertex elimination can be applied selectively. Let us consider algorithm *REDUCE_DELAY* and let us choose W as a minimum-weight vertex separation set in the subgraph induced by U, where the weight of each vertex is its area value, i.e., the increase in area due to its elimination. Let the transformation be the elimination of W. Thus the algorithm will generate a sequence of networks corresponding to different area/delay trade-off points. An area constraint can be easily incorporated as a stopping criterion, i.e., the algorithm terminates when an elimination causes the area estimate to overcome the given bound.

Example 8.6.10. Consider the network of Figure 8.33 (a). Assume that all gates have unit delay and all input data-ready times are 0. Then, (topological) critical paths are $(v_e, v_p, v_s, v_u, v_w)$, $(v_g, v_p, v_s, v_u, v_w)$, $(v_h, v_q, v_s, v_u, v_w)$, $(v_k, v_q, v_s, v_u, v_w)$.

Let us consider the values of the vertices. Those vertices that provide primary output signals, e.g., v_x, v_y, v_w, v_z, are given infinite weight, because they cannot be eliminated. The values of v_t, v_p, v_q, v_u are -1, the value of v_s is 0 and the value of v_r is 1 (using a factored form model).

Hence a critical vertex separation set of minimum cost is $\{v_p, v_q\}$, whose elimination leads to $s = eghk$. Now the critical paths are (v_i, v_r, v_u, v_w) and (v_i, v_s, v_u, v_w), where v_i is some primary input. The value of v_s is now 2. Hence a minimum-cost separation set is $\{v_u\}$, whose elimination yields $w = yrs$.

At this point a few paths with weight 2 still remain, as shown in Figure 8.33 (b). They would be eliminated in the next iteration of the algorithm, which would choose $\{v_r, v_s, v_t\}$ as the separation set, because it is the only choice left. Note that as the algorithm proceeds, the local expression gets larger and the assumption on unit delay gates becomes less reasonable.

Algorithm *REDUCE_DELAY* can support other choices for W as well as other transformations, such as decomposition, substitution and simplification [8, 21]. In addition, general propagation delay models can be used. Note that when the delay model depends on the fanout load, neighboring vertices can be affected by the transformation and should be checked.

Example 8.6.11. Let us consider a delay model where the propagation delay grows with the size of the expression and with the fanout load. Let us analyze the fragment of a network graph shown in Figure 8.34 (a). Let us consider first a critical path (v_k, v_j, v_i). A selective elimination of j into the expression of v_i reduces one stage of delay. Nevertheless, the gain may be offset by the larger propagation delay of v_i, because function f_i has now more literals. In addition, the load on v_k has increased, which can raise the propagation delay d_k. As a result, the data-ready time t_m may increase. Even though the path (v_k, v_j, v_m) was not critical, it may become so if the variation in the data-ready time of t_m is larger than the slack s_m.

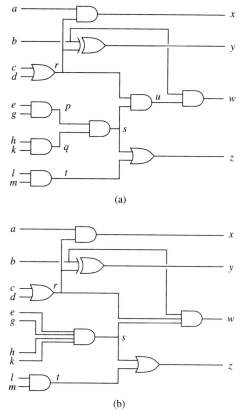

(a)

(b)

FIGURE 8.33
(a) Example of logic network. (b) Example
of logic network after two iterations of the
algorithm.

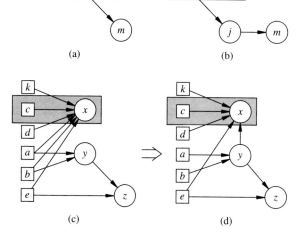

FIGURE 8.34
(a,b) Network fragments before
and after elimination. (c,d)
Network fragments before and after
substitution. (Shaded area is critical.)

Let us analyze the network fragment shown in Figure 8.34 (c) and described by the expressions:

$$x = ka + kb + cde$$

$$y = a + b$$

$$z = ey$$

Let (v_c, v_x) be on a critical path, being t_c is much larger than the other data-ready times of the inputs to v_x. Then it is likely that substituting y for $a + b$ in f_x may reduce its propagation delay d_x and the data-ready time t_x. Note also in this case that the load on v_y increases. Even though this may not affect the critical path (when $t_y < t_c$) it affects t_z. Hence we must ascertain that the variation of t_z is bounded by its slack s_z.

A more elaborate approach is used by the $SPEED\_UP$ algorithm in program MIS [41]. In this case, the network is decomposed beforehand into two-input NAND functions and inverters. This step makes the network homogeneous and simplifies the delay evaluation.

The algorithm can fit into the frame of $REDUCE\_DELAY$ shown before. The set W is determined by first choosing an appropriate vertex separation set in the subgraph induced by U and then adding to it the predecessor vertices separated by no more than d stages, d being a parameter of the algorithm. Then unconstrained elimination is applied to the subnetwork induced by W. Vertices with successors outside the subnetwork are duplicated, and the corresponding cost is called *area penalty*. The *potential speed-up* in the subnetwork is an educated guess on the speed-up after elimination and resynthesis. We can now explain how the separation set was chosen: it is a minimum-weight separation set, where the weights are a linear combination of the *area penalty* and *potential speed-up*.

The subnetwork is then resynthesized by means of a timing-driven decomposition, by selecting appropriate divisors according to the data-ready times. Note that the data-ready times of the inputs to the subnetwork are known. The recursive decomposition process extracts first vertices that are fed by the inputs, and so the data-ready times can be computed correctly bottom up. Eventually the subnetwork is cast into NAND functions, while optimizing the delay. The algorithm is controlled by a few empirical parameters, including ϵ and d described above, the coefficients of the linear combination of the *area penalty* and *potential speed-up* and a choice of propagation delay models, ranging from unit delay to unit delay with fanout dependency or to library cell evaluation. We refer the reader to reference [41] for more details.

Example 8.6.12. Consider the circuit of Figure 8.35 (a). Let all NAND gates have propagation delay equal to 2 units and each inverter equal to 1 unit. Let the data-ready times of all inputs be 0, except for d, where $t_d = 3$. The critical path goes then through v_d and v_x and its delay is 11 units. Let us assume that $\epsilon = 2$ and thus W includes the vertices in the path from v_d to v_x. Assume that v_x is selected as the vertex separation set and the parameter $d = 5$.

Figure 8.35 (b) shows the circuit after the subnetwork feeding v_x has been eliminated into a single vertex, with corresponding expression $a' + b' + c' + d' + e'$. Note

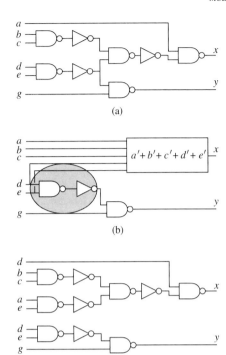

(a)

(b)

(c)

FIGURE 8.35
(a) Logic network. (b) Logic network after elimination.
(c) Logic network after resynthesis by a new decomposition.

that the gates in the shaded area have been duplicated. The cost of that duplication (one NAND gate and one inverter) has been weighted against the possible speed-up of the network due to a more favorable decomposition that balances the skews of the inputs.

Such a decomposition is shown in Figure 8.35 (c). The decomposition has a structure similar to the original circuit, but it reduces the number of stages for the late arriving input d. Almost all input/output paths are now critical, being the critical delay is equal to 8 units.

SYNTHESIS OF MINIMAL-AREA CIRCUITS UNDER DELAY CONSTRAINTS. Another relevant problem in multiple-level logic synthesis is the search for minimal-area implementations subject to timing constraints. The problem can be decomposed into two subproblems:

- compute a timing-feasible solution, i.e., satisfy all timing constraints;
- compute an area-optimal solution.

The former subproblem is more difficult, and sometimes no solution may be found. Unfortunately, due to the heuristic nature of multiple-level logic optimization, the lack of a solution may not be related to an overspecification of the design, because the method itself may not be able to find a solution.

Timing constraints are expressed as input arrival times and output required data-ready times. A circuit is timing feasible if no slack is negative. The larger the magnitude of a negative slack, the more acute is the local problem in satisfying the timing requirements. A first approach to achieve timing feasibility is to apply transformations to the vertices with negative slacks repeatedly until all slacks are non-negative. A pitfall of this approach is that no transformation may be able to make non-critical those vertices with large negative slacks, because a major restructuring is required in the neighborhood of that vertex. Since major changes require a sequence of transformations and most of them lack a look-ahead capability, no transformation may be found and the method may fail.

A gradual approach to achieving timing feasibility is the following. As a first step, different required data-ready times on different outputs are transformed into a single constraint. This can be done by using the largest required data-ready time as an overall constraint and by adding the difference to the actual required data-ready times as "fake" delays on the output ports. In particular, $\bar{t} = \max_{v_i \in V^O} \bar{t}_i$ is the new required data-ready time after having added $\bar{t} - \bar{t}_i$ to the propagation delay of each output port $v_i \in V^O$.

We can then use algorithm $REDUCE\_DELAY$ with a different exit condition. Namely $\tau = \bar{t}$ breaks the loop. At each iteration, some critical vertices are identified and transformed. The goal is to find the transformation that most reduces the delay, rather than the transformation that reduces the delay by a given amount. Hence it is more likely that a transformation is found and that the algorithm succeeds.

When the given circuit is timing feasible, the search for a minimum-area solution can be done by logic transformations. Care has to be taken that no transformation violates the timing constraint. Hence, if the transformation increases the data-ready time at any vertex, this amount should be bounded from above by the slacks. Note that efficient updates of the data-ready times and slacks are of great importance for the overall efficiency of the method.

Example 8.6.13. Consider again the network of Figure 8.33 (a), where gates have unit delay and all input data-ready times are 0. Assume that required output data-ready times are $\bar{\mathbf{t}} = [2332]^T$. We could then assume a single required data-ready time $\bar{t} = 3$ by adding an additional unit delay on the output ports for x and z.

The network is not timing feasible, because the critical path has weight $\tau = 4$. Let us set then the required data-ready time to 4 and compute the slacks. We would detect the same critical paths as in Example 8.6.10. Assume that set $W = \{v_p, v_q\}$ is selected and that these vertices are eliminated.

Then, a successive computation of the longest path would result in $\tau = 3$. Since the timing constraints are now met, we can look for other transformations that reduce the area while preserving timing feasibility. For example, we could eliminate vertices with negative value, such as v_t.

It is important to remember that the transformations are straightforward in this example, because the unit delay model was purposely chosen to show the underlying mechanism of the algorithm. When considering more realistic propagation delay models, the choice of the transformation and their applicability is obviously more involved.

8.7 RULE-BASED SYSTEMS FOR LOGIC OPTIMIZATION

Expert systems are an application of artificial intelligence to the solution of specific problems. A key concept in expert systems is the separation between knowledge and the mechanism used to apply the knowledge, called *inference procedure*. Hence this methodology is applicable to solving a wide class of problems, each characterized by its domain-specific knowledge.

Rule-based systems are a class of *expert systems* that use a set of rules to determine the action to be taken toward reaching a global goal [13]. Hence the inference procedure is a rule interpreter. Rule-based systems evolved over the years to a good degree of maturity, and several *knowledge engineering frameworks* have been designed to support the design and use of such systems.

We consider in this section the application of rule-based systems to multiple-level logic optimization. Hence our view of rule-based systems will be somehow limited in scope and focused on our application. We refer the reader interested in rule-based systems to reference [13] for details.

Rule-based systems for logic optimization have been fairly successful. The first logic synthesis and optimization systems used rules. The most notable example is IBM's Logic Synthesis System (LSS) [18, 19]. Some commercial logic synthesis and optimization systems are based on rules, especially for solving the library binding problem. Therefore some of these considerations will be echoed in Section 10.5. Here we concentrate on the application of the rule-based methodology to the optimization of unbound logic networks.

In a rule-based system, a network is optimized by a stepwise refinement. The network undergoes local transformations that preserve its functionality. Each transformation can be seen as the replacement of a subnetwork by an equivalent one toward the desired goal.

A rule-based system in this domain consists of:

- A rule *database* that contains two types of rules: *replacement rules* and *metarules*. The former abstract the local knowledge about subnetwork replacement and the latter the global heuristic knowledge about the convenience of using a particular strategy (i.e., applying a set of replacement rules).

- A system for entering and maintaining the database.

- A controlling mechanism that implements the inference engine. It is usually a heuristic algorithm.

A rule database contains a family of circuit patterns and, for each pattern, the corresponding replacement according to the overall goal (such as optimizing area, testability or delay). Several rules may match a pattern, and a priority scheme is used to choose the replacement.

Some systems, like LSS, use a common form for representing both the logic network and the rules, for example, interconnections of NAND or NOR gates with a limit on the number of inputs. This facilitates matching the patterns and detecting the applicability of a rule. Other systems use signatures of subcircuits that encode the truth table and can be obtained by simulation.

> **Example 8.7.1.** Consider the rules shown in Figure 8.36. The first rule shows that two cascaded inverters can be replaced by a single inverter. The second rule indicates a direct implementation of a three-input NAND function. The last rule removes a local redundancy, and therefore it increases the testability.

A major advantage of this approach is that rules can be added to the database to cover all thinkable replacements and particular design styles. This feature played a key role in the acceptance of optimization systems, because when a designer could outsmart the program, his or her "trick" (knowledge) could then be translated into a rule and incorporated into the database.

The database must encode the rules in an efficient way to provide for fast execution of the system. Since this task is not simple, specific programs collect the knowledge about the possible transformations and create the database. Even though the generation of the database is time consuming, this task is performed very infrequently.

The most compelling problem in rule-based systems is the order in which rules should be applied and the possibility of look-ahead and backtracking. This is the task of the control algorithm, that implements the inference engine. The algorithm has to measure the quality of the circuit undergoing transformations. The measure of area and delay is usually incremental and involves standard techniques. Then the algorithm enables the rules to identify the subnetwork that is the target for replacement and to choose the replacement. This may involve firing one or more rules.

The problems that the control algorithm has to cope with can be understood by means of the following ideal model. (The adjective "ideal" means that the model is not related to the implementation of the algorithm.) Let us consider the set of all equivalent logic networks that implement the behavior of the original network. We define a *configuration search graph*, where the vertices correspond to the set of all equivalent networks and the edges represent the transformation rules. Assume each vertex is labeled with the cost of the corresponding implementation (e.g., area or

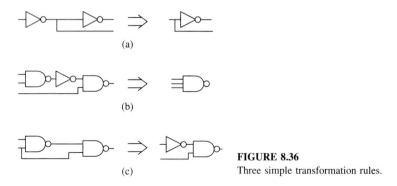

(a)

(b)

(c)

FIGURE 8.36
Three simple transformation rules.

timing cost). A set of rules is *complete* if each vertex is reachable from any other vertex. Needless to say, questioning the completeness of the rule set is a hard problem because of the size of the search configuration graph.

A greedy search is a sequence of rules, each decreasing the circuit cost. Hence it corresponds to a path in the configuration search graph, whose head is the final solution. Obviously it may not be the (global) optimum solution. Indeed the choice of applying a rule affects which rules can be fired next. Some rule-based systems, such as LSS [19], use this kind of strategy for the control algorithm.

Higher quality solutions can be found by using a more complex search for the transformation to apply. The rationale is to explore the different choices and their consequences before applying a replacement. Given a circuit configuration, consider the set of possible sequences of transformations. The branching factor at each configuration is called the *breadth* of the search; the length of the sequences is the *depth*. The larger the breadth and the depth, the better the quality of the solution but also the higher the computing time, which grows exponentially with the depth. Ideally, if the set of rules is complete and the depth and breadth are unbounded, the optimum configuration can be reached.

Some rule-based systems, such as SOCRATES [22, 28], use heuristics to bound the breadth and the depth, as well as for restricting the search to the rules that apply to a small portion of the circuit (*peephole optimization*). Whereas experiments can be used to determine good values for the depth and breadth parameters, a better strategy is to vary them dynamically. *Metarules* are rules that decide upon the value of these parameters, given the state of the optimization [28]. Metarules can be used also to trade off area versus delay or the quality of the solution versus computing time.

From an implementation point of view, the control algorithm iterates the following tasks. First it evaluates the cost functions of interest. Then, it fires the metarules that determine some global parameters of the search. This is followed by firing replacement rules in a specific order. When no rule can improve the cost function of interest, the network is declared optimal and the optimization procedure ends.

The control mechanism is *data driven*, i.e., the moves are determined by how transformations can improve the cost function from a given configuration. This approach, called also *forward chaining* of the rules, is a typical choice when rule-based systems are applied to logic optimization problems.

We shall revisit rule-based systems for the optimization of bound networks in Section 10.5. We shall compare then the rule-based approach to the algorithmic approach in Section 10.5.1.

8.8 PERSPECTIVES

Multiple-level logic synthesis has evolved tremendously over the last two decades. Synthesis and optimization of large-scale circuits is today made possible by CAD tools developed on the basis of the algorithms described in this chapter. The use of automatic circuit optimization has become common practice in industry and in academia. Existing CAD tools support area- and performance-oriented design and are linked to test pattern generators and to programs that ease the testing problem in different ways.

There are still several open issues in this field. First of all, exact methods are few and inefficient. This contrasts the case of two-level minimization, where large problems can be solved exactly today. Even though heuristic logic minimization often satisfies the practical needs of most digital designs, it would be useful to be able to compute at least lower bounds on area and/or delay, to which the solutions computed by heuristic methods could be contrasted. Moreover, it would be very interesting to synthesize (even small) circuits with provably optimum delay properties. Indeed, timing-critical portions of processors are still hand designed or tuned, and they often involve a small amount of logic gates.

Multiple-level logic synthesis and optimization have evolved from minimization methods for *sum of products* expressions and are especially well-suited for manipulating control circuits. When considering data-path design, the optimization of circuits involving exclusive-or primitives, such as adders and multipliers, is often difficult to achieve with presently known heuristic algorithms. Thus an important open problem is the optimal synthesis of data paths, including arithmetic units. Research on optimization of AND-EXOR forms, and of networks of modules with such logic representations, is thus very important and promising.

8.9 REFERENCES

Optimization methods for multiple-level circuits were investigated heavily in the last thirty years. Among the classical methods, it is worth remembering the decomposition techniques of Ashenhurst [2] and Curtis [16]. Davidson [20] proposed exact and approximate algorithms for determining minimal gate-count (i.e., area) implementations of multiple-output networks, subject to stage (i.e., delay), fan-in and fanout constraints. The method, based on a branch-and-bound search (that can be relaxed in the approximate case), has over-polynomial-time complexity, and it is not considered practical for usual circuits. Lawler [35] proposed a minimum-literal factorization algorithm for single-output networks.

Heuristic logic optimization flourished in the 1980s. Most of the contributions are scattered in recent journal papers and conference proceedings, which explains the rich bibliography of this chapter. Algorithms for optimization were pioneered by Brayton and McMullen [7], who developed the algebraic model and the kernel theory. These methods were later perfected by Brayton, Rudell, Wang and Sangiovanni, who introduced and solved the rectangle covering problems [8, 37, 43].

The theory of the *don't care* specifications and computation has been the object of long investigation. Coudert *et al.* [15] proposed a method for image computation that was later perfected by Touati *et al.* [42] and applied to the computation of controllability *don't care* sets. Damiani and De Micheli [17] proposed a method for the exact computation of observability *don't care* sets by means of network traversal and showed the link between the equivalence classes of a Boolean relation and the ODC sets. The concept of ODC sets has been recently extended to observability relations [40].

Algebraic and Boolean algorithms have been used in several programs, such as MIS [8] and BOLD [3]. The transduction method, proposed by Muroga *et al.* [36], has shown to yield good results and it was implemented as part of Fujitsu's design system. The global flow method, proposed by Berman and Trevillyan, was incorporated in IBM's LSS system [18, 19].

Bartlett *et al.* presented first in a seminal paper [3] the links between logic optimality and testability. Recently, Hachtel *et al.* [29] described thoroughly the relations between algebraic transformations and testability for both single and multiple stuck-at faults. Keutzer, Malik and Saldanha [31] addressed the relations between delay optimization and testability.

Whereas methods for timing optimization had been explored since the early stages of logic synthesis [8, 21], the false-path problem was brought back to the community's attention only recently. Brand and Iyengar [5] devised conditions for dynamic sensitization. Devadas *et al.* [24] developed the false-path theory based on static co-sensitization. McGeer and Brayton [32] proposed another theory using the concept of viability. Their book is a good source of information on the problem.

Rule-based methods have been applied to logic optimization for many years. IBM's LSS system was the first logic optimization tool widely used in industry and paved the way to the development of this field. SOCRATES [28] uses both the algorithmic and the rule-based approaches, the latter being mainly applied to the binding problem.

1. P. Abouzeid, K. Sakouti, G. Saucier and F. Poirot, "Multilevel Synthesis Minimizing the Routing Factor," *DAC, Proceedings of the Design Automation Conference*, pp. 365–368, 1990.
2. R. Ashenhurst, "The Decomposition of Switching Functions," *Proceedings of the International Symposium on the Theory of Switching*, pp. 74–116, 1957.
3. K. A. Bartlett, R. K. Brayton, G. D. Hachtel, R. M. Jacoby, C. R. Morrison, R. L. Rudell, A. Sangiovanni-Vincentelli and A. R. Wang, "Multilevel Logic Minimization Using Implicit Don't Cares," *IEEE Transactions on CAD/ICAS*, Vol. CAD-7, No. 6, pp. 723–740, June 1988.
4. L. Berman and L. Trevillyan, "Global Flow Optimization in Automatic Logic Design," *IEEE Transactions on CAD/ICAS*, Vol. CAD-10, No. 5, pp. 557–564, May 1991.
5. D. Barnd and V. Iyengar, "Timing Analysis Using a Functional Relationship," *ICCAD, Proceedings of the International Conference on Computer Aided Design*, pp. 126–129, 1986.
6. R. Brayton, G. Hachtel and A. Sangiovanni-Vincentelli, "Multilevel Logic Synthesis," *IEEE Proceedings*, Vol. 78, No. 2, pp. 264–300, February 1990.
7. R. Brayton and C. McMullen, "The Decomposition and Factorization of Boolean Expressions," *ISCAS, Proceedings of the International Symposium on Circuits and Systems*, pp. 49–54, 1982.
8. R. Brayton, R. Rudell, A. Sangiovanni-Vincentelli and A. Wang, "MIS: A Multiple-level Logic Optimization System," *IEEE Transactions on CAD/ICAS*, Vol. CAD-6, No. 6, pp. 1062–1081, November 1987.
9. R. Brayton, R. Rudell, A. Sangiovanni-Vincentelli and A. Wang, "Multi-Level Logic Optimization and the Rectangular Covering Problem," *ICCAD, Proceedings of the International Conference on Computer Aided Design*, pp. 66–69, 1987.
10. F. Brglez, D. Bryan, J. Calhoun, G. Kedem and R. Lisanke, "Automated Synthesis for Testability," *IEEE Transactions on Industrial Electronics*, Vol. IE-36, No. 2, pp. 263–277, May 1989.
11. F. Brown, *Boolean Reasoning*, Kluwer Academic Publishers, Boston, MA, 1990.
12. D. Bryan, F. Brglez and R. Lisanke, "Redundancy Identification and Removal," *International Workshop on Logic Synthesis*, 1991.
13. B. Buchanan and R. Duda, "Principles of Rule-based Expert Systems," in M. Yovits, Editor, *Advances in Computers*, Vol. 22, Academic Press, San Diego, 1983.
14. H. Chen and N. Du, "Path Sensitization in Critical Path Problem," *TAU90, Proceedings of ACM Workshop on Timing Issues in the Specification and Synthesis of Digital Systems*, 1990.
15. O. Coudert, C. Berthet and J. C. Madre, "Verification of Sequential Machines Based on Symbolic Execution," in J. Sifakis, Editor, *Lecture Notes in Computer Science*, Springer-Verlag, Berlin, Germany, 1990.
16. A. Curtis, *New Approach to the Design of Switching Circuits*, Van Nostrand, Princeton, NJ, 1962.
17. M. Damiani and G. De Micheli, "Don't Care Specifications in Combinational and Synchronous Logic Circuits," *IEEE Transactions on CAD/ICAS*, Vol. CAD-12, pp. 365–388, March 1993.
18. J. Darringer, W. Joyner, L. Berman and L. Trevillyan, "LSS: Logic Synthesis through Local Transformations," *IBM Journal of Research and Development*, Vol. 25, No. 4, pp. 272–280, July 1981.
19. J. Darringer, D. Brand, W. Joyner and L. Trevillyan, "LSS: A System for Production Logic Synthesis," *IBM Journal of Research and Development*, Vol. 28, No. 5, pp. 537–545, September 1984.
20. E. Davidson, "An Algorithm for NAND Decomposition under Network Constraints," *IEEE Transactions on Computers*, Vol. C-18, No. 12, pp. 1098–1109, December 1969.

21. G. De Micheli, "Performance Oriented Synthesis of Large Scale Domino CMOS Circuits," *IEEE Transactions on CAD/ICAS*, Vol. CAD-6, No. 5, pp. 751–764, September 1987.

22. G. De Micheli, A. Sangiovanni-Vincentelli and P. Antognetti (Editors), *Design Systems for VLSI Circuits: Logic Synthesis and Silicon Compilation*, M. Njihoff Publishers, Dordrecht, The Netherlands, 1987.

23. S. Devadas and K. Keutzer, "Synthesis of Robust Delay-Fault Testable Circuits: Theory," *IEEE Transactions on CAD/ICAS*, Vol. 11, CAD-No. 1, pp. 87–101, January 1992.

24. S. Devadas, K. Keutzer and S. Malik, "Delay Computation in Combinational Logic Circuits: Theory and Algorithms," *ICCAD, Proceedings of the International Conference on Computer Aided Design*, pp. 176–179, 1991.

25. D. Dietmeyer, *Logic Design of Digital Systems*, Allyn and Bacon, Needham Heights, MA, 1978.

26. D. Dietmeyer and Y. Su, "Logic Design Automation of Fan-in Limited NAND Networks," *IEEE Transactions on Computers*, Vol. C-18, No. 1, pp. 11–22, January 1969.

27. D. Du, S. Yen and S. Ghanta, "On the General False Path Problem in Timing Analysis," *DAC, Proceedings of the Design Automation Conference*, pp. 555–560, 1989.

28. D. Gregory, K. Bartlett, A. de Geus and G. Hachtel, "Socrates: A System for Automatically Synthesizing and Optimizing Combinational Logic," *DAC, Proceedings of the Design Automation Conference*, pp. 79–85, 1986.

29. G. Hachtel, R. Jacoby, K. Keutzer and C. Morrison, "On Properties of Algebraic Transformations and the Synthesis of Multifault-Irredundant Circuits," *IEEE Transactions on CAD/ICAS*, Vol. CAD-11, No. 3, pp. 313–321, March 1992.

30. B. Kernighan and S. Lin, "An Efficient Heuristic Procedure for Partitioning Graphs," *Bell System Technical Journal*, pp. 291–307, February 1970.

31. K. Keutzer, S. Malik and A. Saldanha, "Is Redundancy Necessary to Reduce Delay?" *IEEE Transactions on CAD/ICAS*, Vol. CAD-10, No. 4, pp. 427–435, April 1991.

32. P. McGeer and R. Brayton, *Integrating Functional and Temporal Domains in Logic Design*, Kluwer Academic Publishers, Boston, MA, 1991.

33. P. McGeer, A. Saldanha, R. Brayton and A. Sangiovanni-Vincentelli, "Delay Models and Exact Timing Analysis," in T. Sasao, Editor, *Logic Synthesis and Optimization*, Kluwer Academic Publishers, Boston, MA, 1993.

34. P. McGeer and R. Brayton, "Efficient, Stable Algebraic Operation on Logic Expressions," in C. Sequin, Editor, *VLSI Design of Digital Systems*, North-Holland, Amsterdam, 1988.

35. E. Lawler, "An Approach to Multilevel Boolean Minimization," *Journal of ACM*, Vol. 11, pp. 283–295, 1964.

36. S. Muroga, Y. Kambayashi, H. Lai and J. Culliney, "The Transduction Method—Design of Logic Networks Based on Permissible Functions," *IEEE Transactions on Computers*, Vol. C-38, No. 10, pp. 1404–1424, 1989.

37. R. Rudell, "Logic Synthesis for VLSI Design," Memorandum UCB/ERL M89/49, Ph.D. Dissertation, University of California at Berkeley, April 1989.

38. A. Saldanha, A. Wang, R. Brayton and A. Sangiovanni-Vincentelli, "Multi-Level Logic Simplification using Don't Cares and Filters," *DAC, Proceedings of the Design Automation Conference*, pp. 277–282, 1989.

39. H. Savoy, R. Brayton and H. Touati, "Extracting Local Don't Cares for Network Optimization," *ICCAD, Proceedings of the International Conference on Computer Aided Design*, pp. 514–517, 1991.

40. H. Savoj and R. Brayton, "Observability Relations and Observability Don't Cares," *ICCAD, Proceedings of the International Conference on Computer Aided Design*, pp. 518–521, 1991.

41. K. Singh, A. Wang, R. Brayton and A. Sangiovanni-Vincentelli, "Timing Optimization of Combinational Logic," *ICCAD, Proceedings of the International Conference on Computer Aided Design*, pp. 282–285, 1988.

42. H. Touati, H. Savoy, B. Lin, R. Brayton and A. Sangiovanni-Vincentelli, "Implicit State Enumeration of Finite State Machines using BDDs," *ICCAD, Proceedings of the International Conference on Computer Aided Design*, pp. 130–133, 1990.

43. A. Wang, "Algorithms for Multilevel Logic Optimization," Memorandum UCB/ERL M89/49, Ph.D. Dissertation, University of California at Berkeley, April 1989.

8.10 PROBLEMS

1. Prove that given two algebraic expressions $f_{dividend}$ and $f_{divisor}$, the algebraic quotient of the former by the latter is empty if any of the following conditions apply:

- $f_{divisor}$ contains a variable not in $f_{dividend}$.
- $f_{divisor}$ contains a cube not contained in any cube of $f_{dividend}$.
- $f_{divisor}$ contains more terms than $f_{dividend}$ (i.e., $m > n$).
- The count of any variable in $f_{divisor}$ is larger than in $f_{dividend}$.

2. Design an algorithm that finds all 0-level kernels of a function.

3. Design an algorithm that finds just one 0-level kernel of a function.

4. Consider the logic network defined by the following expressions:

$$x = ad' + a'b' + a'd' + bc + bd' + ac$$

$$y = a + b$$

$$z = a'c' + a'd' + b'c' + b'd' + e$$

$$u = a'c + a'd + b'd + e'$$

Draw the logic network graph. Outputs are $\{x, y, z, u\}$. Perform the algebraic division f_x/f_y and show all steps. Substitute y into f_x and redraw the network graph. Compute all kernels and co-kernels of z and u. Extract a multiple-cube subexpression common to f_z and f_u. Show all steps. Redraw the network graph.

5. Consider the logic network defined by the following expressions:

$$x = abcf + efc + de$$

$$y = acdef + def$$

$$z = bcd + acf$$

Determine the cube-variable matrix and all prime rectangles. Identify all feasible cube intersections. Determine the minimum-literal network that can be derived by cube extraction.

6. Consider the logic network defined by the following expressions:

$$d = b'$$

$$f = (a + d)'$$

$$e = (ca)'$$

$$x = fe$$

$$y = d \oplus e$$

Inputs are $\{a, b, c\}$ and outputs are $\{x, y\}$. Assume $CDC_{in} = abc'$. Compute CDC_{out}.

7. Consider the logic network of Problem 6. Compute the ODC sets for all internal and input vertices, assuming that the outputs are fully observable.

8. Design an algorithm that computes the exact ODC set by backward traversal and does not use the results of Theorem 8.4.1 but exploits formulae 8.9 and 8.10.

9. Prove formally the following theorem justifying an exact filter. Consider a vertex v_x. Let $DC_x = D \cup E$ such that $|(sup(F) \cup sup(D)) \cap sup(E)| \leq 1$. Then, the cubes of E are useless for optimizing f_x.

10. Give an example of the usefulness of an approximate filter based on the following. Discard those cubes of the *don't care* set whose support is disjoint from that of the function to be minimized. Show also that it is not an exact filter.

11. Prove the following theorem. Let the input/output behavior of the network under consideration be represented by $\mathbf{f}$ and the corresponding external *don't care* set by $\mathbf{DC}_{ext}$. Let us apply simplification to vertex v_x and let g_x be a feasible implementation of the corresponding local function. Then:

$$(\mathbf{f}_{min} \cap \mathbf{f}'|_{x=0}) \cup (\mathbf{f}'_{max} \cap \mathbf{f}|_{x=0}) \subseteq g_x \subseteq (\mathbf{f}_{max} \cup \mathbf{f}'|_{x=1}) \cap (\mathbf{f}'_{min} \cup \mathbf{f}|_{x=1})$$

where $\mathbf{f}_{min} = (\mathbf{f} \cap \mathbf{DC}'_{ext})$ and $\mathbf{f}_{max} = (\mathbf{f} \cup \mathbf{DC}_{ext})$.

12. Derive a recursive formula to compute the controlling set $C_{11}(x)$. Justify the formula.

13. Design an exact algorithm for inverter minimization in a tree-structured network. Determine its time complexity.

14. Formulate the inverter minimization problem as a ZOLP. Consider networks with general topology.

15. Consider the logic network defined by the following expressions:

$$k = a'$$
$$e = k + b$$
$$g = (b + c)'$$
$$f = ag$$
$$h = ab$$
$$i = f \overline{\oplus} d$$
$$j = d + s$$
$$x = e \oplus i$$
$$y = j + h$$

Inputs are $\{a, b, c, d, s\}$ and outputs are $\{x, y\}$. Draw the logic network graph. Assume that the delay of each inverter, AND and OR gates is 1 and that the delay of each NOR, EXOR and EXNOR gate is 2. Compute the data ready and slacks for all vertices in the network. Assume that the input data-ready times are zero except for $t_s = 4$ and that the required data-ready time at the output is 7. Determine the topological critical path.

16. Consider the circuit of Figure 8.31. Is it fully testable for single stuck-at faults? Can you draw the conclusion that a critical true path is not statically sensitizable if the circuit is not fully testable? Why?

CHAPTER 9

SEQUENTIAL
LOGIC
OPTIMIZATION

Non c' è un unico tempo: ci sono molti nastri
che paralleli slittano ...
There isn't one unique time: many
tapes run parallel ...

E. Montale. Satura.

9.1 INTRODUCTION

We consider in this chapter the optimization of sequential circuits modeled by finite-state machines at the logic level. We restrict our attention to *synchronous* models that operate on a single-phase clock. From a circuit standpoint, they consist of inter-connections of combinational logic gates and registers. Thus, synchronous sequential circuits are often modeled by a combinational circuit component and registers, as shown in Figure 9.1. We assume here, for the sake of simplicity, that registers are *edge triggered* and store 1 bit of information. Most of the techniques described in this chapter can be extended, *mutatis mutandis*, to the case of *level-sensitive* latches [32] and that of *multiple-phase* clocks [3].

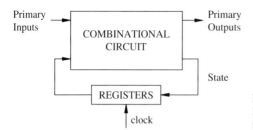

FIGURE 9.1
Block diagram of a finite-state machine implementation.

Sequential circuit optimization has been the subject of intensive investigation for several decades. Some textbooks [20, 21, 23, 29] present the underlying theory and the fundamental algorithms in detail. It is the purpose of this chapter to distill those techniques based on classical methods that have been shown practical for CAD synthesis systems as well as to report on the recent developments in the field.

Sequential circuits can be specified in terms of HDL models or synthesized by means of the techniques described in Chapter 4. In both cases, the sequential circuit model may have either a behavioral or a structural flavor or a combination of both. Whereas the behavior of combinational logic circuits can be expressed by logic functions, the behavior of sequential circuits can be captured by *traces*, i.e., by input and output sequences. A convenient way to express the circuit behavior is by means of finite-state machine models, e.g., state transition diagrams (Section 3.3.3), as shown in Figure 9.2 (a). State transition diagrams encapsulate the traces that the corresponding circuit can accept and produce. Thus, state-based representations have a behavioral flavor.

Classical methods for sequential optimization use state transition diagrams (or equivalent representations, e.g., tables). While many optimization techniques have been proposed, this finite-state machine representation lacks a direct relation between state manipulation (e.g., state minimization) and the corresponding area and delay variations. In general, optimization techniques target a reduction in complexity of the model that correlates well with area reduction but not necessarily with performance

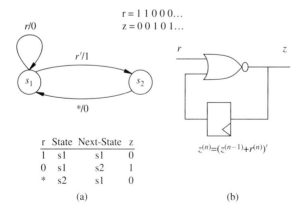

r	State	Next-State	z
1	s1	s1	0
0	s1	s2	1
*	s2	s1	0

(a)

$$z^{(n)} = (z^{(n-1)} + r^{(n)})'$$

(b)

FIGURE 9.2
Sequential circuit models: (a) state transition diagram; (b) synchronous logic network.

improvement. Therefore these optimization methods parallel the behavioral-level transformations for architectural models, described in Section 3.4.2.

An alternative representation of the circuit behavior is by means of logic expressions in terms of time-labeled variables. As in the case of combinational circuits, it is often convenient to express the input/output behavior by means of a set of local expressions with mutual dependencies. This leads to circuit representations in terms of *synchronous logic networks* that express the interconnection of combinational modules and registers, as shown in Figure 9.2 (b). Synchronous logic networks have a structural flavor when the combinational modules correspond to logic gates. They are hybrid structural/behavioral views of the circuit when the modules are associated with logic functions. Some recent optimization algorithms for sequential circuits use the network representation, such as *retiming* [24]. In this case there is a direct relation between circuit transformations and area and/or performance improvement.

State transition diagrams can be transformed into synchronous logic networks by *state encoding* and can be recovered from synchronous logic networks by *state extraction*. State encoding defines the state representation in terms of state variables, thus allowing the description in terms of networks. Unused state codes are *don't care* conditions that can be used for the network optimization. The major task in state extraction is to determine the valid states (e.g., those reachable from the reset state) among those identified by all possible polarity assignments to the state variables. Design systems for sequential optimization leverage optimization methods in both representation domains.

In this chapter we shall first describe sequential circuit optimization using state-based models in Section 9.2. Then we shall consider those methods relying on network models in Section 9.3. Next we shall consider the relations between the network models and state transition diagrams, in particular implicit methods for finite-state machine traversal, in Section 9.4. We shall conclude with some comments on testability of sequential circuits.

9.2 SEQUENTIAL CIRCUIT OPTIMIZATION
USING STATE-BASED MODELS

We consider in this section algorithms for sequential optimization using state-based models as well as transformations into and from structural models. We consider Mealy-type finite-state machines, defined by the quintuple $(X, Y, S, \delta, \lambda)$, as introduced in Section 3.3.3. We denote $|S|$ by n_s in the sequel.

Finite-state machine optimization has been covered by several textbooks [20, 21, 23, 29]. For this reason, we summarize the state of the art in the field, with particular reference to the relation to other synthesis problems and to recent results. In particular, we concentrate on the *state minimization* and *state encoding* problems in Sections 9.2.1 and 9.2.2, respectively.

We shall survey briefly other relevant optimization techniques in Section 9.2.3, including *decomposition*. We comment on the relations among different finite-state machine optimization problems and on issues related to synthesis and optimization of interconnected finite-state machines.

9.2.1 State Minimization

The state minimization problem aims at reducing the number of machine states. This leads to a reduction in size of the state transition graph. State reduction may correlate to a reduction of the number of storage elements. (When states are encoded with a minimum number of bits, the number of registers is the ceiling of the logarithm of the number of states.) The reduction in states correlates to a reduction in transitions, and hence to a reduction of logic gates.

State minimization can be defined informally as deriving a finite-state machine with similar behavior and minimum number of states. A more precise definition relies on choosing to consider completely (or incompletely) specified finite-state machines. This decision affects the formalism, the problem complexity and the algorithms. Hence state minimization is described separately for both cases in the following sections. In addition, a recent technique for state minimization of completely specified finite-state machines using an implicit representation (instead of an explicit state-based model) is described in Section 9.4.2.

STATE MINIMIZATION FOR COMPLETELY SPECIFIED FINITE-STATE MACHINES. When considering completely specified finite-state machines, the transition function δ and the output function λ are specified for each pair (input, state) in $X \times S$. Two states are *equivalent* if the output sequences of the finite-state machine initialized in the two states coincide for any input sequence. Equivalency is checked by using the result of the following theorem [21].

> **Theorem 9.2.1.** Two states of a finite-state machine are equivalent if and only if, for any input, they have identical outputs and the corresponding next states are equivalent.

Since equivalency is symmetric, reflexive and transitive, states can be partitioned into equivalence classes. Such a partition is unique. A minimum-state implementation of a finite-state machine is one where each state represents only one class, or equivalently where only one state per class is retained.

Minimizing the number of states of a completely specified finite-state machine then entails computing the equivalence classes. These can be derived by an iterative refinement of a partition of the state set. Let $\Pi_i, i = 1, 2, \ldots, n_s$, denote the partitions. At first, blocks of partition Π_1 contain states whose outputs match for any input, hence satisfying a necessary condition of equivalence. Then, partition blocks are iteratively refined by further splittings by requiring that all states in any block of Π_{i+1} have next states in the same block of Π_i for any possible input. When the iteration converges, i.e., when $\Pi_{i+1} = \Pi_i$ for some value of i, the corresponding blocks are equivalence classes, by Theorem 9.2.1. Note that convergence is always achieved in at most n_s iterations. In the limiting case that n_s iterations are required, we obtain a *0-partition* where all blocks have one state each, i.e., no pair of states are equivalent. The complexity of this algorithm is $O(n_s^2)$.

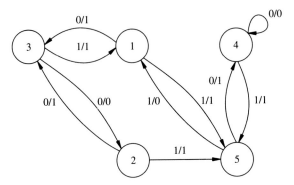

FIGURE 9.3
State diagram.

Example 9.2.1. Consider the state diagram shown in Figure 9.3, whose state table is reported next:

Input	State	Next State	Output
0	s_1	s_3	1
1	s_1	s_5	1
0	s_2	s_3	1
1	s_2	s_5	1
0	s_3	s_2	0
1	s_3	s_1	1
0	s_4	s_4	0
1	s_4	s_5	1
0	s_5	s_4	1
1	s_5	s_1	0

The state set can be partitioned first according to the outputs, i.e.,

$$\Pi_1 = \{\{s_1, s_2\}, \{s_3, s_4\}, \{s_5\}\}$$

Then we check each block of Π_1 to see if the corresponding next states are in a single block of Π_1 for any input. The next states of s_1 and s_2 match. The next states of s_3 and s_4 are in different blocks. Hence the block $\{s_3, s_4\}$ must be split, yielding:

$$\Pi_2 = \{\{s_1, s_2\}, \{s_3\}, \{s_4\}, \{s_5\}\}$$

When checking the blocks again, we find that no further refinement is possible, because the next states of s_1 and s_2 match. Hence there are four classes of compatible states. We denote $\{s_1, s_2\}$ as s_{12} in the following minimal table:

Input	State	Next State	Output
0	s_{12}	s_3	1
1	s_{12}	s_5	1
0	s_3	s_{12}	0
1	s_3	s_{12}	1
0	s_4	s_4	0
1	s_4	s_5	1
0	s_5	s_4	1
1	s_5	s_{12}	0

The corresponding diagram is shown in Figure 9.4.

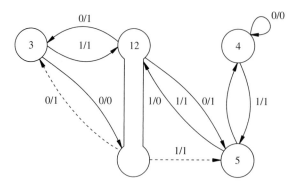

FIGURE 9.4
Minimum-state diagram. (Dotted edges are superfluous.)

In the algorithm described above, the refinement of the partitions is done by looking at the transitions *from* the states in the block under consideration to other states. Hopcroft [19] suggested a partition refinement method where the transitions *into* the states of the block under consideration are considered. Consider a partition block $A \in \Pi_i$ and, for each input, the subset of the states whose next states are in A, which we call P. Then any other block $B \in \Pi_i$ is consistent with A if either $B \subseteq P$ or $B \cap P = \emptyset$. If neither of these conditions are satisfied, block B is split into $B^1 = B \cap P$ and $B^2 = B - B \cap P$, and the two splinters become part of Π_{i+1}. If no block requires a split, then the partition defines a class of equivalent states.

> **Example 9.2.2.** Consider the table of Example 9.2.1. The state set can be partitioned first according to the outputs, as in the previous case:
>
> $$\Pi_1 = \{\{s_1, s_2\}, \{s_3, s_4\}, \{s_5\}\}$$
>
> Then we check each block of Π_1. Let $A = \{s_5\}$. Let us consider input 1. The states whose next state is s_5 are set $P = \{s_1, s_2, s_4\}$. Block $\{s_1, s_2\}$ is a subset of P and requires no further split. Block $B = \{s_3, s_4\}$ is not a subset of P and $B \cap P = \{s_4\}$. Hence the block is split as $\{\{s_3\}, \{s_4\}\}$, yielding:
>
> $$\Pi_2 = \{\{s_1, s_2\}, \{s_3\}, \{s_4\}, \{s_5\}\}$$
>
> No further splits are possible and Π_2 defines four classes of equivalent states.

Hopcroft's method is important because it has a better asymptotic behavior. Note that when a block is considered to split the others, it does not need to be reconsidered unless it is split. In this case, both splinters would yield the same results when considered in turn to split other blocks. Hence we can consider the smaller one in the future iterations. This fact is key in proving that the algorithm can be executed in $O(n_s \log n_s)$ steps. The proof is laborious and reported in references [1] and [19].

STATE MINIMIZATION FOR INCOMPLETELY SPECIFIED FINITE-STATE MACHINES. In the case of incompletely specified finite-state machines, the transition function δ and the output function λ are not specified for some (input, state) pairs. Equivalently, *don't care* conditions denote the unspecified transitions and outputs.

They model the knowledge that some input patterns cannot occur in some states or that some outputs are not observed in some states under some input conditions.

An input sequence is said to be *applicable* if it does not lead to any unspecified transition. Two states are *compatible* if the output sequences of the finite-state machine initialized in the two states coincide whenever both outputs are specified and for any applicable input sequence. The following theorem [21] applies to incompletely specified finite-state machines.

> **Theorem 9.2.2.** Two states of a finite-state machine are compatible if and only if, for any input, the corresponding output functions match when both are specified and the corresponding next states are compatible when both are specified.

This theorem forms the basis of iterative procedures for determining classes of compatible states that are the counterpart of those used for computing equivalent states. Nevertheless two major differences exist with respect to the case of completely specified finite-state machines.

First, compatibility is not an equivalence relation, because compatibility is a symmetric and reflexive relation but not a transitive one. A class of compatible states is then defined to be such that all its states are pairwise compatible. Maximal classes of compatible states do not form a partition of the state set. Since classes may overlap, multiple solutions to the problem may exist. The intractability of the problem stems from this fact.

Second, the selection of an adequate number of compatibility classes to cover the state set is complicated by *implications* among the classes themselves, because the compatibility of two or more states may require the compatibility of others. Hence the selection of a compatibility class to be represented by a single state may imply that some other class also has to be chosen. A set of compatibility classes has the *closure* property when all implied compatibility classes are in the set or are contained by classes in the set.

> **Example 9.2.3.** Consider the finite-state machine of Figure 9.5 (a) and described by the following table, where only the output function λ is incompletely specified for the sake of simplicity:
>
> | Input | State | Next State | Output |
> |:-----:|:-----:|:----------:|:------:|
> | 0 | s_1 | s_3 | 1 |
> | 1 | s_1 | s_5 | * |
> | 0 | s_2 | s_3 | * |
> | 1 | s_2 | s_5 | 1 |
> | 0 | s_3 | s_2 | 0 |
> | 1 | s_3 | s_1 | 1 |
> | 0 | s_4 | s_4 | 0 |
> | 1 | s_4 | s_5 | 1 |
> | 0 | s_5 | s_4 | 1 |
> | 1 | s_5 | s_1 | 0 |

Note first that replacing the *don't care* entries by 1s would lead to the table of Example 9.2.1, which can be minimized to four states. Other choices of the *don't care*

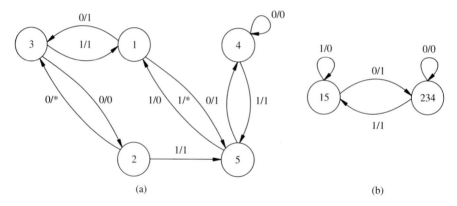

FIGURE 9.5
(a) State diagram. (b) Minimum state diagram.

entries would lead to other completely specified finite-state machines. Unfortunately, there is an exponential number of completely specified finite-state machines in correspondence to the choice of the *don't care* values.

Let us consider pairwise compatibility. The pair $\{s_1, s_2\}$ is compatible. The pair $\{s_2, s_3\}$ is compatible subject to compatibility of $\{s_1, s_5\}$. The pair $\{s_1, s_3\}$ is not compatible. This shows the lack of transitivity of the compatibility relation.

The following table lists the compatible and incompatible pairs:

	Pairs	**Implied Pairs**
Compatible	$\{s_1, s_2\}$	
Compatible	$\{s_1, s_5\}$	$\{s_3, s_4\}$
Compatible	$\{s_2, s_4\}$	$\{s_3, s_4\}$
Compatible	$\{s_2, s_3\}$	$\{s_1, s_5\}$
Compatible	$\{s_3, s_4\}$	$\{s_2, s_4\}, \{s_1, s_5\}$
Incompatible	$\{s_1, s_3\}$	
Incompatible	$\{s_1, s_4\}$	
Incompatible	$\{s_2, s_5\}$	
Incompatible	$\{s_3, s_5\}$	
Incompatible	$\{s_4, s_5\}$	

Maximal compatibility classes are the following:

Classes	**Implied Classes**
$\{s_1, s_2\}$	
$\{s_1, s_5\}$	$\{s_3, s_4\}$
$\{s_2, s_3, s_4\}$	$\{s_1, s_5\}$

Minimizing the number of states of an incompletely specified finite-state machine consists of selecting enough compatibility classes satisfying the closure property so that the states are covered. Hence the state minimization problem can be formulated as a *binate covering* problem and solved exactly or heuristically by the corresponding algorithms. The states of a minimum implementation would correspond then to the selected classes and their number to the cardinality of a minimum cover.

It is worth noticing that the set of maximal compatible classes satisfies always the closure property. Hence, their computation gives always a feasible solution and no check for implications is needed. Unfortunately, its cardinality may be larger than that of a minimum cover and even larger than the original state set cardinality. Minimum covers may involve compatibility classes that are not necessarily maximal. This can be explained informally by noticing that smaller classes may have fewer implication requirements.

> **Example 9.2.4.** Consider the table of Example 9.2.3. There are three maximal classes of compatible states. However, a minimum closed cover involves only $\{s_1, s_5\}$ and $\{s_2, s_3, s_4\}$, therefore with cardinality 2. In this case, the selected classes are maximal, but they are not required to be so in general. The minimum state diagram is shown in Figure 9.5 (b).

Exact methods for state minimization of incompletely specified finite-state machines can take advantage of considering a smaller set of compatibility classes. For example, Grasselli and Luccio [18] suggested restricting the attention to *prime classes*, which are those not included in other classes implying the same set or a subset of classes. The use of prime classes allows us to prune the solution space.

Heuristic methods for state minimization have different flavors. Upper bounds on the optimum solution size are given by the cardinality of the set of maximal compatible classes and by the original state set cardinality. A lower bound is given by a unate cover solution that disregards the implications. It has been noticed [29] that often a unate cover in terms of maximal compatible classes has the closure property and hence it is a valid solution. Otherwise, an approximation to the solution of the binate covering problem can be computed by heuristic covering algorithms.

9.2.2 State Encoding

The *state encoding* (or *assignment*) problem consists of determining the binary representation of the states of a finite-state machine. In the most general case, the state encoding problem is complicated by the choice of register type used for storage (e.g., D, T, JK) [29]. We consider here only D-type registers, because they are the most commonly used.

Encoding affects the circuit area and performance. Most known techniques for state encoding target the reduction of circuit complexity measures that correlate well with circuit area but only weakly with circuit performance. The circuit complexity is related to the number of storage bits n_b used for the state representation (i.e., encoding length) and to the size of the combinational component. A measure of the latter is much different when considering two-level or multiple-level circuit implementations.

For this reason, state encoding techniques for two-level and multiple-level logic have been developed independently. We shall survey methods for both cases next.

STATE ENCODING FOR TWO-LEVEL CIRCUITS. Two-level circuits have been the object of investigation for several decades. The circuit complexity of a *sum of products* representation is related to the number of inputs, outputs and product terms. For PLA-based implementations, such numbers can be used to compute readily the circuit area and the physical length of the longest path, which correlates to the critical path delay.[1] The number of inputs and outputs of the combinational component is the sum of twice the state encoding length plus the number of primary inputs and outputs. The number of product terms to be considered is the size of a minimum (or minimal) *sum of products* representation.

The choice of an encoding affects both the encoding length and the size of a two-level cover. There are $2^{n_b}!/(2^{n_b} - n_s)!$ possible encodings, and the choice of the best one is a formidable task. Note that the size of *sum of products* representations is invariant under permutation and complementation of the encoding bits. Hence the number of relevant codes can be refined to $(2^{n_b} - 1)!/(2^{n_b} - n_s)!n_b!$.

The simplest encoding is *1-hot* state encoding, where each state is encoded by a corresponding code bit set to 1, all others being 0. Thus $n_b = n_s$. The 1-hot encoding requires an excessive number of input/outputs, and it was shown [8, 9] not to minimize the size of the *sum of products* representation of the corresponding combinational component.

Early work on state encoding focused on the use of *minimum-length* codes, i.e., using $n_b = \lceil \log_2 n_s \rceil$ bits to represent the set of states S. Most classical heuristic methods for state encoding are based on a *reduced dependency criterion* [20, 31]. The rationale is to encode the states so that the state variables have the least dependencies on those representing the previous states. Reduced dependencies correlate weakly with the minimality of a *sum of products* representation.

In the early 1980s, symbolic minimization (see Section 7.5) was introduced [8, 9] with the goal of solving the state encoding problem. Minimizing a symbolic representation is equivalent to minimizing the size of the *sum of products* form related to all codes that satisfy the corresponding constraints. Since exact and heuristic symbolic minimizers are available, as well as encoding programs, they can be readily applied to the state encoding problem.

Note that the finite-state machine model requires the solution to both input and output encoding problems. The feedback nature of the finite-state machine imposes a consistency requirement. Namely, symbol pairs (in the input and output fields) corresponding to the same states must share the same code. In other words, the state set must be encoded while satisfying simultaneously both input and output constraints. Exact and heuristic encoding algorithms have been previously reported in Section 7.5.2.

---

[1]The area of a PLA is proportional to the product of the number of I/Os (columns) and the number of product terms (rows). For simplicity we consider as one column any input column pair corresponding to a signal and its complement. The longest physical path is proportional to the sum of twice the number of rows (2 column lengths) plus the number of I/Os (1 row length).

Example 9.2.5. Consider the finite-state machine described by the following state table:

Input	State	Next State	Output
0	s_1	s_3	0
1	s_1	s_3	0
0	s_2	s_3	0
1	s_2	s_1	1
0	s_3	s_5	0
1	s_3	s_4	1
0	s_4	s_2	1
1	s_4	s_3	0
0	s_5	s_2	1
1	s_5	s_5	0

A minimal symbolic cover is the following:

*	$s_1 s_2 s_4$	s_3	0
1	s_2	s_1	1
0	$s_4 s_5$	s_2	1
1	s_3	s_4	1

with the covering constraints:

- s_1 and s_2 cover s_3;
- s_5 is covered by all other states.

The corresponding encoding constraint matrices are then:

$$A = \begin{bmatrix} 1 & 1 & 0 & 1 & 0 \\ 0 & 0 & 0 & 1 & 1 \end{bmatrix} \qquad B = \begin{bmatrix} 0 & 0 & 1 & 0 & 1 \\ 0 & 0 & 1 & 0 & 1 \\ 0 & 0 & 0 & 0 & 1 \\ 0 & 0 & 0 & 0 & 1 \\ 0 & 0 & 0 & 0 & 0 \end{bmatrix}$$

An encoding that satisfies simultaneously both sets of constraints is then:

$$E = \begin{bmatrix} 1 & 1 & 1 \\ 1 & 0 & 1 \\ 0 & 0 & 1 \\ 1 & 0 & 0 \\ 0 & 0 & 0 \end{bmatrix}$$

where each row relates to a state.

An encoded cover of the finite-state machine combinational component is then:

*	1**	001	0
1	101	111	1
0	*00	101	1
1	001	100	1

Whereas symbolic minimization and constrained encoding provide a framework for the solution of the state assignment problem for two-level circuits, we would like to stress a limitation of this approach. The minimum-length solution compatible with a set of constraints may require a number of bits larger than $\lceil \log_2 n_s \rceil$. In practice, experimental results have shown that just a few extra bits may be needed to satisfy all constraints. In addition, constraints can be relaxed to satisfy bounds on the encoding length [2, 9, 35]. Thus, the desired circuit implementation can be searched for by trading off the number of I/Os with the number of product terms.

Example 9.2.6. Consider a finite-state machine whose minimal symbolic cover is:

$$
\begin{array}{llll}
00 & s_1 s_2 & s_3 & 100 \\
01 & s_2 s_3 & s_1 & 010 \\
10 & s_1 s_3 & s_2 & 001
\end{array}
$$

There are no output encoding constraints. Nevertheless, the satisfaction of the input encoding constraints requires at least 3 bits. A feasible encoding is $s_1 = 100$; $s_2 = 010$; $s_3 = 001$. Hence a cover with cardinality 3 and $n_b = 3$ can be constructed. The corresponding PLA would have 3 rows and 11 columns.

Alternatively we may choose not to satisfy one constraint, to achieve a 2-bit encoding. Assume we split the top symbolic implicants into two, namely 00 s_1 s_3 100 and 00 s_2 s_3 100. Then the following 2-bit encoding is feasible: $s_1 = 00$; $s_2 = 11$; $s_3 = 01$. Now we would have a cover with cardinality 4 and $n_b = 2$. The corresponding PLA would have 4 rows and 9 columns.

STATE ENCODING FOR MULTIPLE-LEVEL CIRCUITS. State encoding techniques for multiple-level circuits use the logic network model described in Chapter 8 for the combinational component of the finite-state machine. The overall area measure is related to the number of encoding bits (i.e., registers) and to the number of literals in the logic network. The delay corresponds to the critical path length in the network. To date, only heuristic methods have been developed for computing state encodings that optimize the area estimate.

The simplest approach is first to compute an optimal state assignment for a two-level logic model and then to restructure the circuit with combinational optimization algorithms. Despite the fact that the choice of state codes is done while considering a different model, experimental results have shown surprisingly good results. A possible explanation is that many finite-state machines have shallow transition and output functions that can be best implemented in a few stages.

The difficulty of state encoding for multiple-level logic models stems from the wide variety of transformations available to optimize a network and from the literal estimation problem. Therefore researchers have considered encoding techniques in connection with one particular logic transformation. Devadas et al. [2, 11] proposed a heuristic encoding method that privileges the extraction of common cubes. The method has been improved upon by others in a later independent development [16]. Malik et al. [27] addressed the problem of optimal subexpression extraction and its relation to encoding.

We summarize here a method that targets common cube extraction, and we refer the reader to reference [2] for details. The rationale is to determine desired code proximity criteria. For example, when two (or more) states have a transition to the same next state, it is convenient to keep the distance between the corresponding codes short, because that correlates with the size of a common cube that can be extracted.

> **Example 9.2.7.** Given a finite-state machine with state set $S = \{s_1, s_2, s_3, s_4, s_5\}$, consider two states s_1 and s_2 with a transition into the same state s_3 under inputs i' and i, respectively. Assume that a 3-bit encoding is used. Let us encode both states with adjacent (i.e., distance 1) codes, namely 000 and 001. These correspond to the cubes $a'b'c'$ and $a'b'c$, respectively, where $\{a, b, c\}$ are the state variables. Then the transition can be written as $i'a'b'c' + ia'b'c$, or equivalently $a'b'(i'c' + ic)$. Note that no cube could have been extracted if s_2 were encoded as 111, and a smaller cube could be extracted by choosing 011.

The encoding problem is modeled by a complete edge-weighted graph, where the vertices are in one-to-one correspondence with the states. The weights denote the desired code proximity of state pairs and are determined by scanning systematically all state pairs. State encoding is determined by *embedding* this graph in a Boolean space of appropriate dimensions. Since graph embedding is an intractable problem, heuristic algorithms are used to determine an encoding where pairwise code distance correlates reasonably well with the weight (the higher the weight, the lower the distance).

There are still two major complications. First, when considering the codes of two states with transitions into the same next state, the size of the common cube can be determined by the code distance, but the number of possible cube extractions depends on the encoding of the next states. Therefore, the area gain cannot be related directly to the transitions, and the weights are only an imprecise indication of the possible overall gain by cube extraction. Second, the extraction of common cubes interacts with each other.

To cope with these difficulties, two heuristic algorithms were proposed by Devadas *et al.* [2, 11]. They both use a complete edge-weighted graph, the weights being determined differently. In the first algorithm, called *fanout oriented*, state pairs that have transitions into the same next state are given high edge weights (to achieve close codes). The weights are computed by a complex formula which takes the output patterns into account. This approach attempts to maximize the *size* of common cubes in the encoded next-state function. In the second algorithm, called *fan-in oriented*, state pairs with incoming transitions from the same states are given high weights. Again a complex rule determines the weight while taking the input patterns into account. This strategy tries to maximize the *number* of common cubes in the encoded next-state function.

> **Example 9.2.8.** Consider the table of Example 9.2.5 and the fanout-oriented algorithm. First a complete graph of five vertices is constructed and then the edge weights are determined. For the sake of simplicity, we consider binary-valued weights. Hence a high weight is 1. Take, for example, the state pair $\{s_1, s_2\}$, where either state has a transition to s_3. Then weight 1 is assigned to edge $\{v_1, v_2\}$.

By scanning all state pairs, the edges $\{\{v_1, v_2\}, \{v_2, v_4\}, \{v_1, v_4\}, \{v_4, v_5\}, \{v_3, v_5\}\}$ are assigned unit weight, while the remaining edges are given weight 0.

The encoding represented by the following matrix reflects the proximity requirements specified by the weights:

$$\mathbf{E} = \begin{bmatrix} 1 & 1 & 1 \\ 1 & 1 & 0 \\ 0 & 0 & 0 \\ 0 & 1 & 1 \\ 0 & 0 & 1 \end{bmatrix}$$

By replacing the codes in the table and by deleting those rows that do not contribute to the encoded transition and output function, we obtain:

1	110	111	1
0	000	001	0
1	000	011	1
0	011	110	1
0	001	110	1
1	001	001	0

Now let the input be variable i, the state variables be a, b, c and the encoded transition functions be f_a, f_b, f_c.

Consider expression $f_a = iabc' + i'a'bc + i'a'b'c$, which can be rewritten as $f_a = iabc' + i'a'c(b + b') = iabc' + i'a'c$. The common cube $i'a'c$ is related to the codes of $\{s_4, s_5\}$, which are adjacent. Note that cube $i'a'c$ is not in conjunction with an expression, because in this particular case the primary inputs match in correspondence with the transitions from $\{s_4, s_5\}$ under consideration (rows 4 and 5 of the cover).

For the sake of the example, assume now that s_5 is given code 101, or equivalently $ab'c$. Then $f_a = iabc' + i'a'bc + i'ab'c$, which can be rewritten as $f_a = iabc' + i'c(a'b + ab')$. The expression for f_a is larger in size (i.e., literal count) than the previous one. This is due to the increased distance of the codes of $\{s_4, s_5\}$.

Other rules for detecting potential common cubes have been proposed, as well as formulae for more precise evaluation of literal savings [16]. They all correspond to examining next-state transition pairs that can be represented symbolically as $i_1 s_1 + i_2 s_2$, where i_1, i_2 are input patterns (cubes) and $\{s_1, s_2\} \in S$ is any state pair. Whereas the general case was described before and demonstrated in Example 9.2.7, specific cases apply when $i_1 = i_2$ or $i_1 \subseteq i_2$. Then it is particularly convenient to achieve distance 1 in the encoding of s_1 and s_2, because a larger literal saving can be achieved.

Example 9.2.9. Consider again the first expression of f_a in Example 9.2.8. Since $i_1 = i_2 = i'$ and the distance of the encoding of $\{s_4, s_5\}$ is 1, the common cube $i'a'c(b + b') = i'a'c$ is not in conjunction with any expression.

Let us consider now transitions from states $\{s_1, s_2\}$ to state s_3 under inputs i_1 and i_2, respectively, where $i_1 \subseteq i_2$, e.g., $ij = i_1$ and $i = i_2$. Assume adjacent codes are assigned to s_1 and s_2, e.g., $a'b'c'$ (000) and $a'b'c$ (001), respectively. Then, the transition can be expressed as $ija'b'c' + ia'b'c$ and simplified to $ia'b'(jc' + c) = ia'b'(j + c)$.

It is important to remember that, while cube extraction reduces the number of literals, many other (interacting) transformations can achieve the same goal. Hence, this encoding technique exploits only one method for reducing the complexity of a multiple-level network. By the same token, the estimation of the size of an optimal multiple-level network by considering cube extraction only may be poor. Experimental results have shown this to be adequate, mainly due to the fact that transition functions are shallow ones and do not require many levels.

9.2.3 Other Optimization Methods and Recent Developments*

We review briefly some other optimization techniques for finite-state machines and their interrelations. We comment on the recent developments in the field. This section is intended to give the reader a flavor for the problems involved (for details the reader is referred to the literature [2, 23]).

FINITE-STATE MACHINE DECOMPOSITION. Whereas *collapsing* two (or more) finite-state machines into a single one is straightforward, by merging the transition diagrams, the reverse process is more difficult. Finite-state machine *decomposition* has been a classic problem for many years, and it has been reported in some textbooks [20, 23]. Most techniques have theoretical importance only, because their computational complexity limits their use to small-sized circuits only.

The goal of finite-state machine decomposition is to find an interconnection of smaller finite-state machines with equivalent functionality to the original one. The rationale is that the decomposition may lead to a reduction in area and increase in performance. Decomposition entails a partition, or a cover, of the state set by subsets, each one defining a component of the decomposed finite-state machine. Different decomposition types can be achieved, as shown in Figure 9.6.

As for state minimization and encoding, it is often difficult to relate a decomposition to the final figures of merit of the circuit, because of the abstractness of the state transition graph model. Nevertheless, in the particular case of two-way decomposition with two-level logic models, symbolic minimization and constrained encoding can be used to yield partitions whose figures of merit closely represent those of the resulting circuits. We refer the reader to reference [2] for details.

Recently, decomposition algorithms based on *factorization* have been reported [2]. Factorization consists of finding two (or more) similar disjoint subgraphs in a state transition diagram. An exact factorization corresponds to determining two (or more) subgraphs whose vertices, edges and transitions match. Hence it is possible to achieve a two-way decomposition of the original finite-state machine by implementing the common subgraphs as a single slave finite-state machine that acts as a "subroutine" invoked by a master finite-state machine. This leads to a sharing of states and transitions and therefore to an overall complexity reduction. Exact factors are hard to find in usual designs, unless modeled by state diagrams obtained by flattening hierarchical diagrams. Approximate factors, i.e., similar disjoint subgraphs in a state transition diagram with possibly some mismatches, are more common. Even though

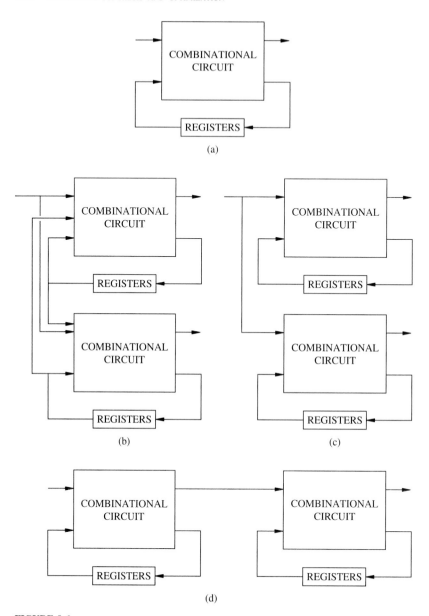

FIGURE 9.6
Different types of finite-state machine decompositions: (a) original machine; (b) general two-way decomposition; (c) parallel decomposition; (d) cascade decomposition.

approximate factorization is more complex and gains are more limited, because mismatches in the common subgraphs must be compensated for, it can still be used as a basis for two-way decomposition and leads to an overall reduction in complexity.

AN OVERALL LOOK AT FINITE-STATE MACHINE STATE-BASED OPTIMIZA-TION. There are many degrees of freedom in optimizing finite-state machines that parallel those existing in multiple-level logic design. Indeed, transformations like *decomposition, factoring* and *collapsing* are defined in both domains and are similar in objectives, even though their nature is different. *State minimization* is the counterpart of *Boolean minimization* and *state encoding* corresponds to *library binding*, i.e., both are transformations into structural representations.

Therefore finite-state machine optimization systems can be built that support these transformations. Unfortunately, as in the case of multiple-level logic optimization, the search for an optimum implementation is made difficult by the degrees of freedom available and the present ignorance about the interrelations among the transformations themselves.

Consider, for example, state minimization and encoding. Whereas it is reasonable to apply both optimization steps (with minimization preceding encoding), counterexamples show that solving exactly, but independently, both problems may not lead to an overall optimum solution. Indeed, encoding non-state-minimal finite-state machines may lead to smaller circuits than encoding state-minimal finite-state machines. State encoding techniques that take advantage of equivalent or compatible states are described in the upcoming section on *symbolic relations.*

The relations of decomposition to state encoding have been investigated [20, 23]. Indeed an encoding can be seen as a set of two-way partitions of the state set, each induced by the choice of a 1 or a 0 in an encoding column. As we mentioned before, in the case of two-level logic models, two-way decomposition can be achieved by symbolic minimization and constrained encoding. The definitions of the generalized prime implicants, encoding problems and evaluation of the objective function differ slightly, but the underlying mechanism is the same. We refer the interested reader to reference [2] for details.

There are also interesting relations between decomposition and state minimization. Indeed, a decomposition yields an interconnection of finite-state machines, and the state minimization problem of each component can take advantage of the decomposition structure. This issue is described in the next section.

STATE MINIMIZATION FOR INTERCONNECTED FINITE-STATE MACHINES. Most classical work on state minimization has dealt with finite-state machines in isolation. Recently, the optimal design of interconnected finite-state machines has prompted scientists to revisit the state minimization problem and extend it. We summarize here the major results, and we refer the reader to reference [2] for more details.

Consider the cascade interconnection of two finite-state machines, as shown in Figure 9.6 (d). The first (driving) machine feeds the input patterns to the second (driven) one. It is then often possible to compute the set of all possible and impossible pattern *sequences* generated by the driving machine. This corresponds to specifying the *input don't care sequences* that can be used for the state minimization of the driven machine. Note that classical state minimization considers only (input, state) pairs, while we are considering here input sequences from a given state. Kim and Newborn [22] proposed a technique for the state minimization of the driven machine. The technique was then improved by Rho *et al.* [30].

Similarly, it is possible to derive the sequences of patterns of the driving machine that do not affect the output of the driven machine, because the latter does not distinguish among them. The knowledge of these *output don't care sequences* can then be used to simplify the driving machines. We shall show in Section 9.3.3 how the input and output *don't care* sequences can be computed in the case of network representations.

SYMBOLIC RELATIONS. *Symbolic relations* are extensions of Boolean relations to the symbolic domain. By replacing the state transition function by a *state transition relation*, we can model the transitions from a state into more than one next state that is indistinguishable as far as these specific transitions are concerned.

Example 9.2.10. Consider the following state table [23]:

$$
\begin{array}{cccc}
0 & s_1 & s_1 & 0 \\
1 & s_1 & s_3 & 0 \\
0 & s_2 & s_2 & 0 \\
1 & s_2 & s_2 & * \\
0 & s_3 & s_2 & 0 \\
1 & s_3 & s_1 & 1
\end{array}
$$

By noticing that compatible classes are $\{s_1, s_2\}$ and $\{s_2, s_3\}$, the table can be minimized to yield [23, 25]:

$$
\begin{array}{cccc}
0 & s_{12} & s_{12} & 0 \\
1 & s_{12} & s_{23} & 0 \\
0 & s_{23} & \{s_{12}, s_{23}\} & 0 \\
1 & s_{23} & s_{12} & 1
\end{array}
$$

The minimal table shows that a transition from s_{23} under input 0 can lead to either s_{12} or s_{23}. Note though that s_{12} and s_{23} are not compatible states.

Symbolic relations allow us to express state tables with more general degrees of freedom for encoding than symbolic functions. The optimization of symbolic relations, as well as the solution to corresponding encoding problems, was proposed by Lin and Somenzi [25], and it has applications to several encoding problems.

9.3 SEQUENTIAL CIRCUIT OPTIMIZATION USING NETWORK MODELS

We already mentioned in the introduction to this chapter that the behavior of sequential circuits can be described by *traces*, i.e., by sequences of inputs and outputs. These sequences correspond to those that the related finite-state machine models can accept and generate. We recall that we restrict our attention to synchronous circuits with single-phase edge-triggered registers for the sake of simplicity. For this reason, it is convenient to discretize time into an infinite set of time points corresponding to the set of integers and to the triggering conditions of the registers. We assume that the observed operation of the network begins at time $n = 0$ after an appropriate initializing (or reset) sequence is applied. By this choice, the reset inputs are applied at some time point $n \leq 0$.

We denote sequences by *time-labeled* variables. For example $x^{(n)}$ denotes variable x at time n. A sequence of values of variable x is denoted by $x^{(n)}$ $\forall n$ in an interval of interest. It is convenient sometimes to have a shorthand notation for variables, without explicit dependency on time but just marking the *synchronous delay offset* with respect to a reference time point. We represent this by appending to the variable the reserved symbol @ and the offset when this is different from zero. Hence $x@k = x^{(n-k)}$ and $x = x@0$. Circuit equations in the shorthand notation are normalized by assuming that the left-hand side has zero offest, e.g., $x^{(n+1)} = y^{(n)}$ is translated to $x = y@1$.

We can extend then the notion of literals, products and sums to time-labeled variables. Thus the behavior of a synchronous network can be expressed by functions (or relations) over time-labeled variables.

> **Example 9.3.1.** Consider the circuit of Figure 9.2 (b), which provides an oscillating sequence when input r is FALSE. Its behavior can be expressed as $z^{(n)} = (z^{(n-1)} + r^{(n)})'$ $\forall n \geq 0$. Input r is a reset condition.
>
> Using the shorthand notation, the circuit can be described by the expression $z = (z@1 + r)'$.

Network models for synchronous circuits are extensions of those used for describing multiple-level combinational circuits. In particular, *logic networks* where modules can represent storage elements can be used to represent both *bound* and *unbound networks*. Bound networks represent gate-level implementation, while unbound networks associate internal modules with logic expressions in terms of time-labeled variables. Hence they represent a structured way of representing the circuit behavior. We consider unbound networks in the sequel.

As in the case of combinational synthesis, we restrict our attention to non-hierarchical logic networks, and we assume that the interconnection nets are split into sets of two-terminal nets. In addition, we represent the registers implicitly, by positive weights assigned to the nets (and to the edges in the corresponding graph), where the weights denote the corresponding synchronous delays. For example, a direct connection between two combinational modules has a zero weight, while a connection through a register has a unit weight. A connection through a k-stage shift register has weight k. Zero weights are sometimes omitted in the graphical representations. We call a *path weight* the sum of the edge weights along that path. (Path weights should not be confused with path delays, which are the composition of vertex propagation delays.) When compared to combinational logic networks, synchronous networks differ in being edge weighted and in not being restricted to acyclic. Nevertheless, the synchronous circuit assumption requires each cycle to have positive weight to disallow direct combinational feedback.

We summarize these considerations by refining the earlier definition of logic network of Section 3.3.2.

Definition 9.3.1. A non-hierarchical **synchronous logic network** is:

- A set of vertices V partitioned into three subsets called primary inputs V^I, primary outputs V^O and internal vertices V^G. Each vertex is assigned to a variable.

- A set of scalar combinational Boolean functions associated with the internal vertices. The support variables of each local function are time-labeled variables associated with primary inputs or other internal vertices. The dependency relation of the support variables corresponds to the edge set E and the synchronous delay offsets, i.e., differences in time labels correspond to the edge weights W.

- A set of assignments of the primary outputs to internal vertices that denotes which variables are directly observable from outside the network.

Note that a local function may depend on the value of a variable at different instances of time. In this case the model requires multiple edges between the corresponding vertices with the appropriate weights. Thus a synchronous network is modeled by a multi-graph, denoted by $G_{sn}(V, E, W)$. Note that synchronous logic networks simplify to combinational networks when no registers are present.

Example 9.3.2. An example of a synchronous circuit and its network model are shown in Figures 9.7 (a) and (b), respectively. The circuit decodes an incoming data stream coded with bi-phase marks, as produced by a compact-disk player. The network model is a multi-graph. For example, there are two edges between v_i and v_a with zero and unit weight. Note that zero weights are not shown. There are two unit-weighted cycles.

As in the case of combinational networks, an alternative representation is possible in terms of logic equations, whose support are variables with explicit dependence on time.

Example 9.3.3. Consider the network of Figure 9.7 (a). It can be described by the following set of equations:

$$a^{(n)} = i^{(n)} \,\overline{\oplus}\, i^{(n-1)}$$

$$b^{(n)} = i^{(n-1)} \,\overline{\oplus}\, i^{(n-2)}$$

$$c^{(n)} = a^{(n)} b^{(n)}$$

$$d^{(n)} = c^{(n)} + d'^{(n-1)}$$

$$e^{(n)} = d^{(n)} e^{(n-1)} + d'^{(n)} b'^{(n)}$$

$$v^{(n)} = c^{(n)}$$

$$s^{(n)} = e^{(n-1)}$$

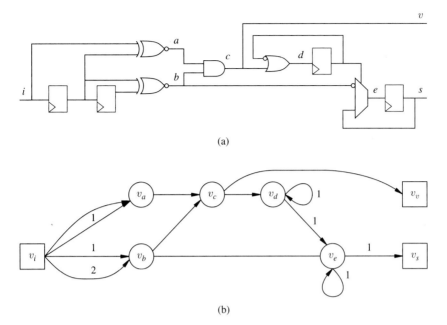

FIGURE 9.7
(a) Synchronous circuit. (b) Synchronous logic network.

or equivalently, using the shorthand notation:

$$a = i \;\overline{\oplus}\; i@1$$

$$b = i@1 \;\overline{\oplus}\; i@2$$

$$c = a\,b$$

$$d = c + d@1'$$

$$e = d\,e@1 + d'\,b'$$

$$v = c$$

$$s = e@1$$

There are different approaches to optimizing synchronous networks. The simplest is to ignore the registers and to optimize the combinational component using techniques of combinational logic synthesis. This is equivalent to deleting the edges with positive weights and to optimizing the corresponding combinational logic network. Needless to say, the removal of the registers from the network segments the circuit and weakens the optimality.

A radically different approach is *retiming*. By retiming a network, we move the position of the registers only. Hence we do not change the graph topology, but we modify the weight set W. Leiserson and Saxe [24] presented polynomially bound algorithms for finding the optimum retiming, which minimizes the circuit cycle-time or area. Unfortunately, retiming may not lead to the best implementation, because only register movement is considered.

The most general approach to synchronous logic optimization is to perform network transformations that blend retiming with combinational transformations. Such transformations can have the algebraic or Boolean flavor. In the latter case, the concept of *don't care* conditions must be extended to synchronous networks.

We present retiming first. Then we survey recent results on synchronous logic transformations as well as on enhancements to the original retiming method. We conclude this section by describing the specification of *don't care* conditions for optimizing synchronous networks.

9.3.1 Retiming

Retiming algorithms address the problem of minimizing the cycle-time or the area of synchronous circuits by changing the position of the registers. Recall that the cycle-time is bounded from below by the critical path delay in the combinational component of a synchronous circuit, i.e., by the longest path between a pair of registers. Hence retiming aims at placing the registers in appropriate positions, so that the critical paths they embrace are as short as possible.

Moving the registers may increase or decrease the number of registers. Thus, area minimization by retiming corresponds to minimizing the overall number of registers, because the combinational component of the circuit is not affected.

MODELING AND ASSUMPTIONS FOR RETIMING. We describe first the original retiming algorithms of Leiserson and Saxe [24] using a graph model that abstracts the computation performed at each vertex. Indeed retiming can be applied to networks that are more general than synchronous logic networks, where any type of computation is performed at the vertices (e.g., arithmetic operations).

When modeling circuits for retiming, it is convenient to represent the environment around a logic circuit within the network model. Hence, we assume that one or more vertices perform combined input/output operations. With this model, no vertex is a source or sink in the graph. Because of the generality of the model, we shall refer to it as a *synchronous network* and denote it by $G_{sn}(V, E, W)$. We shall defer to a later section a discussion of specific issues related to modeling the environment, such as representing distinguished primary input and output ports (e.g., Figure 9.7).

> **Example 9.3.4.** A synchronous network is shown in Figure 9.8 [24]. The numbers above the vertices are the propagation delays.

The retiming algorithms proposed by Leiserson and Saxe [24] assume that vertices have fixed propagation delays. Unfortunately, this is a limitation that may lead to inaccurate results. When registers have input loads different from other gates, shifting the registers in the circuit may indeed affect the propagation delays.

We consider topological critical paths only. Hence, the path delay between two registers (identified by non-zero weights on some edges) is the sum of the propagation

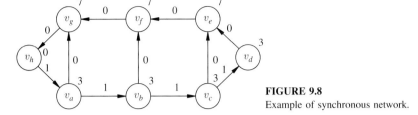

FIGURE 9.8
Example of synchronous network.

delays of the vertices along that path, including the extremal vertices. For a given path $(v_i, \ldots, v_j)$ we define path delay as:

$$d(v_i, \ldots, v_j) = \sum_{k:v_k \in (v_i, \ldots, v_j)} d_k \qquad (9.1)$$

Note the path delay is defined independently from the presence of registers along that path.

The path delay must not be confused with the path weight, which relates to the register count along that path. For a given path $(v_i, \ldots, v_j)$ we define path weight as:

$$w(v_i, \ldots, v_j) = \sum_{k,l:(v_k, v_l) \in (v_i, \ldots, v_j)} w_{kl} \qquad (9.2)$$

Retiming a vertex means moving registers from its outputs to its inputs, or *vice versa*. When this is possible, the retiming of a vertex is an integer that measures the amount of synchronous delays that have been moved. A positive value corresponds to shifting registers from the outputs to the inputs, a negative one to the opposite direction.

Example 9.3.5. Consider the circuit fragment shown in Figure 9.9 (a), whose network is shown in Figure 9.9 (b). A retiming of vertex v_c by 1 leads to the circuit fragment of Figure 9.9 (c), whose network is shown in Figure 9.9 (d).

The retiming of a network is represented by a vector **r** whose elements are the retiming values for the corresponding vertices. Retiming can be formally defined as follows.

Definition 9.3.2. A **retiming of a network** $G_{sn}(V, E, W)$ is an integer-valued vertex labeling $r : V \to Z$ that transforms $G_{sn}(V, E, W)$ into $\widetilde{G}_{sn}(V, E, \widetilde{W})$, where for each edge $(v_i, v_j) \in E$ the weight after retiming $\widetilde{w}_{ij}$ is equal to $\widetilde{w}_{ij} = w_{ij} + r_j - r_i$.

Example 9.3.6. Consider the network fragment shown in Figures 9.9 (b) and (d). The weight $w_{cx} = 1$ before retiming. Since $r_c = 1$ and $r_x = 0$, the weight after retiming $\widetilde{w}_{cx} = 1 + 0 - 1 = 0$.

It is simple to show [24] that the weight on a path depends only on the retiming of its extremal vertices, because the retiming of the internal vertices moves registers within the path itself. Namely, for a given path $(v_i, \ldots, v_j)$:

$$\widetilde{w}(v_i, \ldots, v_j) = w(v_i, \ldots, v_j) + r_j - r_i \qquad (9.3)$$

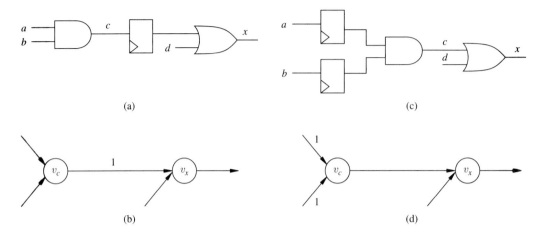

(a)   (c)

(b)   (d)

FIGURE 9.9
(a) Circuit fragment. (b) Network fragment. (c) Circuit fragment after retiming. (d) Network fragment after retiming. (Zero weights are omitted.)

As a consequence, weights on cycles are invariant on retiming. Note that the path delay is invariant with respect to retiming by definition.

A retiming is said to be *legal* if the retimed network has no negative weights. Leiserson and Saxe [24] proved formally that networks obtained by legal retiming are equivalent to the original ones. Moreover they also showed that retiming is the most general method for changing the register count and position without knowing the functions performed at the vertices. Hence, the family of networks equivalent to the original ones can be characterized by the set of legal retiming vectors.

> **Example 9.3.7.** Consider the network of Figure 9.8. An equivalent network is shown in Figure 9.10, corresponding to the retiming vector $\mathbf{r} = -[11222100]^T$, where the entries are associated with the vertices in lexicographic order.

CYCLE-TIME MINIMIZATION BY RETIMING. Retiming a circuit affects its topological critical paths. The goal of this section is to show how an optimum retiming vector can be determined that minimizes the length of the topological critical path in the network.

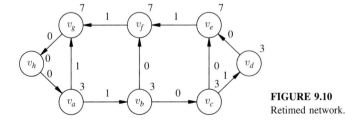

FIGURE 9.10
Retimed network.

Example 9.3.8. Consider the network of Figure 9.8. The topological critical path is $(v_d, v_e, v_f, v_g, v_h)$, whose delay is 24 units. Consider instead the network of Figure 9.10. The topological critical path is (v_b, v_c, v_e), whose delay is 13 units.

For each ordered vertex pair $v_i, v_j \in V$, we define $W(v_i, v_j) = \min w(v_i, \ldots, v_j)$ over all paths from v_i to v_j. Similarly, we define $D(v_i, v_j) = \max d(v_i, \ldots, v_j)$ over all paths from v_i to v_j with weight $W(v_i, v_j)$. The quantity $D(v_i, v_j)$ is the maximum delay on a path with minimum register count between two vertices.

Example 9.3.9. Consider the network of Figure 9.8 and vertices v_a, v_e. There are two paths from v_a to v_e, namely, (v_a, v_b, v_c, v_e) and $(v_a, v_b, v_c, v_d, v_e)$ with weights 2 and 3, respectively. Hence $W(v_a, v_e) = 2$ and $D(v_a, v_e) = 16$.

For a given cycle-time ϕ, we say that the network is *timing feasible* if it can operate correctly with a cycle-time ϕ, i.e., its topological critical path has delay ϕ or less.[2] This condition can be restated by saying that the circuit operates correctly if any path whose delay is larger than ϕ is "broken" by at least one register, or equivalently its weight is larger than or equal to 1. This can be shown to be equivalent to saying that the circuit operates correctly if and only if $W(v_i, v_j) \geq 1$ for all pairs $v_i, v_j \in V$ such that $D(v_i, v_j) > \phi$.

The usefulness of relating the retiming theory to the quantities $W(v_i, v_j)$ and $D(v_i, v_j)$ is that they are unique for each vertex pair and capture the most stringent timing requirement between them. In addition, $\widetilde{W}(v_i, v_j) = W(v_i, v_j) + r_j - r_i$ and $\widetilde{D}(v_i, v_j) = D(v_i, v_j)$. These quantities can be computed by using an all-pair shortest/longest path algorithm, such as Warshall-Floyd's. We denote by **W** and **D** the square matrices of size $|V|$ containing these elements.

We say that a retiming vector is *feasible* if it is legal and the retimed network is timing feasible for a given cycle-time ϕ. We can now state Leiserson and Saxe's theorem [24].

Theorem 9.3.1. Given a network $G_{sn}(V, E, W)$ and a cycle-time ϕ, **r** is a feasible retiming if and only if:

$$r_i - r_j \leq w_{ij} \qquad \forall (v_i, v_j) \in E \qquad (9.4)$$

$$r_i - r_j \leq W(v_i, v_j) - 1 \qquad \forall v_i, v_j : D(v_i, v_j) > \phi \qquad (9.5)$$

Proof. A retiming is legal if and only if $\widetilde{w}_{ij} = w_{ij} + r_j - r_i \geq 0$, which can be recast as inequality (9.4). Given a legal retiming, the topological critical path delay is less than ϕ if and only if the path weight after retiming is larger than or equal to 1 for all those paths with extremal vertices v_i, v_j where $d(v_i, \ldots, v_j) > \phi$. This is equivalent to saying that $\widetilde{W}(v_i, v_j) \geq 1$ for all vertex pairs $v_i, v_j \in V$ such that $\widetilde{D}(v_i, v_j) > \phi$

---

[2]We assume that setup times, clock-skew and register propagation delays are negligible. If this is not the case, they can be subtracted from ϕ.

$RETIME\_DELAY(G_{sn}(V, E, W)){$

    Compute all path weights and delays in $G_{sn}(V, E, W)$ and matrices **W** and **D**;

    Sort the values of the elements of **D**;

    Construct inequalities (9.4);

    foreach ϕ determined by a binary search on the elements of **D** {

        Construct inequalities (9.5);

        Solve inequalities (9.4, 9.5) by the Bellman-Ford algorithm;

    }

    Determine **r** by the last successful solution to the Bellman-Ford algorithm;

$}$

ALGORITHM 9.3.1

and eventually to stating inequality (9.5), because $\widetilde{W}(v_i, v_j) = W(v_i, v_j) + r_j - r_i$ and $\widetilde{D}(v_i, v_j) = D(v_i, v_j) \ \forall v_i, v_j \in V$.

Checking the conditions of Theorem 9.3.1 is fairly straightforward, because the right-hand sides are known constants for a given ϕ. The solution to this linear set of inequalities can be computed efficiently by the Bellman-Ford algorithm, which searches for the longest path in a graph with vertex set V, edges and weights determined by inequalities 9.4 and 9.5). Note that the problem may have no solution when it is overconstrained, i.e., when there exists a topological critical path whose delay exceeds ϕ in all circuit configurations corresponding to a legal retiming.

A naive way to minimize the cycle-time in a network is to check if there exists a feasible retiming for decreasing values of ϕ. For each tentative value of ϕ, the set of inequalities 9.4, and 9.5 is constructed, and a solution is sought by invoking the Bellman-Ford algorithm.

A more efficient search can be done by noticing that the optimum cycle-time must match a path delay for some vertex pair $v_i, v_j \in V$. More specifically, in any network there exists a vertex pair v_i, v_j such that $D(v_i, v_j)$ equals the optimum cycle-time. Hence a binary search among the entries of **D** provides the candidate values for the cycle-time, which are checked for feasibility by the Bellman-Ford algorithm. The overall worst-case computational complexity is $O(|V|^3 \log |V|)$, where the cubical term is due to the Bellman-Ford algorithm and the logarithmic one to the binary search. This is the original algorithm proposed by Leiserson and Saxe [24]. (See Algorithm 9.3.1.)

Example 9.3.10. Consider the network of Figure 9.8. The computation of the path weights and delays is trivial, as well as that of matrices **D** and **W**. (Matrices related to this example are reported in reference [24].) The elements of **D**, once sorted, are (33, 30, 27, 26, 24, 23, 21, 20, 19, 17, 16, 14, 13, 12, 10, 9, 7, 6, 3). The algorithm would then compute the inequalities 9.4, which can be represented by a graph with the same topology as the network and complementing the weights, because $r_i - r_j \leq w_{ij}$ implies $r_j \geq r_i - w_{ij}$ for each edge $(v_i, v_j) \in E$. (See Section 2.4.1.) This is shown by the solid edges of Figure 9.11.

    The binary search selects first $\phi = 19$. Then inequalities 9.5 are constructed, and the Bellman-Ford algorithm is applied. Since no positive cycle is detected, a feasible

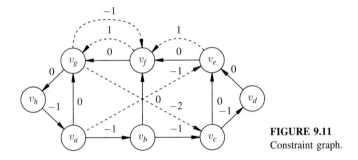

FIGURE 9.11
Constraint graph.

retiming exists for a cycle-time of 19 units. We do not describe the details of this step. We describe instead the case when $\phi = 13$ is selected. Inequalities 9.5 are computed. Some of these are redundant, namely, $D(v_i, v_j) - d(v_i) > 13$ or $D(v_i, v_j) - d(v_j) > 13$. The relevant ones are shown by dotted edges in Figure 9.11.

An inspection of the graph shows no positive cycle, and therefore the constraints are consistent. The Bellman-Ford algorithm can be used to compute the retiming vector. The retiming entries can be thought of as the weights of the longest path from a reference vertex say v_h, with $r_h = 0$. In this case the retiming vector is $-[12232100]^T$. Note that the solution is not unique and that other (larger) retiming values are possible for v_b and v_d. Namely, another feasible retiming vector is $-[11222100]^T$, corresponding to the network shown in Figure 9.10.

Other attempts with a shorter cycle-time would then fail. Note that we could have used the Bellman-Ford algorithm to solve a shortest path problem on a graph obtained from Figure 9.11 by complementing the weights.

Even though this method has polynomial-time complexity, its run time may be high. Computing matrices **W** and **D** may require large storage for graphs with many vertices. Most large networks are sparse, i.e., the number of edges is much smaller than the vertex pairs. Some retiming algorithms exploit the sparsity of the network and are more efficient for large networks. We review here a relaxation method that can be used to check the existence of a feasible retiming for a given cycle-time ϕ [24]. It is called $FEAS$ and it can replace the Bellman-Ford algorithm in $RETIME\_DELAY$.

The algorithm uses the notion of data-ready time at each vertex, which is equal to the the data-ready time of the combinational network obtained by deleting the edges with positive weights. (See Section 8.6.3.) As in the combinational case, the data-ready times are denoted by $\{t_i : v_i \in V\}$. Hence:

$$t_i = d_i + \max_{j:(v_j, v_i) \in E \text{ and } w_{ji} = 0} t_j \qquad (9.6)$$

The circuit operates correctly with cycle-time ϕ when the maximum data-ready time is less than or equal to ϕ.

Algorithm $FEAS$ is iterative in nature. At each iteration it computes the data-ready times and its maximum value. If this value is less than or equal to the given value of ϕ, the algorithm terminates successfully. Otherwise it searches for all those vertices whose data-ready times exceed the cycle-time, i.e., those vertices where the

$FEAS(G_{sn}(V, E, W), \phi)\{$

    Set $\mathbf{r} = \mathbf{0}$;

    Set $\widetilde{G}_{sn}(V, E, \widetilde{W}) = G_{sn}(V, E, W)$;

    for $(k = 1 \ to \ |V|)$ {

        Compute the set of data-ready times;

        if $(\max_{v_i \in V} t_i \leq \phi)$ /* All path delays are bounded by ϕ */

            return ($\widetilde{G}_{sn}(V, E, \widetilde{W})$);

        else {   /* Some path delays exceed ϕ */

            foreach vertex v_i such that $t_i > \phi$

                $r_i = r_i + 1$;   /* Retime vertices with excessive delay */

            $\widetilde{G}_{sn}(V, E, \widetilde{W}) = G_{sn}(V, E, W)$ retimed by $\mathbf{r}$;

        }

    }

    return (FALSE);

}

ALGORITHM 9.3.2

output signal is generated too late to be stored in a register. It then retimes these vertices by 1 unit, i.e., it moves registers backwards along those paths whose delay is too large. Since moving registers may create timing violations on some other paths, the process is iterated until a feasible retiming is found. A remarkable property of the algorithm is that when the algorithm fails to find a solution in V iterations, no feasible retiming exists for the given ϕ. (See Algorithm 9.3.2.)

The computational complexity of the algorithm is $O(|V||E|)$. Only the synchronous network needs to be stored. Note that when the algorithm is incorporated in $RETIME\_DELAY$, matrices $\mathbf{W}$ and $\mathbf{D}$ must still be computed, and the overall complexity may be dominated by their evaluation. However, the algorithm can be used in a heuristic search for a minimum cycle-time, where just a few values of ϕ are tried out in decreasing order. This may correspond to practical design requirements of choosing clock cycles that are multiples of some unit or corresponding to other design choices.

We now show the correctness of the algorithm and then give an example of its application.

Theorem 9.3.2. Given a synchronous network $G_{sn}(V, E, W)$ and a cycle-time ϕ, algorithm $FEAS$ returns a timing-feasible network $\widetilde{G}_{sn}(V, E, \widetilde{W})$ if and only if a feasible retiming exists. If no feasible retiming exists, it returns FALSE [24].

Proof. We first show that the algorithm constructs only legal retiming vectors. Consider any vertex, say v_i, being retimed at any iteration of the algorithm. Since $t_i > \phi$ implies $t_j > \phi$ for each vertex v_j that is head of a zero-weight path from v_i, retiming of vertex v_i as done by the algorithm cannot introduce a negative edge weight, because all its successors on zero-weight paths are retimed as well.

When the algorithm returns $\widetilde{G}_{sn}(V, E, \widetilde{W})$, the returned network is obviously timing feasible for the given ϕ. It remains to prove that when the algorithm returns FALSE, no feasible retiming exists. This can be shown by relating algorithm $FEAS$ to Bellman-Ford's and showing that they are equivalent. We give here a short intuitive proof, as reported in reference [24].

Our goal is to show that we attempt to satisfy inequalities 9.5 in $|V|$ successive iterations by setting $r_j = r_i - W(v_i, v_j) + 1$ for each vertex pair $v_i, v_j : D(v_i, v_j) > \phi$. We then concentrate on vertex pairs $v_i, v_j : D(v_i, v_j) > \phi$, the others being neglected by the algorithm as well as not inducing any constraint. We can also disregard vertex pairs $v_i, v_j : \widetilde{W}(v_i, v_j) = W(v_i, v_j) + r_j - r_i > 0$, because the corresponding constraint in 9.5 is satisfied.

When $D(v_i, v_j) > \phi$ and $\widetilde{W}(v_i, v_j) = 0$, there is some path from v_i to v_j such that $\widetilde{w}(v_i, \ldots, v_j) = \widetilde{W}(v_i, v_j) = 0$ and $d(v_i, \ldots, v_j) = D(v_i, v_j)$. Hence $t_j \geq d(v_i, \ldots, v_j) = D(v_i, v_j) > \phi$, so that the algorithm retimes v_j by 1, i.e., r_j is assigned the value $r_j + 1 = r_i + \widetilde{W}(v_i, v_j) - W(v_i, v_j) + 1 = r_i - W(v_i, v_j) + 1$. Setting $r_j = r_i - W(v_i, v_j) + 1$ is exactly the assignment done by the Bellman-Ford algorithm. Conversely, r_j is incremented only when there exists a path, say from v_k to v_j, with zero weight and delay larger than ϕ, implying that $D(v_k, v_j) \geq d(v_k, \ldots, v_j) > \phi$ and $\widetilde{W}(v_k, v_j) = 0$.

Example 9.3.11. Consider the network of Figure 9.8, reported again in Figure 9.12 (a1). Let $\phi = 13$. Let us compute the data-ready times. Recall that cutting the edges with positive weights determines the combinational component of the circuit, shown in Figure 9.12 (a2). Thus, we can perform a network traversal starting from the vertices that are heads of edges with positive weights, which yields $t_a = 3; t_b = 3; t_c = 3; t_d = 3; t_e = 10; t_f = 17; t_g = 24; t_h = 24$. Hence the subset of vertices to be retimed is $\{t_f, t_g, t_h\}$. Retiming these vertices by 1 corresponds to assigning $w_{ef} = 1; w_{ha} = 0; w_{ag} = 1; w_{bf} = 1$. The retimed network is shown in Figure 9.12 (b1).

We consider again the combinational component of the circuit [Figure 9.12 (b2)] and we then recompute the data-ready times. Thus: $t_a = 17; t_b = 3; t_c = 3; t_d = 3; t_e = 10; t_f = 7; t_g = 14; t_h = 14$. Now the subset of vertices to be retimed is $\{t_a, t_g, t_h\}$. Retiming these vertices by 1 corresponds to assigning $w_{fg} = 1; w_{ab} = 0$. The newly retimed network is shown in Figure 9.12 (c1).

We consider again the combinational component of the circuit [Figure 9.12 (c2)] and determine the data-ready times, which are $t_a = 10; t_b = 13; t_c = 3; t_d = 3; t_e = 10; t_f = 7; t_g = 7; t_h = 7$. The network is timing feasible.

Note that the final network differs from the one in Figure 9.10 by a possible retiming of v_b by 1, which would not alter the feasibility of the network.

AREA MINIMIZATION BY RETIMING.* Retiming may affect the circuit area, because it may increase or decrease the number of registers in the circuit. Since retiming does not affect the functions associated with the vertices of the network, the variation of the number of registers is the only relevant factor as far as area optimization by retiming is concerned.

Before explaining the method, let us recall that the synchronous network model splits multi-terminal nets into two-terminal ones. As a result, synchronous delays are modeled by weights on each edge. Consider a vertex with two (or more) fanout stems. From an implementation standpoint, there is no reason for having independent registers on multiple fanout paths. Registers can be shared, as shown in Figure 9.13 (c).

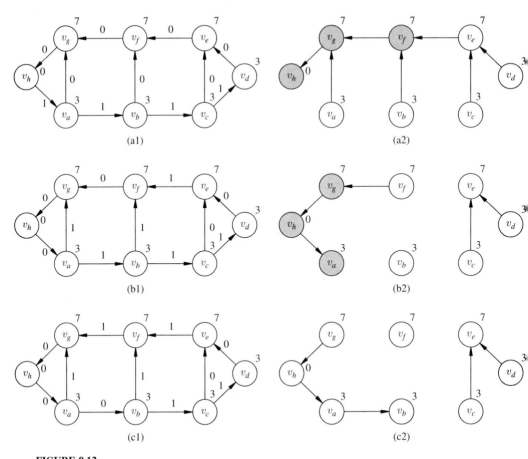

FIGURE 9.12
(a1, b1, c1) Synchronous network being retimed by the *FEAS* algorithm at various steps. (a2, b2, c2) Combinational component of the network at various steps. (Shaded vertices are not timing feasible.)

Nevertheless, for the sake of explanation, we show first how we can achieve minimum area without register sharing. We then extend the method to a more complex network model that supports register sharing.

Consider a generic vertex, say $v_i \in V$, that is retimed by r_i. The local variation in register count is $r_i(indegree(v_i) - outdegree(v_i))$. Therefore, the overall variation of registers due to retiming is $\mathbf{c}^T \mathbf{r}$, where $\mathbf{c}$ is a constant known vector of size $|V|$ whose entries are the differences between the *indegree* and the *outdegree* of each vertex. Since any retiming vector must be legal, i.e., satisfying Equation 9.4, the unconstrained area minimization problem can be stated as follows:

$$\min \quad \mathbf{c}^T \mathbf{r} \quad \text{such that}$$

$$r_i - r_j \leq w_{ij} \quad \forall (v_i, v_j) \in E$$

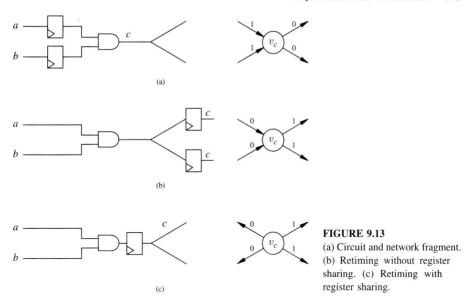

FIGURE 9.13
(a) Circuit and network fragment.
(b) Retiming without register sharing. (c) Retiming with register sharing.

Similarly, area minimization under cycle-time constraint ϕ can be formulated as:

$$\min \quad \mathbf{c}^T \mathbf{r} \quad \text{such that}$$

$$r_i - r_j \leq w_{ij} \qquad \forall (v_i, v_j) \in E$$

$$r_i - r_j \leq W(v_i, v_j) - 1 \qquad \forall v_i, v_j : D(v_i, v_j) > \phi$$

These problems can be solved by any linear program solver, such as the simplex algorithm. Since the right-hand sides of the inequalities are integers, the retiming vector $\mathbf{r}$ has integer entries. This formulation is also the linear programming dual of a minimum-cost flow problem that can be solved in polynomial time [24].

Note that the timing feasibility problem under area constraint $\bar{a}$ can be modeled by appending $\mathbf{c}^T \mathbf{r} \leq \bar{a} - \sum_i \sum_j w_{ij}$ to Equations 9.4 and 9.5 and is solved by linear programming techniques. This provides us with a means of solving the minimum cycle-time problem under area constraints by retiming.

> **Example 9.3.12.** Consider the example of Figure 9.8. Then $\mathbf{c} = [-1, -1, -1, 0, 1, 1, 1, 0]^T$ and $\mathbf{c}^T \mathbf{r} = -r_a - r_b - r_c + r_e + r_f + r_g$. The implementation of Figure 9.8 is area minimal. This can be shown by noticing that there are only four registers and they are all on a cycle. No legal retiming can reduce the cycle weight. Consider the network of Figure 9.10. This implementation requires five registers. The additional register is the price paid for the reduced cycle-time. Note that retiming v_b by $r_b = -1$ would yield a timing-feasible circuit for the same cycle-time, but with an additional register.

To take register sharing into account, we transform the network $G_{sn}(V, E, W)$ into a modified one, called $\widehat{G}_{sn}(\widehat{V}, \widehat{E}, \widehat{W})$. Then we minimize the register count on $\widehat{G}_{sn}(\widehat{V}, \widehat{E}, \widehat{W})$ with a modified cost function. The transformation is such that the overall register count in $\widehat{G}_{sn}(\widehat{V}, \widehat{E}, \widehat{W})$ without register sharing equals the register count in $G_{sn}(V, E, W)$ with register sharing.

The transformation targets only those vertices with multiple direct successors (i.e., where register sharing matters). Without loss of generality, we shall consider only one vertex, say v_i, with k direct successors $v_1, v_2, \ldots, v_k$. The modified network $\widehat{G}_{sn}(\widehat{V}, \widehat{E}, \widehat{W})$ is obtained from $G_{sn}(V, E, W)$ by adding a dummy vertex $\widehat{v}_i$ with zero propagation delay and edges from $v_1, v_2, \ldots, v_k$ to $\widehat{v}_i$. Let $w_{max} = \max_{j=1,2,\ldots,k} w_{ij}$ in $G_{sn}(V, E, W)$. In the modified network, the weights of the new edges are $\widehat{w}_{ji} = w_{max} - w_{ij}$; $j = 1, 2, \ldots, k$.

Let each edge in $\widehat{G}_{sn}(\widehat{V}, \widehat{E}, \widehat{W})$ have a numeric attribute, called *breadth*. We use a modified cost function $\mathbf{c}^T \mathbf{r}$, where the entries of $\mathbf{c}$ are the differences of the sums of the breadths of the incoming edges minus the breadths of the outgoing ones. Note that when all breadths are 1, vector $\mathbf{c}$ represents just the difference of the degrees of the vertices.

We set the breadth of edges (v_i, v_j) and $(v_j, \widehat{v}_i)$, $j = 1, 2, \ldots, k$, to $1/k$. We show now that this formulation models accurately register sharing on the net that stems from v_i when the overall register count in $\widehat{G}_{sn}(\widehat{V}, \widehat{E}, \widehat{W})$ is minimized. For any retiming $\mathbf{r}$, all path weights $w(v_i, \ldots, \widehat{v}_i)$ are the same, because they are equal for $\mathbf{r} = \mathbf{0}$. A minimal-area retimed network is such that the weights $\widehat{w}_{ji}$, $j = 1, 2, \ldots, k$, are minimal, because $\widehat{v}_i$ is a sink of the graph. Thus there exists an edge whose head is $\widehat{v}_i$ with 0 weight after retiming. Then all path weights $w(v_i, \ldots, \widehat{v}_i)$ are equal to the maximum weight of the edges whose tail is v_i after retiming, denoted by $\widehat{w}_{max}$. The overall contribution to the modified cost function by the portion of the network under consideration is thus $\sum_{j=1}^{k} 1/k \ \widehat{w}_{max} = \widehat{w}_{max}$, thus modeling the register sharing effect.

Example 9.3.13. A network fragment is shown in Figure 9.14 (a). Vertex v_i has $k = 3$ outdegree. Dotted edges model the interaction with the rest of the network. The modified network before and after retiming is shown in Figures 9.14 (b) and (c), respectively. Note that the path weights along the paths $(v_i, \ldots, \widehat{v}_i)$ remain equal, even though the edge weights change during retiming. Before retiming $w_{max} = 2$, while after retiming $\widehat{w}_{max} = 3$. The local increase in registers is justified by an overall decrease in other parts of the network. The marginal area cost associated with the modified network is $(3 \cdot 3)/3 = 3$, corresponding to three registers in a cascade connection.

RETIMING MULTIPLE-PORT SYNCHRONOUS LOGIC NETWORKS AND OPTI-MUM PIPELINING. A practical problem associated with retiming synchronous logic networks is the determination of an initial state of the retimed network such that the new network is equivalent to the original one. This is always possible when the reset functions of the registers are modeled explicitly by combinational circuits, i.e., the registers perform only synchronous delays.

On the other hand, it is often convenient to associate the reset function with the registers themselves. This simplifies the network and often matches the circuit specification of the register cells. In this case, the determination of the initial state of a retimed network may be problematic. Consider the negative retiming of a vertex, where

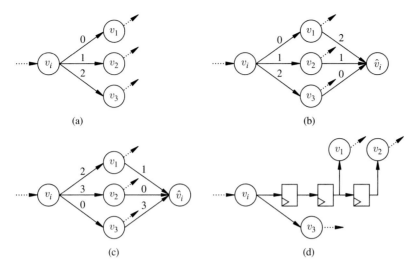

FIGURE 9.14
(a) Network fragment. (b) Modified network before retiming. (c) Modified network after retiming. (d) Circuit fragment after retiming.

registers are moved from the input to the output of the corresponding combinational gate. Then the initial state can be determined by applying the local function to the initial states of the inputs. Conversely, when considering a positive retiming of a vertex, registers are moved from the output to the input of the local gate. Consider the case when no register sharing is used and the gate under consideration drives more than one register with different initial states. Then, an equivalent initial state in the retimed circuit cannot be determined.

Example 9.3.14. An example of a register with explicit external synchronous reset is shown in Figure 9.15 (b), which can be compared to a register with internal reset, as shown in Figure 9.15 (a). In practical circuits, the difference lies in the different internal transistor-level model and in the possible support for asynchronous resets. As far as we are concerned, the difference is in the level of detail of the objects of our optimization technique.

Consider the circuit of Figure 9.15 (c), where registers have internal reset and are initialized to different values. It is impossible to find an equivalent initial state in the circuit retimed by 1.

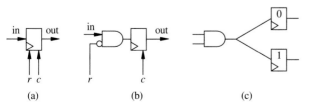

FIGURE 9.15
(a) Register model with internal reset. (b) Register model with explicit external reset circuit. (c) Circuit fragment before retiming with initial conditions.

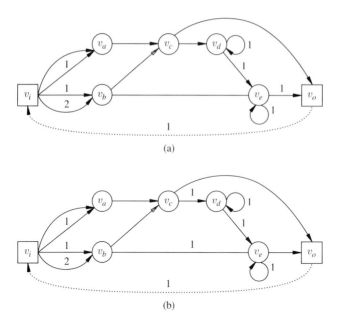

(a)

(b)

FIGURE 9.16
(a) Modified synchronous network. (b) Retimed synchronous network.

Let us consider now the input/output modeling issue. Leiserson's model for retiming synchronous networks assumes that input/output is modeled by one or more vertices of the network. Practical examples of networks have distinguished primary input and output ports. The most common assumption is that the environment provides inputs and samples outputs at each cycle. Hence we can lump all primary input vertices into a single vertex (v_I) and all primary output vertices into another one (v_O). An edge of unit weight between the output vertex and the input vertex models the synchronous operation of the environment. Without loss of generality, the input and output vertices are assumed to have zero propagation delay. Finite delays on any port can be modeled by introducing extra vertices with the propagation delay of the ports.

With this simple transformation, no vertex is a source or a sink in the graph. Retiming preserves the relative timing of the input and output signals, because retiming the input (or the output) merged vertex implies adding/removing the same amount of weight to all edges incident to it. This motivates the use of distinguished vertices for the inputs and outputs. A pitfall of this modeling style is that the synchronous delay of the environment may be redistributed inside the circuit by retiming. This problem can be obviated by the additional constraint $r_I = r_O$.

> **Example 9.3.15.** The network of Figure 9.7 is reproposed in Figure 9.16, after having combined the primary output vertices and added the edge representing the environment. Assume all gates have unit delay. The critical delay is 3 units.
>
> An optimum retiming is shown in Figure 9.16 (b), where the critical path length is 2 units.

A different modeling style for retiming multiple-port networks is to leave the network unaltered and require zero retiming at the ports, i.e., $r_I = r_O = 0$, which is a stronger condition. When using algorithms that allow retiming of a single polarity, like the $FEAS$ algorithm that retimes vertices iteratively by +1 and moves registers from output to input, this modeling style may in some cases preclude finding the optimum solution. Consider, for example, the case of a network where the critical path is an input/output path with a (high) weight on the tail edge. The corresponding registers cannot be distributed by the algorithm along that path.

Let us now consider the following optimum pipelining problem. Given a combinational circuit, insert appropriate registers so that the cycle-time is minimum for a given latency $\bar{\lambda}$. The circuit can be modeled by merging all primary inputs and outputs in two distinguished vertices and joining the output vertex to the input vertex by means of an edge with weight $\bar{\lambda}$. Note that in this case the synchronous delay of the environment is meant to be distributed inside the network. If our pipeline model is such that primary inputs and outputs should be synchronized to the clock, then we may require that the weight on the edge representing the environment is at least 1 after retiming. This can be modeled by the additional constraint $r_O - r_I \leq \bar{\lambda} - 1$.

By applying algorithm $RETIME\_DELAY$ to the circuit for different values of $\bar{\lambda}$, we can derive the exact latency/cycle-time trade-off curve. Similarly, we can minimize the number of registers in a pipeline, under latency and cycle-time constraints, by using the linear programming formulation of the previous section.

9.3.2 Synchronous Circuit Optimization by Retiming and Logic Transformations

When considering the overall problem of optimizing synchronous circuits, it is desirable to perform retiming in conjunction with optimization of the combinational component. It is obviously possible to alternate retiming and combinational optimization. Indeed, we can think of optimization of synchronous circuits as the application of a set of operators, where retiming is one of them. Other operators are the combinational transformations described in Sections 8.3 and 8.4 and applied to the combinational component.

Retiming and combinational optimization are interrelated. Retiming is limited in optimizing the performance of a circuit by the presence of vertices with large propagation delays, which we call combinational bottlenecks. Conversely, combinational optimization algorithms are limited in their ability to restructure a circuit by the fragmentation of the combinational component due to the presence of registers interspersed with logic gates.

Dey *et al.* [15] proposed a method that determines timing requirements on combinational subcircuits such that a retimed circuit can be made timing feasible for a given ϕ when these conditions are met. The timing constraints on the propagation delays of combinational subcircuits can guide algorithms, such as $SPEED\_UP$ (see Section 8.6.3), to restructure the logic circuit to meet the bounds. In other words, the method attempts to remove the combinational bottlenecks to allow the retiming algorithm to meet a cycle-time requirement.

Malik *et al.* [28] proposed a complementary approach that removes the registers temporarily from (a portion of) a network, so that combinational logic optimization can be most effective in reducing the delay (or area). Since the registers are temporarily placed at the periphery of the circuit, the method is called *peripheral retiming*. Peripheral retiming is followed by combinational logic optimization and by retiming. Logic optimization benefits from dealing with a larger unpartitioned circuit.

Other approaches merge local transformation with local retiming, or equivalently perform local transformations across register boundaries. As in the case of combinational circuits, they can have the algebraic or Boolean flavor.

PERIPHERAL RETIMING. Peripheral retiming applies to acyclic networks. When synchronous logic networks have cycles, these can be cut by removing temporarily some feedback edges. Vertices of synchronous logic networks can be partitioned into peripheral vertices, defined as $V^P = V^I \cup V^O$, and internal ones, denoted by V^G. Similarly, edges can be classified as peripheral or internal; the former are incident to a peripheral vertex while the latter are the remaining ones.

In peripheral retiming, peripheral vertices cannot be retimed. When peripheral retiming can be applied, the internal vertices are retimed so that no internal edge has a positive weight, i.e., all registers are moved to the periphery of the circuit.

> **Definition 9.3.3.** A **peripheral retiming** of a network $G_{sn}(V, E, W)$ is an integer-valued vertex labeling $p : V \rightarrow Z$ that transforms $G_{sn}(V, E, W)$ into $\widetilde{G}_{sn}(V, E, \widetilde{W})$, where $p(v) = 0 \ \forall v \in V^P$, and for each internal edge $(v_i, v_j) \in E$ the weight after peripheral retiming $\widetilde{w}_{ij} = w_{ij} + p_j - p_i = 0$.

Note that a *legal* retiming (see Section 9.3.1) requires that $w_{ij} + r_j - r_i \geq 0$ for each edge, while a *peripheral* retiming requires $w_{ij} + p_j - p_i = 0$ for each internal edge. As a consequence of the fact that peripheral retiming sets all weights to zero (removes all registers) on internal edges, negative weights are allowed on peripheral edges. The presence of negative weights means that we are allowed to borrow time from the environment temporarily while applying combinational optimization.

> **Example 9.3.16.** Consider the synchronous circuit of Figure 9.17 (a) [28]. One AND gate is redundant, but combinational techniques cannot detect it when operating on the network fragments obtained by removing the registers. Figure 9.17 (b) shows the circuit after peripheral retiming, where the registers have been pushed to the periphery of the circuit. Note that one register has negative weight, corresponding to borrowing time from the environment.
>
> The result of applying combinational optimization to the peripherally retimed circuit is shown in Figure 9.17 (c), where the redundancy is detected and eliminated. The circuit after retiming is then shown in Figure 9.17 (d). Note that no negative delay is left in the circuit.

The goal of peripheral retiming is to exploit at best the potentials of combinational logic optimization. Registers are temporarily moved to the circuit periphery, so that combinational logic optimization algorithms can be applied to networks that are

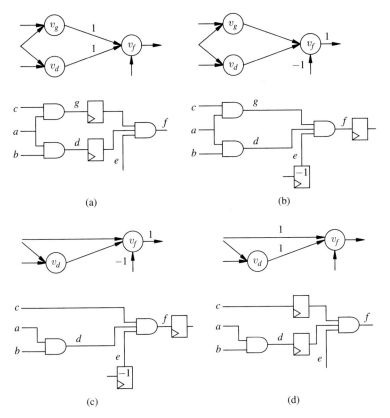

FIGURE 9.17
(a) Synchronous network and circuit. (b) Modified network and circuit by peripheral retiming. (c) Optimized network and circuit after combinational optimization. (d) Synchronous network and circuit after final retiming.

not segmented by the presence of registers. Eventually registers are placed back into the circuit by regular retiming.

Two issues are important to determine the applicability of this method: first, the determination of the class of circuits for which peripheral retiming is possible; second, the characterization of the combinational transformations for which there exist a valid retiming (i.e., so that negative weights can be removed).

The circuits that satisfy the following conditions have a peripheral retiming:

- The network graph is acyclic.
- There are no two paths from an input to an output vertex with different weights.
- There exists integer vectors $\mathbf{a} \in Z^{|V^I|}$ and $\mathbf{b} \in Z^{|V^O|}$ such that $w(v_i, \ldots, v_j) = a_i + b_j$ for all paths $(v_i, \ldots, v_j) \; : \; v_i \in V^I, v_j \in V^O$.

The correctness of this condition is shown formally in reference [28]. Pipelined circuits, where all input/output paths have the same weight, fall in the class of circuits that can be peripherally retimed. Vectors $\mathbf{a}$ and $\mathbf{b}$ represent the weights on the input and output peripheral edges of a retimed circuit. Verifying the condition and computing **a** and **b** (which is equivalent to performing peripheral retiming) can be done in $O(|E||V|)$ time [28].

Let us consider now the applicable logic transformations. Malik *et al.* [28] showed that a legal retiming of a peripherally retimed network requires non-negative input/output path weights. Obviously, it is always possible to retime a peripherally retimed circuit whose topology has not changed by restoring the original position of the registers. On the other hand, some combinational logic transformations may introduce input/output paths with negative weights. When this happens, the transformations must be rejected.

Example 9.3.17. Consider the synchronous circuit of Figure 9.18 (a) [28]. Figure 9.18 (b) shows the circuit after peripheral retiming, and Figure 9.18 (c) shows the circuit after combinational optimization, where a three-input OR gate has been replaced by a two-input OR. Unfortunately, this change induces an input/output path with negative weight, impeding a valid retiming and hence a feasible implementation. Hence the combinational transformation is not acceptable.

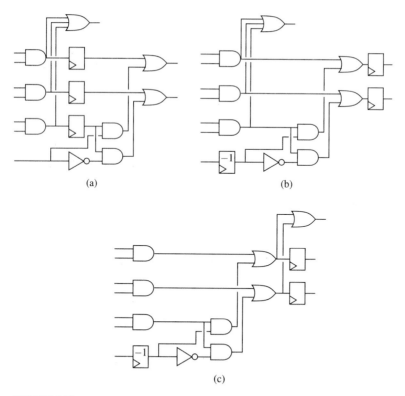

FIGURE 9.18
(a) Synchronous circuit. (b) Synchronous circuit modified by peripheral retiming. (c) Modified circuit after combinational optimization.

Peripheral retiming may be applied to acyclic networks when the above conditions are satisfied or to portions of acyclic networks when partitioning is necessary to satisfy them by breaking some paths. By the same token, peripheral retiming can be applied to cyclic networks by deleting temporarily enough edges to satisfy the assumptions for peripheral retiming. Unfortunately, partitioning the network weakens the power of the method.

TRANSFORMATIONS FOR SYNCHRONOUS NETWORKS. We consider now logic transformations that are specific to synchronous circuits, called *synchronous logic transformations*. They are extensions of combinational logic transformations and can be seen as transformations across register boundaries. In other words, they blend local circuit modification and retiming.

Before describing the transformations, let us consider retiming as a transformation on synchronous logic networks described by time-labeled expressions. Retiming a variable by an integer r is adding r to its time label. Retiming a time-labeled expression is retiming all its variables. Retiming a vertex of a synchronous logic network by r corresponds to either retiming by r the corresponding variable or retiming by $-r$ the related expression.

> **Example 9.3.18.** Consider the network fragment of Figure 9.9, reported for convenience in Figure 9.19. Retiming of vertex v_c by 1 corresponds to changing expression $c^{(n)} = a^{(n)}b^{(n)}$ into $c^{(n+1)} = a^{(n)}b^{(n)}$ or equivalently to $c^{(n)} = a^{(n-1)}b^{(n-1)}$. (With the shorthand notation, $c = ab$ is retimed to $c = a@1\ b@1$.) Therefore it corresponds to retiming variable c by 1, i.e., to replacing $c^{(n)}$ by $c^{(n+1)}$. Alternatively, it corresponds to retiming expression ab by -1.

Synchronous logic transformations can have the algebraic or Boolean flavor. In the algebraic case, local expressions can be viewed as polynomials over time-labeled variables [10]. Combinational logic transformations can be applied by considering the time-labeled variables as new variables, while exploiting the flexibility of using

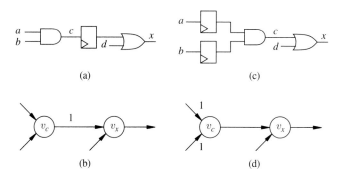

FIGURE 9.19
(a) Circuit fragment. (b) Network fragment. (c) Circuit fragment after retiming. (d) Network fragment after retiming. (Zero weights are omitted.)

retimed variables and expressions to simplify the circuit. Thus, the shorthand notation is particularly convenient for representing synchronous transformations.

Example 9.3.19. Consider the circuit of Figure 9.20 (a), described by expressions $c = ab$; $x = d + c@1$. Variable c is eliminated after retiming the corresponding vertex by 1. This is equivalent to saying that $c = ab$ implies $c@1 = a@1\ b@1$ and replacing $c@1$ in the expression for x, leading to $x = d + a@1\ b@1$. The elimination saves one literal but adds one register.

Example 9.3.20. Consider the circuit of Figure 9.21 (a), described by expressions $x = a@1 + b$; $y = a@2\ c + b@1\ c$. The local expression for y is simplified by algebraic substitution by adding to its support set variable x retimed by -1, i.e., $x@1$. This is possible because the local expression for x, i.e., $a@1 + b$, is an algebraic divisor of $a@2\ c + b@1\ c$ once retimed by -1. The algebraic substitution leads to $x = a@1 + b$; $y = x@1\ c$ while saving two literals and a register.

It is important to note that synchronous transformations affect the circuit area by modifying the number of literals and registers in the network model. Similarly, they affect the network delay by changing the propagation delays associated with the local functions as well as the register boundaries. Hence, the computation of the objective functions is more elaborate than in the combinational case.

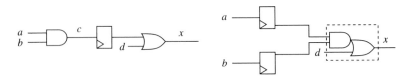

FIGURE 9.20
(a) Fragment of synchronous network. (b) Example of synchronous elimination.

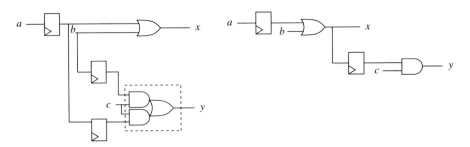

FIGURE 9.21
(a) Fragment of synchronous network. (b) Example of synchronous substitution.

The synchronous Boolean transformations require the use of logic minimizers, as in the case of combinational circuits. In the synchronous case, the minimizers operate on function representations based on time-labeled variables.

Example 9.3.21. Consider the network $\mathcal{N}_2$ of Figure 9.22. Let us consider the simplification of function $u \oplus v$ associated with variable y. The local *don't care* conditions for this function are not straightforward and will be derived in Example 9.3.25. As far as this example is concerned, we assume that the *don't care* conditions include $u@1' \, u' + u@1 \, v'$ and thus involve literals at different time points (different from the combinational case).

Note that $y = u \oplus v = uv + u'v'$ and that $u'v' = u'v'u@1' + u'v'u@1$ is included in the local *don't care* set. Hence, the local function can be replaced by $y = uv$, with a savings of two literals, by Boolean simplification. The resulting network is shown in Figure 9.23.

9.3.3 *Don't Care* Conditions in Synchronous Networks

The power of Boolean methods relies on capturing the degrees of freedom for optimization by *don't care* conditions. Since the behavior of synchronous networks can be described in terms of traces, the *don't care* conditions represent the degrees of freedom in associating output with input sequences. Hence the most general representation for synchronous networks and *don't care* conditions is in terms of Boolean relations. Nevertheless some *don't care* conditions can be expressed by *sum of products* expressions over time-labeled literals and computed by an extension of the methods for *don't care* set computations of combinational circuits.

When the *don't care* conditions of interest are expressed by a *sum of products* form, we can perform simplification by applying any two-level logic minimizer. Oth-

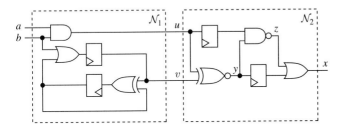

FIGURE 9.22
Interconnected networks.

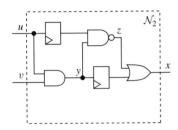

FIGURE 9.23
Optimized network $\mathcal{N}_2$.

erwise, specific relation minimizers or optimization methods are required. We describe next the computation of *don't care* conditions expressed explicitly in *sum of products* forms. We shall then consider implicit specifications of synchronous networks and their *don't care* conditions as well as the corresponding optimization methods.

EXPLICIT *DON'T CARE* CONDITIONS IN SYNCHRONOUS NETWORKS. External *don't care* conditions of synchronous logic networks are related to the embedding of the network in the environment, as in the combinational case. Differently from combinational circuits, sequential networks have a dynamic behavior. As a result, *sum of products* representations of *don't care* sets may involve literals at one or more time points.

External *don't care* conditions consist of a controllability and an observability component.

> **Definition 9.3.4.** The **input controllability** *don't care* set includes all input sequences that are never produced by the environment at the network's inputs.

> **Definition 9.3.5.** The **output observability** *don't care* sets denote all input sequences that represent situations when an output is not observed by the environment at the current time or in the future.

We denote the input controllability *don't care* set by CDC_{in}. We use the vector notation $\mathbf{ODC}_{out}$ for the output observability *don't care* set by associating each entry of the vector with the observability conditions at each output. Moreover, output observability may be restricted to be at a given time point. When referring to the observability at time n, we denote the observability *don't care* set as $\mathbf{ODC}_{out}^{(n)}$. The overall observability *don't care* set $\mathbf{ODC}_{out}$ is the intersection of the ODC sets at the time points of interest. The *external don't care* conditions are denoted by $\mathbf{DC}_{ext} = \mathbf{CDC}_{in} \cup \mathbf{ODC}_{out}$, where $\mathbf{CDC}_{in}$ is a vector with n_o entries equal to CDC_{in}.

> **Example 9.3.22.** Consider the circuit of Figure 9.22. Let us consider the input controllability *don't care* conditions for network $\mathcal{N}_2$. Assume that the network is initialized by the sequence $(b^{(-4)}, b^{(-3)}, b^{(-2)}) = (1, 0, 1)$.
>
> The limited controllability of the inputs of $\mathcal{N}_2$ is reflected by the set of its impossible input sequences. For example, $u^{(n)}v'^{(n+1)}$ is an impossible input sequence for $\mathcal{N}_2$. Indeed for $u^{(n)}$ to be 1, it must be that $a^{(n)} = b^{(n)} = 1$, but $b^{(n)} = 1$ implies $v^{(n+1)} = 1$. Hence, for $\mathcal{N}_2$:
>
> $$u^{(n)}v'^{(n+1)} \subseteq CDC_{in}; \quad \forall n \geq -4$$
>
> As a consequence of the initializing sequence, output v cannot assume the value 0 at time $-3, -1$. Hence:
>
> $$v'^{(-3)} + v'^{(-1)} \subseteq CDC_{in}$$
>
> Similarly, because of the circuit initialization, v cannot take value 0 twice consecutively, as will be shown in detail in Example 9.3.26. Consequently:
>
> $$v'^{(n)}v'^{(n+1)} \subseteq CDC_{in} \quad \forall n \geq -2$$

The interconnection of the two networks limits the observability of the primary outputs of $\mathcal{N}_1$. We now compute the output observability *don't care* conditions for $\mathcal{N}_1$. We concentrate on the scalar component related to output v. In particular, the output of $\mathcal{N}_2$ can be expressed in terms of u and v as:

$$x^{(n)} = y^{(n-1)} + y'^{(n)} + u'^{(n-1)} = u'^{(n-1)} + u^{(n-1)} \overline{\oplus} v^{(n-1)} + u^{(n)} \oplus v^{(n)}$$

$$= u'^{(n-1)} + v^{(n-1)} + u^{(n)} \oplus v^{(n)}$$

(by expanding the $\overline{\oplus}$ operator and removing redundant terms and literals).

The value of $v^{(n)}$ can be observed at the output of $\mathcal{N}_2$ only at time n or at time $n + 1$. In particular, $v^{(n)}$ is observable at time n if $y^{(n-1)} = 0$ and $u^{(n-1)} = 1$. The observability *don't care* of v at time n can thus be described by the function:

$$ODC^{(n)}_{out, v^{(n)}} = u'^{(n-1)} + y^{(n-1)} = u'^{(n-1)} + v^{(n-1)}$$

while the observability *don't care* set at time $n + 1$ is described by:

$$ODC^{(n+1)}_{out, v^{(n)}} = (y^{(n+1)} u^{(n)})' = y'^{(n+1)} + u'^{(n)} = u^{(n+1)} \oplus v^{(n+1)} + u'^{(n)}$$

Conditions for never observing $v^{(n)}$ at the primary output of $\mathcal{N}_2$ are described by:

$$ODC_{out, v^{(n)}} = ODC^{(n)}_{out, v^{(n)}} ODC^{(n+1)}_{out, v^{(n)}}$$

in particular containing the cube $u'^{(n-1)} u'^{(n)}$. Since $u^{(n)} = a^{(n)} b^{(n)}$, then $(a'^{(n-1)} + b'^{(n-1)})(a'^{(n)} + b'^{(n)})$ belongs to the component of the ODC set of $\mathcal{N}_1$ associated with output v. Thus $a'^{(n-1)} a'^{(n)}$ is an input sequence for $\mathcal{N}_1$ that represents a situation when output v is not observed by the environment.

An intuitive explanation of the ODC set is that two consecutive FALSE values at u suffice to keep both z and x TRUE, in particular being x independent of the value of v. Two consecutive FALSE values at a cause two consecutive FALSE values at u.

It is interesting to note that synchronous *don't care* conditions contain a *time-invariant* and a *time-varying* component. The latter may have a *transient* subcomponent, due to the initialization of the circuit, and a *periodic* subcomponent, due to some periodic behavior. The transient component may not be used for circuit optimization. It may be difficult to exploit the periodic component. Therefore, most approaches to Boolean simplification using synchronous *don't care* conditions take advantage of the time-invariant component only.

Example 9.3.23. Consider again the circuit of Figure 9.22. Let us consider the input controllability *don't care* conditions computed in Example 9.3.22. Then $(u^{(n)} + v'^{(n)})v'^{(n+1)}$ $\forall n \geq 0$ is a time-invariant *don't care* condition, while $v'^{(-3)} + v'^{(-1)}$ is a transient condition.

An example of a periodic *don't care* component can be found in a network comprising a two-input OR gate and an oscillator feeding one input of the OR gate. Hence the value at the other input of the OR gate is irrelevant at every other time point.

When considering the internals of a synchronous network, it is possible to define the corresponding satisfiability, controllability and observability *don't care* sets. The satisfiability *don't care* set can be modeled by considering, for each time point n and variable x, the term $x^{(n)} \oplus f_{x^{(n)}}$. The internal controllability and observability sets can be defined by considering them as external *don't care* sets of a subnetwork of the synchronous logic network under consideration.

The complete computation of internal controllability and observability *don't care* conditions is complex and not detailed here. When compared to combinational circuits, additional difficulties stem from the interaction of variables at different time points and from the presence of feedback connections. It is the purpose of the following sections to give the reader a flavor of the issues involved in deriving and using *don't care* conditions for synchronous network optimization, without delving into the technical details, for which we refer the interested reader to reference [6]. We shall first comment on acyclic networks and then extend our considerations to cyclic networks.

ACYCLIC NETWORKS.* The acyclic structure of the network makes it possible to express its input/output behavior at any time point n by a synchronous Boolean expression $\mathbf{f}^{(n)}$ in terms of time-labeled primary input variables. Similarly, all internal variables can be associated with a local expression in terms of time-labeled primary input variables. Thus the internal controllability *don't care* sets can be computed by network traversal by extending the methods presented in Section 8.4.1.

The notion of a *perturbed network* (Section 8.4.1) can also be extended to acyclic synchronous networks, and it is useful to define observability *don't cares* and feasible replacements of local functions. Let us consider an arbitrary vertex $v_x \in V$ and let us denote by $\mathbf{f}^x$ the behavior of the network at time n when perturbed at v_x at time $m \leq n$. Let p denote the maximum weight on a path from v_x to any primary output vertex. Since the output of the network can be affected by perturbations at different time points, we consider sequences of perturbations $\delta^{(n)}, \ldots, \delta^{(n-p)}$ and we denote the behavior of the perturbed network explicitly as $\mathbf{f}^x(\delta^{(n)}, \ldots, \delta^{(n-p)})$. Note that $\mathbf{f} = \mathbf{f}^x(0, \ldots, 0)$.

Definition 9.3.6. We call **internal observability** *don't care* conditions of variable $x^{(m)}$ at time $n \geq m$ the function:

$$\mathbf{ODC}_{x^{(m)}}^{(n)} = \mathbf{f}^x(\delta^{(n)}, \ldots, \delta^{(m+1)}, 1, \delta^{(m-1)}, \ldots, \delta^{(n-p)})$$

$$\overline{\oplus}\ \mathbf{f}^x(\delta^{(n)}, \ldots, \delta^{(m+1)}, 0, \delta^{(m-1)}, \ldots, \delta^{(n-p)}) \tag{9.7}$$

Note that in the present case $\mathbf{ODC}_{x^{(m)}}^{(n)}$ may depend on perturbations $\delta^{(l)}$ at time l, $l \neq m$.

Example 9.3.24. Let us compute the ODC sets for vertex v_y in Figure 9.22. Since $x^{(n)} = y^{(n-1)} + y'^{(n)} + u'^{(n-1)}$, then:

$$ODC_{y^{(n)}}^{(n)} = (y^{(n-1)} + y'^{(n)} + u'^{(n-1)})\ \overline{\oplus}\ (y^{(n-1)} + y^{(n)} + u'^{(n-1)}) = y^{(n-1)} + u'^{(n-1)}$$

and:

$$ODC_{y^{(n-1)}}^{(n)} = (y^{(n-1)} + y'^{(n)} + u'^{(n-1)})\ \overline{\oplus}\ (y'^{(n-1)} + y'^{(n)} + u'^{(n-1)}) = y'^{(n)} + u'^{(n-1)}$$

and thus:

$$ODC_{y^{(n)}}^{(n+1)} = y'^{(n+1)} + u'^{(n)}$$

Variable $y^{(n)}$ is never observed at the primary output of $\mathcal{N}_2$ when:

$$ODC_{y^{(n)}} = ODC_{y^{(n)}}^{(n)} ODC_{y^{(n)}}^{(n+1)} = (y^{(n-1)} + u'^{(n-1)})(y'^{(n+1)} + u'^{(n)})$$

When performing simplification of a combinational function in a synchronous logic network, the effect of replacing a local function on the network behavior can also be expressed by means of perturbations. Namely, the behavior of the network obtained by replacing the local function f_x with g_x is described by $\mathbf{f}^x(\delta^{(n)}, \dots, \delta^{(n-p)})$, where $\delta^{(l)} = f_{y^{(l)}} \oplus g_{y^{(l)}}; l = n, n-1, \dots, n-p$.

The problem of verifying the feasibility of a replacement of a local function bears similarity to multiple-vertex optimization in the combinational case. Instead of having multiple perturbations at different vertices at the same time point, we now have multiple perturbations at the same vertex at different time points.

It is possible to show [6] that sufficient conditions for network equivalence are:

$$\mathbf{e}^{(n)} \subseteq \mathbf{DC}_{ext}^{(n)} \quad \forall n \geq 0 \tag{9.8}$$

where $\mathbf{e}^{(n)} = \mathbf{f}^x(\delta^{(n)}, \dots, \delta^{(n-p)}) \oplus \mathbf{f}^x(0, \dots, 0)$. In other words, $\mathbf{e}^{(n)}$ represents the error introduced at time $l = n, n-1, \dots, n-p$, by replacing f_x by g_x as measured at the output at time n. A sufficient condition for equivalence is that the error is contained in the external *don't care* set, which represents the tolerance on the error.

In general it is not possible to transform Equation 9.8 into unilateral bounds on the perturbation δ, such as $(f_x \oplus g_x) \mathbf{1} \subseteq \mathbf{DC}_x$, where $\mathbf{DC}_x$ is a complete *local don't care* set, as in the case of single-vertex combinational logic optimization. This should not be surprising, because of the similarity to multiple-vertex optimization. Nevertheless we can always determine a set $\widetilde{\mathbf{DC}}_x$ such that a constraint of type $(f_x \oplus g_x)\mathbf{1} \subseteq \widetilde{\mathbf{DC}}_x$ is sufficient to guarantee the validity of Equation 9.8, or equivalently the validity of the replacement.

On the other hand, it is possible to derive unilateral bounds on the perturbation when all paths from v_x to the primary output have the same weight p. In this case, $\mathbf{e}^{(n)}$ depends only on $\delta^{(n-p)}$. It has been shown formally, and analogously to Theorem 8.4.2 for the combinational case, that Equation 9.8 holds if and only if the following inequality holds [6]:

$$\delta^{(n-p)}\mathbf{1} \subseteq \mathbf{ODC}_{x^{(n-p)}}^{(n)} + \mathbf{DC}_{ext}^{(n)} \quad \forall n \geq 0 \tag{9.9}$$

A network models a *pipeline* when, for each vertex, all paths from the vertex to any primary output have the same weight. Equation 9.9 shows that these *don't care* conditions fully describe the degrees of freedom for the optimization of these networks. Note also that the computation of *don't care* sets in pipelined networks is a straightforward extension of the combinational case, represented by $p = 0$ in Equation 9.9. As a result, Boolean optimization of pipelined networks is conceptually no more difficult than Boolean optimization of combinational networks.

In the general case, when considering arbitrary acyclic networks, a vertex may have multiple paths with different weights to a primary output, so that the function $\mathbf{e}^{(n)}$ has multiple dependencies upon $\delta^{(n)}, \ldots, \delta^{(n-p)}$, and the associated *don't care* conditions expressed by Equation 9.8 are correspondingly more complex.

Techniques similar to those used for computing CODC sets in combinational circuits (see Section 8.4.2) can be applied to the present situation. Indeed we search for subsets of the observability *don't care* sets that are independent of the multiple perturbations $\delta^{(n)}, \ldots, \delta^{(n-p)}$. A possible approach is to derive bounds on each perturbation, make each bound independent on the other perturbations and then consider their intersection. We refer the interested reader to reference [6] for the details of the method. We report here an example.

Example 9.3.25. Consider again the circuit of Figure 9.22 and the optimization of v_y. The conjunction of the observability *don't care* sets at times n and $n + 1$ yields $ODC_{y^{(n)}} = (y^{(n-1)} + u'^{(n-1)})(y'^{(n+1)} + u'^{(n)})$, which contains cube $u'^{(n-1)}u'^{(n)}$, which is independent of variable y.

The time-invariant controllability *don't care* set contains cube $u^{(n-1)}v'^{(n)}$. Hence the set $\widetilde{DC}_y$ contains $u'^{(n-1)}u'^{(n)} + u^{(n-1)}v'^{(n)}$ $\forall n \geq 0$. Replacing the EXNOR gate by an AND gate is equivalent to the following perturbation: $\delta^{(n)} = (u^{(n)}v^{(n)} + u'^{(n)}v'^{(n)}) \oplus (u^{(n)}v^{(n)}) = u'^{(n)}v'^{(n)} = u^{(n-1)}u'^{(n)}v'^{(n)} + u'^{(n-1)}u'^{(n)}v'^{(n)}$ $\forall n \geq 0$. The first term is included in the CDC component and the second in the ODC component of $\widetilde{DC}_y$. Hence the replacement is possible. The optimized network is shown in Figure 9.23. Note that $\widetilde{DC}_y$ can be expressed as $u@1' \, u' + u@1 \, v'$ in the shorthand notation, as used in Example 9.3.21.

CYCLIC NETWORKS.* The computation of internal controllability and observability *don't care* conditions is more complex for cyclic networks, due to the feedback connection. The simplest approach to dealing with cyclic networks is to consider the subset of internal *don't care* conditions induced by any acyclic network obtained by removing a set of feedback edges. Note that the acyclic network may be combinational or not. In the affirmative case, this simplification is equivalent to resorting to combinational *don't care* set computation techniques.

In the general case, controllability and observability conditions are induced by the feedback connection and are inherent to the cyclic nature of the network. For example, some feedback sequences may never be asserted by the network and may therefore be considered as external controllability *don't care* conditions for the acyclic component. Similarly, some values of the feedback input may never be observed at the primary outputs, thereby resulting in external observability *don't care* conditions of the feedback outputs of the acyclic component.

Iterative methods can be used for computing the *don't care* conditions related to the feedback connection. The principle is to consider at first the feedback connection fully controllable (or observable). This is equivalent to assuming a corresponding empty CDC (or ODC) set. Then, the impossible feedback sequences (or unobservable feedback sequences) are computed for the acyclic network, possibly using the external *don't care* conditions. If the corresponding CDC (or ODC) set is different than before, the process can be iterated until the set stabilizes [6]. A particular form of iterative method is used for state extraction, described later in Section 9.4.1.

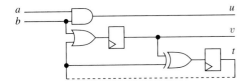

FIGURE 9.24
Network $\mathcal{N}_1$ redrawn to highlight the feedback connection.

Example 9.3.26. Consider again network $\mathcal{N}_1$ of Figure 9.22, redrawn in Figure 9.24 to highlight the feedback connection. Assume that the initialization sequence $(b^{(-4)}, b^{(-3)}, b^{(-2)}) = (1, 0, 1)$ is applied to the network. It follows that $v^{(-3)} = 1$ and $t^{(-2)} = t'^{(-3)}$. Thus, only $(0, 1)$ and $(1, 0)$ are possible sequences for $t^{(-3)}, t'^{(-2)}$.

Consider now the acyclic component of the network obtained by removing the feedback connection. We denote the dangling feedback connection as $\tilde{t}$. External (transient) *don't care* conditions on the subnetwork are $DC_{ext} = \tilde{t}^{(-2)}\tilde{t}^{(-3)} + \tilde{t}'^{(-2)}\tilde{t}'^{(-3)}$. Moreover, if we assume that $\tilde{t}'^{(n-1)}, \tilde{t}'^{(n)}$ can never be an input sequence, it follows that $t'^{(n-1)}, t'^{(n)}$ cannot be produced by the acyclic component at the feedback output $\forall n$.

Since $\tilde{t}'^{(-2)}\tilde{t}'^{(-3)} \in DC_{ext}$, $\tilde{t}'^{(n-1)}\tilde{t}'^{(n)} \; \forall n \geq -2$ is an impossible sequence. Hence $v'^{(n-1)}v'^{(n)} \; \forall n \geq -1$ is also an impossible sequence and it is part of the output controllability *don't care* set of network $\mathcal{N}_1$.

Note that similar considerations apply when we neglect the initialization sequence, but we assume that at least one register has an initial condition equal to TRUE.

IMPLICIT *DON'T CARE* CONDITIONS IN SYNCHRONOUS NETWORKS.* We stated previously that traces model best the behavior of sequential circuits. Therefore the most general representation of synchronous networks and their corresponding *don't care* conditions is in terms of relations that associate the possible input sequences with the corresponding possible output sequences. Consider the optimization of a synchronous network confined to Boolean simplification of the local function associated with a single-vertex or a single-output subnetwork. It may be necessary to resort to a relational model if we want to be able to consider all possible degrees of freedom for optimization. Unfortunately, *don't care* sets expressed as *sum of products* of time-labeled literals are functional (and not relational) representations, and they may fall short of representing all degrees of freedom for optimization.

Example 9.3.27. Consider the circuit fragment of Figure 9.25 (a). Assume that the external *don't care* set is empty, for the sake of simplicity. It can be easily verified that the inverter can be replaced by a direct connection, leading to the simpler and equivalent circuit of Figure 9.25 (b).

We try to interpret this simplification with the analysis methods learned so far. First, note that the circuit fragment has two input/output paths of unequal weight, ruling out the use of (peripheral) retiming techniques. Second, note that removing the inverter yielding variable x is equivalent to a perturbation $\delta^{(n)} = a'^{(n)} \oplus a^{(n)} = 1 \; \forall n$. This could lead us to the erroneous conclusion that $DC_{x^{(n)}} = 1 \; \forall n$ and hence that x could be replaced by a permanent TRUE or FALSE value.

Whereas the inclusion of the perturbation in a local *don't care* set is a sufficient condition for equivalence, it is by no means necessary in the case of synchronous networks. Therefore we must search for more general conditions for stating feasible transformations.

When considering single-vertex optimization, say at v_x, the most general method for verifying the feasibility of the replacement of a local function f_x by another one g_x is to equate the terminal behavior of the network in both cases. A feasible replacement must yield indistinguishable behavior at all time points (possibly excluding the external *don't care* conditions of the network).

Example 9.3.28. Consider again the circuit fragment of Figure 9.25 (a) and assume that the external *don't care* set is empty. The input/output behavior of the network is:

$$z^{(n)} = a'^{(n)} \oplus a'^{(n-1)} \quad \forall n \geq 0$$

Consider the subnetwork $\mathcal{N}_1$ shown inside a box in the figure. Any implementation of $\mathcal{N}_1$ is valid as long as the following holds:

$$x^{(n)} \oplus x^{(n-1)} = a'^{(n)} \oplus a'^{(n-1)} \quad \forall n \geq 0$$

The above equation represents the constraints on the replacement for subnetwork $\mathcal{N}_1$. Possible solutions are the following:

- $x^{(n)} = a'^{(n)} \; \forall n \geq 0$. This corresponds to the original network, shown in Figure 9.25 (a).
- $x^{(n)} = a^{(n)} \; \forall n \geq 0$. This corresponds to removing the inverter, as shown in Figure 9.25 (b). (It can be derived by noticing that the parity function is invariant to complementation of an even number of inputs.)
- $x^{(n)} = a^{(n)} \oplus a^{(n-1)} \oplus x^{(n-1)} \; \forall n \geq 0$. This solution can be derived by adding the term $\oplus x^{(n-1)}$ to both sides of the constraint equation after having complemented a'^n and $a'^{(n-1)}$. The corresponding circuit is shown in Figure 9.25 (c).

Note that the last implementation of the network introduces a feedback connection.

A few issues are important in selecting a replacement for a local function. First, feasibility must be ensured. Second, we may restrict our attention to replacements

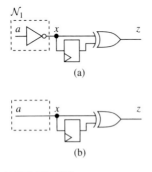

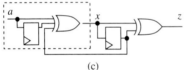

FIGURE 9.25
(a) Circuit fragment. (b) Optimized circuit fragment. (c) Other implementation of circuit fragment.

that do not introduce further cycles in the network. Third, the substitution of the local function must satisfy some optimality criteria.

A method for computing a minimum *sum of products* implementation of an acyclic replacement subnetwork is described in reference [7]. We present here the highlights of the method by elaborating on an example.

The first step is equating the terminal behavior of the original network and of the network embedding the local replacement. This yields an implicit *synchronous recurrence equation* that relates the network variables at different time points. A tabulation of the possible values of the variables leads to the specification of a *relation table* describing the possible input and output traces of the subnetwork that satisfy the equation.

Example 9.3.29. Consider again the circuit fragment of Figure 9.25 (a). The corresponding synchronous recurrence equation is:

$$x^{(n)} \oplus x^{(n-1)} = a'^{(n)} \oplus a'^{(n-1)} \quad \forall n \geq 0$$

which can be tabulated as follows:

$a^{(n)}$	$a^{(n-1)}$	$x^{(n)}$ $x^{(n-1)}$
0	0	{00, 11}
0	1	{01, 10}
1	0	{01, 10}
1	1	{00, 11}

The right part of the table shows the possible traces for x in conjunction with the input traces.

The synchronous recurrence equation, and equivalently the relation table, specifies implicitly the subnetwork and all degrees of freedom for its optimization. Hence *don't care* conditions are represented implicitly.

An implementation compatible with the relation table is a specification of $x^{(n)}$ as a function of the network inputs at different time points. Note that the relation table expresses the possible traces for x and thus the values of the same variable at different time points. This differs from the specification of a combinational Boolean relation, specifying the values of different variables at the same time. Therefore, combinational Boolean relation minimizers (as described in Section 7.6) cannot be applied *tout court* to the solution of this problem. A specific synchronous relation minimizer [33] has been developed to solve this problem. Alternatively, the solution can be sought for by representing the desired replacement by a truth table in terms of unknown coefficients [7]. Constraints on the feasible values of the coefficients can be inferred from the relation table. As a result, truth tables satisfying the constraints correspond to feasible network replacements. Among these, an optimal solution may be chosen.

Example 9.3.30. Assume that network $\mathcal{N}_1$ of Figure 9.25 (a) is replaced by a function specified by the following truth table:

$a^{(n)}$	$a^{(n-1)}$	f
0	0	f_0
0	1	f_1
1	0	f_2
1	1	f_3

We can now re-express the constraints of the relation table in terms of the coefficients. Consider traces $x^{(n)}, x^{(n-1)} = \{00, 11\}$ implying $x^{(n)} = x^{(n-1)}$, or equivalently $f^{(n)}(a^{(n)}, a^{(n-1)}) = f^{(n-1)}(a^{(n-1)}, a^{(n-2)})$. Such traces imply also that $z^{(n)}$ is TRUE, which happens in correspondence with the following input traces $a^{(n)}, a^{(n-1)}, a^{(n-2)}$: $\{000, 001, 110, 111\}$. For input trace 000, $f^{(n)}(0, 0) = f^{(n-1)}(0, 0)$ implies $f_0 = f_0$ or $(f_0' + f_0)(f_0 + f_0') = 1$. For input trace 001, $f^{(n)}(0, 0) = f^{(n-1)}(0, 1)$ implies $f_0 = f_1$ or $(f_0' + f_1)(f_0 + f_1') = 1$. Similar considerations apply to the remaining input traces $\{110, 111\}$ and to those related to output traces $\{01, 10\}$. The resulting constraints on the coefficients, excluding the tautological ones and duplications, are:

$$(f_0' + f_1)(f_0 + f_1') = 1$$

$$(f_3' + f_2)(f_3 + f_2') = 1$$

$$(f_1' + f_2')(f_1 + f_2) = 1$$

$$(f_1' + f_3')(f_1 + f_3) = 1$$

$$(f_2' + f_0')(f_2 + f_0) = 1$$

Solutions are $f_0 = 1$; $f_1 = 1$; $f_2 = 0$; $f_3 = 0$ and $f_0 = 0$; $f_1 = 0$; $f_2 = 1$; $f_3 = 1$. The first solution corresponds to selecting $x^{(n)} = a'^{(n)}$ and the second to selecting $x^{(n)} = a^{(n)}$. The second situation is preferable, because it does not require an inverter. Note that this method does not consider solutions with additional feedbacks, as shown in Figure 9.25 (c).

It is interesting to note that approaches to solving exactly the simplification problem for synchronous networks involve the solution of binate covering problems. This fact is rooted in the need for considering, even for single-vertex optimization, the values of the variable at multiple time points and therefore modeling the problem by relations rather than by functions.

9.4 IMPLICIT FINITE-STATE MACHINE TRAVERSAL METHODS

Traversing a finite-state machine means executing symbolically its transitions. If a state transition diagram representation is available, an *explicit* traversal means following all directed paths whose tail is the reset state, thus detecting all reachable states. States that are not reachable can be eliminated and considered as *don't care* conditions. If the finite-state machine is described by a synchronous logic network, a traversal means determining all possible value assignments to state variables that can be achieved, starting from the assignment corresponding to the reset state. In this case, reachable and unreachable state sets are represented *implicitly* by functions over the state variables.

Implicit methods are capable of handling circuit models with large state sets (e.g., 10^{60}). The potential size of the state space is large, because it grows exponentially with the number of registers. Consider, for example, a circuit with 64 state variables, $\{x_1, x_2, \ldots, x_{64}\}$. The condition $x_1 + x_2$ denotes all states with one of the first two state variables assigned to 1. There are $3 \cdot 2^{62} \approx 10^{19}$ states that can be represented implicitly by this compact expression.

Finite-state machine traversal is widely applicable to verification of the equivalence of two representations of sequential circuits [4, 5, 26, 34]. We limit our comments here to those applications strictly related to synthesis and to the topics described in this chapter, namely, we consider state extraction and we revisit state minimization.

9.4.1   State Extraction

The extraction of a state transition diagram from a synchronous logic network requires us to identify the state set first. If the network has n registers, there are at most 2^n possible states. In general only a small fraction are valid states, i.e., reachable from the reset state under some input sequence. Thus, finite-state machine traversal is performed to extract a valid set of states.

> **Example 9.4.1.** Consider the synchronous network of Figure 9.26 (a), with one input (x), one output (z) and two state variables (p, q). Let the reset state correspond to the state assignment $p = 0; q = 0$. There are at most four states, corresponding to the different polarity assignments of p and q. We question if all four states are reachable from the reset state. Next we shall attempt to construct a consistent state transition diagram.

We first consider an intuitive method for reachability analysis and then comment on some refinements. Let the network under consideration have n_i inputs and n state

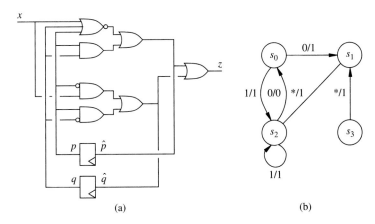

FIGURE 9.26
(a) Synchronous network. (b) Extracted state transition diagram.

variables. Let $\mathbf{f} : \mathbf{B}^{n+n_i} \to \mathbf{B}^n$ define the state transition function (denoted elsewhere as δ) which defines the next-state variables in terms of the present-state variables and primary inputs. Let r_0 represent the reset state in terms of the state variables.

The states directly reachable from r_0 are the image of r_0 under $\mathbf{f}$ and can be expressed implicitly as a function of the state variables. Similarly, the states directly reachable from any state set represented implicitly by r_k are the image of r_k under $\mathbf{f}$. By defining $r_{k+1} : k \geq 0$ to be the union of r_k and the image of r_k under $\mathbf{f}$, we specify an iterative method for implicit reachability computation. The iteration terminates when a fixed point is reached, i.e., when $r_k = r_{k+1}$ for some value of $k = k^*$. It terminates in a finite number of steps, because the functions $r_k, k = 0, 1, \ldots, k^*$, denote monotonically increasing state sets and the number of states is finite. The expression for r_{k^*} encapsulates all reachable states, whereas the complement of r_{k^*} denotes unreachable states and represents *don't care* conditions that can be used to optimize the circuit.

Example 9.4.2. Consider again the synchronous network of Figure 9.26 (a), whose reset state is represented by $(p = 0; q = 0)$. Thus $r_0 = p'q'$. The state transition function is $\mathbf{f} = \begin{bmatrix} f^1 \\ f^2 \end{bmatrix}$, where $f^1 = x'p'q' + pq$ and $f^2 = xp' + pq'$. The image of $p'q'$ under $\mathbf{f}$ can be derived intuitively by considering that when $(p = 0; q = 0)$, the function $\mathbf{f}$ reduces to $\begin{bmatrix} x' \\ x \end{bmatrix}$, whose range is represented by vectors $[01]^T$ and $[10]^T$. (The formal computation is shown in Example 9.4.4.) Equivalently, the states reachable from the reset state are encoded as $(p = 1; q = 0)$ and $(p = 0; q = 1)$. Therefore $r_1 = p'q' + pq' + p'q = p' + q'$.

The image of r_1 under $\mathbf{f}$ is represented by vectors $[00]^T$, $[01]^T$ and $[10]^T$. (The detailed computation is shown also in Example 9.4.4.) Thus the reachable states can be encoded as $(p = 0; q = 0)$, $(p = 1; q = 0)$ and $(p = 0; q = 1)$. Since $r_2 = p' + q' = r_1$, the iteration has converged. Thus $p' + q'$ represents implicitly all reachable states. Each state can be associated with a minterm of $p' + q'$. Namely s_0 corresponds to $p'q'$, s_1 to pq' and s_2 to $p'q$. The state corresponding to pq is unreachable and thus pq can be considered as a *don't care* condition.

Once the state set has been identified, the state transition diagram can be constructed by determining the transitions which correspond to the edges of the graph. The transition edges and the qualifying inputs on the edges can be derived by computing the inverse image of the head state representation under $\mathbf{f}$.

Example 9.4.3. Transitions into s_0 corresponding to $p'q'$ are identified by those patterns that make $\mathbf{f} = [00]^T$. Equivalently, they satisfy $(f^1 = 0)(f^2 = 0) = (x'p'q' + pq)'(xp' + pq')' = x'p'q$. Hence there is a transition into state s_0 from state s_2 (encoded as $p'q$) under input x'. The corresponding primary output can be derived by evaluating the network in a straightforward way. In this case, input $x = 0$ and state $p'q$ yield $z = 0$.

All other transitions can be extracted in a similar fashion. The extracted state transition diagram is shown in Figure 9.26 (b).

The image computation can be performed efficiently using ROBDDs [4, 5, 26, 34], as in the case of controllability *don't care* computation of combinational circuits (see Section 8.4.1). Some variations and improvements on the aforementioned scheme have been proposed.

Reachability analysis can be performed using the state transition relation, which is the characteristic equation of the state transition function and can be efficiently represented as an ROBDD. Namely, let $\mathbf{y} = \mathbf{f}(\mathbf{i}, \mathbf{x})$ denote the state transition function. Then $\chi(\mathbf{i}, \mathbf{x}, \mathbf{y}) = 1$ is the corresponding relation, linking possible triples $(\mathbf{i}, \mathbf{x}, \mathbf{y})$ of inputs, present states and next states. Let $r_k(\mathbf{x})$ be a function over the state variables $\mathbf{x}$ denoting implicitly a state set. The image of r_k under the state transition function can be computed as $\mathcal{S}_{\mathbf{i}, \mathbf{x}}(\chi(\mathbf{i}, \mathbf{x}, \mathbf{y}) \cdot r_k(\mathbf{x}))$ [26, 34].

The inverse image of a state set under the state transition function is also of interest, because it allows us to derive all states that have transitions into a given state subset. The inverse image can also be computed with the same ease. Namely, let $r_k(\mathbf{y})$ be a function over the state variables $\mathbf{y}$ denoting implicitly a (next) state set. Then, the inverse image of r_k under the state transition function can be computed as $\mathcal{S}_{\mathbf{i}, \mathbf{y}}(\chi(\mathbf{i}, \mathbf{x}, \mathbf{y}) \cdot r_k(\mathbf{y}))$. With this formalism, it has been possible to develop algorithms for finite-state machine traversal in both directions, applicable to sequential circuits with potential large state sets (e.g., 10^{60}).

Example 9.4.4. Consider again Example 9.4.1. Let $\hat{p}$ and $\hat{q}$ denote the next-state variables. The state transition relation is represented by the characteristic equation:

$$\chi(x, p, q, \hat{p}, \hat{q}) = (\hat{p} \ \overline{\oplus} \ (x'p'q' + pq)) \ (\hat{q} \ \overline{\oplus} \ (xp' + pq')) = 1$$

By expanding the expression, we get:

$$\chi(x, p, q, \hat{p}, \hat{q}) = \hat{p}\hat{q}'(x'p'q' + pq) + \hat{p}'\hat{q}(xp' + pq') + \hat{p}'\hat{q}'(x'p'q)$$

The states reachable from $r_0 = p'q'$ are those denoted by:

$$\mathcal{S}_{x,p,q}(\chi(x, p, q, \hat{p}, \hat{q}) \ p'q') \ = \ \mathcal{S}_{x,p,q}(\hat{p}\hat{q}'(x'p'q') + \hat{p}'\hat{q}(xp'q')) \ = \ \hat{p}\hat{q}' + \hat{p}'\hat{q}$$

corresponding to s_1 and s_2. Similarly, the states reachable from $r_1 = p' + q'$ are those denoted by:

$$\mathcal{S}_{x,p,q}(\chi(x, p, q, \hat{p}, \hat{q}) \ (p' + q')) = \mathcal{S}_{x,p,q}(\hat{p}\hat{q}'(x'p'q') + \hat{p}'\hat{q}(xp' + pq') + \hat{p}'\hat{q}'(x'p'q))$$

$$= \hat{p}'\hat{q}' + \hat{p}'\hat{q} + \hat{p}\hat{q}'$$

representing states $\{s_0, s_1, s_2\}$.

The inverse image of $r_0 = \hat{p}'\hat{q}'$ can be computed as:

$$\mathcal{S}_{x,\hat{p},\hat{q}}(\chi(x, p, q, \hat{p}, \hat{q})) \ \hat{p}'\hat{q}' \ = \ \mathcal{S}_{x,\hat{p},\hat{q}}(\hat{p}'\hat{q}'(x'p'q)) \ = \ p'q$$

Therefore state s_0 can be reached from state s_2.

Example 9.4.5. Consider again the control unit of the complete differential equation integrator, described in Example 4.7.6 as a synchronous logic network. This description uses a *1-hot* encoding and it entails 46 literals and 6 registers.

The corresponding state transition diagram, extracted from the network, is shown in Figure 9.27. (The output signals are omitted in the figure.) There are six states and

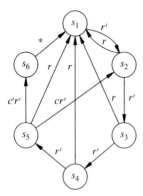

FIGURE 9.27
State transition diagram for the control unit of the complete differential equation integrator.

state s_1 is the reset state. No state can be eliminated by state minimization. State encoding requires 3 bits. The encoding produced by program NOVA, called by program SIS, yields the network reported in Example 8.2.10. Such a network entails 93 literals and 6 registers. After logic optimization, the network can be reduced to 48 literals and 3 registers.

9.4.2 Implicit State Minimization*

State traversal methods can be used for finite-state machine state minimization using an implicit model, which is appealing because it can handle much larger state sets than other explicit methods do. We consider here state minimization of completely specified finite-state machines, in particular the derivation of the equivalence classes as suggested by Lin *et al.* [26]. Classes of equivalent states are determined by considering pairwise state equivalence by exploiting the transitivity of the equivalence relation.

Given a sequential logic network representation of a finite-state machine, we duplicate the network to obtain a *product machine* representation, as sketched in Figure 9.28. The states of the product machine are identified by the state variables. Each state of the product machine corresponds to a pair of states in the original finite-state machine. As a result, all states of the product machine whose output is TRUE for all inputs denote state pairs in the original finite-state machine whose outputs match for all inputs. This is a necessary, but not sufficient, condition for equivalence. So, we call these state pairs *candidate equivalent pairs.*

Candidate state pairs are represented implicitly by a characteristic function. Since state pairs correspond to a single state in the product machine, the inverse image of any state of the product machine under the transition function yields the state pairs that are mapped into the state under consideration. As a result, it is possible to check if a state pair has next states that are candidate equivalent pairs for all inputs. This can be performed by multiplying the characteristic function by the inverse image of the candidate set under all input conditions.

An iterative procedure can thus be defined that updates the characteristic function by its product with the inverse image. Equivalently, the state pair information stored in the characteristic function is refined by requiring that such pairs have next states that are also represented by the characteristic function. It is possible to show [26]

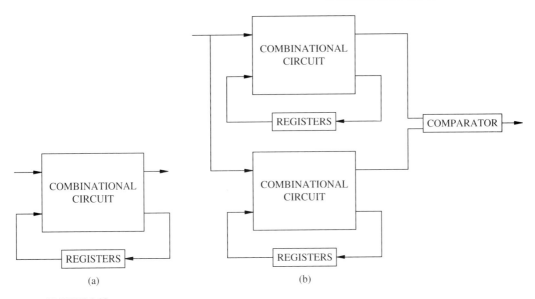

FIGURE 9.28
(a) Original finite-state machine. (b) Product finite-state machine.

that this iteration converges in a finite number of steps to a characteristic function representing implicitly all equivalent pairs.

It is interesting to compare this method with the classical methods described in Section 9.2.1. Classical methods perform iterative refinements of state partitions. They require explicit representations of the states, and thus they are limited by the state set cardinality. The implicit method performs an iterative refinement of a set of state pairs that are described implicitly in terms of state variables of the product machine (which are just twice as many as those of the original machine). Since implicit methods leverage efficient set manipulation based on ROBDD representations and operations, they can handle problems of much larger size than classical methods do. We refer the interested reader to reference [26] for details.

9.5 TESTABILITY CONSIDERATIONS FOR SYNCHRONOUS CIRCUITS

The synthesis of testable sequential circuits is a broad area of research. We limit ourselves to a few comments, and we refer the reader to specialized books on the subject [17, 29]. To be precise, we use single *stuck-at* fault models and we consider fully testable those sequential circuits where all faults can be detected.

Different testing strategies are used for synchronous sequential circuits. *Scan* techniques were introduced and divulged by IBM Corporation and are used today with different flavors. In this design and test methodology, registers can be configured to be linked in a chain so that data can be introduced and/or observed directly while testing the circuit. Therefore, scan methods provide full controllability and observability of the registers. As a consequence, the sequential testing problem reduces to that of the corresponding combinational component.

The overhead of using scan techniques varies according to the circuit technology and clocking policy. In some cases, a considerable area penalty is associated with the replacement of regular registers with those supporting scan. To alleviate this problem, partial scan techniques have been proposed where only a fraction of the registers are in the scan chain. The controllability and observability of the remaining registers is either achieved by exploiting features of the combinational logic component or compromised partially.

Testing sequential circuits without scan registers may require long sequences of test vectors. The circuit has to be driven first to a state that excites the fault, and then the effects of the fault have to be propagated to the primary outputs. This is caused by the lack of direct controllability and observability of the state registers.

Recent approaches to design for testability of sequential logic circuits have explored the possibility of using the degrees of freedom in logic synthesis to make the circuit fully testable with relatively short sequences of test vectors. The degrees of freedom are, for example, *don't care* conditions and/or the choice of state encoding. The starting points for these methods are either state transition diagram [12] or network [17] representations of sequential circuits. The advantages of these methods are that both eliminate the area overhead due to the scan registers and that they provide support for faster testing by avoiding to load and unload the scan chain.

There is a wide spectrum of methods for designing fully testable sequential circuits not based on scan. On one side of the spectrum, there are techniques that constrain the implementation of the sequential circuit. On the other side, we find synthesis techniques that eliminate untestable faults by using optimization methods with appropriate *don't care* sets. Both approaches have advantages and disadvantages. The first family of methods requires simpler synthesis tasks but possibly an overhead in terms of area. Conversely, the second group requires computationally expensive (or prohibitive) synthesis techniques but no implementation overhead [14]. We shall consider two representative examples of these approaches.

We explain first a method for achieving full testability using an appropriate state encoding and requiring a specific implementation style based on partitioning (and possibly duplicating) a portion of the combinational component [12]. This technique is applicable to sequential circuits modeled as both Moore-style and Mealy-style finite-state machines. We comment on the first case only, for the sake of simplicity.

Consider the Moore-style finite-state machine shown in Figure 9.29 (a). Assume first that the circuit is partitioned into registers, a combinational output-logic block and a combinational next-state logic block (feeding the registers). To detect a fault in the output-logic circuit, the finite-state machine has to be driven by a sequence of vectors into the state that propagates the effect of the fault to the primary outputs. To detect a fault in the next-state logic block, we must find first a state and an input vector which propagate the effects of the fault to the state registers, which would represent a faulty state instead of the corresponding non-faulty next state. We must then be able to distinguish the faulty state from the non-faulty state by observing the primary outputs. In general, this problem is complex unless the state variables are

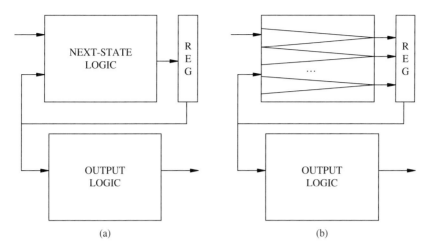

FIGURE 9.29
(a) Original Moore-style finite-state machine. (b) Moore-style finite-state machine with partitioned next-state logic.

directly observable at the outputs. Moreover, if the faulty state is equivalent to the non-faulty state, the fault is not detectable.

A fully testable implementation of a Moore-style finite-state machine can be achieved as follows. Assuming that n registers are used, all possible 2^n states must be made reachable from the reset state, possibly by adding transitions to the state transition diagram. The states are encoded so that two state codes have distance 2 (differ in 2 bits) when they assert the same output. Then the next-state logic block is further partitioned into independent combinational circuits each feeding a state register, as shown in Figure 9.29 (b). The combinational circuits are made prime and irredundant using the techniques of Section 8.5. As a result, any single fault in the next-state logic will affect one state variable and its effect will be observable at the primary outputs. Thus, the testing method requires just constructing sequences of input test vectors that set the machine in the desired state. This task can be done by reachability analysis.

We consider next a synthesis method to achieve testable sequential circuits that relies on optimization techniques. For this purpose, we need to characterize faults in sequential circuits.

A fundamental problem in sequential synthesis for testability is relating faults to the sequential behavior of the circuit. Untestable faults can be divided into two categories: combinational and sequential faults. Combinational untestable faults are those that are untestable in the combinational component of the circuit, even if the registers were fully observable and controllable. They can be detected by using combinational logic optimization methods and removed by making the combinational component prime and irredundant. (See Section 8.5.) Sequential untestable faults are faults in the combinational component of the circuit that cannot be tested because of the sequential behavior of the circuit. Sequential untestable faults can be grouped in the following three classes:

- *Equivalent-sequential* untestable faults cause the interchange/creation of equivalent states.
- *Invalid-sequential* untestable faults cause transitions from invalid states corresponding to unused state codes.
- *Isomorphic-sequential* untestable faults cause the circuit to behave as modeled by a different, but isomorphic, state transition diagram, with a different state encoding.

Devadas *et al.* [13, 17] showed that these are all possible types of untestable sequential faults.

> **Example 9.5.1.** This example of untestable faults is due to Devadas *et al.* [13]. Consider the state transition diagram of Figure 9.30 (a), whose network implementation is shown in Figure 9.30 (b). The circuit has one primary input (i), one primary output (o) and five states encoded by variables p_1, p_2, p_3. The states encoded as 010 and 110 are equivalent.
>
> Consider a *stuck-at*-0 fault on input $w1$, as shown in Figure 9.30 (b). The fault changes the state transition diagram to that of Figure 9.30 (c). The corrupted transition is shown by a dotted edge in the figure. Since states 010 and 110 are equivalent, the fault causes an interchange of equivalent states and so is an equivalent-sequential untestable fault.
>
> Consider next a *stuck-at*-1 fault on input $w2$. The fault changes the state transition diagram to that of Figure 9.30 (d). The fault creates an extra state, encoded as 111, that was originally invalid (unreachable), and that is equivalent to state 110. This is again an example of an equivalent-sequential untestable fault.
>
> An example of an invalid-sequential fault is a fault that affects the circuit when in an invalid state, e.g., 101. An example of an isomorphic-sequential untestable fault is shown in Figure 9.30 (e), where the faulty machine represents an equivalent machine with a different encoding. Namely the states with codes 000 and 001 have been swapped.

Designing a fully testable sequential circuit involves checking for and removing untestable sequential faults as well as removing combinational untestable faults. Devadas *et al.* [13] proposed a method that uses three major steps: state minimization, state encoding and combinational logic optimization with an appropriate *don't care* set that captures the interplay of equivalent and invalid states. A special state encoding is used that guarantees that no single stuck-at fault can induce isomorphic sequential untestable faults. *Don't care* set extraction and combinational logic optimization are iterated, and the overall method is guaranteed to converge to a fully testable sequential circuit. We refer the reader to references [13] and [17] for the details.

9.6 PERSPECTIVES

Synthesis and optimization of sequential synchronous circuits has been a research playground for a few decades. There are a wealth of techniques for optimizing state-based representations of finite-state machines, including algorithms for decomposition,

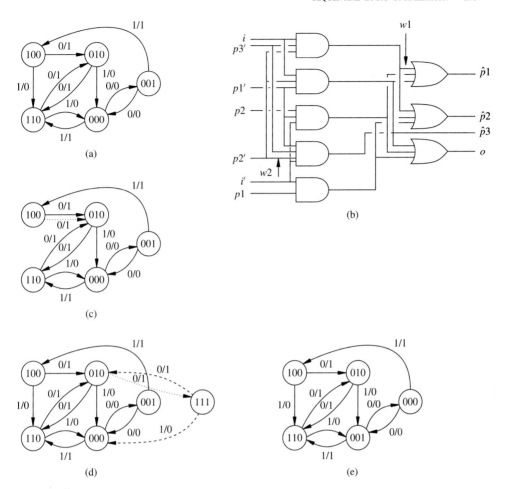

FIGURE 9.30

(a) Original state transition diagram. (b) Network implementation of the combinational component. (c) State transition diagram of the faulty circuit due to an equivalent-sequential untestable stuck-at-0 fault at $w1$. (d) State transition diagram of the faulty circuit due to an equivalent-sequential untestable stuck-at-1 fault at $w2$. (e) State transition diagram of a faulty circuit with an isomorphic-sequential untestable fault.

state minimization and assignment. Most classical methods are not practical for circuits of usual size. Heuristic algorithms have been shown useful, but the quality of their results is not well assessed yet. In particular, when considering the state minimization problem for incompletely specified finite-state machines, we can solve small or medium-sized problems and we cannot compare heuristic to optimum solutions. This should be contrasted to two-level combinational logic minimization, where approximate solutions are known to be close in quality to the exact solutions in many instances. One reason for this discrepancy is the harder underlying covering problem (i.e., binate versus unate). When considering state assignment for multiple-level logic implementations, the quality of the results is still often unpredictable, because of the inability of current algorithms to forecast precisely the effects of the choice of the codes on the area and performance of the resulting circuits.

Sequential optimization methods using network models evolved from retiming and multiple-level combinational logic synthesis. Once again, the flexibility in implementing multiple-level circuits allows us to develop a rich set of circuit transformations but makes it hard to devise algorithms yielding optimum implementations. Thus, most network-based optimization algorithms are heuristic in nature, with the exception of retiming, which, on the other hand, exploits only one degree of freedom (i.e., register positioning) for circuit optimization.

Network transformations affect the encoding of the states of the circuit being optimized. Whereas state transition diagrams can be easily extracted from network models, the implication of network transformations on the corresponding properties of the state transition diagrams are not yet fully understood and exploited. To date, most sequential circuit optimization programs use different modeling paradigms (e.g., state transition diagrams or networks) without exploiting the synergism between the two representations. Even though these programs are used routinely and successfully for digital design, we believe that this field still has many open problems for further research and that progress in sequential logic synthesis and optimization will lead to even more powerful design tools in the years to come.

9.7 REFERENCES

Finite-state machine optimization using state models has been described in several classical textbooks, e.g., Hartmanis and Stearns' [20], Kohavi's [23], Hill and Peterson's [21] and McCluskey's [29]. Hopcroft's algorithm for state minimization was first presented in reference [19], but a generalization of the method is described in detail in reference [1].

The symbolic approach to the state encoding problem for two-level circuits was first presented in reference [8] and then extended in references [9] and [35]. State encoding for multiple-level circuits was addressed by Devadas *et al.* [2, 11] and perfected by Du *et al.* [16]. Malik *et al.* [27] developed a solution to an input encoding problem that optimizes the extraction of common subexpressions and that can be used for state encoding of multiple-level circuits.

Recent work on finite-state machine optimization, with particular emphasis on encoding and decomposition, is reported in a recent monograph on sequential synthesis [2]. Machine optimization using *don't care* sequences was first postulated by Paull and Unger [36]. Kim and Newborn [22] described a state minimization method using input *don't care* sequences. The technique was perfected later by Rho *et al.* [30]. Symbolic relations were introduced by Lin and Somenzi [25].

Retiming was proposed by Leiserson and Saxe in the early 1980s. A complete and accurate description was reported in a later paper [24]. Several extensions to retiming have been proposed [10, 15, 28]. An analysis of *don't care* conditions for sequential circuits modeled with state transition graphs is reported in reference [2], while specifications of *don't care* sets using network models are reviewed in reference [6]. Recently, Damiani and De Micheli [7] investigated modeling and optimization of sequential circuits using synchronous recurrence equations.

The implicit finite-state machine traversal problem was tackled by Coudert *et al.* [5] and other authors [4, 26, 34], with the goal of verifying equivalence between sequential circuit models. The ramifications in the synthesis field have been several, including the efficient derivation of controllability *don't care* conditions for both combinational and sequential circuits. Devadas *et al.* [12–14] investigated several aspects of design for testability of sequential circuits. A recent book describing the overall problem is reference [17].

1. A. Aho, J. Hopcroft and J. Ullman, *The Design and Analysis of Computer Algorithms*, Addison-Wesley, Reading, MA, 1974.

2. P. Ashar, S. Devadas and A. Newton, *Sequential Logic Synthesis*, Kluwer Academic Publishers, Boston, MA, 1992.

3. K. Bartlett, G. Borriello and S. Raju, "Timing Optimization of Multiphase Sequential Logic," *IEEE Transactions on CAD/ICAS*, Vol. CAD-10, No. 1, pp. 51–62, January 1991.

4. J. Burch, E. Clarke, E. McMillan and D. Dill, "Sequential Circuit Verification Using Symbolic Model Checking," *DAC, Proceedings of the Design Automation Conference*, pp. 46–51, 1990.

5. O. Coudert, C. Berthet and J. C. Madre, "Verification of Sequential Machines based on Symbolic Execution," in J. Sifakis, Editor, *Lecture Notes in Computer Science*, Springer-Verlag, Berlin, Germany, 1990.

6. M. Damiani and G. De Micheli, "*Don't care* Set Specifications in Combinational and Synchronous Logic Circuits," *IEEE Transactions on CAD/ICAS*, Vol. CAD-12, pp. 365–388, March 1993.

7. M. Damiani and G. De Micheli, "Recurrence Equations and the Optimization of Synchronous Logic Circuits," *DAC, Proceedings of the Design Automation Conference*, pp. 556–561, 1992.

8. G. De Micheli, R. Brayton and A. Sangiovanni-Vincentelli, "Optimal State Assignment for Finite State Machines," *IEEE Transactions on CAD/ICAS*, Vol. CAD-4, No. 3, pp. 269–284, July 1985.

9. G. De Micheli, "Symbolic Design of Combinational and Sequential Logic Circuits Implemented by Two-level Logic Macros," *IEEE Transactions on CAD/ICAS*, Vol. CAD-5, No. 4, pp. 597–616, October 1986.

10. G. De Micheli, "Synchronous Logic Synthesis: Algorithms for Cycle-Time Minimization," *IEEE Transactions on CAD/ICAS*, Vol. 10, No. 1, pp. 63–73, January 1991.

11. S. Devadas, T. Ma, R. Newton and A. Sangiovanni-Vincentelli, "MUSTANG: State Assignment of Finite-State Machines Targeting Multi-level Logic Implementations," *IEEE Transactions on CAD/ICAS*, Vol. CAD-7, No. 12, pp. 1290–1300, December 1988.

12. S. Devadas, T. Ma, R. Newton and A. Sangiovanni-Vincentelli, "Synthesis and Optimization Procedures for Fully and Easily Testable Sequential Machines," *IEEE Transactions on CAD/ICAS*, Vol. 8, No. 10, pp. 1100–1007, October 1989.

13. S. Devadas, T. Ma, R. Newton and A. Sangiovanni-Vincentelli, "Irredundant Sequential Machines Via Optimal Logic Synthesis," *IEEE Transactions on CAD/ICAS*, Vol. CAD-9, No. 1, pp. 8–18, January 1990.

14. S. Devadas and K. Keutzer, "A Unified Approach to the Synthesis of Fully Testable Sequential Machines," *IEEE Transactions on CAD/ICAS*, Vol. CAD-10, No. 1, pp. 39–50, January 1991.

15. S. Dey, M. Potkonjak and S. Rothweiler, "Performance Optimization of Sequential Circuits by Eliminating Retiming Bottlenecks," *ICCAD, Proceedings of the International Conference on Computer Aided Design*, pp. 504–509, 1992.

16. X. Du, G. Hachtel, B. Lin and R. Newton, "MUSE: A MUltilevel Symbolic Encoding Algorithm for State Assignment," *IEEE Transactions on CAD/ICAS*, Vol. CAD-10, No. 4, pp. 28–38, January 1991.

17. A. Ghosh, S. Devadas and A. Newton, *Sequential Logic Testing and Verification*, Kluwer Academic Publishers, Boston, MA, 1992.

18. A. Grasselli and F. Luccio, "A Method for Minimizing the Number of Internal States in Incompletely Specified Sequential Networks," *IEEE Transactions on Electronic Computers*, Vol. EC-14, pp. 350–259, June 1965.

19. J. Hopcroft, "An $n \log n$ Algorithm for Minimizing States in a Finite Automaton," in Z. Kohavi, (Editor) *Theory of Machines and Computations*, Academic Press, New York, pp. 189–198, 1971.

20. J. Hartmanis and R. Stearns, *Algebraic Structure Theory of Sequential Machines*, Prentice-Hall, Englewood Cliffs, NJ, 1966.

21. F. Hill and G. Peterson, *Switching Theory and Logical Design*, Wiley, New York, 1981.

22. J. Kim and M. Newborn, "The Simplification of Sequential Machines with Input Restrictions," *IEEE Transactions on Computers*, Vol. C-21, pp. 1440–1443, December 1972.

23. Z. Kohavi, *Switching and Finite Automata Theory*, McGraw-Hill, New York, 1978.

24. C. Leiserson and J. Saxe, "Retiming Synchronous Circuitry," *Algorithmica*, Vol. 6, pp. 5–35, 1991.

25. B. Lin and F. Somenzi, "Minimization of Symbolic Relations," *ICCAD, Proceedings of the International Conference on Computer Aided Design*, pp. 88–91, 1990.

26. B. Lin, H. Touati and A. Newton, "Don't Care Minimization of Multi-Level Sequential Logic Networks," *ICCAD, Proceedings of the International Conference on Computer Aided Design*, pp. 414–417, 1990.

27. S. Malik, L. Lavagno, R. Brayton and A. Sangiovanni-Vincentelli, "Symbolic Minimization of Multiple-level Logic and the Input Encoding Problem," *IEEE Transactions on CAD/ICAS*, Vol. CAD-11, No. 7, pp. 825–843, July 1992.

28. S. Malik, E. Sentovich, R. Brayton and A. Sangiovanni-Vincentelli, "Retiming and Resynthesis: Optimizing Sequential Networks with Combinational Techniques," *IEEE Transactions on CAD/ICAS*, Vol. CAD-10, No. 1, pp. 74–84, January 1991.

29. E. McCluskey, *Logic Design Principles*, Prentice-Hall, Englewood Cliffs, NJ, 1986.

30. J. Rho, G. Hachtel and F. Somenzi, "Don't Care Sequences and the Optimization of Interacting Finite-State Machines," *ICCAD, Proceedings of the International Conference on Computer Aided Design*, pp. 418–421, 1991.

31. G. Saucier, M. Crastes de Paulet and P. Sicard, "ASYL: A Rule-based System for Controller Synthesis," *IEEE Transactions on CAD/ICAS*, Vol. CAD-6, No. 6, pp. 1088–1097, November 1987.

32. N. Shenoy, R. Brayton and A. Sangiovanni-Vincentelli, "Retiming of Circuits with Single Phase Transparent Latches," *ICCD, Proceedings of the International Conference on Computer Design*, pp. 86–89, 1991.

33. V. Singhal, Y. Watanabe and R. Brayton, "Heuristic Minimization for Synchronous Relations," *ICCD, Proceedings of the International Conference on Computer Design*, pp. 428–433, 1993.

34. H. Touati, H. Savoy, B. Lin, R. Brayton and A. Sangiovanni-Vincentelli, "Implicit State Enumeration of Finite State Machines using BDDs," *ICCAD, Proceedings of the International Conference on Computer Aided Design*, pp. 130–133, 1990.

35. T. Villa and A. Sangiovanni-Vincentelli, "NOVA: State Assignment for Finite State Machines for Optimal Two-level Logic Implementation," *IEEE Transactions on CAD/ICAS*, Vol. CAD-9, No. 9, pp. 905–924, September 1990.

36. S. Unger, *Asynchronous Sequential Switching Circuits*, Wiley, New York, 1972.

9.8 PROBLEMS

1. Consider the state table of Example 9.2.3. Derive a completely specified cover by replacing the *don't care* entries by 0s. Minimize the machine using the standard and Hopcroft's algorithms. Repeat the exercise with the *don't care* entries replaced by 0 and 1.

2. Consider the state table of Example 9.2.3. Derive a minimum symbolic cover and the corresponding encoding constraints. Then compute a feasible encoding. Can you reduce the encoding length by using constraints derived from a non-minimum cover? Show possible product term/encoding length trade-off.

3. Consider the state encoding problem specified by two matrices $\mathbf{A}$ and $\mathbf{B}$ derived by symbolic minimization. Assume that matrix $\mathbf{B}$ specifies only covering constraints (i.e., exclude disjunctive relations). Let S be the state set. Prove the following. A necessary and sufficient condition for the existence of an encoding of S satisfying both the constraints specified by $\mathbf{A}$ and $\mathbf{B}$ is that for each triple of states $\{r, s, t\} \subseteq S$ such that $b_{rs} = 1$ and $b_{st} = 1$ there exists no row k of $\mathbf{A}$ such that $a_{kr} = 1, a_{ks} = 0, a_{kt} = 1$.

4. Consider the network of Figure 9.8. Draw the weighted graph modeling the search for a legal retiming with cycle-time of 22 units using the Bellman-Ford method. Compute the retiming and draw the retimed network graph.

5. Suppose you want to constrain the maximum number of registers on a path while doing retiming. How do you incorporate this requirement in the retiming formalism? Would the complexity of the algorithm be affected? As an example, assume that you require at most one register on the path (v_d, v_h) in the network of Figure 9.8. What would the

minimum cycle-time be? Show graphically the constraints of the Bellman-Ford algorithm in correspondence with the minimum cycle-time as well as the retimed network graph.

6. Give examples of synchronous substitution and elimination as applied to the network of Figure 9.7.

7. Show that for a pipelined network, for any single perturbation at any vertex v_x representing the replacement of a local function f_x by g_x, necessary and sufficient conditions for the feasibility of the replacement are:

$$\delta^{(n-p)}\mathbf{1} \subseteq \mathbf{ODC}^{(n)}_{x^{(n-p)}} + \mathbf{DC}^{(n)}_{ext} \quad \forall n \geq 0 \tag{9.10}$$

for a suitable value of p.

8. Consider the acyclic synchronous logic network obtained from that of Figure 9.7 by cutting the loops and using additional input variables for the dangling connections. Determine the observability *don't care* sets at all internal and input vertices.

CHAPTER

10

CELL-LIBRARY BINDING

Cum legere non possis quantum habueris, satis est habere quantum legas.
Since you cannot read all the books which you may possess,
it is enough to possess only as many books as you can read.

Seneca. Epistulae ad Lucilium.

10.1 INTRODUCTION

Cell-library binding is the task of transforming an unbound logic network into a bound network, i.e., into an interconnection of components that are instances of elements of a given library. This step is very important for standard-cell and array-based semicustom circuit design, because it provides a complete structural representation of the logic circuit which serves as an interface to physical design tools. Library binding allows us to retarget logic designs to different technologies and implementation styles. Hence it is of crucial importance for updating and customizing circuit designs.

Library binding is often called *technology mapping*. The origin of this term is due to the early applications of semicustom circuits, which re-implemented circuits, originally designed in TTL SSI bipolar technologies, in LSI CMOS technologies. Circuit technological parameters are fully represented by the characterization of a library in terms of the area and speed parameters of each cell. Therefore we prefer the name *library binding*, because the essential task is to relate the circuit representation to that of the cell library and to find a cell interconnection.

The library contains the set of logic primitives that are available in the desired design style. Hence the binding process must exploit the features of such a library in the search for the best possible implementation. Optimization of area and/or delay, as well as testability enhancement, is always associated with the binding process. We shall show that the optimization tasks are difficult, because they entail the solution of intractable problems.

Practical approaches to library binding can be classified into two major groups: *heuristic* algorithms and *rule-based* approaches. In this chapter we consider both methods in detail and highlight their advantages and limitations. Whereas most binding algorithms are limited to single-output combinational cells, rule-based approaches can handle arbitrarily complex libraries, including multiple-output, sequential and interface elements (e.g., Schmitt triggers and three-state drivers). The drawbacks of the latter methods are the creation and maintenance of the set of rules and the speed of execution. Most commercial tools use a combination of algorithms and rules in library binding to leverage the advantages of both.

Even though digital circuits are often sequential and hierarchical in nature, the most studied binding problems deal with their combinational components, because the choice of implementation of registers, input/output circuits and drivers in a given library is often straightforward and binding is done by direct replacement. Therefore we consider here the library binding problem for multiple-level combinational circuits. Two-level logic representations are decomposed into multiple-level networks before library binding unless they are implemented with a specific macro-cell style (e.g., PLAs).

After having formulated and analyzed the library binding problem in Section 10.2, we describe algorithms for standard libraries of combinational gates in Section 10.3. Next, in Section 10.4, we consider algorithms for specific design styles, such as *field-programmable gate arrays* (FPGAs), where the library can be defined implicitly, instead of by enumeration. We describe rule-based library binding in Section 10.5 and we compare it to the algorithmic approach.

10.2 PROBLEM FORMULATION AND ANALYSIS

A *cell library* is a set of primitive logic gates, including combinational, sequential and interface (e.g., driver) elements. Each element is characterized by its function, terminals and some macroscopic parameters such as area, delay and capacitive load. In the case of standard-cell libraries, a cell layout is associated with each element. For array-based design, information is provided (to the physical design tools) on how to implement the cell with the pre-diffused patterns.

We are concerned in this chapter with the logic abstraction level of cell libraries. We also restrict our attention to libraries of combinational single-output cells. Even though this assumption may seem restrictive, practical approaches to library binding involve several techniques. Specific multiple-output functions (e.g., adders, encoders, etc.), as well as registers, may be identified, and bound to the corresponding cells, using simple replacement rules. For the sake of simplicity, we restrict our attention to non-hierarchical combinational logic networks.

We assume that the library is a set of cells, each one characterized by:

- a single-output combinational logic function;
- an area cost;
- the input/output propagation delays (that are generally specified as a function of the load, or fanout, for rising/falling transitions at the output, or by worst-case values).

Binding for other design styles, such as look-up-table-based FPGAs where the library is best characterized otherwise, is presented in Section 10.4.

Let us consider a combinational logic network that may have been optimized by means of the algorithms described in Chapter 8. The library binding problem entails finding an equivalent logic network whose internal modules are instances of library cells. Note that cell sharing is not possible at the logic level, as done for resources at the architectural level, because cells evaluate logic functions concurrently.

It is usual to search for a binding that minimizes the area cost (possibly under some delay constraints) or the maximum delay (possibly under an area constraint). The binding problem is computationally hard. Indeed, even checking the equivalence of an unbound and a bound network (tautology problem) is intractable.

A common approach for achieving library binding is to restrict binding to the replacement of subnetworks (of the original unbound network) with cell-library instances [19, 23, 26]. This is called *network covering* by library cells. Covering entails recognizing that a portion of a logic network can be replaced by a library cell and selecting an adequate number of instances of library elements to cover the logic network while optimizing some figure of merit, such as area and/or delay. A simple example of network covering is shown in Figure 10.1.

We say that a cell *matches* a subnetwork when they are functionally equivalent. Note that a cell may match a subnetwork even if the number of inputs differs and some of these are shorted together or connected to a fixed voltage rail. For example, a three-input AND cell can be used to implement a two-input AND function. We neglect these cases in the sequel, because libraries usually contain those cells that can be obtained from more complex ones by removing (or bridging) inputs.

An unbound logic network, where each local function matches a library cell, can be translated into a bound network in a straightforward way by binding each vertex to an instance of a matching library cell. We call this binding trivial. Even when the unbound network is area and/or timing optimal, a trivial binding may not be so, because optimization of unbound networks involves simplified models and ignores the actual cell parameters. Thus, library binding often involves a restructuring of the logic network.

We describe now the covering problem of an unbound network $G_n(V, E)$ in more detail. A rooted subnetwork is a subgraph of $G_n(V, E)$ that has a single vertex with zero outdegree. In the covering problem, we associate with each internal vertex

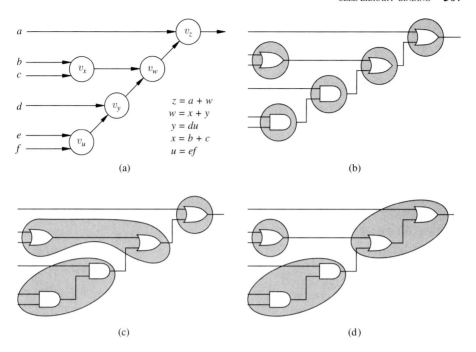

FIGURE 10.1
(a) Simple network. (b) Trivial binding by using one gate per vertex. (c) Network cover with library cells including two-input and three-input AND and OR gates. (d) Alternative network cover.

v of $G_n(V, E)$ the subset M_v of library cells that match some subnetwork rooted at v. We say that a cell in M_v covers v and the other vertices of the matching subnetwork. Let M be the set of all matching cells associated with the internal vertices of the network, i.e., $M = \bigcup_{v \in V^G} M_v$.

The covering problem can be modeled by selecting enough matches in M that cover all internal vertices of the unbound network. For each selected match, we must ensure that the vertices bound to the inputs of the corresponding cell are also associated with the outputs of other matching cells. Thus the choice of a match implies the choice of some other matches, and the network covering problem can be classified as a binate covering problem. An example of an optimal solution is one that minimizes the total cost associated with the selected matches. For example, the cost may be the area taken by the individual cells.

A necessary condition for the network covering problem to have a solution is that each internal vertex is covered by at least one match. This condition is usually satisfied by decomposing the unbound network prior to covering so that all local functions have at least one match in the library.

> **Example 10.2.1.** Consider the simple library shown in Figure 10.2 (a) and the unbound network of Figure 10.2 (b). We consider the problem of finding a network cover that minimizes the total area cost associated with the chosen cells.

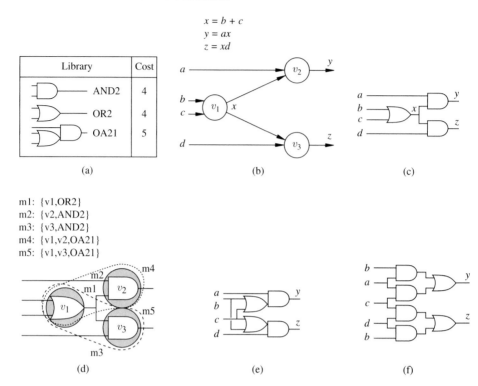

$$x = b + c$$
$$y = ax$$
$$z = xd$$

m1: {v1,OR2}
m2: {v2,AND2}
m3: {v3,AND2}
m4: {v1,v2,OA21}
m5: {v1,v3,OA21}

FIGURE 10.2
(a) Simple library. (b) Unbound network. (c) Trivial binding. (d) Match set. (e) Network cover. (f) Alternative bound network which is not a cover of the unbound network shown in (b).

There are several possibilities for covering this network. For example, a trivial binding is shown in Figure 10.2 (c). A more interesting binding can be found by considering the possible matches. Consider, for example, vertex v_1, which can be bound to a two-input OR gate (OR2), and vertex v_2, which can be bound to a two-input AND gate (AND2). Moreover, the subnetwork consisting of $\{v_1, v_2\}$ can be bound to a complex gate (OA21). We can associate binary variables to denote the matches. Variable m_1 is TRUE whenever the OR2 gate is bound to v_1, variable m_2 is TRUE whenever the AND2 gate is bound to v_2, and m_4 represents the use of OA21 for covering the subnetwork $\{v_1, v_2\}$. Similar considerations apply to vertex v_3. We use variable m_3 to denote the choice of a two-input AND gate (AND2) for v_3 and m_5 to represent the choice of OA21 for $\{v_1, v_3\}$. The possible matches are shown in Figure 10.2 (d).

Therefore we can represent the requirement that v_1 be covered by at least one gate in the library by the unate clause $m_1 + m_4 + m_5$. Similarly, the covering requirements of v_2 and v_3 can be represented by clauses $m_2 + m_4$ and $m_3 + m_5$, respectively.

In addition to these requirements, we must ensure that the appropriate inputs are available to each chosen cell. For instance, binding an AND2 gate to v_2 requires that its inputs are available, which is the case only when an OR2 gate is bound to v_1. The former choice is represented by m_2 and the latter by m_1. This implication can be expressed by $m_2 \rightarrow m_1$, or alternatively by the binate clause $m_2' + m_1$. Similarly, it can be shown that

$m_3 \rightarrow m_1$, or $m_3' + m_1$. Therefore, the following overall clause must hold:

$$(m_1 + m_4 + m_5)(m_2 + m_4)(m_3 + m_5)(m_2' + m_1)(m_3' + m_1) = 1$$

The clause is binate. An exact solution can be obtained by binate covering, taking into account the cell costs. For each cube satisfying the clause, the least-cost one denotes the desired binding. In this case, the optimum solution is represented by cube $m_1' m_2' m_3' m_4 m_5$, with a total cost of 10, corresponding to the use of two OA21 gates. The optimal bound network is shown in Figure 10.2 (e).

The bound network obtained by covering depends on the initial decomposition. For example, Figure 10.2 (f) shows another bound network, which is equivalent to the network of Figure 10.2 (b), but not one of its covers.

An exact algorithm for solving the network covering problem must cope with binate covering. The branch-and-bound algorithm of Section 2.5.3 can be used for this purpose, but experimental results have shown that this approach is viable only for small-scale circuits, which are not of practical interest [26].

It is important to remember that library binding could be solved exactly by methods other than covering. Unfortunately, the difficulty of the covering problem stems from its binate nature, due to the fact that the choice of any cell requires selecting other cells to provide correct connectivity. Any other formulation of library binding will face the same problem, and thus we conjecture that solving library binding is at least as complex as solving it by network covering.

For this reason, heuristic algorithms have been developed to approximate the solution of the network covering problem. Alternatively, rule-based systems can perform network covering by stepwise replacement of matching subnetworks. We shall review heuristic algorithms next.

10.3 ALGORITHMS FOR LIBRARY BINDING

Algorithms for library binding were pioneered at AT&T Bell Laboratories by Keutzer [19], who recognized the similarity between the library binding problem and the code generation task in a software compiler. In both cases, a matching problem addresses the identification of the possible substitutions and a covering problem the optimal selection of matches.

There are two major approaches to solving the matching problem which relate to the representation being used for the network and the library. In the *Boolean* approach, the library cells and the portion of the network of interest are described by Boolean functions. In the *structural* approach, graphs representing algebraic decompositions of the Boolean functions are used instead. Since the algebraic representation of an expression can be cast into a graph, expression pattern matching approaches can be classified as structural techniques. The structural approach was used by programs DAGON [19], MIS [10, 26] and TECHMAP [24], while the Boolean approach was implemented first in program CERES [23] and in Fujitsu's binder [29].

The structural and Boolean approaches differ mainly in the matching step. We define formally the matching of two scalar combinational functions, representing a cell and a subnetwork, as follows.

Definition 10.3.1. Given two single-output combinational functions $f(\mathbf{x})$ and $g(\mathbf{y})$, with the same number of support variables, we say that f **matches** g if there exists a permutation matrix $\mathbf{P}$ such that $f(\mathbf{x}) = g(\mathbf{P}\,\mathbf{x})$ is a tautology.

The degrees of freedom in associating input pins is modeled by the permutation matrix. We refer to this type of matching as *Boolean matching* because it is based on the property of the function and to distinguish it from another weaker form of matching. Given a structural representation of two functions by two graphs in a pre-defined format (e.g., *sum of products* representations or networks of two-input NAND gates and inverters), there is a *structural match* if the graphs are isomorphic. A structural match implies a Boolean match, but the converse is not true.

> **Example 10.3.1.** Consider the following two functions: $f = ab + c$ and $g = p + qr$. Since it is possible to express g as a function of $\{a, b, c\}$ with a suitable input variable permutation so that $f = g$ is a tautology, it follows that f and g have a Boolean match.
>
> Functions f and g can be represented by their OR-AND decomposition graphs, as shown in Figures 10.3 (a) and (b). Since the two graphs are isomorphic, f and g have a structural match.
>
> Consider now the following two functions over the same support set: $f = xy + x'y' + y'z$ and $g = xy + x'y' + xz$. They are logically equivalent and hence they yield a Boolean match. Nevertheless, they are entirely different in the expression patterns and in their structural representation. Note that different structures for a given function arise because there exist different possible ways of factoring an expression and there are even different *sum of products* representations of the same function.

Matching algorithms are described in Sections 10.3.1 and 10.3.2. The Boolean matching problem is intractable, because the complement of the tautology problem belongs to the $\mathcal{NP}$-complete class [12]. The structural matching problem for general functions, represented by dags, is also conjectured to be intractable, because it is transformable into the graph isomorphism problem [12]. Nevertheless, efficient algorithms for matching have been developed, because the size of the matching problem is usually small, since it is related to the maximum number of inputs of the library cells.

The major difficulty in solving the library binding problem lies in the network covering problem, as we have shown in the previous section. To render the problem solvable and tractable, most heuristic algorithms apply two pre-processing steps to the network before covering: *decomposition* and *partitioning*.

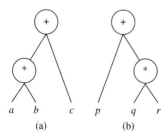

FIGURE 10.3
(a) Representative graph for $f = ab + c$. (b) Representative graph for $g = p + qr$.

Decomposition is required to guarantee a solution to the network covering problem by ensuring that each vertex is covered by at least one match. The goal of decomposition in this context is to express all local functions as simple functions, such as two-input NORs or NANDs, that are called *base functions*. The library must include cells implementing the base functions to ensure the existence of a solution. Indeed, a trivial binding can always be derived from a network decomposed into base functions. Conversely, if no library cell implements a base function f, there may exist a vertex of the network whose expression is f and that is not included in any subnetwork that matches a library element. The choice of the base functions is important, especially for those approaches based on structural matching, and it is obviously dependent on the library under consideration.

Different heuristic decomposition algorithms can be used for this purpose, but attention must be paid because network decompositions into base functions are not unique and affect the quality of the solution. Therefore heuristics may be used to bias some features of decomposed networks. For example, while searching for a minimal-delay binding, a decomposition may be chosen such that late inputs traverse fewer stages.

The second major pre-processing step in heuristic binding is partitioning, which allows the covering algorithm to consider a collection of multiple-input single-output networks in place of a multiple-input multiple-output network. The subnetworks that are obtained by partitioning the original network are called *subject graphs* [19]. Subject graphs are then covered by library elements one at a time.

The rationale for partitioning is twofold. First, the size of each covering problem is smaller. Second, the covering problem becomes tractable under some additional assumptions, as described in Section 10.3.1. In the case of general networks, partitioning is also used to isolate the combinational portions of a network from the sequential elements and from the I/Os.

Different schemes can be used for partitioning. When considering a combinational network, vertices with multiple outdegrees can be marked, and the edges whose tails are marked as vertices define the partition boundary. This method implies that library binding will not try to improve the previous structure of the network as far as changing multiple-fanout points. Another approach is to iteratively identify a partition block consisting of all vertices that are tails of paths to a primary output and to delete them from the graph [10]. While this approach privileges the formation of larger partition blocks, it suffers from the dependency on the choice of output.

It is important to stress that, although the partitioning and decomposition steps are heuristics that help reduce the problem difficulty, they can hurt the quality of the solution.

Finally each subject graph is covered by an interconnection of library cells. For selected portions of the subject graph, all cells in the library are tried for a match, and when one exists and is selected, that portion of the subject graph is labeled with the matching cell along with its area and timing attributes. The choice of a match is done according to different covering schemes, as described in detail in the following sections. The decomposition, partitioning and covering steps are illustrated in Figures 10.4, 10.5 and 10.6.

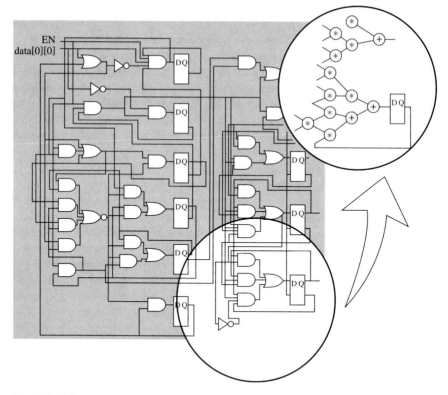

FIGURE 10.4
Network decomposition into base functions (two-input ORs and two-input ANDs).

10.3.1 Covering Algorithms Based on Structural Matching

Structural matching relies on the identification of common patterns. For this reason, both the subject graph and the library functions must be cast into a form that is comparable. Hence they are both decomposed in terms of the same set of base functions. The graphs associated with the library elements are called *pattern* graphs. Some library cells may have more than one corresponding pattern graph, because the decomposition of their representative functions may not be unique.

The subject and pattern graphs are acyclic and rooted, the root being associated with the subnetwork and cell outputs. We assume in the sequel that decomposition yields either trees or dags where paths that stem from the root reconverge only at the input vertices (leaves). Such dags are called in jargon *leaf-dags*. When the decomposition yields a tree, all corresponding input variables are associated with different vertices. Otherwise, an input variable may be associated with more than one vertex.

We also assume in the sequel that inverters are explicitly modeled in the subject and pattern graphs. In some cases, we shall represent these graphs by their

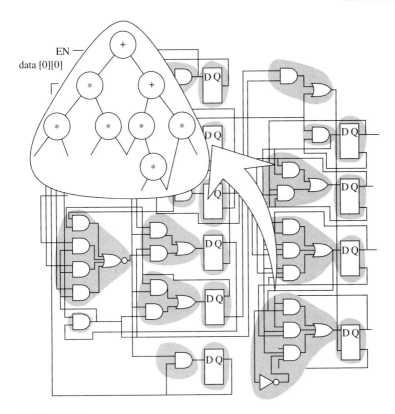

FIGURE 10.5
Network partitioning into subject graphs.

trivial binding, because operations on this representation are easier to visualize and understand.

> **Example 10.3.2.** Consider the following cells in a library: a two-input AND gate, a two-input EXOR gate and a four-input AND gate whose representative functions are $f_1 = ab$, $f_2 = a \oplus b$ and $f_3 = abcd$, respectively. The pattern graphs are expressed using a decomposition in two-input NANDs and inverters.
>
> The pattern graphs are shown in Figure 10.7. In the first case the graph is a tree; in the second it is a leaf-dag. In the last case, there are two pattern graphs associated with the cell (excluding isomorphic graphs). Non-terminal vertices are labeled by the letter "N" to denote a two-input NAND and by the letter "I" to denote an inverter.

Structural matching can be verified by checking the isomorphism between two rooted dags. Even though this problem is conjectured to be intractable [12], experimental results have shown that the computation time is negligible for problems of practical size [10]. Nevertheless, the matching problem can be simplified by noticing that most cells in any library can be represented by rooted trees and that *tree matching* and *tree covering* problems can be solved in linear time. For those cells, such as EXOR

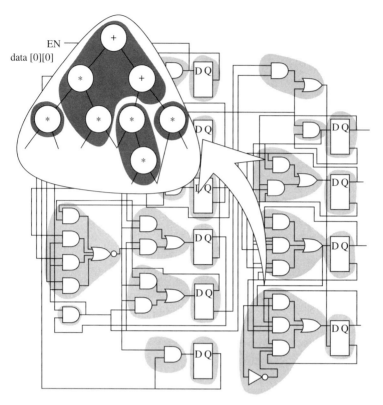

FIGURE 10.6
Covering of a subject graph.

and EXNOR gates, that do not have a tree-like decomposition, it is still possible to use a tree representation (by splitting the leaves of the leaf-dag) with an additional notation for those input variables associated with more than one leaf vertex. An example of a simple library and the corresponding pattern trees are shown in Figures 10.8 (a) and (b).

We consider now structural matching and covering using the tree-based representation as proposed by Keutzer [19]. We assume that the subject graph is represented by a rooted tree, obtained by splitting the leaves of the corresponding leaf-dag. We describe first tree-based matching and then tree-based covering. In particular, we describe two methods for tree matching. The first is a simple and intuitive approach that applies to trees obtained by decomposition into base functions of one type only. The second approach is more general in nature, because it supports arbitrary sets of base functions and it uses an automaton to capture the library and check for matching.

SIMPLE TREE-BASED MATCHING. We describe tree matching in the case that only one type of base function is used in the decomposition. We consider two-input NANDs,

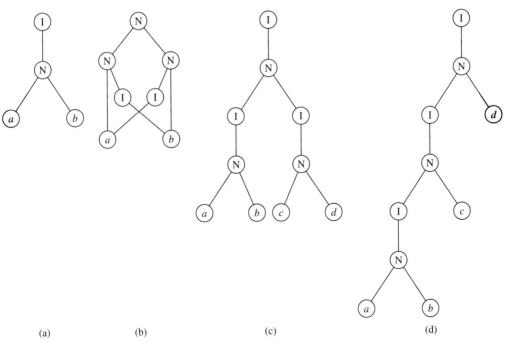

(a) (b) (c) (d)

FIGURE 10.7
(a) Pattern graph for $f_1 = ab$. (b) Pattern graph for $f_2 = a \oplus b$. (c,d) Pattern graphs for $f_3 = abcd$.

but the same considerations are applicable to two-input NORs. Thus, each vertex of the subject and pattern trees is associated with either a two-input NAND and has two children or an inverter and has one child. Note that an inverter can be seen as a single-input NAND. Since only one type of base function is used, vertices need not be labeled by the base function. The type of a vertex (e.g., NAND, inverter or leaf) is easily identified by its *degree*, i.e., by the number of its children.[1]

There are several algorithms for tree matching. We describe a simple algorithm that determines if a pattern tree is isomorphic to a subgraph of the subject tree. This is performed by matching the root of the pattern tree to a vertex of the subject tree and visiting recursively their children. The isomorphism can be easily verified by comparing the degrees of pairs of vertices in both the subject and the pattern trees, starting from the initial vertices and proceeding top down until the leaves of the pattern tree are reached. If there is a mismatch, the algorithm terminates with an unsuccessful match. Otherwise, the corresponding children are recursively visited.

Algorithm 10.3.1 is invoked with v as a vertex of the subject graph and u as the root of the pattern graph.

---

[1] It is customary to call the degree of a vertex in a rooted tree the number of its children rather than the number of edges incident to it.

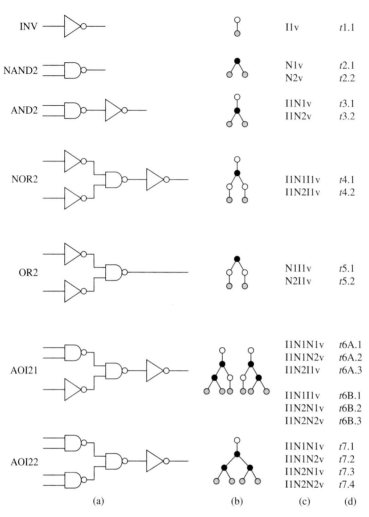

INV	I1v	t1.1
NAND2	N1v N2v	t2.1 t2.2
AND2	I1N1v I1N2v	t3.1 t3.2
NOR2	I1N1I1v I1N2I1v	t4.1 t4.2
OR2	N1I1v N2I1v	t5.1 t5.2
AOI21	I1N1N1v I1N1N2v I1N2I1v	t6A.1 t6A.2 t6A.3
	I1N1I1v I1N2N1v I1N2N2v	t6B.1 t6B.2 t6B.3
AOI22	I1N1N1v I1N1N2v I1N2N1v I1N2N2v	t7.1 t7.2 t7.3 t7.4

(a)   (b)   (c)   (d)

FIGURE 10.8
(a) Simple cell library. (b) Pattern trees (I = white, N = black, v = gray.) (c) Pattern strings. (d) Pattern tree identifiers.

When visiting a vertex pair, if the vertex of the pattern graph is a leaf, then a path from the root to that leaf in the pattern graph has a match in the subject graph. Conversely, when the vertex of the subject graph is a leaf, and the vertex of the pattern graph is not a leaf, a match is impossible. When both vertices are not leaves, they must have the same number of children which must recursively match for a match to be possible. The algorithm is linear in the size of the graphs.

Example 10.3.3. Consider the subject tree for function $x = a + b'$ in terms of a two-input NAND decomposition, as shown in Figure 10.9 (a,b). Let us consider the pattern trees of

```
MATCH(u, v) {
        if (u is a leaf) return (TRUE);                          /* Leaf of the pattern graph reached */
        else {
                if (v is a leaf) return (FALSE);                /* Leaf of the subject graph reached */
                if (degree(v) ≠ degree(u)) return(FALSE);       /* Degree mismatch */
                if (degree(v) == 1) {                           /* One child each: visit subtree recursively */
                        u_c = child of u ; v_c = child of v ;
                        return (match(u_c, v_c) )
                }
                else {                                          /* Two children each: visit subtrees recursively */
                        u_l = left-child of u ; u_r = right-child of u ;
                        v_l = left-child of v ; v_r = right-child of v ;
                        return (MATCH(u_l, v_l) · MATCH(u_r, v_r) + MATCH(u_r, v_l) · MATCH(u_l, v_r));
                }
        }
}
```

ALGORITHM 10.3.1

the simple library shown in Figure 10.8. Two patterns are also reported in Figures 10.9 (c) and (d) for convenience.

We use algorithm *MATCH* to determine if a pattern tree matches a subtree of the subject tree with the same root. First, let us apply the algorithm to the subject tree and the inverter (INV) pattern tree. Since the root of the subject tree has two children and that of the INV pattern tree has only one, there is no match. Next, let us apply the algorithm to the subject tree and to the NAND2 pattern tree. Both roots have two children. In both recursive calls, the children of the pattern tree are leaves and the calls return TRUE. Hence the NAND2 pattern tree matches a subtree of the subject tree with the same root.

TREE-BASED MATCHING USING AUTOMATA.* We consider now a method for tree matching that uses an automaton to represent the cell library. This approach is based on an encoding of the trees by strings of characters and on a stringrecognition algorithm.

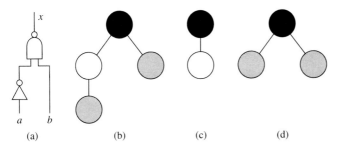

(a)   (b)   (c)   (d)

FIGURE 10.9
(a) NAND2 decomposition of $x = a + b'$. (b) Subject tree. (c) Pattern tree for INV. (d) Pattern tree for NAND2.

This method supports library descriptions in terms of arbitrary base functions, each one encoded by a single character. Thus, it is more general in nature than the simple algorithm of the previous section.

Tree matching using string matching is based on the Aho-Corasick algorithm, which was devised to recognize strings of characters in a given text [1]. It constructs an automaton that detects the matching strings. In our context, there is one automaton for the entire cell library, i.e., all pattern trees contribute to the automaton. The automaton processes strings that encode paths in the subject tree and recognizes those that match paths in the pattern trees. This mechanism, detailed later in this section, is used to detect all matching patterns.

A tree can be represented by a set of strings, where each string corresponds to a path from the root to a leaf. Since a path is a sequence of vertices and edges, a string has a character denoting each vertex in the path and its type (i.e., corresponding base function) alternated with a number denoting the edge.

Example 10.3.4. Consider the subject tree of Figure 10.9 (b). Let us label edges according to this convention. When a vertex has two children, the left edge is labeled by 1 and the right edge is labeled by 2. When a vertex has one child, the edge to its child is labeled by 1.

There are two paths from the root to the leaves. They can be labeled by strings {N1I1v,N2v}, where "v" denotes a leaf.

We summarize now the algorithm for constructing the automaton, which is fully detailed in reference [1]. The automaton consists of a set of states, a set of transitions that are partitioned into *goto* and *failure* transitions corresponding to the status of detection of a character in a string and an output function that signals the full detection of a string. The automaton is constructed once for all elements of a given library.

The automaton is constructed incrementally, while considering the strings corresponding to the pattern trees. Initially the automaton contains only the *reset* state. Then each string is processed in turn one character at a time, a new state being added for each of the characters not recognized by the automaton under construction. These characters are used as labels of the *goto* transitions into the states being added. The output function that signals the full detection of a string is specified in conjunction with each transition to a state corresponding to the last character of the string itself. Once all strings have been considered, the automaton transition diagram has a tree structure. The states can be assigned a level equal to the number of transitions from the *reset* state and a parent/child relation can be established among state pairs.

Next the automaton is revisited to add the *failure* function, which indicates which next state to reach when an incoming character does not match the label of any outgoing edge from the current state. Initially the *failure* function of states from level 1 is set to the *reset* state. Next, while traversing the automaton by breadth-first search, the failure function for a state s reached by state r under input character c is set to the state reached by a transition under c from the failure state for r. The output function is updated while deriving the failure functions. In particular, we append to the output function of a given state the output function associated with the transition into the failure state under the corresponding character.

The automaton may recognize arbitrary subtrees of a given tree. In practice, it is convenient to detect subtrees that have the same root or the same leaves. This can be easily achieved by using a distinguished character for the root or the leaves.

Example 10.3.5. Consider a library of two cells: a three-input NAND and a two-input NAND. The corresponding pattern trees are shown in Figure 10.10, where we assume that the base functions are a two-input NAND and an inverter. Hence, vertices are labeled by "N," "I" and "v," where "v" denotes a leaf.

The following strings encode the first pattern tree: {N1v, N2I1N1v, N2I1N2v}. These strings are labeled by $t1.1, t1.2, t1.3$, respectively. Similarly, the second pattern is encoded by {N1v, N2v}, denoted as $t2.1, t2.2$. Note that the first string in both patterns is the same.

We want to construct an automaton that recognizes the strings related to the two cells just mentioned. This automaton is shown in Figure 10.11. We consider now the individual steps for assembling the automaton.

While processing the first string, i.e., N1v, three states are added to the reset state, forming a chain with transitions labeled by N, 1 and v. The last transition is coupled with the output value $t1.1$, denoting the detection of the first string. Next the second string is considered: N2I1N1v. Starting from the reset state, the existing automaton would recognize the first character (e.g., N), but not the following ones. Hence new states and transitions are added to the automaton: the first transition to be added is from state 1 to state 4, and so on.

When all strings have been considered and the state set finalized, the remaining transitions are determined based on the *failure* function. The *failure* function for state 1 is the reset state. This means that when in state 1, if an input character c does not yield a *goto* transition (i.e., $c \neq 1$ and $c \neq 2$), the next state is determined by the transition from the reset state under c. In particular, when $c = $ N, the next state is state 1. And so on.

Let us now consider the use of the automaton for finding the pattern trees that match a given subject tree. We shall consider the particular case where we search for pattern trees isomorphic to subtrees of the subject graph and with the same leaves.

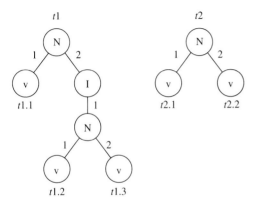

FIGURE 10.10
Two pattern graphs.

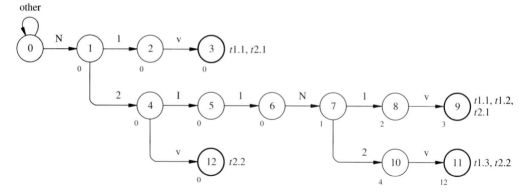

FIGURE 10.11
Aho-Corasick matching automaton. The *goto* transitions are indicated by edges and the *failure* function is denoted by subscripts.

The subject tree is visited bottom up, and for each vertex all strings that correspond to paths to the leaves are processed. A match exists if the strings are recognized by the automaton as belonging to the same pattern tree. The algorithm complexity is linear in the size of the subject graph.

Example 10.3.6. We consider here the simple library and the patterns trees of Figure 10.8. The corresponding automaton is shown in Figure 10.12.

Let us assume that the subject tree models the function $x = a + b'$, as shown in Figure 10.9. The subject tree can be modeled by strings {N1I1v,N2v}.

Let us feed these strings (representing the subject tree) to the automaton of Figure 10.12 (representing the library). It can be seen that the automaton would recognize both strings, leading to output functions $t5.1$ and $t2.2$, respectively (states 30 and 32 in Figure 10.12). Since these functions belong to different pattern graphs, there is no cell in the library that can implement the subject graph directly.

Nevertheless, substring I1v can be recognized by the automaton, yielding output $t1.1$ (state 3 in Figure 10.12). This means that an inverter covers a part of the subject tree. After pruning the corresponding part of the tree, the subject tree is reduced to the strings {N1v,N2v}, which can be recognized by the automaton (states 27 and 32 in Figure 10.12). The corresponding output functions are $t1.1$ and $t1.2$, denoting to two-input NAND cell. Thus, a NAND2 cell and an INV cell provide a cover.

When comparing this example with Example 10.3.3, the reader may be surprised to notice that the *MATCH* algorithm finds a match for the NAND2 cell, while the automaton finds a match for the INV cell. Note first that these matches cover part of the subject tree and that in both cases a tree cover can be achieved by using both a NAND2 cell and an INV cell. Second, the difference in the matches detected by the two algorithms is due to the fact that the former algorithm looks for a pattern tree isomorphic to a subtree of the subject graph with the same root. The automaton-based algorithm looks instead for a pattern tree isomorphic to a subtree of the subject graph with the same leaves. It is possible to modify the automaton algorithm by using a distinguished character for the root instead for the leaves and by reversing the strings, so that it detects if a pattern tree is isomorphic to a subtree of the subject graph with the same root.

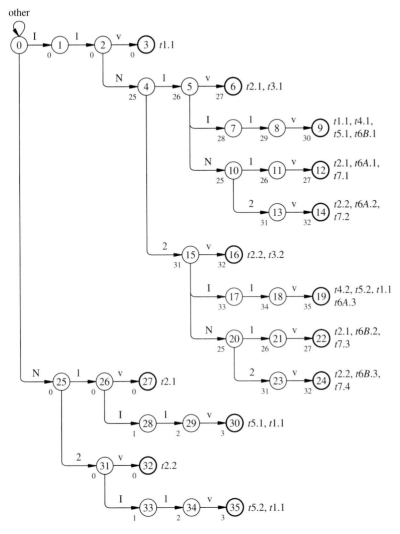

FIGURE 10.12
Aho-Corasick automaton for the simple library.

In the previous examples, the subject and pattern trees where represented in terms of NAND2 and INV base functions. Thus, all strings contained only three characters, including the terminator (e.g., "v"). The automaton recognizer can support tree recognition with decompositions into arbitrary sets of base functions by just associating each base function with a character. The automaton construction and recognition algorithms remain the same.

When compared to the simple *MATCH* algorithm, the automaton-based approach has the additional advantage that it considers all patterns available for matching at the same time, while the *MATCH* algorithm compares one tree at a time. This advantage is offset by the increased complexity of handling trees as separate strings.

The set of base functions useful for library binding is usually small. The choice of the base functions affects the quality of the solution as well as the number of pattern graphs and consequently the computing time for binding. There are some arguments and experimental results favoring the choice of using one base function only (e.g., NAND2 or NOR2 plus inverters) [26]. In this case, the *MATCH* algorithm is preferable due to its simplicity.

TREE-BASED COVERING. Optimum tree covering can be computed by dynamic programming [2, 19]. Here we consider the algorithm described in Section 2.3.4 in the context of solving the library binding problem.

We describe first the minimum-area covering problem. Each cell has an area cost, representing its area usage. The total area of the bound network is the objective to minimize. The tree covering algorithm traverses the subject graph in a bottom-up fashion. For the sake of this explanation, we consider matching of pattern trees whose roots correspond to the vertex of the subject tree under consideration. Other approaches yield equivalent results.

For all vertices of the subject tree, the covering algorithm determines the matches of the locally rooted subtrees with the pattern trees. There are three possibilities for any given pattern tree:

1. The pattern tree and the locally rooted subject subtree are isomorphic. Then, the vertex is labeled with the corresponding cell cost.
2. The pattern tree is isomorphic to a subtree of the locally rooted subject subtree with the same root and a set of leaves L. Then, the vertex is labeled with the corresponding cell cost plus the labels of the vertices L.
3. There is no match.

If we assume that the library contains the gates implementing the base functions, then for any vertex there exists at least one cell for which one of the first two cases applies, and we can label that vertex. Therefore, it is possible to choose for each vertex in the subject graph the best labeling among all possible matches. At the end of the tree traversal, the vertex labeling corresponds to an optimum covering. Note that overall optimality is weakened by the fact that the total area of a bound network depends also on the partitioning and decomposition steps. The complexity of the algorithm is linear in the size of the subject tree.

> **Example 10.3.7.** Consider the network shown in Figure 10.13 with the library and patterns of Figure 10.8. (The network is represented by its trivial binding after a NAND2 decomposition, for the sake of clarity.) While visiting the subject tree bottom up, only one match is found for vertices x, y, z and w. The corresponding pattern is recorded

Network	Subject graph	Vertex	Match	Gate	Cost
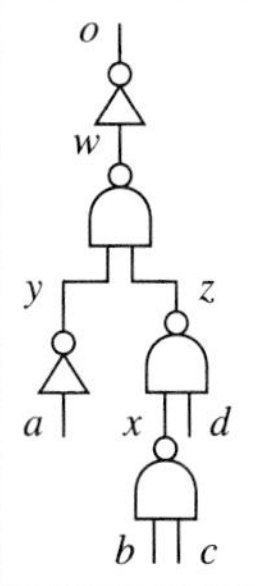	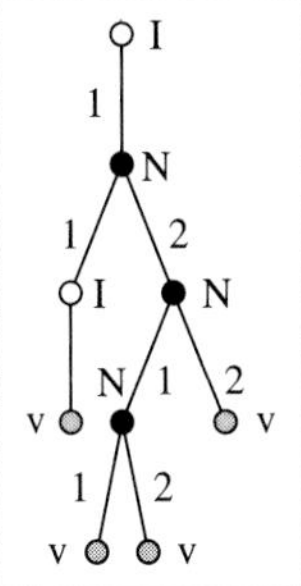	x	$t2$	NAND2(b,c)	NAND2
		y	$t1$	INV(a)	INV
		z	$t2$	NAND2(x,d)	2 NAND2
		w	$t2$	NAND2(y,z)	3 NAND2+INV
		o	$t1$	INV(w)	3 NAND2+2INV
			$t3$	AND2(y,z)	2 NAND2+AND2+INV
			$t6B$	AOI21(x,d,a)	NAND2+AOI21

FIGURE 10.13
Example of structural covering: network; subject graph; possible matches at each vertex and corresponding costs.

along with the cost of the network bound to the rooted subtree. Let us consider then the root vertex o. Three matches are possible:

- One INV gate. Hence the cost is that of an inverter plus that of the subtree rooted in w.
- One AND2 gate with inputs y, z. Hence the cost is that of an AND2 plus those of the subtrees rooted in y and z.
- One AOI21 gate with inputs x, d, a. Hence the cost is that of the AOI21 gate plus that of the subtree rooted in x.

Assume now that the area costs of the cells {INV, NAND2, AND2, AOI21} are { 2, 3, 4, 6 } respectively. Then, the optimum binding is given by the third option, i.e., by choosing gate AOI21 fed by a NAND2 gate (Figure 10.14).

Let us now consider delay minimization in conjunction with library binding. When considering the network partitioned into subject graphs, the problem reduces to minimizing the data-ready time (see Section 8.6.3) at the output corresponding to the root of the subject graph. In this case, the cost of each cell is its input/output

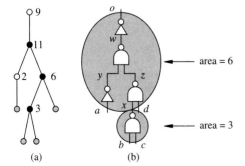

area = 6

area = 3

FIGURE 10.14
(a) Subject graph with area-cost annotation.
(b) Optimum cover for area.

propagation delay. For the sake of simplicity, we consider the worst-case delay for the cells' rising and falling transitions at the outputs, and we assume that the cell delay from each input to the output is the same. The algorithms still apply, with some slight modifications, if these assumptions are removed.

Therefore the propagation delay of a cell can be measured in terms of a constant plus a load-dependent term, which depends on the cell (or cells) being driven and the wiring load. First, we consider the case in which the propagation delay is constant, and then we look at the more general case.

When the propagation delay is constant, the cost associated with a cell is just a single positive number. The overall cost is the maximum path delay, assuming that all paths can be sensitized. The data-ready time at a cell output is then the sum of its propagation delay plus the maximum of the data-ready time at the cell inputs. A bottom-up traversal of the subject tree would allow us to determine the binding that minimizes the data-ready time at each vertex and hence the minimum time at the root. The input data-ready times can be easily taken into account. Note again that the delay optimality is valid within the tree model and is dependent on the chosen decomposition.

> **Example 10.3.8.**   Consider again the network of Figure 10.13 with the library and patterns of Figure 10.8. Assume that a minimum-delay cover is searched for, that the delays of the cells {INV, NAND2, AND2, AOI21} are {2, 4, 5, 10}, respectively, and that all inputs are available at time 0 except for input d, which is available at time 6. While visiting the subject tree bottom up, the data-ready times of vertices x, y, z and w are then 4, 2, 10 and 14. Three matches are again possible for vertex o. If an inverter is chosen, the output data-ready time is 16. If an AND2 gate is chosen, it is 5 plus the maximum data-ready time at its inputs y and z, i.e., $5 + \max\{2, 10\} = 15$. The choice of an AOI21 gate leads to an output data-ready time of $10 + \max\{0, 4, 6\} = 16$. Hence the binding corresponding to the fastest network corresponds to the second choice. It entails the interconnection of an AND2, two NAND2 and an INV gate (Figure 10.15).

Let us now consider the general case, where the propagation delay depends on the load. This model is highly desirable for an accurate delay representation. Most libraries have multiple gates with different drive capabilities for the same logic function. The higher the driving capability, the shorter the propagation delay for the same load and the higher the input capacitance, because larger devices are employed.

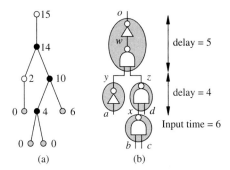

(a)   (b)

FIGURE 10.15

(a) Subject graph with vertex data-ready annotation.

(b) Optimum cover for delay.

The problem of selecting a cell in a bottom-up traversal of the subject graph is not straightforward, because the input capacitance loads of the following stages are unknown when matching. Indeed the following stages correspond to vertices closer to the root and therefore that are yet to be bound.

Rudell realized that for most libraries the values of input capacitances are a finite and small set [26]. Therefore he used a *load binning* technique, which consists of labeling each vertex with all possible total load values or by an approximation. This can be done as a pre-processing step.

The tree covering algorithm is extended by computing an array of solutions for each vertex, corresponding to the loads under consideration. For each match, the arrival time is computed for each load value. For each input to the matching cell, the best match for driving the cell (for any load) is selected and the corresponding data-ready time is used. If all possible load values are considered, then the algorithm guarantees an optimum solution for the delay model within the tree modeling assumption. Otherwise, it represents a heuristic method approximating the exact solution. The computational complexity of the tree covering approach is linear in the size of the subject tree and in the number of load values being considered.

Example 10.3.9. Consider again the network of Figure 10.13 with the library and patterns of Figure 10.8. Assume that a minimum-delay cover is searched for, that the delays of the cells {INV, NAND2, AND2, AOI21} are now $\{1+1\cdot l, 3+1\cdot l, 4+1\cdot l, 9+1\cdot l\}$, where l is the load at the output of the gate. All cells load the previous stage by $l = 1$. If the output load is also 1, the problem reduces to the previous one described in Example 10.3.8.

Assume next that a super-inverter cell is available, SINV, with load $l = 2$ and delay $\{1 + 0.5 \cdot l\}$. Hence, the possible loads on the cells are 1 and 2. The algorithm would compute the best binding for each load. In this case, the best match at vertices x, y and z are the same as in Example 10.3.8. Nevertheless, at vertex w a load of either 1 or 2 may be possible, with corresponding data-ready times of either 14 or 15.

Assume the output load is 1. There are four choices for matching. The regular inverter INV would have a propagation delay of $1+1\cdot 1$ and a load of 1, yielding an output data-ready time of $14 + 2 = 16$. The super-inverter SINV would have a propagation delay of $1 + 0.5\cdot 1$ and a load of 2, yielding an output data-ready time of $15 + 1 + 0.5\cdot 1 = 16.5$. The AND2 would yield an output data-ready time of $4 + 1 \cdot 1 + \max\{2, 10\} = 15$ and the AOI21 would yield an output data-ready time of $9 + 1 \cdot 1 + \max\{0, 6, 4\} = 16$. Hence, the AND2 solution would be preferred as in the previous example.

Assume now that the output load is 5. The regular inverter INV would have a propagation delay of $1 + 1 \cdot 5$ and a load of 1, yielding an output data-ready time of $14 + 6 = 20$. The super-inverter SINV would have a propagation delay of $1 + 0.5 \cdot 5$ corresponding to an output data-ready time of $15 + 1 + 0.5 \cdot 5 = 18.5$. The AND2 would yield an output data-ready time of $4 + 1 \cdot 5 + \max\{2, 10\} = 19$ and the AOI21 would yield an output data-ready time of $9 + 1 \cdot 5 + \max\{0, 6, 4\} = 20$. Hence, the SINV solution would now be preferred.

Whereas the tree covering and matching approach is very appealing for its simplicity and efficiency, there are three pitfalls. First, there are multiple non-isomorphic patterns for some cells, because the decomposition into given base functions is not necessarily unique. Therefore, a library cell may correspond to more than one pattern graph. As a result, each vertex of the subject graph must be tested for matching against a potentially larger number of pattern graphs, increasing the computational burden of the algorithm.

Second, some cells, such EXOR and EXNOR gates, cannot be represented by trees. A partial solution is to extend tree covering to use leaf-dags [10, 26], which does not change the complexity of the algorithm. Pattern leaf-dags can match vertices of the subject graphs as long as the corresponding leaves match. Thus, a limited use of these cells can be achieved.

Lastly, and more importantly, structural matching can only detect a subset of the possible matches and it does not permit the use of the *don't care* information in library binding. This can lead to solutions of inferior quality.

10.3.2 Covering Algorithms Based on Boolean Matching

Boolean matching can overcome the aforementioned pitfalls of structural matching. Boolean matching requires an equivalence check between two functions, one representing a portion of the network and called *cluster function* and the other representing a cell and named *pattern function*. Boolean covering consists of identifying subnetworks whose corresponding cluster functions have matches in the library and selecting an adequate set of matches that optimize the area and/or delay of the bound network.

Whereas Boolean covering and matching are potentially more computationally expensive than structural covering and matching, recent implementations have shown that computing times may be comparable. In addition, Boolean matching can find matches that are not detected by structural matching and it may exploit the degrees of freedom provided by *don't care* conditions. As a consequence, it may lead to better quality solutions.

BOOLEAN MATCHING. We consider Boolean matching of completely specified functions in this section, and we defer consideration of the use of *don't care* conditions to Section 10.3.4. We represent the cluster function by $f(\mathbf{x})$ and the pattern function by $g(\mathbf{y})$, where $\mathbf{x}$ and $\mathbf{y}$ denote the input variables associated with the subnetwork and library cell, respectively. We assume that $\mathbf{x}$ and $\mathbf{y}$ have the same size n.

Boolean matching addresses the question of whether functions $f(\mathbf{x})$ and $g(\mathbf{y})$ match according to Definition 10.3.1, i.e., if a permutation matrix $\mathbf{P}$ exists such that $f(\mathbf{x}) = g(\mathbf{P}\ \mathbf{x})$ is a tautology. Thus, Boolean matching requires solving a factorial number of tautology checks. Fortunately, this number can be greatly reduced by applying the techniques described next.

The equivalence between two functions can be detected using *ordered binary decision diagrams* (OBDDs) in various ways (see Section 2.5.2 and reference [23]). The order of the variables of one OBDD can be arbitrary and fixed, while different orderings of the variables in the other OBDD are tried until a match is found. Boolean matching can be made practical by using filters that drastically reduce the number of variable permutations to be considered. Similarly, filters can prune the set of pattern functions that need to be tested against a cluster function. Filters are based on properties of the Boolean functions that represent necessary conditions for matching and that can be easily verified. Namely:

- Any input permutation must associate a unate (binate) variable in the cluster function with a unate (binate) variable in the pattern function.
- Variables or groups of variables that are interchangeable in the cluster function must be interchangeable in the pattern function.

The first condition implies that the cluster and pattern functions must have the same number of unate and binate variables to have a match. In addition, if there are b binate variables, then an upper bound on the number of variable permutations to be considered in the search for a match is $b! \cdot (n - b)!$.

Example 10.3.10. Consider the following pattern function from a commercial library: $g = s_1 s_2 a + s_1 s_2' b + s_1' s_3 c + s_1' s_3' d$ with $n = 7$ variables. Function g has 4 unate variables and 3 binate variables.

Consider a cluster function f with $n = 7$ variables. First, a necessary condition for f to match g is to have also 4 unate variables and 3 binate variables. If this is the case, only $3! \, 4! = 144$ variable orders and corresponding OBDDs need to be considered in the worst case. (A match can be detected before all 144 variable orders are considered.) This number must be compared to the overall number of permutations, $7! = 5040$, which is much larger.

The second condition allows us to exploit symmetry properties to simplify the search for a match [23, 24]. Consider the support set of a function $f(\mathbf{x})$. A *symmetry set* is a set of variables that are pairwise interchangeable without affecting the logic functionality. A *symmetry class* is an ensemble of symmetry sets with the same cardinality. We denote a symmetry class by C_i when its elements have cardinality i, $i = 1, 2, \ldots, n$. Obviously classes can be void.

Example 10.3.11. Consider the function $f = x_1 x_2 x_3 + x_4 x_5 + x_6 x_7$. The support variables of $f(\mathbf{x})$ can be partitioned into three symmetry sets: $\{x_1 x_2 x_3\}, \{x_4 x_5\}, \{x_6 x_7\}$. There are two non-void symmetry classes, namely, $C_2 = \{\{x_4, x_5\}, \{x_6, x_7\}\}$ and $C_3 = \{\{x_1, x_2, x_3\}\}$.

Most libraries have pattern functions exhibiting symmetries, and the symmetry classes can be used to simplify the search for a match in different ways [23]. First, they can be used to restrict the set of pattern functions that can match a given cluster function as follows. The symmetry classes of the pattern functions can be computed beforehand, and they provide a signature for the patterns themselves. The symmetry classes of the cluster function can be quickly determined before comparing the OBDDs. A necessary condition for two functions to match is having symmetry classes of the same cardinality for each $i = 1, 2, \ldots, n$. Thus, pattern functions not satisfying this test can be weeded out.

Second, the symmetry classes are used to determine non-redundant variable orders. Indeed, variables can be paired only when they belong to symmetry sets of the same size, and since all variables in any given symmetry set are equivalent, the ordering of the variables within the set is irrelevant. As a result, only permutations over symmetry sets of the same size need to be considered. Thus another upper bound on the number of permutations required to detect a match is $\prod_{i=1}^{n}(|C_i|!)$.

> **Example 10.3.12.** Consider again function $f = x_1x_2x_3 + x_4x_5 + x_6x_7$. There are two non-void symmetry classes, one of cardinality 2 and one of cardinality 1. Thus, when considering the library pattern functions, only those with $|C_2| = 2$ and $|C_3| = 1$ are retained. For each retained pattern function, only $2! = 2$ variable orders and corresponding OBDDs need to be considered.
>
> For example, assuming that the OBDD of f has variables ordered as $(x_1, x_2, x_3, x_4, x_5, x_6, x_7)$, the relevant variable orders for the OBDDs of g are $(x_1, x_2, x_3, x_4, x_5, x_6, x_7)$ and $(x_1, x_2, x_3, x_6, x_7, x_4, x_5)$.

In addition, the unateness information and symmetry classes can be used together to derive a tighter bound on the number of non-redundant variable orders. Unate and binate symmetry sets are disjoint, since both unateness and symmetry properties have to be the same for two variables to be interchangeable. Thus we can write $C_i = C_i^b \cup C_i^u$, $i = 1, 2, \ldots, n$, and $|C_i| = |C_i^b| + |C_i^u|$, where the superscripts b and u denote binate and unate, respectively. Thus the number of non-redundant permutations are at most $\prod_{i=1}^{n} |C_i^b|! \cdot |C_i^u|! = \prod_{i=1}^{n} |C_i^b|! \cdot (|C_i| - |C_i^b|)!$.

BOOLEAN COVERING. We describe here a procedure for Boolean covering of a subject graph. We assume that the network has been partitioned into subject graphs and decomposed into base functions beforehand. The procedure is reminiscent of the tree-covering algorithm, because it uses a bottom-up traversal of the subject graph. However, the subject graph is not required to be a tree, but just a rooted dag. As with structural covering, the library is required to include the base functions.

We define a *cluster* as a rooted connected subgraph of the subject graph. It is characterized by its *depth* (longest path from the root to a leaf). The associated *cluster function* is the Boolean function obtained by collapsing the logic expressions associated with the vertices into a single expression.

> **Example 10.3.13.** Consider the subject graph shown in Figure 10.16. The base functions for the decomposition of the subject graph are 2-input AND and OR functions. The expressions associated with the subject graph are the following:
>
> $$j = xy; \quad x = e + z; \quad y = a + c; \quad z = c' + b$$
>
> We consider the clusters rooted at vertex v_j of the subject graph and shown by different shadings in the picture. The corresponding cluster functions are:
>
> $$f_j^1 = xy; \quad f_j^2 = x(a + c); \quad f_j^3 = (e + z)y;$$
>
> $$f_j^4 = (e + z)(a + c); \quad f_j^5 = (e + c' + b)y; \quad f_j^6 = (e + c' + b)(a + c)$$

Let us consider first the minimum-area covering problem. The covering algorithm attempts to match each cluster function to a library element. For each match,

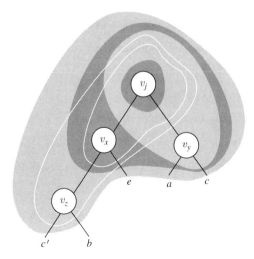

FIGURE 10.16
Clusters of the Boolean covering algorithm.

the area cost of a cover is computed by adding the area cost of the matching cell to the area cost of the covers of the subgraphs rooted at the vertices providing the inputs to the matching cell. There is always at least one match for each vertex, because the library contains the base functions. When matches exist for multiple clusters rooted at a given vertex, the algorithm selects the cluster that minimizes the cost of the cover of the locally rooted subgraph.

A similar procedure is used to minimize the data-ready time at the root of the subject graph under the assumption of constant propagation delays. The data-ready time at any vertex is computed by adding the propagation delay of the matching cell to the maximum of the data-ready times at the vertices providing the inputs to the matching cell. When matches exist for multiple clusters, the algorithm selects again the cluster that minimizes the local data-ready time.

The Boolean covering algorithm is based on the same dynamic programming paradigm used for structural matching. However, its complexity is higher, because many clusters exist for each vertex of the subject graph. An exact pruning procedure disregards those cluster functions whose support cardinality is larger than the maximum number of inputs in any library cell. A heuristic method used in CERES [23] limits the clusters to those whose depth is bounded from above by a pre-defined parameter. Unfortunately, this may prevent reaching an optimum solution.

The Boolean covering algorithm still yields optimum solutions for tree or leaf-dag decompositions of the subject graph when the depth of the clusters is unbounded. As in the case of structural covering, the solution depends on the particular decomposition and partition into subject graphs. Therefore the global optimality of the covering step *per se* has limited practical value, and near-optimal covering solutions are often more than adequate to obtain results of overall good quality.

Example 10.3.14. We consider once more the control unit of the complete differential equation integrator described by the optimized logic network in Example 8.2.10. The result of binding the network to a commercial library with program Ceres is shown next:

```
.model control ;
      .inputs   reset c CLK ;
      .outputs s1 s2 s3 s4 s5 s6 s1s3 s2s3 s2s4 s1s2s3 s1s3s4
               s2s3s4 s1s5s6 s2s4s5s6 s2s3s4s6 s3s4s5s6 ;

      .call DF1_LatchOut_v2 DF1 (v5.0, CLK ;;LatchOut_v2 ) ;
      .call DF1_LatchOut_v3 DF1 (v5.1, CLK ;;LatchOut_v3 ) ;
      .call DF1_LatchOut_v4 DF1 (v5.2, CLK ;;LatchOut_v4 ) ;
      .call NOR2_s1 NOR2 (LatchOut_v4, LatchOut_v3 ;;s1 ) ;
      .call NOR2B_s2 NOR2B (LatchOut_v4, LatchOut_v3 ;;s2 ) ;
      .call AND2A_s3 AND2A (LatchOut_v2, LatchOut_v4 ;;s3 ) ;
      .call AND2A_s4 AND2A (LatchOut_v4, LatchOut_v2 ;;s4 ) ;
      .call AND2A_s5 AND2A (LatchOut_v2, LatchOut_v3 ;;s5 ) ;
      .call AND2A_s6 AND2A (LatchOut_v3, LatchOut_v2 ;;s6 ) ;
      .call OR2_s1s3 OR2 (s3, s1 ;;s1s3 ) ;
      .call AND2A_s2s3 AND2A (s6, LatchOut_v4 ;;s2s3 ) ;
      .call OR2_s2s4 OR2 (s4, s2 ;;s2s4 ) ;
      .call OR2_s1s2s3 OR2 (s2s3, s1s3 ;;s1s2s3 ) ;
      .call OA1B_s1s3s4I OA1B (LatchOut_v4, s5, s1s3 ;;s1s3s4I ) ;
      .call OR2_s2s3s4 OR2 (s2s4, s2s3 ;;s2s3s4 ) ;
      .call OA1B_s1s5s6I OA1B (LatchOut_v3, s2s3, s5 ;;s1s5s6I ) ;
      .call OR2_s2s4s5s6 OR2 (LatchOut_v3, LatchOut_v2 ;;s2s4s5s6 ) ;
      .call OR2_s2s3s4s6 OR2 (LatchOut_v4, LatchOut_v2 ;;s2s3s4s6 ) ;
      .call AO1A_s3s4s5s6 AO1A (s2s3, s2s4s5s6, s3 ;;s3s4s5s6 ) ;
      .call NOR2_v5.0 NOR2 (LatchOut_v2, reset ;;v5.0 ) ;
      .call AO2A_v5.1 AO2A (reset, s4, zz_7, zz_4 ;;v5.1 ) ;
      .call AO1A_v5.2 AO1A (s2s3s4s6, v5.0, zz_8 ;;v5.2 ) ;
      .call AND3B_zz_4 AND3B (reset, LatchOut_v2, c ;;zz_4 ) ;
      .call NOR2_zz_7 NOR2 (s2s4s5s6, reset ;;zz_7 ) ;
      .call AND2A_zz_8 AND2A (reset, s2 ;;zz_8 ) ;
      .call INV_s1s3s4 INV (s1s3s4I ;;s1s3s4 ) ;
      .call INV_s1s5s6 INV (s1s5s6I ;;s1s5s6 ) ;
.endmodel control ;
```

The network is represented as a module-oriented netlist. Each record corresponds to a cell instance: the first label is an arbitrary instance identifier, the second label denotes the cell type and the argument in parentheses denotes the names of the nets that are inputs and outputs to each cell. Note that the original signal names of Example 4.7.6 have been preserved.

10.3.3 Covering Algorithms and Polarity Assignment

In this section we describe the polarity assignment problem in conjunction with the matching problem, because together they impact the cost of an implementation. There-fore we consider the possibility of matching a subnetwork to a library cell that implements the complement of the subnetwork's function and/or uses complemented input signals.

The goal of considering the polarity assignment problem together within library binding is to find the best cover, regardless of the polarity of the signals. Hence a cover of lower cost can be found as compared to approaches that disregard the flexibility in choosing the signal polarities.

STRUCTURAL COVERING. In the case of structural covering, the optimal polarity assignment can be achieved by using a clever trick [10, 26]. Consider the network and the library cells after decomposition into base functions. All connections between base gates are replaced by inverter pairs which do not affect the network and cell behavior. (Connections to or from inverters do not need to be replaced.) This transformation can be easily applied to both the subject graph and the pattern graphs by replacing selected edges by edge pairs with an internal vertex labeled as an inverter.

The dynamic programming covering algorithm can now take advantage of the existence of both polarities for each signal in the subject graph in the search for a minimum (area or delay) cost solution. It is important that the newly introduced inverters are removed when they do not contribute to lowering the overall cost. For this reason, a fake element is added to the library. It consists of an inverter pair whose actual implementation is a direct connection and whose cost is zero. Because of the optimality of the covering algorithm, the cost of the bound network starting from an unbound network with inverter pairs has lower (or at most equal) cost than the cost of a bound network derived without using inverter pairs. Note again that the optimality is within a tree. Heuristics can be used to minimize the inverters across different subject graphs [26]. The only drawback of using the inverter-pair heuristic is a slightly increased computational cost due to the larger size of the subject and pattern graphs.

Example 10.3.15. Consider the network of Figure 10.13, repeated for convenience in Figure 10.17 (a). Figure 10.17 (b) shows the network after the insertion of inverter pairs and Figure 10.17 (c) the subject graph. We assume that input signals are available in both polarities and this is modeled by adding a zero-cost inverter on all inputs. Let us assume that library cells available are those shown in Figure 10.8. (Note that inverter pairs should be added to the pattern trees of the last two cells of Figure 10.8.)

We search for a minimum-area cover. Let us visit the subject graph bottom up. The best cover at v_1 and v_2 is provided by a zero-cost INV gate, because input signals are available in both polarities. The best cover at v_3 is provided either by a NAND2 gate (plus two zero-cost INV gates) or an OR2 gate. When considering vertex v_4, there are three possible covers: (1) an INV gate plus a NAND2 gate (and two zero-cost INV gates); (2) an AND2 gate (and two zero-cost INV gates); (3) a NOR2 gate. And so on. The choice depends on the actual cost of the cells.

Assume that the area costs of the cells {INV, NAND2, AND2, AOI21} are {2, 3, 4, 6} respectively, as in Example 10.3.7. In addition, assume that the cost of a NOR2 cell is 2.5 and the cost of an inverter pair is 0. Then, a minimum-area cost cover can be derived that uses three NOR2 cells, with a total cost of 7.5. Note that the cost is inferior to that of the cover computed in Example 10.3.7, which was 9. Note also that inputs b, c, d have been used with the negative polarity.

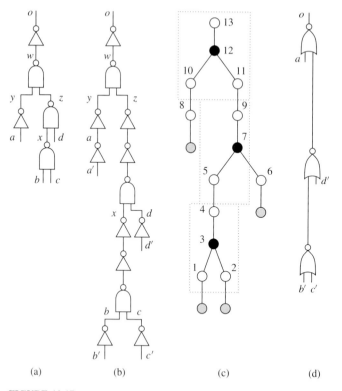

FIGURE 10.17

(a) Original network. (b) Network after the insertion of inverter pairs. (c) Subject tree. (The numbers are vertex identifiers.) (d) Network cover.

BOOLEAN COVERING. In the case of Boolean covering, the polarity assignment problem can be explained with the help of a formalism used to classify Boolean functions. Consider all scalar Boolean functions over the same support set of n variables. Two functions $f(\mathbf{x})$ and $g(\mathbf{x})$ are said to belong to the same $\mathcal{NPN}$ class if there are a permutation matrix $\mathbf{P}$ and complementation operators $\mathbf{N}_i, \mathbf{N}_o$ such that $f(\mathbf{x}) = \mathbf{N}_o\, g(\mathbf{P}\, \mathbf{N}_i\, \mathbf{x})$ is a tautology [15]. The complementation operators specify the possible negation of some of its arguments. In other words, two functions belong to the same $\mathcal{NPN}$ class, and are called $\mathcal{NPN}$-*equivalent*, when they are equivalent modulo the negation ($\mathcal{N}$) of the output, the permutation ($\mathcal{P}$) of the inputs and the negation ($\mathcal{N}$) of the inputs.

When considering the Boolean covering problem jointly with the polarity assignment, the definition of Boolean match is extended as follows. Two single-output combinational functions $f(\mathbf{x})$ and $g(\mathbf{y})$ (with the same number of support variables) match when they are $\mathcal{NPN}$-equivalent.

Note that the extension to Definition 10.3.1 entails the use of the complementation operator that models the freedom in choosing the polarity of the input and output signals. As a result of the more general definition of a match, more library cells can match a given cluster function and *vice versa*.

Example 10.3.16. Any cluster function $f(a, b)$ in the set:

$$\{a + b, a' + b, a + b', a' + b', ab, a'b, ab', a'b'\}$$

can be matched by the pattern function $g(x, y) = x + y$ (OR2 cell). Note, however, that cluster functions in the set:

$$\{a'b + ab', a'b' + ab\}$$

cannot be matched to that pattern function.

The covering algorithm described in Section 10.3.2 can still be used, provided that the search for a match is extended to consider all input/output polarities and input permutations. Because of the extension to the definition of a Boolean match, the polarity information of the unate input variables is irrelevant. This has two consequences. First, the search for a match can be simplified by complementing the negative unate variables. Second, the search for a match can be reduced to a sequence of equivalence tests while considering the possible polarity assignments to the binate variables [23].

Filters that quickly check necessary conditions for matching and that are based on the unateness and symmetry properties (Section 10.3.2) can still be used to reduce further the number of equivalence checks [23].

Example 10.3.17. Consider the cluster function $f_1 = a' + b$ and the pattern function $g_1 = p + q$, which have a match provided that input a is complemented. It would be wasteful to consider the polarity assignments of all input variables of f_1, because there are no binate variables in f_1. The polarity of a can be changed, and $\widetilde{f_1} = a + b$ is compared to $g_1 = p + q$.

Consider now the cluster function $f_2 = a'b' + ac$ and the pattern function $g_2 = pq + p'r$, which also have a match provided that inputs a and b are complemented. Let us consider the equivalence tests required to detect the match. First, the polarity of the unate variable b can be changed and $\widetilde{f_2} = a'b + ac$ be considered instead of f_2. Second, since there is only one binate variable (i.e., a), the polarity assignments to be considered are those of a, corresponding to functions $\widetilde{f_2^1} = a'b + ac$ and $\widetilde{f_2^2} = ab + a'c$. These two functions are compared to the pattern function $g_2 = pq + p'r$. (Note that a permutation of the unate variables, i.e., exchanging b with c, would lead to the same functions in this case.)

10.3.4 Concurrent Logic Optimization and Library Binding*

It is customary to perform library-independent logic optimization and library binding as two separate tasks. Multiple-level optimization techniques based on logic transformations do not impose constraints on the local expressions, such as having to match some library element. For example, the extraction of a subexpression is done regardless of whether this subexpression has a match. It is conjectured that if logic transformations were constrained to provide only valid matches, the solution space would be reduced and the results would be poor.

We consider now the possibility of combining logic transformations and library binding in a single step. In particular, we concentrate on the relations between logic simplification (see Section 8.4.2) and binding. Since simplification takes advantage of the degrees of freedom expressed by *don't care* conditions, we consider those *don't care* sets that are specified at the network boundary and those that arise from the network interconnection itself. Since the topology of the network changes during the covering stage, *don't care* conditions must be computed dynamically.

We consider first the problem of extracting the *don't care* information of a logic network during binding. A sketch of a partially bound network is shown in Figure 10.18. The interconnection of the cells in the bound portion of the network induces a satisfiability *don't care* (SDC) set (see Section 8.4.1). For example, if vertex v_a, corresponding to variable a, is bound to an OR2 cell with inputs b and c, the assignment $a = b + c$ implies that the relations among the variables given by $a \oplus (b + c) = a'b + a'c + ab'c'$ can never happen and belong to the SDC set.

Consider now the problem of matching a cluster to a cell using the same local inputs or a subset thereof. The degrees of freedom in matching the corresponding cluster function f are the impossible local input patterns and those whose effect is unobservable at the network outputs. The former are given by its local *controllability don't care* (CDC) set and the latter by the local *observability don't care* (ODC) set. (See Section 8.4.1.) Note that by the nature of the binding algorithms that progress from the network inputs to the outputs, it is easier to use CDCs than ODCs, because CDCs depend on a bound portion of the network while ODCs depend on an unbound

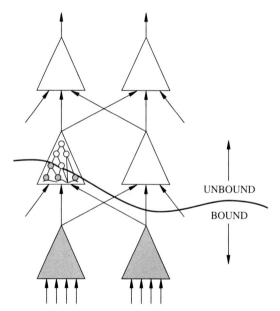

UNBOUND

BOUND

FIGURE 10.18
Example of a partially bound network.

part, and hence are subject to change. Therefore ODCs must be either updated after each cell binding or approximated (e.g., by using compatible ODCs).

The purpose of using the *don't care* conditions in library binding is to increase the possible number of matches, as allowed by the *don't cares*, in the search for a lower cost cover. To exploit *don't care* conditions in library binding, the concept of matching needs to be generalized further. Let us consider a cluster function $f(\mathbf{x})$ and a pattern function $g(\mathbf{y})$. The *don't care* conditions are represented here as a function $d(\mathbf{x})$. Thus, they are expressed in terms of the variables that are in support of $f(\mathbf{x})$. The pattern function can replace the cluster function if there exists a completely specified function $\tilde{f}(\mathbf{x})$ that matches $g(\mathbf{y})$ such that $f(\mathbf{x}) \cdot d'(\mathbf{x}) \subseteq \tilde{f}(\mathbf{x}) \subseteq f(\mathbf{x}) + d(\mathbf{x})$.

> **Example 10.3.18.** Consider again the subject graph of Figure 10.16. Let us assume that vertex v_x has been bound to a three-input OR gate, i.e., $x = \text{OR } 3(c', b, e)$. Hence $x \oplus (c' + b + e) = x'c' + x'b + x'e + xcb'e'$ belongs to the satisfiability *don't care* set. Consider now the cluster function for vertex v_j, namely $f = x(a + c)$, and its corresponding controllability *don't care* set. The controllability *don't care* set expresses that portion of the SDC that is independent of the values of the variables that are not in $sup(f)$. Namely, $CDC = \mathcal{C}_{b,e}(x'c' + x'b + x'e + xcb'e') = x'c'$. Pattern $x'c'$ cannot be an input to the cluster represented by cluster function $f = x(a + c)$.
>
> Let us consider now the matching of $f = x(a + c)$ with *don't care* conditions $d = x'c'$. Equivalently, we can consider the possible matches for function $\tilde{f}$ such that $f \cdot d' \subseteq \tilde{f} \subseteq f + d$, or equivalently $xa + xc \subseteq \tilde{f} \subseteq xa + xc + x'c'$. Whereas f can match $g_1 = p(q + r)$, it can also match $g_2 = qr + q's$. (This is possible because $\tilde{f} = cx + c'a$ satisfies the bounds and matches g_2.) Pattern function g_2 represents a multiplexer. The choice of g_2 is preferable to g_1 in some libraries, due to efficient implementation of the multiplexing function with pass transistors (Figure 10.19).

By using the controllability (and possibly observability) *don't care* conditions of the cluster function f while trying to match it to a cell, we combine Boolean simplification with library binding. Indeed we search for the local best match allowed by the degrees of freedom specified by the *don't care* conditions. A further extension to this method is to allow the matching cell to use as inputs any output of a bound cell. In this case, the CDC set is replaced by the SDC set induced by the portion of the network already bound. This approach is then analogous to the use of Boolean division, which is now performed concurrently with library binding.

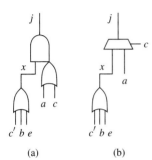

(a)   (b)

FIGURE 10.19
(a) Bound network. (b) Bound network exploiting *don't care* conditions.

Let us consider now Boolean matching with *don't care* conditions. The simplest approach to using *don't care* conditions in library binding is to simplify the cluster functions before matching. This entails the computation of the *don't cares* induced by the partially bound network and invoking a two-level logic minimizer. This approach has a potential pitfall. Minimizers for multiple-level designs target the reduction of the number of literals of a Boolean expression. Whereas such an approach leads to a smaller (and faster) implementation in the case of a design style based on cell generators [4], it may not improve the local area and timing performance in a standard-cell based or array-based design style. For example, cell libraries exploiting pass transistors may be faster and/or smaller than other gates having fewer literals.

> **Example 10.3.19.** Consider again the network of Example 10.3.18. After binding vertex v_x to a three-input OR gate, the CDC set for cluster function $f = x(a + c)$ is $d = x'c'$. The simplification of f with this *don't care* set would leave f unaltered. Thus f would be bound to a cell modeled by pattern function $g_1 = p(q + r)$ and not to a multiplexer. This choice is unfortunate in the cases in which the multiplexer implementation is smaller and faster.

Example 10.3.19 has just shown that applying Boolean simplification before matching may lead to inferior results, as compared to merging the two steps in a single operation. Thus, the degrees of freedom provided by *don't care* sets is best used in selecting an appropriate Boolean match that minimizes area (or delay).

Boolean matching with *don't care* conditions can be verified using ROBDDs by performing the containment tests $f(\mathbf{x}) \cdot d'(\mathbf{x}) \subseteq \tilde{f}(\mathbf{x}) \subseteq f(\mathbf{x}) + d(\mathbf{x})$ while considering all possible polarities and input assignments of $g(\mathbf{x})$ to $\tilde{f}(\mathbf{x})$. Savoj *et al.* [28] proposed a Boolean matching method based on ROBDDs that exploits the symmetry information to shorten the search of a match. Mailhot *et al.* [23] proposed an alternative formulation based on a traversal of a *matching compatibility* graph representing all $\mathcal{NPN}$ classes of functions of n variables and annotated with the library information. Both methods have been shown to yield bound networks of superior quality when compared to methods that do not exploit *don't care* conditions.

10.3.5 Testability Properties of Bound Networks

The testability of a network is affected by all logic synthesis and optimization techniques, including library binding. We restrict our attention in this section to testability of bound combinational networks with *stuck-at* fault models. We use the definition of testability introduced in Sections 7.2.4 and 8.5, where a fully testable circuit is one that has test patterns which can detect all faults. We assume that the library cells are individually fully testable.

Let us consider first a circuit implementing a subject tree. Tree covering techniques which replace subtrees by logically equivalent cells preserve the circuit's testability. This is true for both structural matching and Boolean matching without the use of *don't cares*.

Let us now consider multiple-input, multiple-output networks. The following theorem, due to Hachtel *et al.* [14], gives conditions for full testability for multiple faults.

> **Theorem 10.3.1.** Consider a logic network decomposed in terms of NAND base functions and partitioned into trees. Let the network be bound to library cells in such a way that each tree is replaced by a tree of cells that is fully testable for single stuck-at faults, and let the fanout points of the unbound network match the fanout points of the bound network. Then, if a circuit implementing a trivial binding of the original decomposed network is fully testable for multiple faults, so is a circuit realizing the bound network.

The practical significance of this theorem is the following. If we can design an unbound network which is fully testable (using logic synthesis methods such as those shown in Section 8.5), then the covering methods shown in Sections 10.3.1 and 10.3.2 yield fully testable networks for multiple stuck-at faults when applied while satisfying the assumptions of Theorem 10.3.1.

Lastly, let us consider the testability properties of bound networks constructed using Boolean matching techniques with *don't care* conditions and non-tree-based decompositions. Two extreme cases can be detected while matching a cluster rooted at a vertex, say v_x. The first is when the cluster function f is included in the *don't care* set and the second is when the disjunction (i.e., union) of the function f and the *don't care* set is a tautology. In these two cases, vertex v_x is associated with either a FALSE or a TRUE value, which is then propagated in the network. No cell is bound to v_x. When the complete controllability *don't care* set can be computed for each cluster function f, Boolean matching binds a cluster to a cell when there exist test patterns at the circuit input that can set the output of that cell to FALSE and to TRUE. Therefore, the bound network is fully controllable. If redundancy removal is applied to the bound network to remove unobservable cells, then the resulting network is fully testable for single stuck-at faults regardless of the properties of the original network.

10.4 SPECIFIC PROBLEMS AND ALGORITHMS FOR LIBRARY BINDING

Standard-cell and mask-programmable gate array libraries can be described well as collections of cells, and so the binding algorithms of Section 10.3 are directly applicable. We consider now other design styles where libraries are represented implicitly, i.e., without enumerating their elements. This is the case of some macro-cell and *field-programmable gate array* (FPGAs) design styles where special techniques for library binding are required.

There are several types of macro-cell design styles where *module generators* construct the physical view of a macro-cell from an (optimized) unbound logic network. Some macro-cell generators construct the macro-cell by placing and wiring *functional cells*, which are similar in principle to standard cells but are not stored in a library. Their physical layout is automatically synthesized from logic expressions. To satisfy area and performance requirements related to the corresponding physical view, the logic expressions must satisfy some constraints. Typical constraints are related to the maximum number of inputs and/or the maximum number of transistors in series or in parallel in any given cell. Logic expressions and cells that satisfy these functional constraints constitute a *virtual library*. Library binding consists of manipulating the logic network until all its expressions satisfy the functional constraints and the network has some desired optimality properties. Berkelaar and Jess [4] proposed a heuristic binding algorithm for this task where functional constraints are the maximum number of transistors in series or in parallel in any cell. When the functional constraint is only the maximum number of cell inputs, the problem is similar to a binding problem for FPGAs, described in Section 10.4.1.

Field-programmable gate arrays are pre-wired circuits that are programmed by the users (on the field and after chip fabrication) to perform the desired functions. (See Section 1.3.) Today, there are several FPGA providers, and circuits differ widely in their programming technology and implementation style. We consider here just their major features, without delving into implementation details, to explain the relation to the library binding problem. Broadly speaking, FPGAs can be classified as either *soft* or *hard* programmable. Circuits in the first class are implemented by a programmable interconnection of registers and *look-up tables*, which store the personality of local combinational logic functions. They can be programmed by loading an on-chip memory that configures the tables and the interconnection. Circuits in the second class consist of an array of programmable logic modules, each implementing a logic function and that can be personalized and connected by programming the *antifuses*.

Library binding for FPGAs is an important and complex problem. Its difficulty stems from the fact that binding pre-wired circuits is deeply related to physical design issues [27]. Bound networks must be routable, and routability depends on the binding itself. In addition, path delays are heavily conditioned by wiring, because of the programmable interconnect technology. Thus iterative improvement techniques have been shown to be important in achieving good quality solutions. At present, this topic is still the subject of ongoing research. A review of the state of the art is reported in reference [27].

Since FPGA architectures are novel and quickly evolving, it is hard to present the binding problem in a comprehensive way. We therefore consider here the fundamental problems and solutions at the logic level of abstraction. We neglect purposely physical design issues, as these are beyond the scope of this book and also because they are often dependent on the specific architecture.

10.4.1 Look-Up Table FPGAs

The *virtual library* of look-up table FPGAs is represented by all logic functions that can be realized by the tables. Even when considering just those tables implementing single-output functions of a few variables (e.g., 5), enumerating the library cells is not practical.

> **Example 10.4.1.** Let us consider the FPGAs marketed by Xilinx Inc. In the 3000 series, each look-up table implements any single-output function of five variables or any two-output function of five variables, where each component has no more than four variables in its support.

Let us concentrate on the first option only. There are 2^{2^n} different functions of n variables. Hence, there are 2^{2^5}, or about 4 billion, possible single-output functions.

The organization of the look-up tables differs in various FPGA products. For this reason, we concentrate here on binding combinational networks and in particular on the following problem that is important for all look-up-table-based FPGAs. We assume that look-up tables can implement any scalar combinational function of n input variables. Given a combinational logic network, binding consists of finding an equivalent logic network, with a minimum number of vertices (or minimum critical path delay) such that each vertex is associated with a function implementable by a look-up table, or equivalently with a local function having at most n input variables.

Example 10.4.2. Consider the network of Figure 10.20 (a). Assume that the look-up tables can implement any combinational function of $n = 5$ inputs. Then the network can be covered by three tables, as shown by Figure 10.20 (b).

The algorithms of Section 10.3 are not applicable to this problem because the library cannot be enumerated. Some specialized binding algorithms have been recently proposed, but at the time of writing this topic is still under investigation. Thus, we present only the flavor of two approaches [6, 25] and refer the reader to specific articles [17, 22] for the others.

Usually, the starting point for binding is a logic network decomposed into base functions, such as ANDs and ORs. When considering n-input look-up tables, the base functions are required to have at most n inputs, but it is usually convenient to use two-input base functions to achieve a finer network granularity.

The tree covering paradigm has been adapted to this problem by Francis *et al.* [25], whose heuristic approach to binding involves covering a sequence of subject graphs into which the decomposed network has been partitioned. Covering is driven by the principle of packing as much functionality as possible into each look-up table subject to its input size constraint n. This leads to the need for solving the following subproblem.

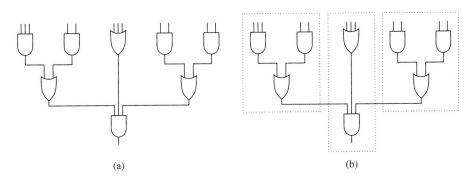

(a)   (b)

FIGURE 10.20
(a) Subject tree. (b) Cover by three five-input look-up tables.

Consider a *sum of products* representation of a single-output function whose product terms have at most n variables. Assume first that the product terms have disjoint support, as in the case corresponding to a tree-like decomposition. If the function has at most n inputs, it can be trivially implemented by one table. Otherwise, groups of product terms must be assigned to different tables. This problem bears a strong resemblance to the *bin packing* problem, which consists of packing a given set of objects (here product terms) of bounded size into bins (here tables) of a given capacity. Unfortunately bin packing does not model the problem precisely, because partitioning the set of product terms into b tables requires more than b look-up tables to implement the function, the additional ones being devoted to performing the sum of the partial sums.

> **Example 10.4.3.** Let the table size be $n = 3$. Let the function to be implemented into tables be $f = ab + cd$. Even though the two product terms have $2 < n = 3$ variables each, function f has $4 > n = 3$ inputs and cannot be implemented by a single table. If two look-up tables are used to implement each product term, then one additional table should be devoted to performing their sum. This is equivalent to a decomposition into three tables, implementing $f_1 = ab$; $f_2 = cd$; $f = f_1 + f_2$. Note that two tables can cover the function, according to the following decomposition: $f = ab + f_2$; $f_2 = cd$.

This problem can be solved by modifying a bin packing procedure as follows [25]. The algorithm selects iteratively the product term with most variables and places it into any table where it fits. (Recall that each table handles at most n variables.) If no table has enough capacity, a new table is added to the current solution containing the selected product term. When all product terms have been assigned to tables, the following steps are iterated. The table with the fewest unused variables is declared final, and a variable is associated with this table and assigned to the first table that can accept it. The procedure terminates when one table is left; this table yields the desired output. Even though this algorithm is heuristic, it can be shown that when the product terms are disjoint (i.e., for subject trees), the solution has a minimum number of tables when $n \leq 6$.

> **Example 10.4.4.** Consider the problem of Example 10.4.3. The algorithm assigns a table to each product term. Without loss of generality, let the first table correspond to product term cd. Then, this table is declared final and a variable, say z, is associated with it. It represents the assignment $z = cd$ implemented by the table. Then the algorithm tries to fit the single-variable product term z into the other table. Since the second table has an unused input, product term z can be added to yield $f = ab + z$.

This algorithm has been implemented in program CHORTLE-CRF [25] with a few extensions. First, subject graphs are not restricted to be trees and as a consequence product-term pairs may share variables. To cope with this extension, CHORTLE-CRF exhaustively explores all possible assignments of product terms with intersecting support to the same table. The second extension addresses the inefficiencies due to the network partition. CHORTLE-CRF attempts to duplicate product terms in exchange for reducing the total number of look-up tables. We refer the interested reader to references [25] and [27] for further details.

A second remarkable approach to the look-up table binding problem was recently proposed by Cong and Ding [6]. They considered the problem of minimizing the critical path delay in a bound network. Since all tables are alike and display the same propagation delay, this problem is equivalent to minimizing the maximum number of tables on any input/output path. This problem can be solved exactly in polynomial time, given a network decomposed into base functions with no more than n inputs. Unfortunately, the bound network depends on the particular network decomposition that is used as the starting point for this procedure.

The algorithm is based on a transformation of the binding problem into a network flow computation that can be solved exactly by standard algorithms. The transformation is fully detailed in reference [6]. Recently, the same authors proposed another algorithm that minimizes the number of look-up tables required under a bound on the critical path delay [7]. The algorithm assumes that overlapping look-up tables in the cover are not beneficial as far as minimizing the number of tables and avoids it. This algorithm has been implemented in program FLOWMAP-R, which can be used to obtain area-delay trade-off curves for a given network.

10.4.2 Anti-Fuse-Based FPGAs

The *virtual library* of anti-fuse-based FPGAs is represented by all logic functions that can be implemented by personalizing the logic module. This is usually achieved by shorting inputs either to a voltage rail or together, by programming the antifuses.

Example 10.4.5. Let us consider the FPGAs marketed by Actel Inc. In the *Act1* series, the module implements the function $m_1 = (s_0 + s_1)(s_2 a + s_2' b) + s_0' s_1' (s_3 c + s_3' d)$, while in the *Act2* series it implements the function $m_2 = (s_0 + s_1)(s_2 s_3 a + (s_2 s_3)' b) + s_0' s_1' (s_2 s_3 c + (s_2 s_3)' d)$. In both cases, the module is a function of $n = 8$ inputs.

As an example of programming, by setting $s_0 = s_1 = 1$, function m_1 implements the multiplexer $s_2 a + s_2' b$. This is achieved by providing a path from inputs s_0 and s_1 to the power rail through an antifuse.

There are about 700 functions that can be derived by programming either module.

The organization of the FPGAs and the type of logic module differ in various products. For this reason, we make a simplifying assumption that defines a fundamental problem common to different anti-fuse-based FPGAs. We assume that all programmable modules implement the same single-output combinational function, called the *module function*. We concentrate also on the binding problem of combinational logic networks. Namely, given a combinational logic network, binding consists of finding an equivalent logic network with a minimum number of vertices (or minimum critical path delay) such that each expression can be reduced to a personalization of the module function. Note that a module personalization can be seen as inducing a stuck-at or a bridging fault on its inputs.

In some cases it is practical to derive the library explicitly, by considering all possible personalizations of the module function, because the size of the library is limited (e.g., fewer than 1000 functions). This approach has some advantages. First, standard binding algorithms can be used. Second, a library subset can be considered that excludes gates with undesirable delays or pin configurations. Third, the precise area and delay cost of each cell can be established.

We comment here on the general case where the virtual library is so large that it is not practical to enumerate it. Specialized algorithms for binding have been proposed based on structural and Boolean representations [11, 18, 22, 27].

Structural approaches exploit the specific nature of the uncommitted module. For example, commercial implementations of FPGAs use programmable modules based on signal multiplexing. In this case, it is convenient to decompose the subject graph by choosing a multiplexer as the base function. The entire library can then be implicitly represented by pattern graphs that use a similar decomposition. Binding can then be done by structural covering using the dynamic programming paradigm [22, 27].

> **Example 10.4.6.** Consider the module function $m = s_1(s_2a + s_2'b) + s_1'(s_3c + s_3'd)$. There are four corresponding pattern graphs that represent the functions that m can implement. The pattern graphs are leaf-dags and are shown in Figure 10.21.
>
> When the module differs from a cascade of multiplexers, the number of pattern graphs increases and may include dags that are not leaf-dags. In particular, this applies to module function $m_1 = (s_0 + s_1)(s_2a + s_2'b) + s_0's_1'(s_3c + s_3'd)$. By using a restricted set of patterns to represent m_1, namely eight leaf-dag patterns, good-quality solutions were achieved using program MIS-PGA [22].

The Boolean covering algorithm of Section 10.3.2 can be combined with a specialized Boolean matching algorithm that detects whether a cluster function can be implemented by personalizing the module function and that determines the personalization at the same time. Let us consider only personalizations by input stuck-ats, for the sake of simplicity. Then, the module function can implement any cluster function that matches any of its cofactors.

ROBDD representations can be very useful to visualize and solve this matching problem. Indeed, given an order of the variables of the module function and a corresponding ROBDD representation, its cofactors with respect to the first m variables in the order are represented by subgraphs of the ROBDD. These subgraphs are rooted at those vertices reachable from the root of the module ROBDD along m edges corresponding to the variables with respect to which the cofactors have been taken, or equivalently to those variables that are stuck at a fixed value by the personaliza-

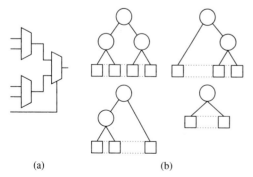

(a)   (b)

FIGURE 10.21
(a) Module function. (b) Pattern graphs.

tion. Unfortunately, to consider all possible personalizations, all variable orders of the module function and the corresponding ROBDDs must also be considered. This can be done by constructing a shared ROBDD that encapsulates the virtual library corresponding to the module function. Extensions to cope with personalization by bridging have also been proposed [11].

Example 10.4.7. Consider the module function $m = s_1(s_2 a + s_2' b) + s_1'(s_3 c + s_3' d)$ and cluster function $f = xy + x'z$, shown in Figures 10.22 (a) and (d), respectively. Figure 10.22 (b) shows the ROBDD of m for variable order $(s_1, s_2, a, b, c, s_3, d)$, and Figure 10.22 (c) shows the ROBDD of f for variable order (x, y, z). Since the ROBDD of f is isomorphic to the subgraph of the ROBDD of m rooted in the vertex labeled s_2 (which is the right child of s_1), the module function can implement f by sticking s_1 at 1.

Note that other cluster functions that can be implemented by the module function may have ROBDDs that are not isomorphic to any subgraph of the ROBDD of Figure 10.22 (b). This is due to the fact that a specific variable order has been chosen to construct this ROBDD.

Recently, Burch and Long [5] developed canonical forms for representing functions under input negation and permutation that can be used efficiently for Boolean matching as well as for matching under a stuck-at constant and/or bridging of some inputs. These forms have been applied to the development of efficient algorithms for library binding of anti-fuse-based FPGAs [5].

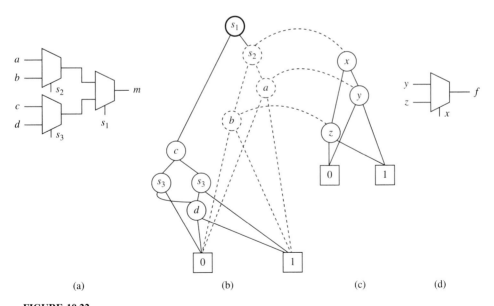

FIGURE 10.22
(a) Programmable module. (b) Module ROBDD. (c) Cluster ROBDD. (d) Representation of the cluster function.

10.5 RULE-BASED LIBRARY BINDING

Rule-based library binding is a widely used alternative and a complement to the algorithmic approach described in the previous sections. Some of the early logic synthesis systems, such as LSS [8, 9], used rules for both logic optimization and library binding. Some current commercial and proprietary design systems use rule-based binding, sometimes in conjunction with algorithmic binding [13, 16, 30].

In a rule-based system, a network is bound by a stepwise refinement. The network undergoes local transformations that preserve its behavior. Each transformation can be seen as the replacement of a subnetwork by an equivalent one that best exploits the cell library. Since rule-based library binding is similar to rule-based optimization of unbound networks, which was presented in Section 8.7, we shall describe only those issues that are specific to the library binding problem.

Each entry in the rule database contains a circuit pattern, along with an equivalent replacement pattern in terms of one or more library elements. Entries may represent simple or complex rules. Simple rules propose just the best match for a subnetwork. Complex rules address situations requiring a restructuring of the network. As an example, a complex rule may be applicable to a cell with a high load, which may require the use of a high-drive cell, or insertion of buffers, or even duplication of some gates.

> **Example 10.5.1.** Consider the rules shown in Figure 10.23. The first two can be called simple, the third complex.
>
> The first rule shows that two cascaded two-input AND gates can be bound to a three-input AND gate.
>
> The second rule indicates that a two-input AND gate with an inverted input and output can be bound to a two-input NOR gate with an inverted input.
>
> The third rule shows that an AND gate on a critical path with a high load can be replaced by a NAND gate followed by two inverters in parallel. The first inverter drives the critical path and the second the remaining gates, which amount to a large load.

The execution of the rules follows a priority scheme. For each rule in a given order, the circuit patterns that match the rules are detected and the corresponding replacements are applied. The overall control strategy of rule-based systems for binding is similar to that used for optimizing unbound logic networks and described in Section 8.7. Some rule-based systems, such as LSS [9] and LORES/EX [16], use a greedy

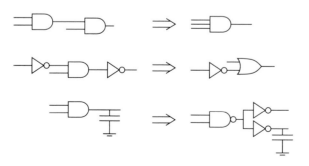

FIGURE 10.23
Two simple transformation rules.

search. Namely, the rule-based binder repeatedly fires the rules corresponding to the local best improvement of the network cost according to some metric.

Other systems, such as SOCRATES [13], use a more refined search for choosing the transformations in an attempt to explore the different choices and their consequences before applying a replacement. For example, SOCRATES tries a set of sequences of rules in the search for the best move. Recall from Section 8.7 that the size of this set is the *breadth* of the search and the length of the sequences is its *depth*. The results depend often on the breadth and depth parameters, which can be varied dynamically by the binder during its execution. *Metarules* decide upon the breadth and depth of the search and on the trade-off between area and delay or between the quality of the solution and the computing time.

10.5.1 Comparisons of Algorithmic and Rule-Based Library Binding

Let us consider first the technical differences and then the overall merit of the two approaches. Present algorithms for library binding use a covering approach, where the network is systematically scanned in a prescribed order and covered. On the other hand, most rule-based systems execute rules in a given order, and replace all circuit patterns that match the rules, either in the entire circuit or in a selectively chosen subcircuit. Whereas covering algorithms bind a subnetwork to a cell only once, iterative re-binding and stepwise improvement are supported by rule-based systems.

Let us consider the generality of both approaches. Rules can be thought of for all kinds of library cells without any restriction. Rules can be complex at times, but all cases can in principle be covered. By contrast, algorithms for library binding have been conceived to only handle single-output combinational logic cells. Extensions to other types of cells involve *ad hoc* methods.

The overall quality of the solutions is comparable for both approaches. However, for restricted classes of circuits, provable optimality and testability properties can be claimed by some algorithmic approaches. On the other hand, it is hard to prove that networks bound by rule-based systems have similar attributes, unless the rule set satisfies some completeness property and the order in which rules are applied follows some particular discipline.

Some binding algorithms can execute in a short time, because of their low computational complexity, but the speed of running a rule-based system varies. The number of rules under consideration and the metarules can be tuned so that a desired quality can be achieved with a predictable execution time. Therefore, the same rule-based system can provide for better or worse solutions, according to the amount of time allowed to perform a binding and possibly improve upon it.

Whereas the library description is straightforward in the case of the algorithmic-based approach, rule databases are large and complex. As a result, programs, often assisted by human experts, are used to compile the rule database. Maintaining a rule set is difficult, because rule sets must be continuously updated to reflect any change in the library, as cells are added or deleted and library cell parameters are updated when faster fabrication processes become available.

In summary, both approaches have advantages and disadvantages. Complex design systems for library binding often couple algorithms and rules. Binding algorithms are usually applied to a large portion of the circuit and provide a first solution, which can be improved upon by the application of rules.

10.6 PERSPECTIVES

Library binding is the key link between logic synthesis and the physical design of semicustom circuits. Library binding tools are widely available and successfully used. Nevertheless, the present techniques leave space for improvements, which are always highly desired because they are directly coupled to the circuits' quality. In particular, binding algorithms are dependent on the initial network decomposition into base functions. It would be highly desirable to develop algorithms whose solutions depend only on the network behavior. Similarly, it would be useful to be able to compute precise lower bounds on the area and/or speed of bound networks to evaluate the closeness of the solutions provided by heuristic binding algorithms and rule-based binders.

There is a wealth of techniques that are applicable to binding which have not been presented in this chapter for various reasons. Algorithms and strategies for rule-based systems of most commercial and proprietary binders are described only in documents with restricted access. Other techniques which have been shown to be promising are applicable to specific subproblems of library binding. We mention three examples. Spectral analysis of Boolean functions is useful for determining criteria for Boolean matching and for filtering probable matches. Iterative binding techniques have been used for performance-oriented binding, where gates along critical paths are repeatedly identified and re-bound. Algorithms have been studied for determining the optimal usage and sizing of buffers to drive highly loaded nets. As in the case of multiple-level circuit optimization of unbound networks, there are still interesting open research challenges to be solved in the years to come.

10.7 REFERENCES

Early work on automating library binding was done at IBM as part of the LSS project [8, 9]. Library binding in LSS was based on rules. Most commercial [13] and proprietary [16, 30] programs use rules, combined with algorithms, for binding.

At AT&T Bell Laboratory, Keutzer [19] proposed the first algorithm for binding that leveraged the work on string matching [1, 2]. He developed program DAGON, which binds logic networks using a dag model and uses program TWIG for string matching. Detjens *et al.* [10] and Rudell [26] expanded Keutzer's ideas and implemented the binder of program MIS [10] at the University of California at Berkeley. They introduced the inverter pair heuristic and developed algorithms for delay optimization. Morrison *et al.* [24] developed a binding algorithm using expression pattern matching and implemented it in program TECHMAP. Other programs based on structural covering with various flavors are reported in the literature [3, 20, 21].

Boolean matching and covering using *don't care* conditions was proposed by Mailhot and De Micheli [23], who developed program CERES, and by Sato *et al.* [29] at Fujitsu. Savoj *et al.* [28] extended the use of filters based on symmetries to enhance Boolean matching with *don't care* conditions. Burch and Long [5] developed canonical forms for Boolean matching.

Library binding techniques for FPGAs have been the subject of intensive investigation in recent years, due to the novelty and rapid distribution of FPGA technologies. Francis *et al.* [25] studied heuristic methods for binding table look-up architectures. Cong and Ding developed first exact methods for minimum-delay binding [6] and for determining the area/delay trade-off [7]. Karplus [17, 18] and independently Murgai *et al.* [22] studied methods for binding both soft- and hard-programmable FPGAs. A good survey of the state of the art is given in reference [27].

1. A. Aho and M. Corasick, "Efficient String Matching: An Aid to Bibliographic Search," *Communications of ACM*, Vol. 18, No. 6, pp. 333–340, June 1975.

2. A. Aho and S. Johnson, "Optimal Code Generation for Expression Trees," *Journal of ACM*, Vol. 23, No. 3, pp. 488–501, June 1976.

3. R. Bergamaschi, "SKOL: A System for Logic Synthesis and Technology Mapping," *IEEE Transactions on CAD/ICAS*, Vol. CAD-10, No. 11, pp. 1342–1355, November 1991.

4. M. R. C. M. Berkelaar and J. Jess, "Technology Mapping for Standard-Cell Generators," *ICCAD, Proceedings of the International Conference on Computer Aided Design*, pp. 470–473, 1988.

5. J. Burch and D. Long, "Efficient Boolean Function Matching," *ICCAD, Proceedings of the International Conference on Computer Aided Design*, pp. 408–411, 1992.

6. J. Cong and Y. Ding, "An Optimal Technology Mapping Algorithm for Delay Optimization in Lookup-table Based FPGA Designs," *ICCAD, Proceedings of the International Conference on Computer Aided Design*, pp. 48–53, 1992.

7. J. Cong and Y. Ding, "On Area/Depth Trade-off in LUT-based FPGA Technology Mapping," *DAC, Proceedings of the Design Automation Conference*, pp. 213–218, 1993.

8. J. Darringer, W. Joyner, L. Berman and L. Trevillyan, "LSS: Logic Synthesis through Local Transformations," *IBM Journal of Research and Development*, Vol. 25, No. 4, pp. 272–280, July 1981.

9. J. Darringer, D. Brand, W. Joyner and L. Trevillyan, "LSS: A System for Production Logic Synthesis," *IBM Journal of Research and Development*, Vol. 28, No. 5, pp. 537–545, September 1984.

10. E. Detjens, G. Gannot, R. L. Rudell, A. Sangiovanni-Vincentelli and A. R. Wang, "Technology Mapping in MIS," *ICCAD, Proceedings of the International Conference on Computer Aided Design*, pp. 116–119, 1987.

11. S. Ercolani and G. De Micheli, "Technology Mapping for Electrically Programmable Gate Arrays," *DAC, Proceedings of the Design Automation Conference*, pp. 234–239, 1991.

12. M. Garey and D. Johnson, *Computers and Intractability*, W. Freeman, New York, 1979.

13. D. Gregory, K. Bartlett, A. de Geus and G. Hachtel, "Socrates: A System for Automatically Synthesizing and Optimizing Combinational Logic," *DAC, Proceedings of the Design Automation Conference*, pp. 79–85, 1986.

14. G. Hachtel, R. Jacobi, K. Keutzer and C. Morrison, "On Properties of Algebraic Transformations and the Synthesis of Multi-fault Irredundant Circuits," *IEEE Transactions on CAD/ICAS*, Vol. CAD-11, No. 3, pp. 313–321, March 1992.

15. S. Hurst, D. Miller and J. Muzio, *Spectral Techniques in Digital Logic*, Academic Press, London, 1985.

16. J. Ishikawa, H. Sato, M. Hiramine, K. Ishida, S. Oguri, Y. Kazuma and S. Murai, "A Rule-based Reorganization System: Lores/ex," *ICCD, Proceedings of the International Conference on Computer Design*, pp. 262–266, October 1988.

17. K. Karplus, "Xmap: A Technology Mapper for Table-lookup Field-Programmable Gate Arrays," *DAC, Proceedings of the Design Automation Conference*, pp. 240–243, 1991.

18. K. Karplus, "Amap: A Technology Mapper for Selector-based Field-Programmable Gate Arrays," *DAC, Proceedings of the Design Automation Conference*, pp. 244–247, 1991.

19. K. Keutzer, "Dagon: Technology Binding and Local Optimization by Dag Matching," *DAC, Proceedings of the Design Automation Conference*, pp. 341–347, 1987.

20. M. C. Lega, "Mapping Properties of Multi-level Logic Synthesis Operations," *ICCD, Proceedings of the International Conference on Computer Design*, pp. 257–261, 1988.

21. R. Lisanke, F. Brglez and G. Kedem, "Mcmap: A Fast Technology Mapping Procedure for Multi-level Logic Synthesis," *ICCD, Proceedings of the International Conference on Computer Design*, pp. 252–256, 1988.

22. R. Murgai, Y. Nishizaki, N. Shenoy, R. Brayton and A. Sangiovanni-Vincentelli, "Logic Synthesis for Programmable Gate Arrays," *DAC, Proceedings of the Design Automation Conference*, pp. 620–625, 1990.

23. F. Mailhot and G. De Micheli, "Technology Mapping with Boolean Matching," *IEEE Transactions on CAD/ICAS*, Vol. CAD-12, No. 5, pp. 599–620, May 1993.

24. C. R. Morrison, R. M. Jacoby and G. D. Hachtel, "Techmap: Technology Mapping with Delay and Area Optimization," in G. Saucier and P. M. McLellan (Editors), *Logic and Architecture Synthesis for Silicon Compilers*, North-Holland, Amsterdam, The Netherlands, pp. 53–64, 1989.

25. R. Francis, J. Rose and Z. Vrasenic, "Chortle-crf: Fast Technology Mapping for Look-up Table-based FPGAs," *DAC, Proceedings of the Design Automation Conference*, pp. 227–233, 1991.

26. R. Rudell, "Logic Synthesis for VLSI Design," Memorandum UCB/ERL M89/49, Ph.D. Dissertation, University of California at Berkeley, April 1989.

27. A. Sangiovanni-Vincentelli, A. El Gamal and J. Rose, "Synthesis Methods for Field Programmable Gate Arrays," *Proceedings of the IEEE*, Vol. 81, No. 7, pp. 1057–1083, July 1993.

28. H. Savoj, M. Silva, R. Brayton and A. Sangiovanni-Vincentelli, "Boolean Matching in Logic Synthesis," *EURODAC, Proceedings of the European Design Automation Conference*, pp. 168–174, 1992.

29. H. Sato, N. Takahashi, Y. Matsunaga and M. Fujita, "Boolean Technology Mapping for Both ECL and CMOS Circuits Based on Permissible Functions and Binary Decision Diagrams," *ICCAD, Proceedings of the International Conference on Computer Aided Design*, pp. 286–289, 1990.

30. S. Suzuki, T. Bitoh, M. Kakimoto, K. Takahashi and T. Sugimoto, "TRIP: An Automated Technology Mapping System," *DAC, Proceedings of the Design Automation Conference*, pp. 523–529, 1987.

10.8 PROBLEMS

1. Consider the binding of a network implementing the conjunction of 10 variables. Available cells are only two-input AND gates with cost 2, three-input AND gates with cost 3 and four-input AND gates with cost 4. Find an optimum cover of a balanced tree decomposition of the network using two-input AND gates as base functions. Is this the best implementation of the given network with the available cells? Is there another decomposition into the same base function leading to a lower cost solution?

2. Derive a formula that yields the number of distinct decompositions of a function f, implementing the conjunction of n variables, into two-input AND gates. Tabulate the number of distinct decompositions for $n = 2, 3, \ldots, 10$.

3. Consider the simple library of cells of Figure 10.8 (a). Derive all pattern trees and corresponding strings for a decomposition into NOR2 and INV base functions. Repeat the exercise by considering NOR2, NAND2 and INV as base functions.

4. Consider the simple library of cells of Figure 10.8 (a) and the pattern trees according to a decomposition in NOR2 and INV (see Problem 3). Compute the automaton that represents the library.

5. Consider a library including the following cells: {AND2 with cost 4; OR2 with cost 5; INV with cost 1}. Draw the pattern trees for these cells using NAND2 and INV as base functions. Then consider function $f = ab' + c'd'b$. Determine the subject graph for f using the same base functions. Find a minimum cost cover of the subject graph using the inverter-pair heuristic. *Hint*: use the following decomposition: $f = $ NAND2 (p, q); $p = $ NAND2 (a, b'); $q = $ NAND2 $(c'r')$; $r = $ NAND2 (d', b).

6. Consider a scalar function of n variables with a symmetry set of cardinality m. How many of the cofactors of f, w.r.t. all variables in the set, differ?

7. Consider the library of virtual gates corresponding to static CMOS implementations of single-output functions with at most s transistors in series and p transistors in parallel. What is the size of these libraries for $s = 1, 2, \ldots, 5$, $p = 1, 2, \ldots, 5$?

8. Enumerate the different $\mathcal{NPN}$ classes of functions of three variables, and show one representative function for each class.

PART
IV

CONCLUSIONS

CHAPTER
11

STATE OF THE
ART AND
FUTURE TRENDS

I libri non sono fatti per crederci, ma per essere sottoposti a indagine.
Books are not made to be believed, but to be subjected to inquiry.

U. Eco. Il nome della rosa.

11.1 THE STATE OF THE ART IN SYNTHESIS

In the introduction to this book we commented on the importance of computer-aided synthesis and optimization methods for the advancement of the electronic industry. We also presented a concise history of the major breakthroughs in this field in Section 1.8. We now critically review the use of synthesis techniques in the support of microelectronic circuit design. For this reason, we consider the implementation of the ideas presented in this book in current computer-aided design systems and their use for digital circuit design. We refer to synthesis and optimization methods as synthesis for brevity.

The success of many ideas in CAD can often be measured by their use in design systems. Some algorithms failed to be applied because they either were not practical

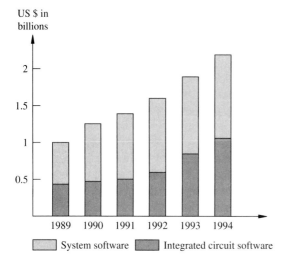

US $ in
billions

FIGURE 11.1
Electronic CAD software market.
(Source: VLSI Research Inc.)

☐ System software   ■ Integrated circuit software

or did not address problems relevant to the design methodologies being used. Some techniques disappeared because they were superseded by newer ones. As an example, algorithms with lower computational complexity or with more powerful heuristics displaced others that executed in longer time or provided lower-quality solutions. Generally speaking, most of the techniques presented in this book are currently used by CAD systems.

On the other hand, CAD systems are very complex and their usability depends not only on the algorithms being supported but also on the user interface, database and adaptability to the user environment and needs. Software engineering techniques play a major role in the development of robust and user-friendly CAD systems. In other words, this book has described some of the algorithms that constitute the inner engine of CAD systems.

Overall, the impact of synthesis methods on microelectronic design has been extremely positive. Some microelectronic circuits could never have been designed at the required performance levels without the aid of CAD systems. Many synthesis and optimization tools are now commercially available, and most designers of digital circuits use them for at least some part of the designs. Recently, sales of synthesis systems have soared, as shown in Figure 11.1 [7]. The market growth has been impressive, especially for synthesis tools at the architectural and logic levels.

As in many other fields, the directions in growth have been driven by trends. For example, the use of specific hardware description languages and design formats is often dictated by company policies or by marketing issues. Standards in design representations have evolved through the years and design synthesis tools have followed the evolution by trying to match the standards.

In this chapter, we focus on the present state of the art and the likely future directions. We present first a taxonomy of synthesis systems and describe some representative ones. We shall take then a broader view of the problem and put into perspective the growth of circuit synthesis as a fundamental design method. Eventually, we shall consider requirements for electronic circuit and system design in the close and distant future and mention relevant open and unresolved problems.

11.2 SYNTHESIS SYSTEMS

Synthesis systems can be classified according to different criteria. Therefore we consider different factors that are useful for evaluating the fitness of a synthesis system for a given design task and for comparing different synthesis systems.

A first differentiation among synthesis systems can be done by considering their primary goal, i.e., according to whether they are used for commercial circuit production or for research and exploratory reasons.

Production-level synthesis systems are conceived to be used for designing circuits to be manufactured and either marketed as off-the-shelf components or incorporated into electronic systems. Since the reliability of the CAD tools is of primary importance, they are based upon mature (i.e., well-experienced) techniques. They can be classified further as *commercial* or *internal* tools. The former are available for sale, while the latter are designed within companies for proprietary use. The difference between internal and commercial tools is beginning to blur, as some companies are now selling CAD programs originally developed for internal use.

Research synthesis systems are designed to explore new ideas in synthesis and optimization. Hence they tend to incorporate the most novel and advanced algorithms, which in turn may not have reached an adequate degree of maturity and stability. They are prototypes of production-level synthesis tools and they are occasionally used for designing circuits that will be fabricated. Universities, research centers and industries are usually involved in designing such systems. Some research synthesis systems developed in universities are available for free or for a nominal charge.

Design systems can be classified according to the circuit abstraction level they support, namely as *architectural*, *logic* and *geometrical* design tools. Some synthesis systems can claim full top-down synthesis, i.e., support synthesis at all levels. Other systems are limited to particular levels and tasks. In addition, specific synthesis systems have been designed for some application domains such as digital signal processing.

Synthesis systems should be classified according to the quality of the designs they produce, measured in terms of area, performance and testability. While such a figure would be extremely useful, specific comparative data are still not available. On the other hand, indicative measures support the belief that successful systems have targeted either full top-down synthesis in a restricted application domain or a restricted set of synthesis tasks for general circuits.

Users of CAD synthesis systems also evaluate them on the basis of their integration in the current design flow, which varies from site to site according to the circuit technology, implementation style and overall design methodology. The user is often confronted with the problem of mixing and matching tools from different sources and/or blending them with other tools developed in-house for specific purposes. Thus, an important feature of synthesis systems is the possibility and ease of augmenting them by incorporating user-designed or other programs. Standard data representation formats play a key role in tool integration. Fast and easy access to internal data representation is very important for combining tools efficiently.

CAD *frameworks* have gained much attention in recent years. Frameworks support the integration of CAD systems and isolated tools, by providing guidelines for links to a common user interface, data representation formats and inter-tool communication. The scope of CAD frameworks goes beyond circuit synthesis at all levels, and it includes circuit simulation, verification and testing support. When thinking of the increasingly difficult challenge of designing larger and larger microelectronic circuits, the key role played by CAD tool development, integration and management is immediately recognized.

11.2.1 Production-Level Synthesis Systems

Limited published information is available on details of the algorithms used by production-level synthesis systems. The scarcity of detailed information is due to the attempt to protect proprietary ideas. Therefore we shall restrict our comments to the major features of these systems.

Proprietary internal production-level synthesis tools have been developed by several companies, such as AT&T, Fujitsu, HP, IBM, Intel, NEC, NTT, Philips and Siemens among others. IBM developed the first production-level logic synthesis and optimization system: LSS [4]. LSS was a major success not only for IBM but for the entire field, because it showed the practicality of using synthesis and optimization techniques for large-scale designs. Recently, a novel implementation of IBM's architectural and logic synthesis system, called BOOLEDOZER, has been put on the market. Also other companies (e.g., Philips) are selling design systems that were originally developed as internal tools. Most ASIC vendors, especially field-programmable gate array suppliers, provide their customers with synthesis tools that are targeted to their libraries.

Several commercial synthesis tools are now available [13]. Most systems accept circuit specifications in hardware description languages like Verilog or VHDL. For architectural synthesis, specific synthesis policies are mandated to make unambiguous the interpretation of behavioral models. Some companies provide physical design tools that are directly coupled to the corresponding logic synthesis and optimization programs. Others provide synthesis from HDL models to the specification of bound networks in standard netlist formats, thus providing a well-defined interface to external physical design tools.

Commercial CAD systems, their current features and their costs are often described in trade magazines [1, 2]. We report in Table 11.1 a summary of the offerings of some synthesis vendor companies in 1993. This table is only indicative, as these data evolve with time.

TABLE 11.1
Some commercial synthesis systems in 1993.

Organization	System	Main features
Cadence Design Systems	SYNERGY	Synthesis from VHDL and Verilog. Logic synthesis and optimization. Library binding.
Compass	ASIC SYNTHESIZER	Separate synthesis of data path and control from VHDL, Verilog and graphical inputs. Resource sharing. Logic synthesis and optimization. Library binding.
Dazix/Intergraph	ARCHSYN	Synthesis from VHDL and Verilog. Resource sharing. Logic synthesis and optimization. Library binding.
Exemplar Logic	CORE	Synthesis from VHDL and Verilog. Logic optimization and binding for FPGAs.
Mentor Graphics	AUTOLOGIC	Synthesis from VHDL and M. Logic synthesis and optimization. Library binding.
Synopsys	HDL/DESIGN COMPILER DESIGNWARE	Synthesis from VHDL and Verilog. Resource sharing. Logic synthesis and optimization. Library binding.
Viewlogic	SILCYSN VIEWSYNTHESIS	Synthesis from VHDL. Resource sharing. Control synthesis for loops. Logic synthesis and optimization. Library binding.

11.2.2   Research Synthesis Systems

Several research synthesis systems have been developed, and it is impossible to comment on all of them here. Some specialized books [4, 5, 15] describe these systems in detail, and thus we just summarize their major features in Table 11.2.

   We describe instead the most salient features of four synthesis systems, which have been selected because they represent archetypes of different research ideas and directions.

THE SYSTEM ARCHITECT'S WORKBENCH. Several programs for architectural synthesis and exploration have been developed at Carnegie-Mellon University for over one decade. The SYSTEM ARCHITECT'S WORKBENCH (or SAW) [14] is a design system that encapsulates some of these architectural synthesis tools. Inputs to this design system are circuit specifications in the ISPS or Verilog languages that are compiled into an intermediate data-flow format called *value trace*. The value trace can be edited graphically to perform manual operations such as partitioning and expansion of

TABLE 11.2
Some research synthesis systems in 1993 (FSM = finite-state machine specification, ASO = architectural synthesis and optimization, LSO = logic synthesis, optimization and binding, MG = module generation).

Organization	System	Input	Scope
AT&T	BRIDGE/CHARM	FDL2	ASO
Carleton University	HAL	Graph models	ASO
CMU	SAW	Verilog, ISPS	ASO
IBM	HIS	VHDL	ASO
IMAG	ASYL	FSM, networks	LSO
IMEC	CATHEDRAL I-IV	Silage	ASO, MG
Philips	PYRAMID, PHIDEO	Silage	ASO, MG
Princeton University	PUBSS	FSM	ASO, LSO
Stanford University	OLYMPUS	HardwareC	ASO, LSO
U.C. Berkeley	SIS	FSM, networks, BDS	LSO
U.C. Berkeley	HYPER, LAGER	Silage	ASO, MG
U.C. Boulder	BOLD	FSM, networks	LSO
U.C. Irvine	VSS	VHDL	ASO
U. Karlsrühe/Siemens	CADDY/CALLAS	DSL, VHDL	ASO
USC	ADAM	SLIDE, DDS	ASO

selected blocks. The synthesis outcome is a structural circuit representation in terms of a network of hardware resources and a corresponding control unit.

The workbench includes the following tools. APARTY is an automatic partitioner based on a cluster search. CSTEP is responsible for deriving the hardware control portion: it is based on a list scheduling algorithm which supports resource constraints. EMUCS is a global data allocator that binds resources based on the interconnection cost. BUSSER synthesizes the bus interconnection by optimizing the hardware using a clique covering algorithm. SUGAR is a dedicated tool for microprocessor synthesis which recognizes some specific components of a processor (e.g., an instruction decode unit) and takes advantage of these structures in synthesis. CORAL maintains the correspondence between the behavioral and structural views. All tools are interfaced to each other and share a common data structure and user interface.

THE SEQUENTIAL INTERACTIVE SYNTHESIS SYSTEM. The SEQUENTIAL INTER-ACTIVE SYNTHESIS system (or SIS), developed at the University of California at Berkeley, is a program that supports sequential and combinational logic synthesis, including library binding. The SIS program evolved from the MULTILEVEL INTERACTIVE SYNTHESIS program (or MIS), which was limited to synthesis and optimization of combinational circuits.

The MIS program has been very popular and widely distributed. Some commercial tools have drawn ideas from MIS, and some proprietary internal CAD systems have directly incorporated parts of the MIS program. MIS supports logic optimization by means of both algebraic and Boolean transformations. It uses the ESPRESSO program for Boolean simplification and it performs library binding by using a structural covering approach.

Program SIS has now replaced MIS. SIS supports all features of MIS as well as sequential logic optimization using either state-based or structural models. SIS is an interactive program with batch capability. Script files can be used to sequence a set of commands, each related to a set of logic transformations of a given type. Example 8.2.9 of Chapter 8 reports the *rugged script*, often used for optimization of combinational circuits.

Specific transformations for sequential circuits include state minimization and encoding as well as retiming to reduce the cycle-time or area. The state transition graph can be extracted from a structural representation of the circuit. *Don't care* conditions can also be computed for sequential circuits. In addition, SIS supports a variety of transformations among network representations. The output of SIS can be transferred to the OCTTOOLS suite, a package that supports physical design, also developed at the University of California at Berkeley.

THE CATHEDRAL SYNTHESIS SYSTEMS. The *Cathedral* project was developed at IMEC in connection with the Catholic University of Leuven in Belgium and other partners under the auspices of project Esprit of the European Community. One of the guiding principles of the project is the tailoring of synthesis tools to specific application domains and implementation styles. Therefore, different CATHEDRAL programs have been designed to transform behavioral models of a particular class of designs, namely digital signal processors (DSP), into circuits with particular design styles. The Silage language is used for circuit modeling.

CATHEDRAL-I [5] is a hardware compiler for bit-serial digital filters. CATHEDRAL-II [3, 5] is a synthesis system for DSP applications using concurrent bit-parallel processors on a single chip. Typical applications are speech synthesis and analysis, digital audio, modems, etc. CATHEDRAL-III [11] targets hard-wired bit-sliced architectures intended for the implementation of algorithms in the real-time video, image and communication domains. CATHEDRAL-IV is planned for implementing very repetitive algorithms for video processing. Commercial versions of CATHEDRAL-I and CATHEDRAL-II are also available. CATHEDRAL-II has been recently extended to cope with retargetable code generation. We describe here CATHEDRAL-II and CATHEDRAL-III because of their relevance in connection with the topics described in this book.

The general design methodology in CATHEDRAL-II is called *meet in the middle* strategy. There are two major tasks in the system: architectural synthesis and module generation. Architectural synthesis maps behavioral circuit models into interconnections of instances of primitive modules, such as data paths, memories, I/O units and controllers. Architectural optimization includes the following tasks: system partitioning into processes and protocols, data-path synthesis (i.e., mapping the partition blocks into execution units while minimizing the interconnection busses) and control synthesis based on a microcode style. Data-path synthesis is done with the aid of an architecture knowledge database. Control synthesis is based on a heuristic scheduling algorithm. The physical layout is achieved by invoking module generators which can be seen as a library of high-level cells described in a procedural style.

Module generators are designed to be portable across different, but similar, circuit technologies.

CATHEDRAL-III exploits the concept of *application-specific units*, which are clusters of resources tailored to specific tasks [11], such as sorting an array, performing a convolution or computing the minimum/maximum. Thus, data-path synthesis is centered on the optimal use of application-specific units. Architectural synthesis provides operation clustering in the signal-flow graph representation into which the behavioral model is compiled. Clusters identify application-specific units. Optimizing transformations include distributivity and associativity exploitation as well as retiming. Memory management is provided in the CATHEDRAL-III environment by supporting different storage models and the synthesis of the addresses of the memory arrays where data are held.

THE OLYMPUS SYNTHESIS SYSTEM. The OLYMPUS synthesis system, developed at Stanford University, is a vertically integrated set of tools for the synthesis of digital circuits, with specific support for ASIC design. Circuits are modeled (at the architectural level) in a hardware description language called *HardwareC*, which has both procedural and declarative semantics and a C-like syntax [10].

The OLYMPUS system supports architectural and logic synthesis. A front-end tool called HERCULES performs parsing and behavioral-level optimization. Program HEBE executes architectural optimization. It strives to compute a minimal-area implementation subject to performance requirements, modeled as relative timing constraints. HEBE applies the relative scheduling algorithm after having bound resources to operations. If a valid schedule cannot be found that satisfies the timing constraints, a new resource binding is tried. Binding and scheduling are iterated until a valid solution is found, unless HEBE determines that the constraints cannot be met and need to be relaxed. Details are reported in reference [10].

Logic synthesis and library binding are used in OLYMPUS for estimating area and delay of application-specific logic blocks during architectural synthesis as well as for generating the system's output in terms of a hierarchical logic network bound to library elements. Program CERES performs library binding using a Boolean matching approach. Program MERCURY supports some logic transformations, an interface to the SIS program and gate-level simulation. Programs THESEUS and VENUS provide a waveform and a sequencing graph display facility, respectively. OLYMPUS does not support physical design, but it provides netlist translation into a few standard formats.

11.2.3 Achievements and Unresolved Issues

The major achievements of synthesis and optimization techniques are an improvement in the quality of circuit implementations (smaller, faster, more testable) and a reduction in their design time. These two factors are so important for circuit design that synthesis systems have become pervasive. The continuous increase in complexity and improvement in performance of microelectronic circuits could not have been achieved without the use of automated synthesis systems.

Several authors described examples of VLSI circuits that have been synthesized in full from architectural models. Thomas and Fuhrman [3] reported on the industrial use of the SYSTEM ARCHITECT'S WORKBENCH in connection with a commercial physical design tool for the design of applications for the automotive market. Nakamura [3] described the full synthesis of a 32-bit RISC processor (called FDDP because it was designed in four days) and of a *very long instruction word* (VLIW) vector processor for DSP. Both designs were achieved with the PARTHENON synthesis system developed at NTT and required about 14,000 and 400,000 transistors, respectively. Application-specific circuits for the consumer industry, such as compact-disk controllers and interfaces, have been fully synthesized by several systems, e.g., CATHEDRAL-II, PYRAMID and OLYMPUS. Moreover, many examples of applications of logic synthesis and optimization to chip design, including popular processors, have been reported.

Hurdles have been encountered in applying synthesis techniques to the engineering design flow. Designers had to be educated to think of circuits in a more abstract way and to rely on the tools for decisions they used to make themselves. To gain acceptance, synthesis systems had to support all (or almost all) circuit features that handcrafted designs have.

Other difficulties are intrinsic to the nature of the design problem. Even though heuristic algorithms are often used to cope with the computational complexity of many synthesis tasks, circuit designs can always be found that are too large for existing design systems. This has limited the use of some optimization techniques to portions of VLSI circuits. Unfortunately, synthesis and optimization methods are needed the most where the circuit scale is so large that human design is unlikely to be effective. Moreover, due to the heuristic nature of the algorithms, there are also pathological circuit examples where optimization leads to poor results.

There are still some problems that limit the use of synthesis systems and that are due to the lack of maturity of this field. Some specific design tasks have not been satisfactorily addressed by automated synthesis methods. An example is logic optimization of data-path units, including arithmetic functions. Most logic optimization algorithms have been devised for coping with sparse or control logic and are unable to exploit the special structure of arithmetic functions that is key to their optimization.

Today, the use of architectural synthesis is still limited by a few factors. The synthesis of efficient pipelined circuits, which supports all features of handcrafted designs, is not available yet. Similarly, the design of efficient data storage in memory hierarchies is crucial for the design of some circuits, but it is supported only by a few synthesis tools and to a limited extent. We believe that these problems will be overcome with time due to their technical, rather than fundamental, nature.

11.3 THE GROWTH OF SYNTHESIS IN THE NEAR AND DISTANT FUTURE

The ideas presented in this book are typical of an evolving science. Computer-aided design methods started as applications of algorithms to circuit design. Today, CAD techniques have matured and acquired the strength of an independent discipline. This is corroborated by the fact that the CAD research and development community has grown in size through the years.

Synthesis of digital circuits is part of *design science*. Design science encodes the engineering design flow in a rigorous framework and allows us to reason formally about design problems and their solutions. Improvements in science are either *evolutionary* or *revolutionary*. The former consists of compounding little steps, each related to perfecting the solutions to some problems. The latter involves radical changes in the way in which problems are modeled and solved.

Architectural and logic synthesis have been revolutionary, because they changed the way in which we reason about circuits and address their optimization. Describing and synthesizing circuits from HDL models instead of using gate-level schematics can be paralleled to the replacement of assembly languages with high-level programming languages in the software domain.

An evolutionary growth of synthesis is likely in the near future. Many techniques need to be perfected, and many subproblems, which were originally considered of marginal interest, need now to be solved. In particular, the evolutionary growth will involve the horizontal extension of present techniques to less conventional design styles and circuit technologies as well as the full integration of present synthesis methods.

Examples of the horizontal growth of logic synthesis are the application of library binding to novel field-programmable gate arrays or to circuit families with specific connection rules, such as *emitter-coupled logic* (ECL) circuits (which support dotting). The horizontal growth of architectural synthesis can be related to extending its application domains. To date, architectural synthesis has been most used for signal processing circuits. We expect the growth of its use in the application-specific circuit and instruction-set processor domains.

Whereas synthesis and optimization of the geometric features of integrated circuits is a mature field, the integration of logic and architectural synthesis techniques with physical design is still only a partially solved problem. In the future its importance will grow as circuit density increases and wiring delays tend to dominate, thus affecting cell selection (in library binding) and resource selection and sharing. Hence, logic and architectural synthesis programs will need to access accurate information of the physical layout. The integration of different synthesis tasks is not simple, because architectural and logic synthesis are performed before physical design. Estimation techniques have been used for this purpose, but their level of accuracy needs to be raised to cope with forthcoming circuits with an increasingly higher level of integration.

As circuits become more and more complex, validation of properties by means of formal verification methods becomes even more important. By the same token, implementation verification by comparing synthesized representations at different levels will be relevant to ensure that the circuit has no flaws introduced by the synthesis process. Indeed, even though synthesis and optimization algorithms have guaranteed properties of correctness, their software implementation may have bugs. The coupling of synthesis and verification for circuit design will become common practice in the future.

Revolutionary changes in synthesis methods will be required when considering circuits with both synchronous and asynchronous operation modes as well as with analog components. Due to the wide disparity of design paradigms, objective functions and constraints, new modeling, synthesis and optimization methods will be necessary. At present, isolated synthesis and optimization algorithms exist for some asynchronous styles and for some analog components. On the other hand, design systems in these domains are not yet available. The integration of synthesis techniques for synchronous, asynchronous and analog components, as well as the support for the concurrent design of heterogeneous chip sets, is a major challenge for the years to come.

In the longer term, the vertical growth of synthesis methods will lead to the extension of the scope of synthesis beyond the design of integrated circuits. Synthesis of composite electrical (and electromechanical) systems, possibly involving both hardware and software components, will be a forthcoming major revolutionary conquest in the sphere of CAD and it will change the practice of engineering jobs in several sectors.

11.3.1 System-Level Synthesis

System-level design is a broad term, as different meanings can be associated with the word *system*. It is customary to refer to computers as information systems. The scale of systems may vary widely. Consider, for example, a laptop computer and a distributed computing environment or a telephone and a telephone switching network. To be concrete, we consider systems that are single physical objects and that have an electrical component with, possibly, interfaces and/or a mechanical component. In particular, we shall consider here issues related to the synthesis of the electrical component of a system that can be thought of as an interconnection of integrated circuits. Computer-aided design of electromechanical systems is a subject of ongoing research, but synthesis of such systems is still far on the horizon.

Computer-aided design of multiple-chip systems involves several tasks, including system specification, validation and synthesis. System specification may be multiform, because different system components may be *heterogeneous* in nature. Consider, for example, the specification of a cellular telephone, handling both digital and analog signals at different frequencies. Distinct functional requirements can be best described with different modeling paradigms, e.g., hardware description languages, circuit schematics and constraints. System validation may be solved (at least in part) by mixed-mode simulation. Whereas many mixed-mode simulators are available on the market, few address the real problems of modeling heterogeneous systems. Among these, PTOLEMY is a research design environment and simulator [9] for signal processing and communication-system design that provides a means for heterogeneous co-specification by supporting several modeling styles, including data-flow, discrete-event and user-defined models.

System-level synthesis is a challenging field still in its infancy. We shall consider synthesis at different levels of abstraction. Whereas placement and routing tools for electronic boards have been available for a long time, the physical design of electronic systems has evolved as the physical means of composing systems have progressed. Multiple-chip carriers provide a means of connecting efficiently several integrated circuits. A well-known example of a multiple-chip carrier is IBM's *thermal conduction module* (TCM). Important issues for multiple-chip physical design are wiring delay estimation and performance-oriented placement and routing. At present, several research and some production-level tools have been developed for solving these problems.

Logic design of electronic systems must address data communication and clocking problems. Systems may have synchronous and asynchronous components. Synchronous subsystems may operate on different clocks. Communication among the various system components must satisfy protocol requirements and possibly different data rates. As a result, system-level logic design is a very challenging task which is performed by experienced designers due to the lack of CAD tools. With the increasing complexity of system design, system-level logic design will become prohibitively time-consuming and risky in the future, thus motivating the development of specialized tools.

A major task in the architectural synthesis of systems is defining a partition of the system's function over components (i.e., integrated circuits) which can be seen as system resources. Even though this problem seems to be an extension of architectural synthesis methods from the circuit to the system level, the cost functions and constraints are different. In system design, it is often convenient to leverage the use of components that are already available on the market (as off-the-shelf parts) or in-house (from previous system designs). Thus synthesis methods must support the use of pre-designed components which impose constraints on the synthesis of the remaining parts. In this case, the overall cost of an electronic system depends on the cost of designing and manufacturing some components and on the actual cost of the available components. Hence system-level partitioning can heavily affect the cost of the system as well as its performance. Since architectural-level decisions may affect strongly the cost/performance trade-off of a system, CAD tools for this task, though not yet available, will be very important.

System-level design is not confined to hardware. Indeed most digital systems consist of a hardware component and software programs which execute on the hardware platform. A given system can deliver higher performance when the hardware design is tuned to its software applications and *vice versa*. Computer designers have exploited the synergism between hardware and software for many years, while defining hardware architectures and providing architectural support for operating systems. An important problem for architectural-level synthesis of any composite digital system is to find the appropriate balance between hardware and software. This problem falls into the domain of hardware/software co-design, described in the next section.

11.3.2 Hardware-Software Co-Design

The *hardware/software co-design* problem is not new but has been receiving more and more attention in recent years, due to the search for computer-aided design tools for effective co-design.

There are several reasons for using mixed hardware/software solutions in a system design. First, we can search for implementations where the performance provided by customized hardware units can balance the programmability of the software components. Second, the evolution of a system product may be better supported by allowing the software programs to undergo upgrades in subsequent releases of the system. By the same token, the development of a product may be eased by supporting prototypes where most (if not all) functionality is assigned to the software component, reducing (or avoiding) circuit fabrication in the prototyping stage.

Several problems are encompassed by hardware/software co-design. We comment here on those related to *general purpose* computing and to *dedicated computing and control*. The design of instruction-set processors falls in the former class, while design of embedded controllers is an example of the latter.

Relevant hardware/software co-design problems for instruction-set processors are *cache* and *pipeline* design. The design and sizing of a memory cache require a match between circuit performance and the updating algorithm and its parameters. Most cache designs are based on validating the design assumptions through simulation with specialized tools. No tools are yet available for the full automatic synthesis of caches and of the related updating algorithms.

The design and control of a pipeline in a processor requires removing *pipeline hazards*. Hardware or software techniques can be used for this purpose. An example of a hardware mechanism is *flushing* the pipe, while a typical software solution is reordering the instructions or inserting *No-Operations*. The overall processor performance is affected by the choice. Furthermore, performance estimation is not trivial and requires appropriate models for both hardware and software. Computing the most effective number of pipe stages for a given architecture is thus a hardware/software co-design problem. CAD tools can explore the hardware/software trade-off and suggest an implementation which meets the needs of the design problem. The PIPER synthesis program is an example of a co-design tool that addresses this problem [8]. It provides pipe-stage partitioning and pipeline scheduling and determines the appropriate instruction reorder that the corresponding back-end compiler should use to avoid hazards.

We consider now embedded systems that are computing and control systems dedicated to an application (Figure 11.2). The most restrictive view of an embedded system is a micro-controller or a processor running a fixed program. This model can be broadened to a processor, assisted by application-specific hardware and memory, that performs a dedicated function. Sensors and actuators allow the system to communicate with the environment. Embedded systems often fall into the class of *reactive systems*. They are meant to react to the environment by executing functions in response to specific input stimuli. In some cases, their functions must execute within predefined time windows. Hence they are called *real-time systems*. Examples of reactive real-time systems are pervasive in the automotive field (e.g., engine combustion control), in the manufacturing industry (e.g., robot controllers) and in the consumer and telecommunication industries (e.g., portable telephones).

Computer-aided synthesis of embedded systems, called here *co-synthesis*, is the natural evolution of existing hardware architectural synthesis methods. A working

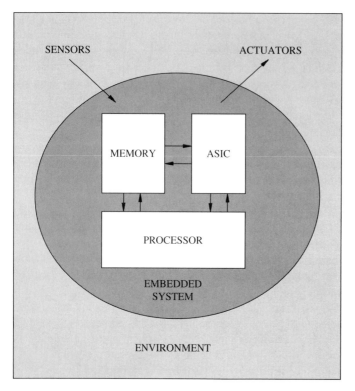

FIGURE 11.2
Embedded system: a simplified structural view.

hypothesis for co-synthesis is that the overall system can be modeled consistently and be partitioned, either manually or automatically, into a hardware and software component. The hardware component can be implemented by application-specific circuits using existing hardware synthesis tools; the software component can be generated automatically to implement the function to which the processor is dedicated. Co-synthesis must support a means for interfacing and synchronizing the functions implemented in the hardware and software components.

The overall system cost and performance are affected by its partition into hardware and software components. At one end of the spectrum, hardware solutions may provide higher performance by supporting parallel execution of operations at the expense of requiring the fabrication of one or more ASICs. At the other end of the spectrum, software solutions may run on high-performing processors available at low cost due to high-volume production. Nevertheless, operation serialization and lack of specific support for some tasks may result in loss of performance. Thus a system design for a given market may find its cost-effective implementation by splitting its functions between hardware and software.

At present, the overall CAD support for co-synthesis is primitive. Nevertheless, the potential payoffs make it an attractive area for further research and development.

Several open problems still impede the rapid growth of the field. First and foremost, there exists a need to define better abstract models for hardware/software systems and to develop consistent languages to express them. Possible solutions range from the extension of existing hardware and software languages to the use of new heterogeneous paradigms. Second, cost and performance evaluation of mixed systems play an important role in driving partitioning and synthesis decisions. The problem is complicated by the remoteness of the abstract system models from the physical implementation. Last, but not least, methods for validating hardware/software systems are very important. Co-simulation provides a simple way of tracing the input/output (and internal) system behavior. However, for large system design, co-simulation may provide insufficient evidence to prove some desired system property. Extending formal verification techniques to the hardware/software domain would thus be desirable.

11.4 ENVOY

Research and development in the CAD field has progressed tremendously in the last three decades. Computer-aided synthesis of digital circuits has become a scientific discipline attracting a large number of researchers, developers and users. The CAD industry, and in particular the digital design sector, has grown in size and occupies an important place in the overall electronic market.

The knowledge accumulated in this field to date should stimulate us to think, as Socrates suggested, of all other important design problems that are not solved yet. The limit to our engineering design capabilities lies in the instruments we have. In the case of electronic design, CAD is one of the keys to the future evolution.

This book could only mention some of the relevant problems in the synthesis field. Nevertheless, we hope it has raised enough interest in the reader to motivate him or her to search for additional information in the referenced articles and books and to follow the evolution of this discipline in the regular conferences and scientific journals.

11.5 REFERENCES

Some research synthesis systems are described in detail in the books edited by Gajski [5] and Camposano with Wolf [3]. Walker and Camposano [15] summarize the major features of several architectural synthesis systems in a specialized book.

The SYSTEM ARCHITECT'S WORKBENCH is described in a dedicated book [14]. A good reference to the CATHEDRAL-II system can be found in Gajski's book [5], while a description of some more recent optimization algorithms used in CATHEDRAL-II and in CATHEDRAL-III are reported in reference [11]. Information about the SIS system can be found in reference [12] and about *Olympus* in Ku's [10] and in Camposano with Wolf's [3] books.

The areas of system-level synthesis and hardware/software co-design are fairly new, and only a few published papers report on the recent developments [9, 6, 8].

1. "Focus Report Summaries," *ASIC & EDA*, pp. 85–114, December 1992.
2. B. Arnold, "Turning to HDLs for Fast Design Relief," *ASIC & EDA*, pp. 54–64, June 1993.
3. R. Camposano and W. Wolf, Editors, *High-Level VLSI Synthesis*, Kluwer Academic Publishers, Boston, MA, 1991.
4. J. Darringer, D. Brand, W. Joyner and L. Trevillyan, "LSS: A System for Production Logic Synthesis," *IBM Journal of Research and Development*, Vol. 28, No. 5, pp. 537–545, September 1984.

5. D. Gajski, Editor, *Silicon Compilation*, Addison-Wesley, Reading, MA, 1987.

6. R. Gupta and G. De Micheli, "Hardware-Software Co-synthesis for Digital Systems," *IEEE Design & Test*, Vol. 10, No. 3, pp. 29–41, September 1993.

7. B. Groves, "1991: What Will It Mean for ASICs, EDA?" *ASIC Technology & News*, Vol. 2, No. 9, pp. 1–9, January 1991.

8. I. Huang and A. Despain, "High-level Synthesis of Pipelined Instruction Set Processors and Back-end Compilers," *DAC, Proceedings of the Design Automation Conference*, pp. 135–140, 1992.

9. A. Kalavade and E. Lee, "A Hardware-Software Codesign Methodology for DSP Applications," *IEEE Design & Test*, Vol. 10, No. 3, pp. 16–28, September 1993.

10. D. Ku and G. De Micheli, *High-Level Synthesis of ASICS under Timing and Synchronization Constraints*, Kluwer Academic Publishers, Boston, MA, 1992.

11. S. Note, F. Chattoor, G. Goossens and H. De Man, "Combined Hardware Selection and Pipelining in High-Level Performance Data-Path Design," *IEEE Transactions on CAD/ICAS*, Vol. CAD-11, pp. 413–423, April 1992.

12. E. Sentovich, K. Singh, C. Moon, H. Savoj, R. Brayton and A. Sangiovanni-Vincentelli, "Sequential Circuit Design Using Synthesis and Optimization," *ICCD, Proceedings of the International Conference on Computer Design*, pp. 328–333, 1992.

13. M. Smith, "More Logic Synthesis for ASICs," *IEEE Spectrum*, No. 11, pp. 44–48, November 1992.

14. D. Thomas, E. Lagnese, R. Walker, J. Nestor, J. Rajan and R. Blackburn, *Algorithmic and Register Transfer Level Synthesis: The System Architect's Workbench*, Kluwer Academic Publishers, Boston, MA, 1990.

15. R. Walker and R. Camposano, *A Survey of High-Level Synthesis Systems*, Kluwer Academic Publishers, Boston, MA, 1991.

INDEX

AHPL - A Hardware Programming Language
ALAP - As Late As Possible scheduling
ALU - Arithmetic Logic Unit
APL - A Programming Language
AFAP - As Fast As Possible
ASAP - As Soon As Possible scheduling
ASIC - Application Specific Integrated Circuit
ATPG - Automatic Test Pattern Generation
BDD - Binary Decision Diagram
BiCMOS - Bipolar Complementary Metal Oxide Semiconductor
BIST - Built-In Self-Test
CAD - Computer Aided Design
CDFG - Control/Data Flow Graphs
CMOS - Complementary Metal Oxide Semiconductors
DSP - Digital Signal Processing
ECL - Emitter Coupled Logic
FPGA - Field Programmable Gate Array
HDL - Hardware Description Language
ILP - Integer Linear Program
ITE - If-Then-Else
LSI - Large-Scale Integration
LSS - Logic Synthesis System
LSSD - Level-Sensitive Scan Design
MIS - Miltiple-level Interactive System
MPGA - Mask Programmable Gate Array
OBDD - Ordered Binary Decision Diagram
ODC - Observability *Don't Care* (sets)
PLA - Programmable Logic Array
RAM - Random Access Memory
RISC - Reduced Instruction Set Computer
ROBDD - Reduced Ordered Binary Decision Diagram
SAW - System's Architech Workbench
SIS - Sequential Interactive Synthesis system
SOS - Silicon on Sapphire
TCM - Thermal Conduction Module
TTL - Transistor-Transistor-Logic
UDL/I - Unified Design Language for Integrated Circuits

VHDL - VHSIC Hardware Description Language
VHSIC - Very High Speed Integrated Circuits
VLIW - Very Long Instruction Word
VLSI - Very Large Scale Integration
YLE - Yorktown Logic Editor
ZOLP - Zero-One Linear Program

1-hot encoding, 323, 450
ADA, 98
AHPL, 98
ALAP scheduling algorithm, 188, 198, 209–210
APL, 98
ASAP scheduling algorithm, 188, 198
absolute constraints, 190
abstract models, 27, 98, 114
activation signals, 166, 167, 171, 175
acyclic graphs, 38, 117, 146, 476
    directed acyclic, 39, 121, 186
acyclic networks, 484
adaptive control synthesis, 174–175, 177
adjacency matrix, 40
adjacent vertices, 38, 40
Aho-Corasick algorithm, 518, 521
algebraic divisor, 361
algebraic model, 360
algebraic transformations, 360
algorithmic approach to network optimization, 356
algorithmic binding, 544, 545–546
algorithms,
    Aho-Corasick, 518, 521
    ALAP scheduling, 188, 198, 210
    ASAP scheduling, 188, 198
    approximation, 43, 45
    Bellman-Ford, 55, 57, 194, 197, 466, 467
    Boolean covering, 526, 541
    Boolean matching, 526, 541
    branch and bound, 47–48, 61, 89, 92, 337, 506
    controllability *don't care* sets, 384–388
    decidable, 43
    delay evaluation and optimization, 418, 426–431

567